Lecture Notes in Engineering

The Springer-Verlag Lecture Notes provide rapid (approximately six months), refereed publication of topical items, longer than ordinary journal articles but shorter and less formal than most monographs and textbooks. They are published in an attractive yet economical format; authors or editors provide manuscripts typed to specifications, ready for photo-reproduction.

The Editorial Board

Lecture Notes in Engineering

Edited by C. A. Brebbia and S. A. Orszag

45

M. Gad-el-Hak (Editor)

Advances in Fluid Mechanics Measurements

Springer-Verlag
Berlin Heidelberg New York
London Paris Tokyo Hong Kong

Series Editors

C. A. Brebbia · S. A. Orszag

Consulting Editors

J. Argyris · K.-J. Bathe · A. S. Cakmak · J. Connor · R. McCrory
C. S. Desai · K.-P. Holz · F. A. Leckie · G. Pinder · A. R. S. Pont
J. H. Seinfeld · P. Silvester · P. Spanos · W. Wunderlich · S. Yip

Editor

Prof. Mohamed Gad-el-Hak
Dept. of Aerospace & Mechanical Engineering
University of Notre Dame
Notre Dame, IN 46556
USA

ISBN-13:978-3-540-51136-6   e-ISBN-13:978-3-642-83787-6
DOI: 10.1007/978-3-642-83787-6

Library of Congress Cataloging-in-Publication Data

Advances in fluid mechanics measurements / [edited by] M. Gad-el-Hak.
(Lecture notes in engineering ; 45)
ISBN-13:978-3-540-51136-6 (U.S.)
1. Fluid mechanics – Measurement.   I. Gad-el-Hak, M.   II. Series.
TA357.A279   1989
620.1'06 – dc20   89-10103

2161/3020-543210

PREFACE

One cannot overemphasize the importance of studying fluids in motion or at rest for a variety of scientific and engineering endeavors. Fluid mechanics as an art reaches back into antiquity, but its rational formulation is a relatively recent undertaking. Newton's (1687) *Principia* provides the cornerstone of modern science, and about 150 years after its publication the complete equations describing fluid motion were derived from first principles. Another 150 years have elapsed with only about 70 particular solutions of these nonlinear, partial differential equations found. Increasingly, therefore, we must rely on experimental and numerical simulations for answers to our most pressing problems.

The speed and capacity of today's supercomputers have improved by roughly a factor of 10^3 in the last 20 years. Nevertheless, direct integration of the equations of fluid motion at realistic Reynolds numbers is still a remote prospect, let alone prohibitively expensive. For high Re turbulent flows, the equations are averaged and the additional unknowns resulting from the nonlinearity of the governing equations are heuristically modeled in terms of the original dependent variables. Experimental input is required for both developing and validating the different turbulence models used to close the problem.

Much of the physics of a particular flow situation can be understood by conducting appropriate experiments. Of course measuring a flow quantity can itself be an end product, providing useful predictive or postdictive information such as flow rate, aerodynamic forces, etc. In an article appearing in <u>Fluid Mechanics Measurements</u>[*] , Roger Arndt provides a useful summary of what we measure and why. Flow visualization techniques offer a useful tool to establish an overall picture of a flow field and to delineate broadly its salient features before embarking on more detailed quantitative measurements. Among the single-point measurements that are particularly difficult are those in separated flows, non-Newtonian fluids, rotating flows, and nuclear aerosols. Pressure, shear stress, vorticity, and heat transfer coefficient are also difficult quantities to measure, particularly for time-dependent flows. These and other special situations are among the topics covered in this volume of Springer's <u>Lecture Notes in Engineering</u>. The book is, in the words of Peter Bradshaw, a collection of twelve noninteracting review articles. Each emphasizes the development of a particular measuring technique. A companion volume entitled "<u>Frontiers in Experimental Fluid Mechanics</u>" contains ten articles with more emphasis on the use of experiments to achieve better physical understanding of various classes of flow problems.

Chapters 1 through 4 describe novel flow visualization methods, including image process-ing. Pressure and shear stress measurements are discussed in Chapters 5 and 6, respectively. Techniques to measure the fluctuating vorticity field in transitional and turbulent flows are reviewed in Chapter 7. Chapter 8 is devoted to heat transfer and skin friction measurements in unsteady flows. Scanning laser velocimetery and other measurement techniques suitable for separated flow

---

[*] R.J. Goldstein, Editor, pp. 1-42, Hemisphere, Washington, 1983.

regions are discussed in Chapter 9. Chapters 10 and 11 give broad overviews of measurement techniques particularly suited for rotating flows and drag-reducing polymer flows, respectively. Measurement techniques for nuclear aerosols are reviewed in the final chapter.

The topics covered were chosen because of their importance to the field, recent appeal, and potential for future development. The articles are comprehensive and coverage is pedagogical with a bias towards recent developments. Whenever possible, the manuscripts were each reviewed by two independent referees. The reviewers' constructive comments were incorporated in the final version.

The editor wishes to express his deep appreciation to all the authors and the reviewers who unselfishly gave much of their time and effort. Their reward is in helping the reader interested in a particular topic through the ever expanding maze of available literature.

Mohamed Gad-el-Hak
Notre Dame, Indiana
October 1988

LIST OF CONTRIBUTORS

Professor T.E. Diller
Department of Mechanical Engineering
VPI & SU
Blacksburg, VA 24061

Professor P.F. Dunn
Department of Aerospace
  and Mechanical Engineering
University of Notre Dame
Notre Dame, IN 46556

Professor J.F. Foss
Department of Mechanical Engineering
Michigan State University
East Lansing, MI 48824

Professor P. Freymuth
Department of Aerospace Engineering Sciences
University of Colorado
Boulder, CO 80309

Professor M. Gad-el-Hak
Department of Aerospace
  and Mechanical Engineering
University of Notre Dame
Notre Dame, IN 46556

Professor M. Gharib
Department of Applied Mechanics
  and Engineering Sciences
University of California
La Jolla, CA 92093

Dr. R.A. Handler
Laboratory for Computational Physics
  and Fluid Dynamics
Naval Research Laboratory
Washington, D.C. 20375

Dr. J.H. Haritonidis
Department of Aeronautics
  and Astronautics
MIT
Cambridge, MA 02139

Mr. E.W. Hendricks
Laboratory for Computational Physics
  and Fluid Dynamics
Naval Research Laboratory
Washington, D.C. 20375

Dr. M.P. Horne
Laboratory for Computational Physics
  and Fluid Dynamics
Naval Research Laboratory
Washington, D.C. 20375

Professor A. Krothapalli
Department of Mechanical Engineering
Florida State University
Tallahassee, FL 32306

Dr. J.V. Lawler
Research Division
Hoechst Celanese Co.
Summit, NJ 07901

Professor P. Leehey
Department of Mechanical Engineering
MIT
Cambridge, MA 02139

Professor L.M. Lourenco
Department of Mechanical Engineering
Florida State University
Tallahassee, FL 32306

Professor R.B. Miles
Department of Mechanical
  and Aerospace Engineering
Princeton University
Princeton, NJ 08544

Professor D.M. Nosenchuck
Department of Mechanical
  and Aerospace Engineering
Princeton University
Princeton, NJ 08544

Dr. V.J. Novick
Engineering Division
Argonne National Laboratory
Argonne, IL 60439

Dr. B.J. Schlenger
Reactor Analysis & Safety Division
Argonne National Laboratory
Argonne, IL 60439

Professor R.L. Simpson
Dept. of Arospace and
  Ocean Engineering
VPI & SU
Blacksburg, VA 24061

Dr. C.A. Smith
Aeromechanics Branch
NASA-Ames
Moffett Field, CA 94035

Dr. J.D. Swearingen
Laboratory for Computational Physics
  and Fluid Dynamics
Naval Research Laboratory
Washington, D.C. 20375

Professor D.P. Telionis
Department of Engineering Science
  and Mechanics
VPI & SU
Blacksburg, VA 24061

Professor J.M. Wallace
Department of Mechanical Engineering
University of Maryland
College Park, MD 20742

Professor P.D. Weidman
Department of Mechanical Engineering
University of Colorado
Boulder, CO 80309

Mr. C. Willert
Department of Applied Mechanics
  and Engineering Sciences
University of California
La Jolla, CA 92093

TABLE OF CONTENTS

AIR FLOW VISUALIZATON USING TITANIUM TETRACHLORIDE

Peter Freymuth
Department of Aerospace Engineering Sciences,
University of Colorado
Boulder, Colorado 80309-0429

Abstract

The titanium tetrachloride method of air flow visualization is traced to its beginnings. The method is described in detail to prospective users. Recent progress in visualizing unsteady separated vortex flows by means of this method is documented by many examples.

Introduction and Historical Development

Titanium tetrachloride ($TiCl_4$) is a clear liquid which gives off dense white fumes when brought into contact with moist air. The brightness of this smoke consisting, according to Freymuth, Bank and Palmer (1985), of micron size TiO_2 particles, is an advantage in flow visualization when compared to other smokes. The ability of the liquid to adhere to aerodynamic model surfaces where it forms a non-intrusive, smoke producing liquid film has become a decisive factor in recent approaches to vortex visualization.

Unfortunately, the $TiCl_4$ fumes are toxic and corrosive since hydrochloric acid is generated as a byproduct of the reaction with moisture. Freymuth, Bank and Palmer (1985) recomend some precautions, mainly careful handling of the liquid, good ventilation and the wearing of safety glasses. This does not pose any unsurmountable obstacles.

The remainder of this section traces the $TiCl_4$-method from the early 30's to the present time, followed by a section which describes the method in detail for prospective users, followed by a section with numerous examples of visualizations obtained in our laboratory.

Flow visualization by means of $TiCl_4$ seems to have been introduced by Simmons and Dewey (1931) for boundary layer visualizations as has been mentioned in the well known textbook by Prandtl and Tietjens (1934). Nearly simultaneously, it was used by Farren (1932) for visualization of steady and unsteady flow around small aerodynamic models. These and later developments in visualization at Cambridge have recently been reviewed by Head (1982) who used $TiCl_4$ to visualize separation in the 50's. Brown at the University of Notre Dame also used the $TiCl_4$ method in the late 30's according to Mueller (1988). The book by Pankhurst and Holder (1952) on wind tunnel techniques and the review by Maltby (1962) mention this method. These authors recommend thinning of $TiCl_4$ with carbon tetrachloride in order to avoid clogging of a smoke rake driven by $TiCl_4$ fumes. This suggestion has most recently been taken up again by Bienkiewicz and Cermak (1987) who describe such a smoke rake system in detail. $TiCl_4$ has been used to visualize flow around model buildings in a large closed return wind tunnel at least since 1974, as the report by Peterka and Cermak (1974) attests to. In the 60's, $TiCl_4$ was used to wet the walls of a round nozzle from which a laminar jet issued. The vortex rings forming outside the nozzle were visualized this way and reported by Wille (1963) (without, however, mentioning the $TiCl_4$ method). The $TiCl_4$ method has been mentioned by Taneda (1977) for the study of unsteady separated flows, but he showed mostly visualizations in water using an electro-chemical method, presumably because they were of higher quality. Such experience may have led Werlé (1973) to express the view that many aerodynamic tests are more easily done in water. Subsequently, it was shown by Freymuth, Bank and Palmer (1985), Freymuth (1985), and Freymuth, Bank and Finaish (1985) that the $TiCl_4$ method can be very successful in visualizing unsteady vortex flow around airfoils and other bluff cylinders when $TiCl_4$ is painted as a centerstrip on the airfoil surface from the leading to the trailing edge. In this way smoke was generated wherever vorticity was generated which allowed visualization of vortex development in side view. Primary, secondary, and even higher order vortices were visualized this way as two-dimensional cuts.

This method of "vortex tagging" or "vorticity tagging" in two dimensions was extended by Freymuth, Finaish and Bank (1985) to three-dimensional global visualization of unsteady flow around airfoils and by Freymuth (1986, 1988 a) and by Freymuth, Finaish and Bank (1986, 1987) to three-dimensional wings. $TiCl_4$ was painted along wing perimeters where vorticity production was strongest. Global visualization of transition and of entire three-dimensional vortex systems around lifting surfaces became available this way for the first time. A logical extension of global visualization would be the "flooding" of an entire wing surface by $TiCl_4$ prior to visualization, which is in the planning stage.

$TiCl_4$ can also visualize steady separation bubbles as has recently been shown by Mueller (1987) who took advantage of the brightness of the fumes. While recent progress in visualization by uniform vortex tagging has primarily taken place in airflow, similar progress in water flow seems promising and would allow the use of color as an additional experimental dimension. The reader is referred to Werlé (1971, 1986) and to Gad-el-Hak (1986, 1987), for the state of the art in three-dimensional water flow visualization.

Description of the TiCl$_4$ Method

The TiCl$_4$ method used in our laboratory for visualization of unsteady flows is rather elementary. We progressed by experience and research needs rather than by refinements. For instance, TiCl$_4$ is introduced to airfoil surfaces manually by a small brass pipette rather than by means of elaborate tubing systems. We used conventional flood lighting and a conventional 16 mm movie camera. Airfoil or wing motions were manually controlled for maximum flexibility and low cost at modest accuracy. The description of our practices should, therefore, be considered a baseline reference only, allowing much flexibility, leeway and upgrading when applied by other researchers.

We obtain liquid TiCl$_4$ in purified form in 500 ml bottles from a commercial source. We decant 60 ml into a test tube which rests in a foam rubber cushion as shown in Fig. 1. The content of the test tube lasts easily through a one-hour movie session for a 100 ft long film reel. After a session, the test tube is replenished and sealed with two layers of heavy plastic tape, then stored in al locked cabinet, along with the TiCl$_4$ bottle. During a photographic session, the test tube is placed on top of the wind tunnel together with a commercial brass pipe which we use as a pipette (inner diameter of the pipe is 2.5 mm, outer diameter is 3 mm, length is 80 cm). Fumes from the test tube and during withdrawal of the pipette are blown away from the experimenter by means of a small blower, as shown in Fig. 1.

The test model located in the wind tunnel is then painted by means of the pipette with TiCl$_4$, inserting it through a hole in the top tunnel wall and removing it prior to visualization. The tunnel is of the open return type as shown in Fig. 2 and located in a large ventilated hall. For further improvement in ventilation, two fans are placed in an outside door adjacent to the tunnel exhaust and other doors in the hall are opened. Fig. 3 shows introduction of the pipette into the tunnel and a three-dimensional model to be painted by the TiCl$_4$ at its perimeters. It was found that gravity flow of the liquid TiCl$_4$ along surfaces aids greatly in providing uniformity of the liquid film where needed. For visualization of two-dimensional flow, the airfoil is mounted horizontally across the tunnel. A 2 cm wide center strip of TiCl$_4$ is provided by means of the pipette from the leading to the trailing edge of the airfoil surface for visualiztion of vortex patterns in side view. Since the pipette is removed prior to visualization our method is nearly nonintrusive.

Floodlighting from the top is by several 500 W projector lamps. Photographing is from outside of the tunnel through a Plexiglas wall. The backwall of the tunnel and the test model are painted flat black for contrast to the white TiCl$_4$ smoke. Test models need frequent cleaning with soap and water to avoid buildup of sticky residues on its surfaces. The surface needs then to be dried with paper towels and airflow before reapplication of TiCl$_4$. Also, the pipette needs to be cleaned after

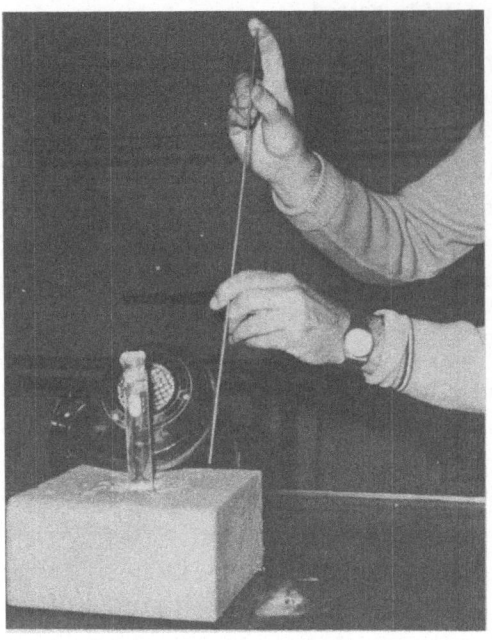

Fig. 1   View of TiCl$_4$ filled test tube
and pipette

Fig. 3.   View of test model with
inserted pipette; the pipette is removed
prior to documenting visualization by a
movie sequence.

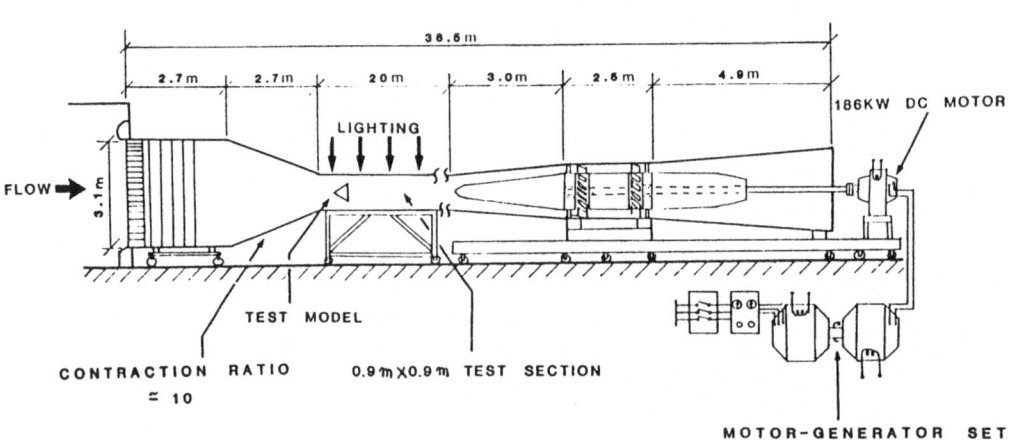

Fig. 2   Experimental setup

a session by letting water flow through it, then to be dried by means of compressed air flow.

The experiment was usually run by the experimenter who introduced the smoke and who controlled the desired unsteadiness in the dxperiment and by a photographer. A Bolex 16 mm movie camera with a frame rate of 64 frames/sec was used. The f-stop varied between 2 and 5.6 Kodak 16 mm 4x negative film was utilized.

The $TiCl_4$ method has also beeen used by us outside the wind tunnel in still air, in conjunction with a pickup carriage which allowed impulsive start of an airfoil from rest. Details of this experimental arrangement have been described by Finaish (1987) and by Finaish and Freymuth (1988).

For safe handling of $TiCl_4$ we had available safety goggles, rubber gloves, and acid filters inserted into masks. In six years of experience, we have not had an accident.

$TiCl_4$ can also be utilized for a smoke-wire placed upstream of the test model. $TiCl_4$ is released from the pipette to the wire and allowed to run down and wet the wire. Because of inhomogeneities in the wire surface closely spaced streaklines emanate from the wire for visualization. The smoke-wire method is of particular value to visualize attached flow where little vorticity separates and therefore the vorticity tagging method has little to show. This is best illustrated by an example. Fig. 4,left, shows smoke-wire visualization of flow over a carefully stall controlled airfoil at a 40° angle of attack. The overall streakline pattern is very lively, clearly showing flow attachment and illustrating the potential flow streakline pattern surounding the boundary layer and the wake behind the airfoil. Fig. 4, right , shows corresponding visualization by the vorticity tagging method. Since there is no separation over the airfoil a single smoke line leaves the airfoil with rather limited information being conveyed. Stall control and how it is achieved will be considered in more detail in the section "Stall control using an airfoil with rotating nose."

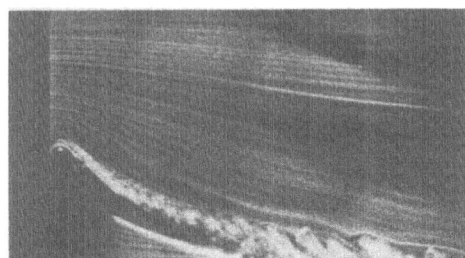 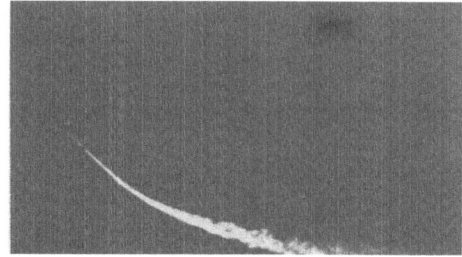

Fig. 4   Left: Streakline visualization for a   stall   controlled airfoil at
$\alpha = 40^{\circ}$, c = 12.4 cm, U_0 = 61 cm/sec, Re = 4200.
Right: Visualization by vorticity tagging for the same airfoil, for details of control see Fig. 11, top.

Limitations and Comparison with Other Techniques

Some reservations in interpreting visualization results are in order. In practicing the vorticity tagging technique $TiCl_4$ smoke inadvertently is introduced in flow regions where no vorticity is produced and is not introduced completely homogeneously. These flaws mask vortex patterns to some degree and introduce "optical noise." Smoke patterns foremost visualize vortex shapes but not vortex intensities. Another problem area is the difference in diffusion of smoke and vorticity. Usually smoke diffusion is much smaller than vorticity diffusion. Furthermore, while vortices of opposite rotation can annihilate each other by diffusion smoke streaks can not. Streakline methods like the smoke-wire method are prone to the same problems. Diffusion problems are most severe for downstream of the regions of vorticity and smoke production. No general quantitative analysis is available presumably because the above effects are highly configuration dependent.

Streamline techniques where particle paths are traced out in the flow field are not prone to the above effects and have good potential for obtaining quantitative information. On the other hand, streamline techniques do not yield vortex patterns without extensive data processing, are hard to utilize in three-dimensional flows and lack the immediate appeal of the vorticity tagging and streakline methods.

Computer simulation and visualization represents a new tool with a very high potential for "outvisualizing" physical flow visualization and filling in quantitative information not available by other methods.

At present it seems best to have different types of visualization (streamline, streakline, vortex tagging) in different media (air, water, oil) by different researchers available, for mutual reinforcement, confirmation and complementation. For general overviews of flow visualization the reader is referred to Mueller (1983), Settles (1986) and Merzkirch (1987).

Progress in visualizing unsteady separation: Examples

Introduction

At the first Workshop on Unsteady Flow at the Air Force Academy, Freymuth, Palmer and Bank (1984) presented flow visualization examples of accelerated starting flow around various bodies. By applying smoke producing liquid titanium tetrachloride to vorticity producing body surfaces, vortices were visualized in remarkable detail. In the meantime Freymuth (1985) has progressed from initial examples of vortex tagging to an extensive parametric study of accelerated starting flow around airfoils and other two-dimensional bodies. In the course of this work many new vortex interactions like splitting, shredding, squeezing, and others were identified by Freymuth, Bank and Palmer (1985) and by Finaish, Freymuth and Bank (1986).

From this basis Freymuth (1988 b,c) progressed to visualizations of vortex developments for pitching and plunging airfoils. The visualization of turbulent spots and of the closed three-dimensional vortex systems associated with finite wings was also achieved by Freymuth, Finaish and Bank (1985), by Freymuth (1986) and by Freymuth, Finaish and Bank (1986, 1987).

In this overview we document our progress in visualization by selected examples. The plan is to commence from two-dimensional flows and to proceed to three-dimensional configurations.

Impulsively Started Airfoil

Figure 5 sequences the development of dynamic vortical separation for an NACA 0015 airfoil set impulsively into motion. Movie frames are ordered into columns from top to bottom and from left to right. Flow is from left to right while the airfoil is stationary with respect to the camera. The chord length of the airfoil is c = 5.1 cm; speed after start from rest is U_o = 37 cm/sec at an angle of attack α = 50°. The resulting Reynolds number is Re = $U_o c/\nu$ = 1000 , where ν is the kinematic viscosity in air. The sequence is taken from the thesis by Finaish (1987) which contains a parametric investigation of this flow configuration. Flow starts with the first frame, which shows random smoke due to smoke introduction on the airfoil by means of a pipette. The next few frames clearly show the development of leading and trailing edge starting vortices. The first trailing edge vortex leaves the airfoil to the right, while the leading edge vortex rolls over the suction surface, inducing creation of a secondary vortex. While the leading edge vortex rolls over the trailing edge, it induces the generation of a second trailing edge vortex. This vortex jumps up the rear section of the airfoil, aided by the leading edge vortex to the right in Column 2. This encourages the secondary vortex to slip down where it gets incorporated into the trailing edge vortex before this reinforced vortex finally moves to the right. This sequence of events was first observed on the basis of streamline visualization by Monnet et al (1985). The process repeats in Column 3 and gets more turbulent in Column 4. The peculiarities of timing of vortex generation allowed the first leading edge vortex to escape unscathed by the second trailing edge vortex which is reinforced by secondary vortex slipdown. In contrast, an airfoil started at constant acceleration showed splitting of the first leading edge vortex by the second trailing edge vortex and incorporation of the secondary vortex into the first leading edge vortex, a process which Freymuth, Bank and Palmer (1985) termed vortex shredding. We conclude that starting flow history can have dramatic and previously unknown effects on dynamic separation.

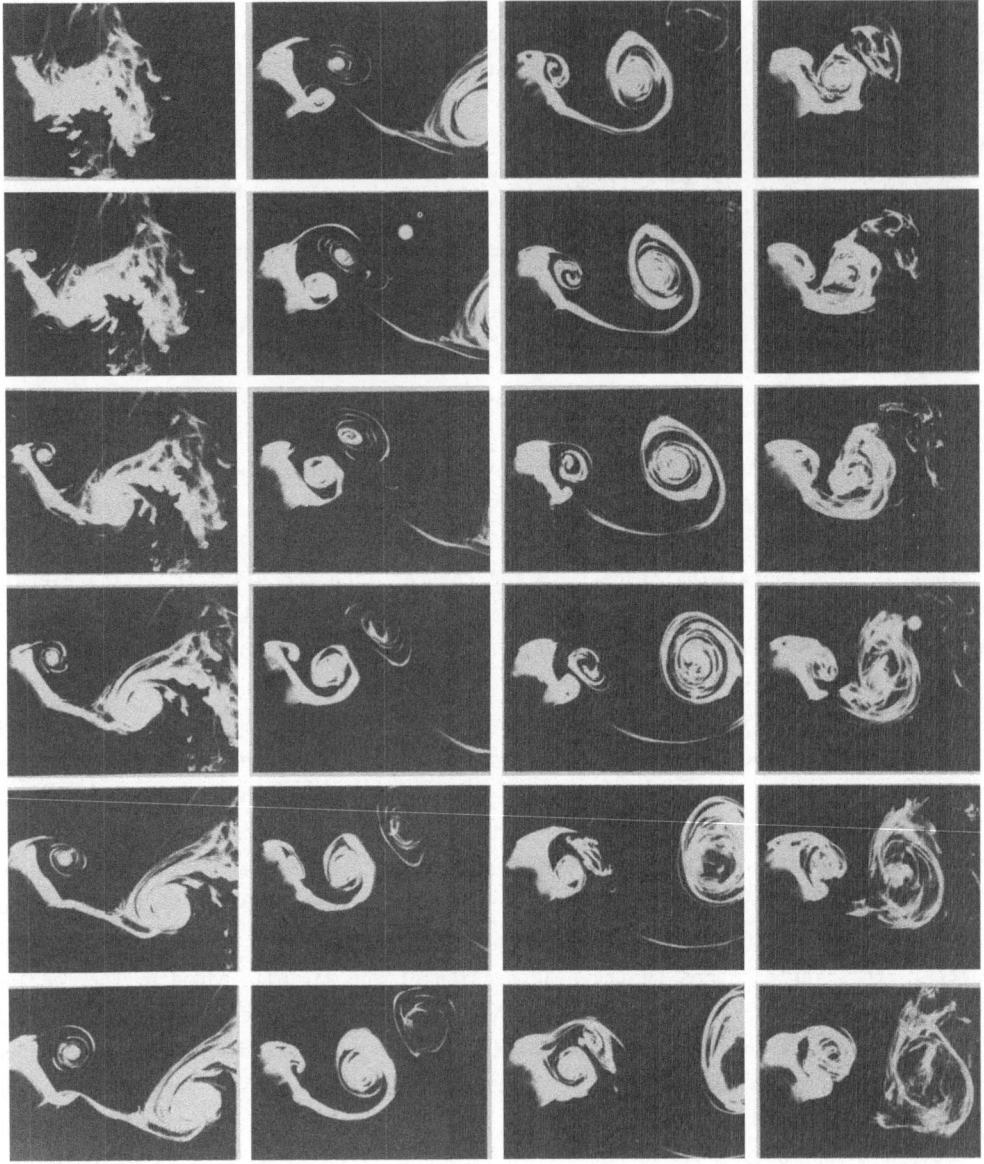

Fig. 5   Impulsive flow over an airfoil,
$\alpha = 50^{o}$, c = 5.1 cm, U_{o} = 37 cm/sec, Re = 1000,
time between consecutive frames is Δt = 1/16 sec.

Pitching Airfoil

A classic smoke rake visualization of an airfoil in pitch up and exposed to steady wind is shown in Fig. 6. It was published by Lippisch (1935) over 50 years ago and has been reproduced recently in the definitive historical review by Bublitz (1986).   In Fig. 7 we complement this classic with increased vortical detail obtained by our vortex tagging technique in the Rynolds number range best suited for visualization.

According to Freymuth (1988 b,c,d), this technique also yialded detailed results for airfoils in periodic pitch; an example is shown in Fig. 8. Periodic pitching was between angles of attack 20° and -20° at a frequency f = 1.6 Hz resulting in a reduced frequency k = 2.9, where k = $\pi fc/U_o$.

Alternating vortices are shed from the trailing edge of the airfoil and form a vortex street behind the airfoil. The vortex street has vortices with reverse sense of rotation if compared to the well known drag indicating Karman vortex street in flow behind stationary cylinders. Mutual induction of these vortices is in the downstream direction and the resulting reacting force on the pitching airfoils is therefore in the upstream direction, i.e. the pitching airfoil generates thrust, similar to a pitching fishtail. Some interesting single frame vortex patterns in the wake of an oscillating airfoil in water have recently been published by Koochesfahani (1987).

Plunging Airfoil

Pitching and plunging airfoils, as active elements, allow the generation of thrust and this is at the root of sustained animal flight. Obviously, bird and insect flight mechanics can be very complex since pitching, flapping, and other motions are employed in combination and since wings are finite. Almost nothing has been done in terms of flow visualiztion to elucidate basic unsteady propulsion mechanisms relevant to bird flight. Simple thrust generation by an airfoil in pure periodic plunging motion and exposed to a steady free stream velocity, U_o = 61 cm/sec, is shown in Fig. 9. We have c = 15.2 cm, α = 5°, Re = 5200. The plunging frequency is f = 4 Hz resulting in a reduced frequency k = $\pi fc/U_o$ = 2.7 , the plunging amplitude was h = 3 cm. The sequence which should crudely model the flapping of a bird wing, mainly shows the generation of trailing edge alternate vortices which move backward in jet-like fashion, thus generating thrust similar as in Fig. 8.

An interesting variation of the reverse vortex street was obtained at a higher angle of attack α = 10° as shown in Fig. 10. here the vortices team up in vortex pairs or jetlets which propel themselves backward and downward, thus thrusting the

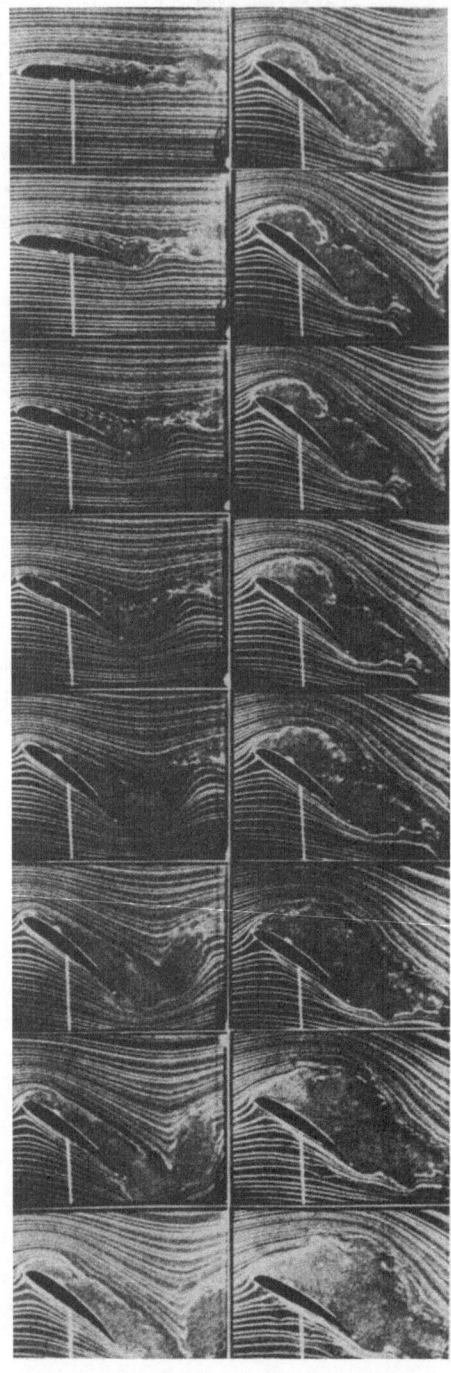

Fig. 6   Large amplitude pitchup of an airfoil in steady wind; from a streakline
         sequence by Lippisch (1935).

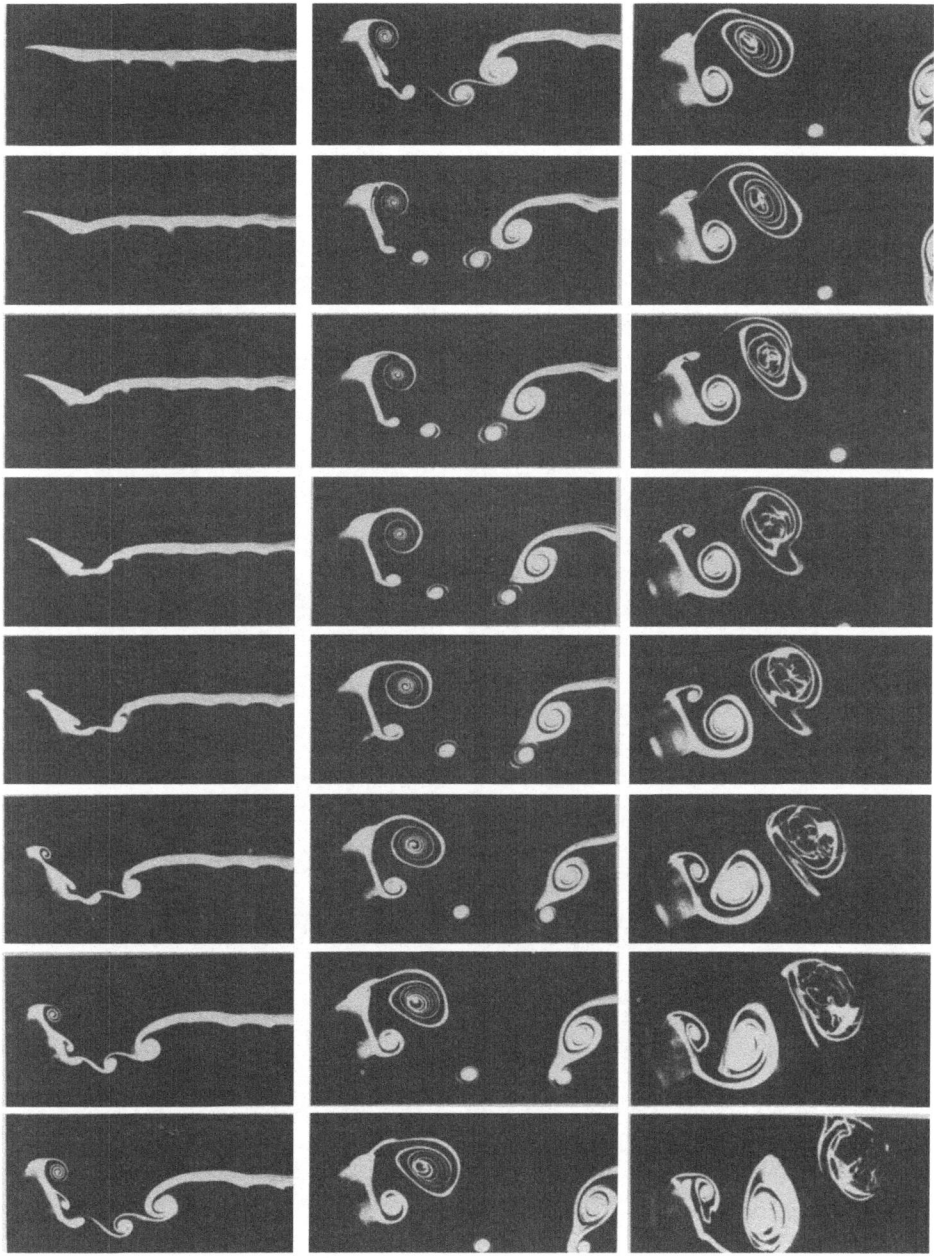

Fig. 7 Large amplitude pitchup of an airfoil in steady wind from 0° to 60° around the c/4 axis; c = 5.1 cm, U_0 = 61 cm/sec, Re = 1700, $\alpha^+ = \dfrac{\dot{\alpha}c}{U_0}$ = 0.26, Δt = 1/32 sec.

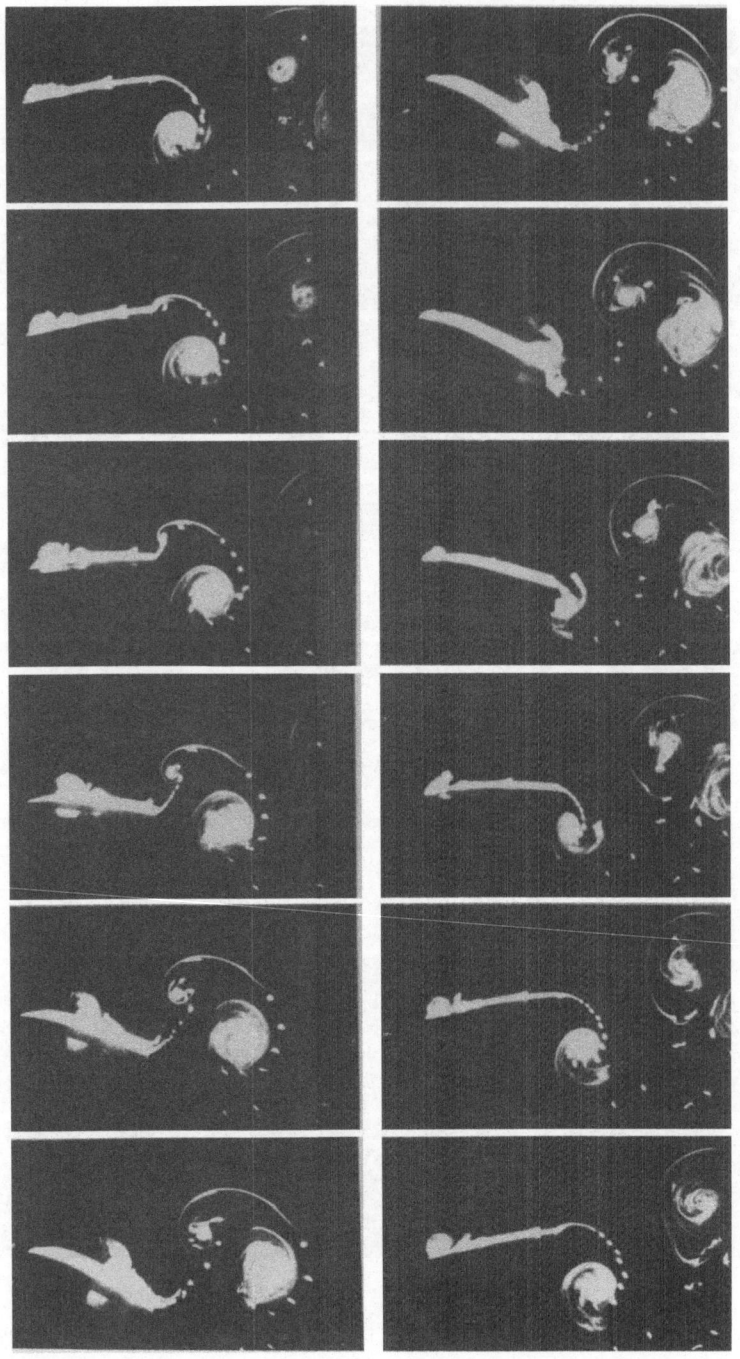

Fig. 8   Airfoil in periodic pitch between -20° and +20°.
c = 35.6 cm, U_0 = 61 cm/sec, Re = 12000, f = 1.6 Hz around the c/4 axis,
k = 2.9, Δt = 1/16 sec.

13

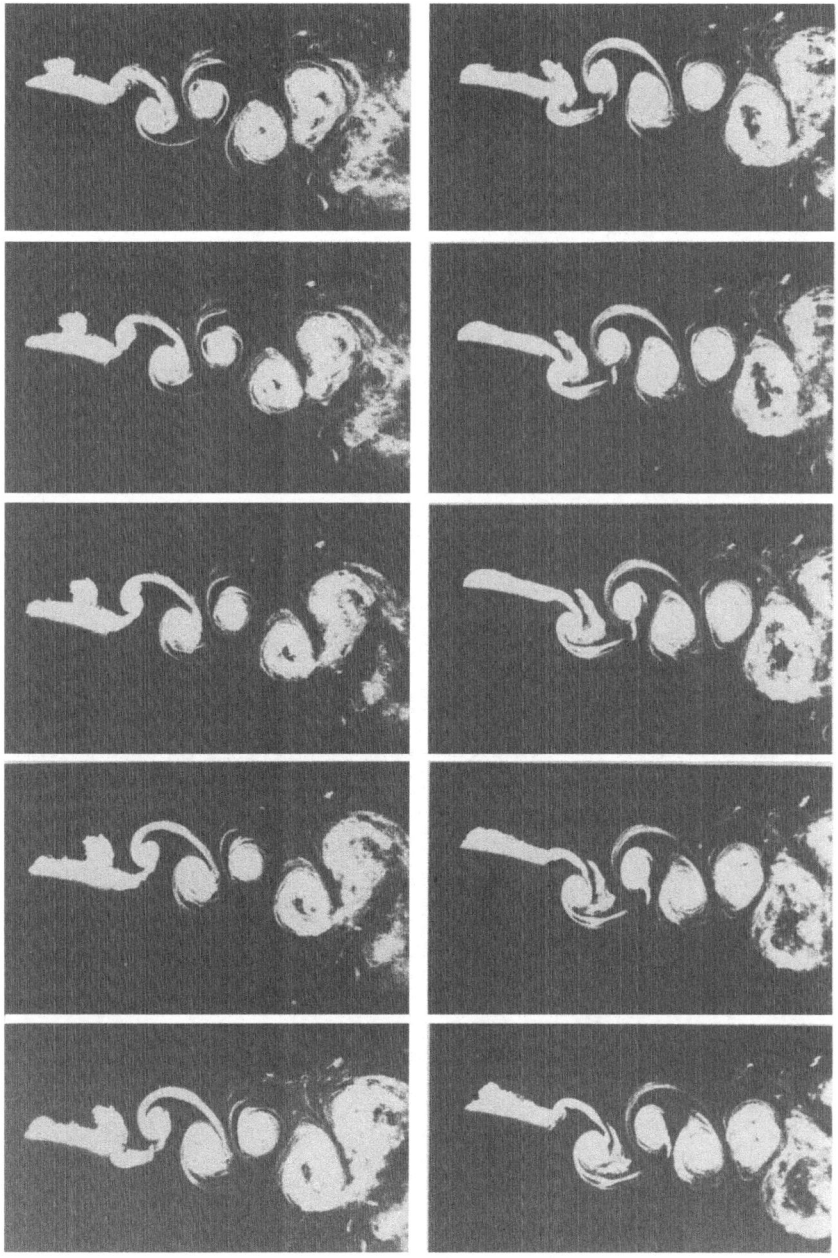

Fig. 9   Airfoil in periodic plunge,
α = 5°, c = 15.2 cm, U_0 = 61 cm/sec, Re = 5200, , plunge amplitude
h = 3 cm, f = 3.5 Hz, k = 2.7, Δt = 1/32 sec.

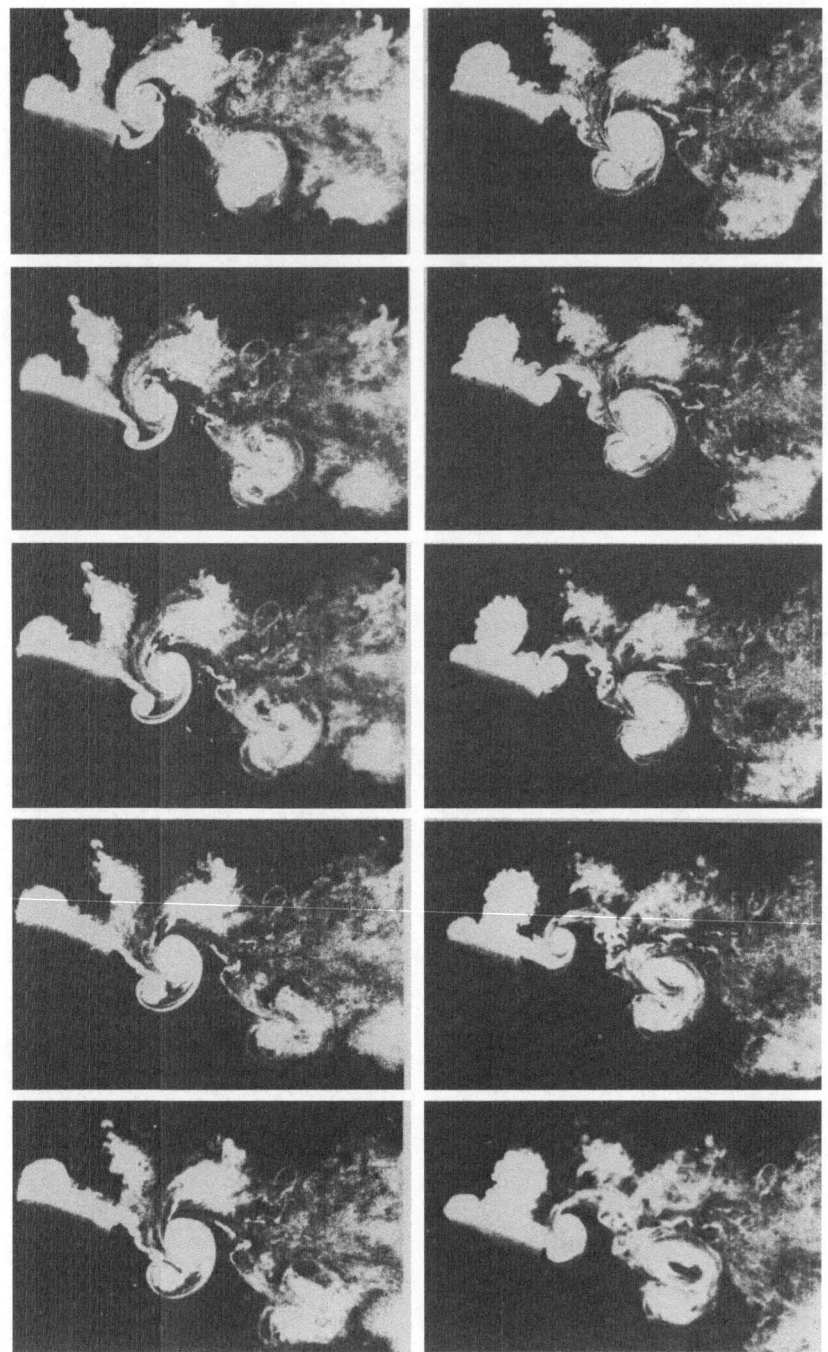

Fig. 10 Airfoil in periodic plunge,
$\alpha = 10^O$, c = 15.2 cm, U_o = 61 cm/sec, Re = 5200,
h = 2.5 cm, f = 3.5 Hz, k = 2.7, Δt = 1/32 sec.

airfoil forward and upward.

More needs to be done to closely model bird flight, but the basic signatures of thrust generating mechanisms has already been revealed by our simple examples in Figs. 8 to 10, from Freymuth (1988 b).

Stall Control Using an Airfoil with Rotating Nose

Rotating nose airfoils have been under discussion and experimentation for a long time according to Modi et al (1981). There seems to exist no significant flow visualization of the stall control achieved by this type of device. Furthermore, there is no mention of this device in the context if supermaneuverability.

An airfoil with rotating nose has been designed by Bush et al (1987) as a senior design laboratory project and flow reattachment has been successfully visualized for high angles of attack. Fig. 11, top shows the basic design sketch and the photographic frames below it show the process of reattachment for the airfoil at $\alpha = 40^0$, at a free stream velocity $U_0 = 91$ cm/sec, c = 12.4 cm, Re = 6000. Circumferential speed of the rotating cylinder was 4.2 times the free stream velocity. The cleanness of the final reattachment is sensational and was achievable at even higher angles of attack.

Fig. 12 shows the airfoil with rotating nose in pitch from 0° to 40° in the first 4 frames, then remaining at 40° in the remaining frames. Visualization is by smoke wire wetted with $TiCl_4$. Vortex shedding from the trailing edge is readily apparent during and after pitchup but no leading edge vortex is generated. This new concept of dynamic stall control over the leading edge, while allowing dynamic separation from the trailing edge, may lead to important applications in fast maneuvering aircraft. Fig. 13 shows the same airfoil during pitchup to 40° but without rotation of the nose. Development of the well known dynamic stall vortex behind the leading edge of the airfoil is obvious in this case and is in stunning contrast to the dynamic stall control displayed in Fig. 12.

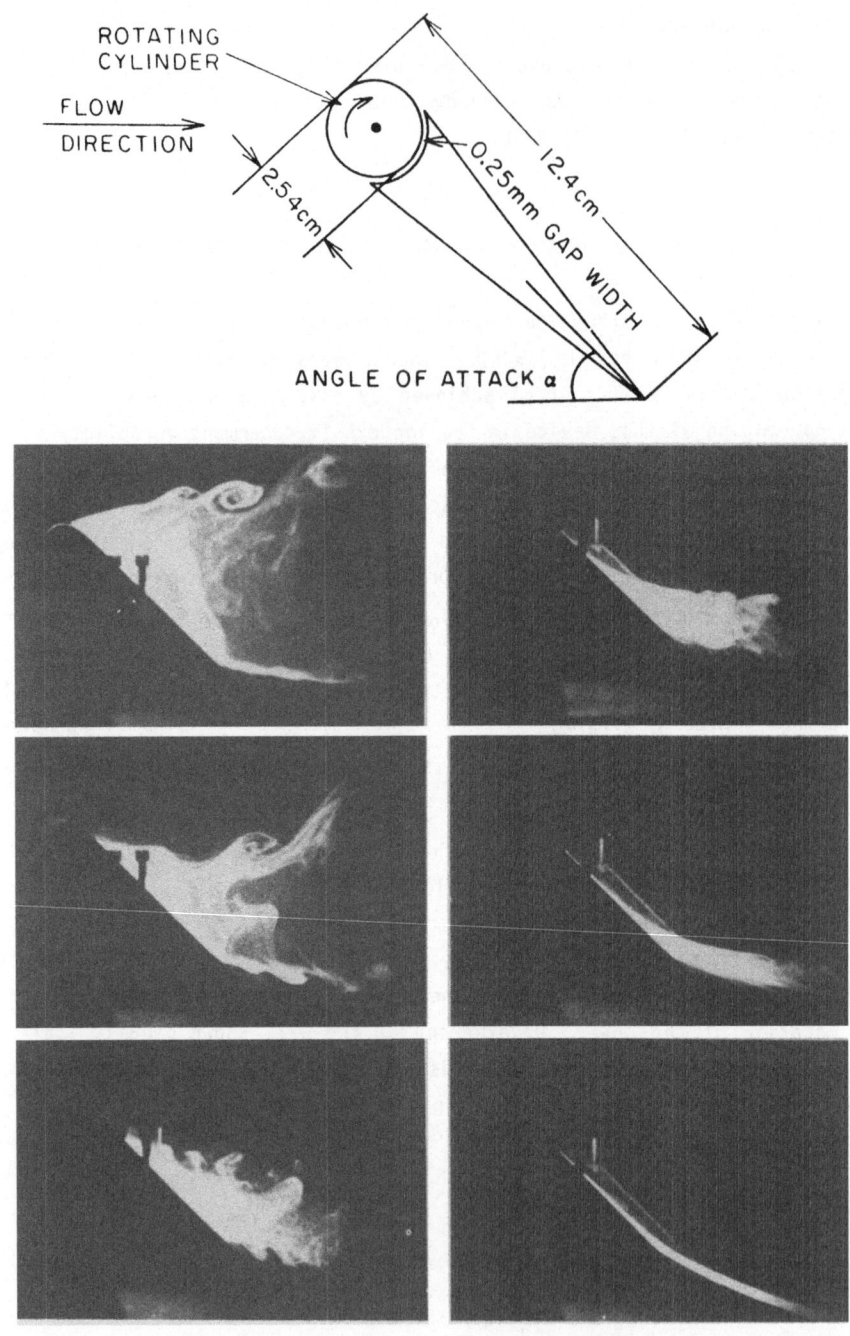

Fig. 11 Rotating nose airfoil. The flow reattachment process is shown after startup of the nose rotation.
$\alpha = 40^{\circ}$, $U_0 = 91$ cm/sec, Re = 6000, $\Delta t = 1/8$ sec.

17

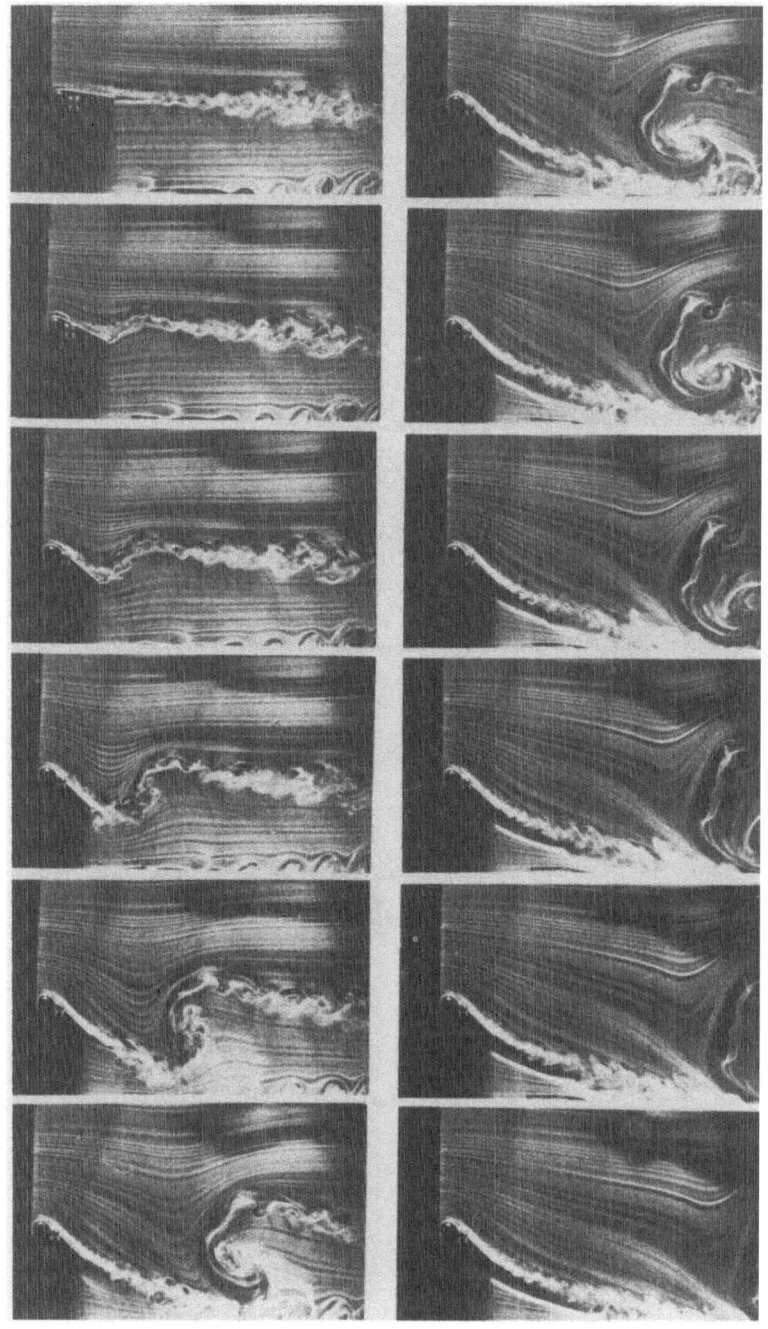

Fig. 12   Rotating nose airfoil pitching from 0° to 40°
$\alpha^+ = \dot\alpha\, c/U_0 = 0.4$, $\dot\alpha$ = angular rotation rate in rad/sec.

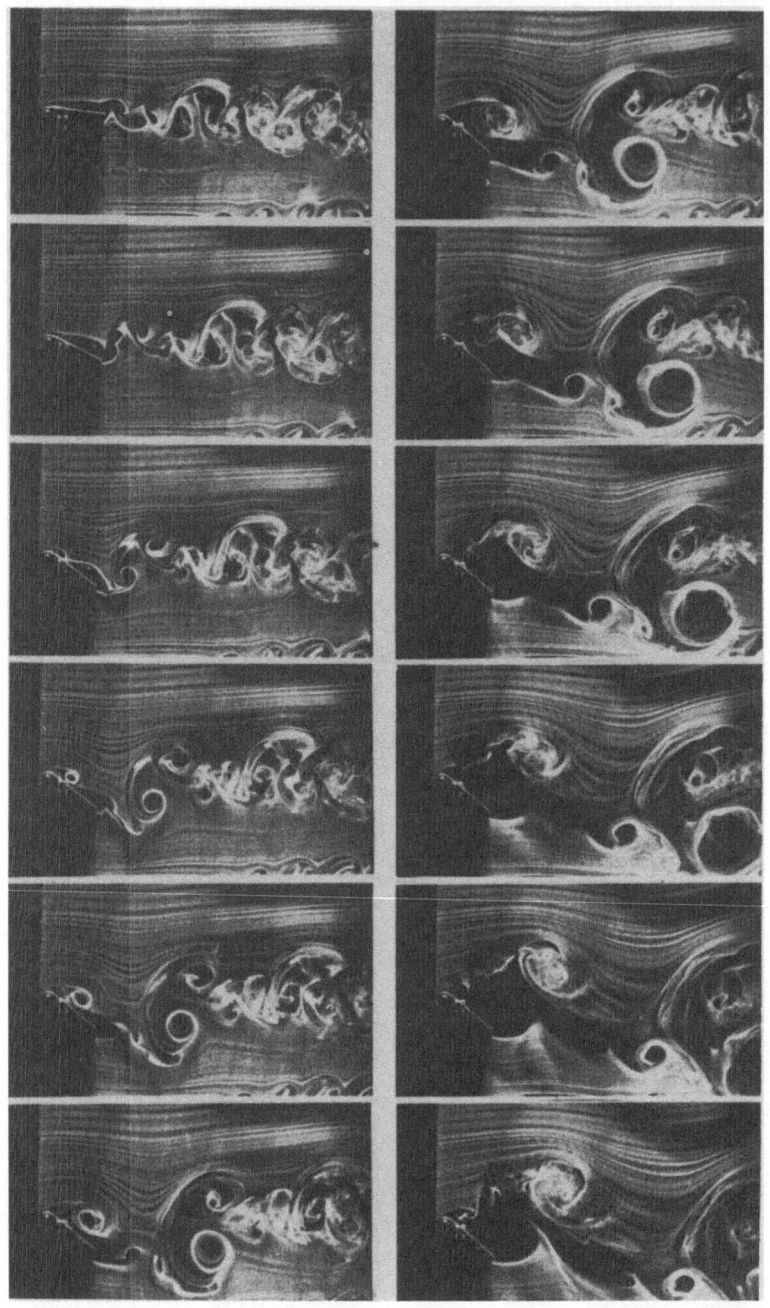

Fig. 13 Same airfoil in pitch as in Fig. 12 but without rotation of the nose.

Three-Dimensional Vortex Systems of Finite Wings

The vortex strands of a finite three-dimensional vortex system must be connected in accordance with Helmholtz's law. While the vortex strands must be connected viscous interaction allows reconnection of strands. A piercing question is how well flow visualization documents connecting and reconnecting of vortices. Two-dimensional and other cuts of streakline visualizations are incapable of addressing the question. Global visualization where we introduce the smoke as homogeneously as possible in areas of vorticity production can document connecting and reconnecting of vortices with moderate success as our subsequent examples will show.

Fig. 14 focuses on the connectivity of vortex strands at the top left corner of a rectangular half wing implusively started from rest with an angle of attack $\alpha = 20^{0}$. Impulsive start was achieved by means of a start cart described in detail by Finaish (1987) and by Finaish and Freymuth (1988). The movie frames depict the tip of a half wing protruding from below. Leading and trailing edges are marked for orientation above the first frame of the sequence, which represents a spanwise view or top or global view. After impulsive start in the first frame a nearly horizontal tip vortex and a series of vertical leading edge vortices develop which join together at the corner of leading edge and tip. Initially these vortices are separated from each other by a dark gap except at the leading edge corner. With ongoing time the two vortex systems approach each other so closely in column 3 that partial viscous annihilation and consequent reconnection of the vortex strands farther downstream seem likely.

Fig. 15 sequences vortex development over a square wing at 30° angle of attack started impulsively from rest, in spanwise view. The entire closed vortex system consisting of the leading and trailing edge starting vortices and of the tip vortices can be seen. The vortex loop formed by the trailing edge starting vortex and the tip vortices has a tendency to neck down in column 3.

Fig. 16 shows a top view or global view of a half wing which in this case protrudes from the ceiling into the center of the wind tunnel. It is mounted in a turntable which allows periodic and other pitchings. The sequence renders global visualization of vortex development in the tip region of the wing. Pitching is ±5° around a 20° mean angle of attack; furthermore, $c = 15.2$ cm, $U_0 = 61$ cm/sec, $f = 0.67$ Hz, $\Delta t = 1/8$ sec. Two counterrotating vortices develop near the tip on the suction side of the wing as Columns 1 and 2 show. Both tip vortices join together at the front corner of the tip and in this way accommodate the Helmholtz law. The farther inboard vortex also connects to the vortices which separate from the leading edge. In Column 3 both tip vortices move so close together that they may partially annihilate each other by viscous diffusion, thus allowing a growth decay cycle of the tip vortices. Growth and decay of the outer tip vortex has been noticed by

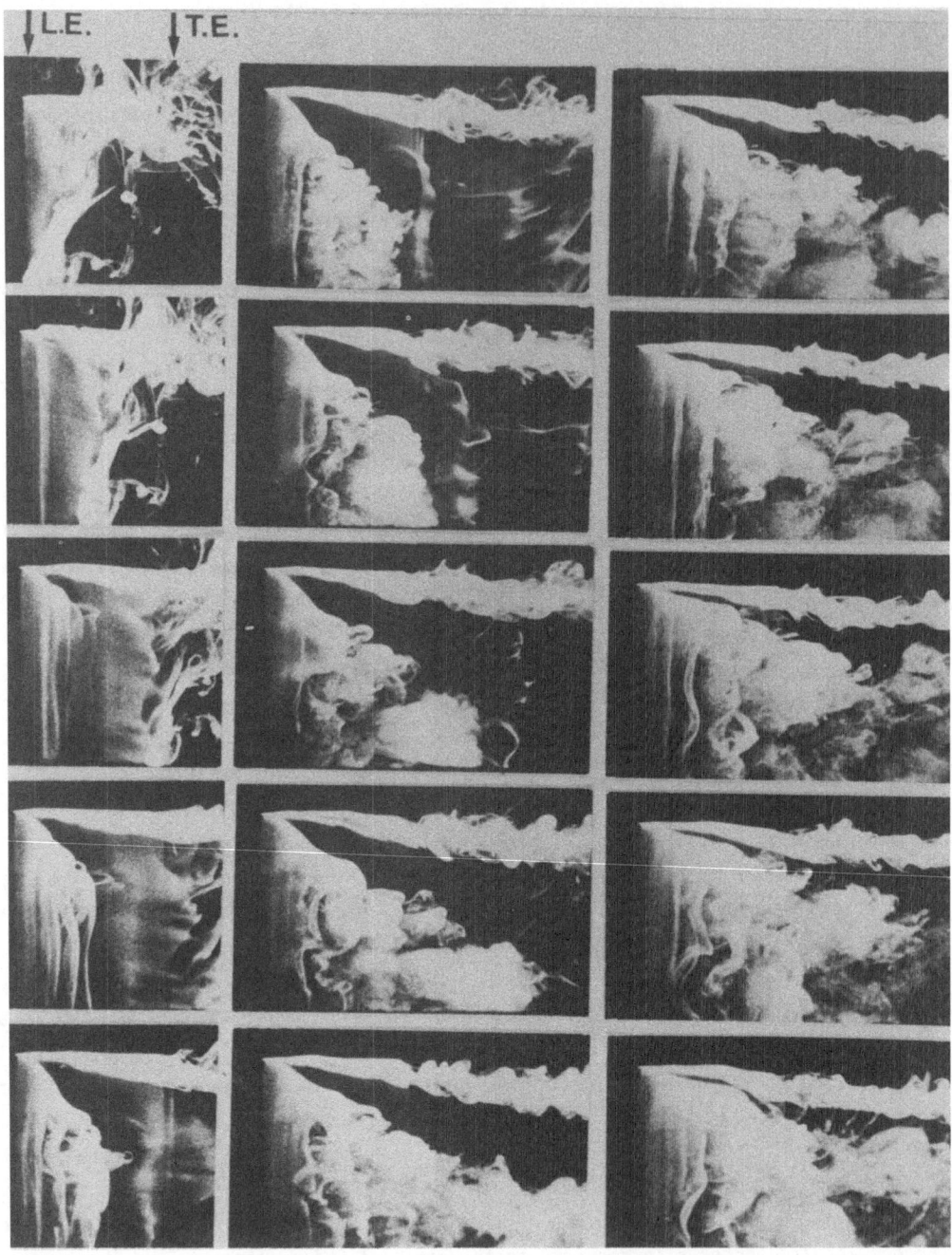

Fig. 14 Impulsive flow over a half-wing, spanwise or top view.
c = 15.2 cm, U_0 = 49 cm/sec, Re = 4100, α = 20^0, Δt = 1/8 sec.

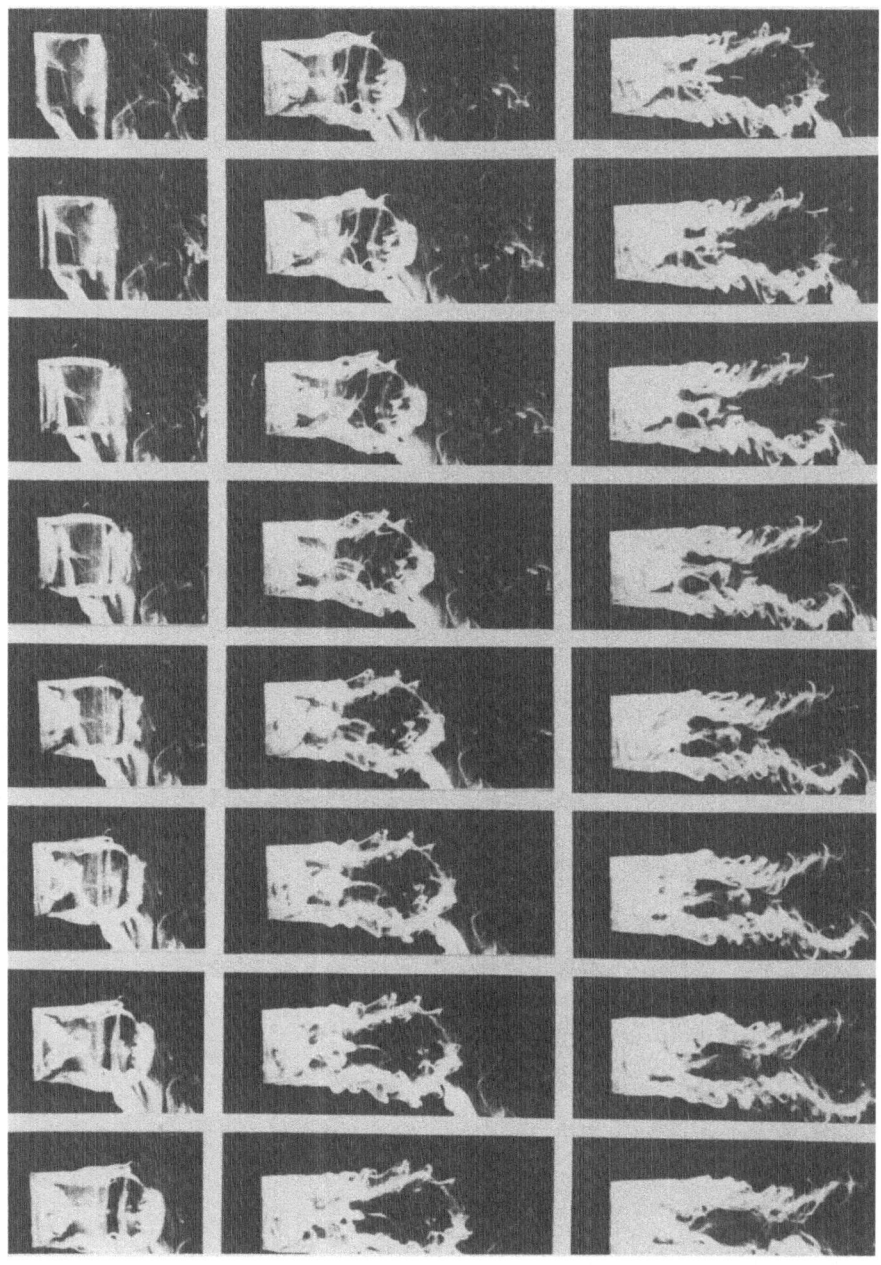

Fig. 15 Impulsive flow over a square airfoil, spanwise or topview,
c = 7.6 cm, U_0 = 37 cm/sec, Re = 1600, α = 30°, Δt = 1/32 sec.

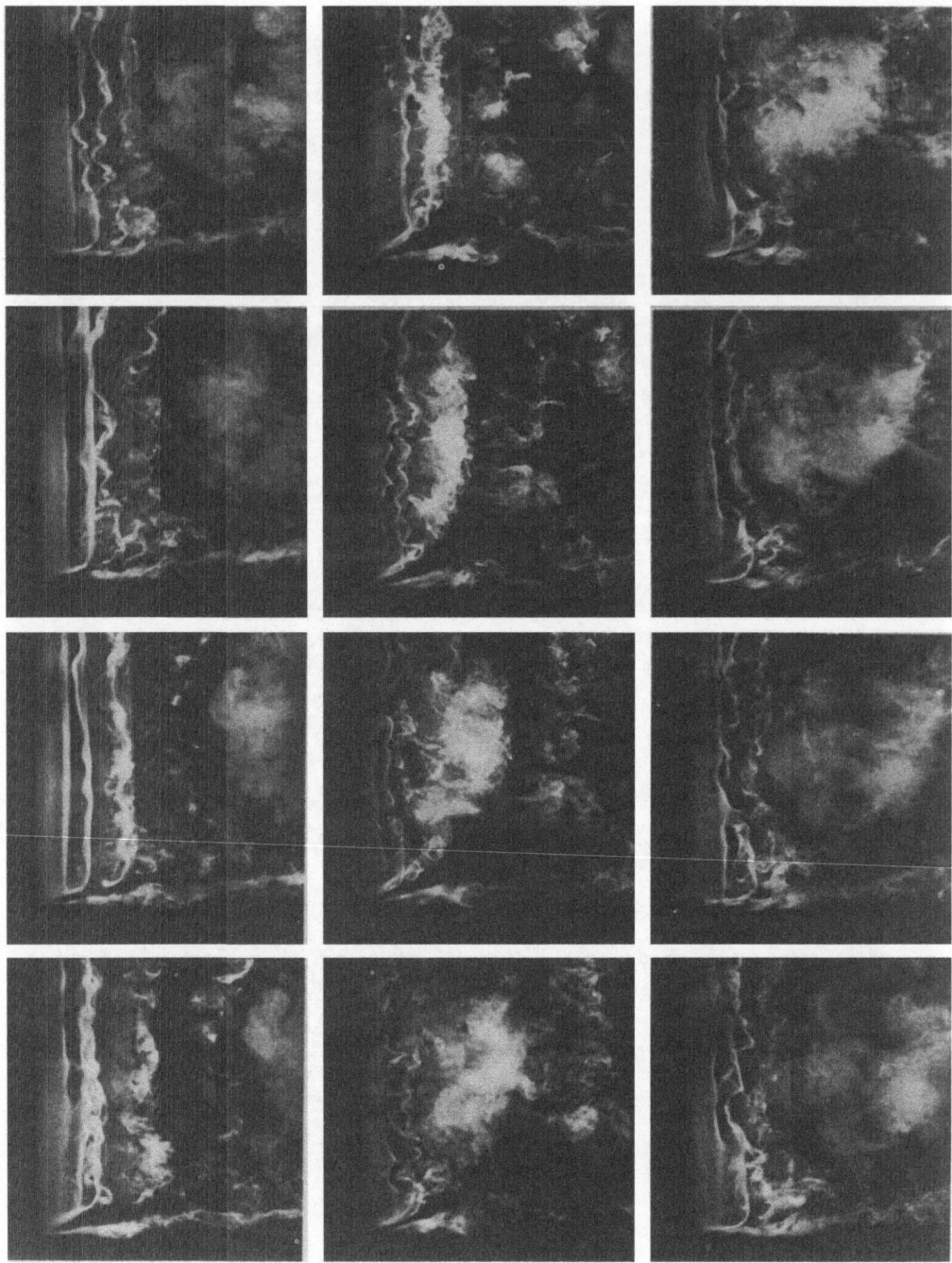

Fig. 16 Rectangular half wing in periodic pitch around the c/2 axis, spanwise or
top view,
$\alpha = 20^{0}$, pitch amplitude $\alpha = 5^{0}$, c = 15.2 cm, U_{o} = 61 cm/sec, Re = 5200,
k = 0.53, Δt = 1/16 sec.

Adler and Luttges (1985) in their smoke wire visualizations. Vortex tagging thus allowed us to obtain a span-wise or three-dimensional or global view of tip vortex development. Furthermore, we have been able to learn how the tip vortices accommodate Helmholtz's law and how a growth-decay cycle can be accommodated.

Fig. 17 shows the situation for a pitching delta wing. Periodic pitching is between angles of attack 0° and 30° at a reduced frequency k = 1.6. In this case we recognize some growth of the conical leading edge vortices in Column 1. They move toward the center line of the delta wing. The parts of the tip vortices close to the front corners keep their laminar and straight appaearance. Farther downstream they take on somewhat irregular spiral shapes (double spirals in this case) in the lower frames of Column 1.This change in appearance has been observed at constant angle of attack for a long time and is termed vortex break-down or bursting by Payne et al. (1986) and by Lugt (1983). Near the front corner the two tip vortices approach each other so closely that they link up and annihilate above the link in the upper frames of Column 2. ·The linked up or reconnected vortex then convects downstream. Simultaneously, new leading edge vortices start to grow inboard of and close to the leading edges also in Column 2 and go through the entire growth-decay cycle. A growth-decay cycle for a pitching delta wing has first been inferred by Gad-el-Hak, Ho and Blackwelder (1984) and by Gad-el-Hak and Ho (1985, 1986) from their two-dimensional cuts of visualization. We now interpret such a cycle by means of three-dimensional vortex principles.

A powerful starting vortex was generated by rapid pitch of a rectangular wing (aspect ratio is 2) from 0° to 60° angle of attack within 0.25 sec. A snapshot of the overall vortex system is shown in Fig. 18. Leading and trailing edge vortices are again linked at the front corners of the wing. The main trailing edge vortex looks very three-dimensional to the eyes which may be captured by stereoscopic or holographic techniques in the future.

Further progress can be achieved in the future by applying our global visualization of vortex development to other three-dimensional configurations and pitch histories.

Quest for Three-Dimensional Details

While the two examples of the previous section give good global views of three-dimensional vortex systems additional vortical detail is desirable in areas where smoke is too dense to fully visualize the vortical strands. To obtain finer detail we diluted the titanium tetrachloride to 30% by volume by means of trichloroethylene with the results of less dense smoke production when applied to the wing. We then took closeup movies of areas of interest. An example is shown in Fig. 19 where developments near the upper leading edge of a delta wing at 40° angle of attack are

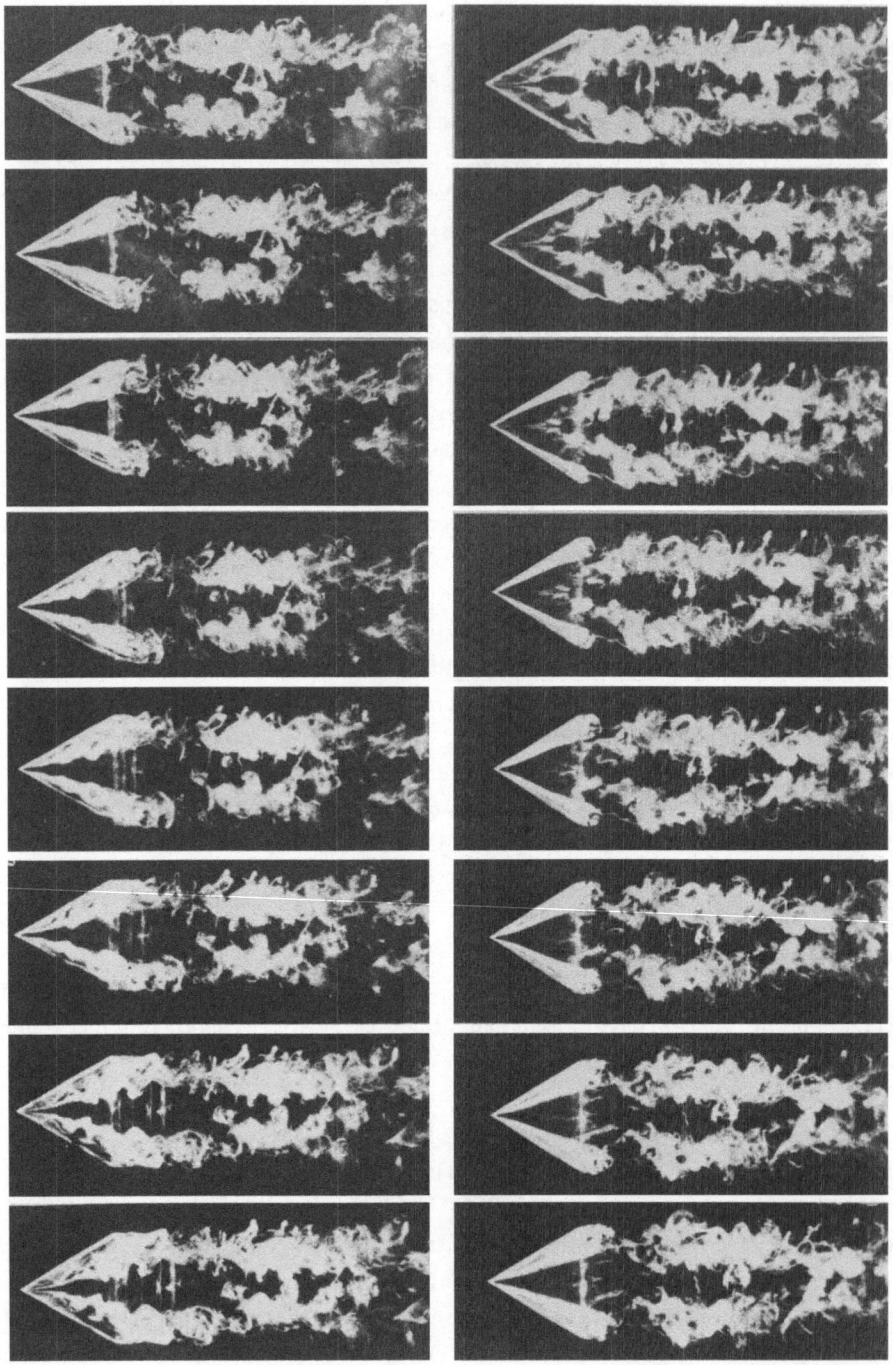

Fig. 17 Equilateral delta wing in periodic pitch around the c/4 axis, top view,
$\alpha = 15^O$, $\hat{\alpha} = 15^O$, c = 15.2 cm = side lengths, U_o = 61 cm/sec,
Re = 5200, f = 2 Hz, k = 1.6, Δt = 1/64 sec.

shown in a starting flow of constant acceleration. The development of the tip vortex and of a trailing edge starting vortex are shown in Column 1. In Column 2 a counterrotating secondary vortex develops more closely to the leading edge and connects to the tip vortex near the upper back corner of the wing, as does the trailing edge vortex and subsequent trailing edge vortices. In Column 3 even this method gets insufficient to more fully resolve the various vortex strands which become numerous and turbulent. The last few frames of Column 3 show the bursting phenomenon of the main leading edge tip vortex close to the front corner of the airfoil.

Freymuth et al. (1987) previously published a global view of this vortex system, which did not produce the fine detail achieved with the dilution method used for Fig. 19. On the other hand, using the dilution method in conjunction with a global view produced inferior overall results.

Fig. 20 shows enlargement of a Nikon shot corresponding to frame 2, Column 3 of Fig. 19. This frame shows the linkage knots of the secondary and of the trailing edge starting vortices with the leading edge starting vortex quite well although details are still not fully resolved. A peculiar vortex bar exists between the two linkage knots.

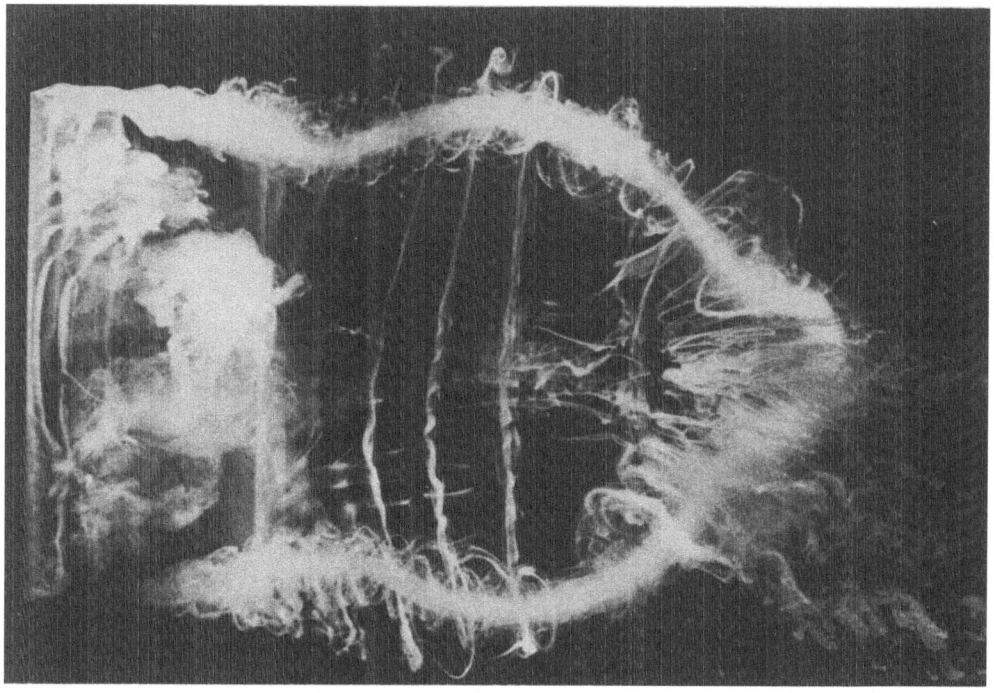

Fig. 18 Nikon shot of vortex system of a rectangular wing with aspect ratio 2 after a rapid pitch from 0° to 60°; c = 15.2 cm, U_0 = 61 cm/sec, Re = 5200.

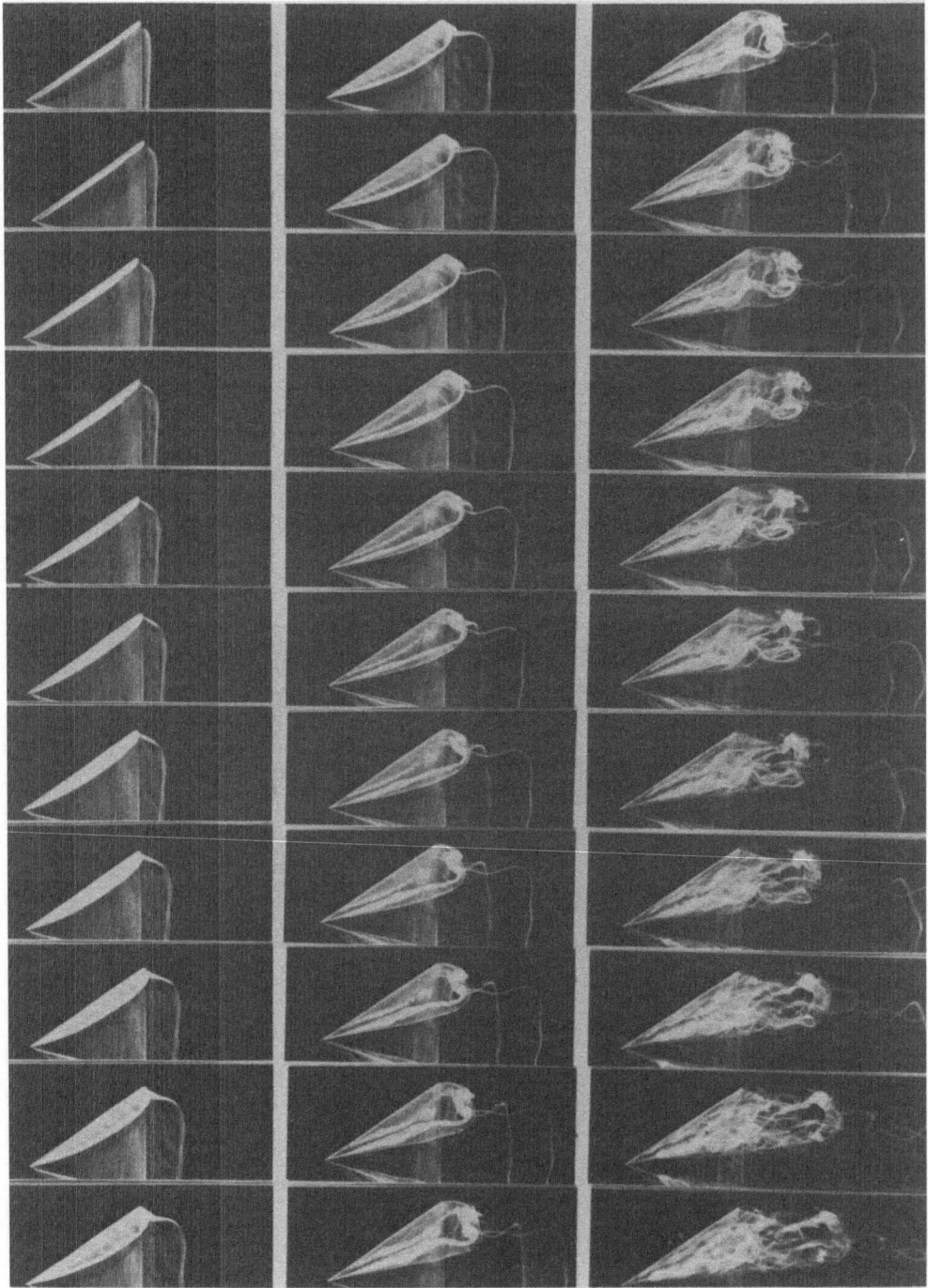

Fig. 19 Closeup top view for equilateral delta wing in accelerated starting flow at
$\alpha = 40^0$, $a = 2.4$ m/sec^2, side length c = 15.2 cm,
$R = a^{1/2}c^{3/2}/\nu = 5200$, $\Delta t = 1/64$ sec.

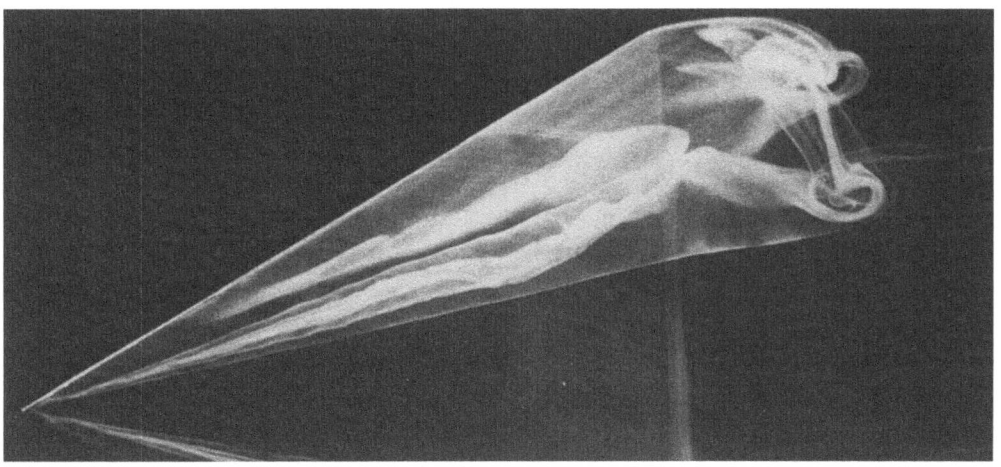

Fig. 20 Enlargement of a Nikon shot corresponding to frame 2, column 3 of Fig. 19.

Our three-dimensional examples show that we are making progress in visualizing the connectivity of vortex strands. Complete resolution is, however, an elusive goal and becomes impossible in a strongly turbulent regime.

Conclusions

The $TiCl_4$ method of flow visualization has been described in detail. We have increased the application range of this method. We have visualized and identified two-dimensional vortical interactions in great detail. The utility of the method in stall control work has been demonstrated. Global views of three-dimensional vortex systems have become accessible.

Future refinements of the method can yield further progress in the quality of visualization. Extending the application range can contribute to the integration and unification of the entire field of unsteady separated flows.

Acknowledgement

The photographic assistance by W. Bank and F. Finaish is gratefully acknowledged. An AFOSR Contract F49620-84-C-0065 supported this work.

References

Adler, J. N., Luttges, M. W. (1985) Three-dimensionality in unsteady flow about a wing, Paper AIAA-85-0132.

Bienkievicz, B., Cermak, J. E. (1987) A flow visualization technique for low-speed wind tunnel studies, Experiments in Fluids 5, pp. 212-214.

Bublitz, P. (1986) History of Aeroelasticity in Germany from the beginning until 1945, (in German), DFVLR-Mitt. 86-25.

Bush, R., Freece, T., Hora, K., Jackson, S., Kallemeyn, P., McNall, S., White, T., (1987) The effects of a rotating cylinder as a leading edge on boundary layer separation over an airfoil assembly. Senior design internal report, U. of Colorado, Dept. Aerospace Engineering Sciences (Available upon rquest), May.

Farren, W. S., (1932) Air flow with demonstrations on the screen by means of smoke, J. Roy. Aero. Soc. 36, pp. 451-472.

Finaish, F., Freymuth, P., Bank., W. (1986) Starting flow over spoilers, double steps and cavities, J. Fluid Mech. 168, pp. 383-392.

Finaish, F. (1987) Experimental study of two-dimensional vortex patterns for impulsively started bodies in comparison with other configurations. Ph.D. Thesis, University of Colorado, 1987.

Finaish, F., Freymuth, P. (1988) Aerodynamic visualization for impulsively started airfoils. Recherche Aerspatiale , accepted for publication.

Freymuth, P., Palmer, M., Bank, W. (1984) Comparative visualization of accelerating flow around various bodies, starting from rest, Workshop on Unsteady Separated Flow, Aug. 10-11, 1983, Air Force Academy, Colorado Springs, pp. 52-57.

Freymuth, P. (1985) The vortex patterns of dynamic separation: A parametric and comparative study. Prog. Aerospace Sci. 22, pp. 161-208.

Freymuth, P., Bank, W., Finaish, F. (1985) Surveying unsteady flows by means of movie sequences: a case study. International Symposium on Physical and Numerical Flow Visualization, Albuquerque, N.M., ASME Proceedings FED-Vol. 22.

Freymuth, P., Bank, W., Palmer, M. (1985) Use of titanium tetrachloride for visualization of accelerating flow around airfoils. Flow Visualization III, Hemisphere Publishing Corp., pp. 99-105.

Freymuth, P., Bank, W., Palmer, M. (1985) Further experimental evidence of vortex splitting, J. Fluid Mech. 152, pp. 289-299.

Freymuth, P., Finaish, F., Bank., W. (1985) Three-dimensional vortex patterns in a starting flow, J. Fluid Mech. 161, pp. 239-248.

Freymuth, P. (1986) Visualizing the combined system of wing tip and starting vortices, TSI Flow Lines, Premier Issue, May.

Freymuth, P., Finaish, F., Bank. W. (1986) The wing tip vortex system in a starting flow, Zeitschrift·fuer Flugwissenschaften und Weltraumforschung (ZFW) 10, pp. 116-118.

Freymuth, P., Finaish, F, Bank, W. (1987) Further visualization of combined wing tip and starting vortex systems, AIAAJ 25, pp. 1153-1159.

Freymuth, P. (1988 a) Visualizing the connectivity of vortex systems for pitching wings, First National Fluid Dynamics Congress, July 24-28, Cincinnati. To appear as conference paper.

Freymuth, P. (1988 b) Propulsive vortical signatures of plunging and pitching airfoils. Paper AIAA-88-0323, a shorter version also to appear in the AIAAJ.

Freymuth, P. (1988 c) Vortices, To appear in Handbook of Flow Visualization, Ed. Wen Jei Yang, Hemisphere Publ. Corp.

Freymuth, P. (1988 d) Vortex patterns of dynamic separation: New directions. To appear in Encyclopedia of Fluid Mechanics 8, Ed. N.P. Cheremisinoff, Gulf Publishing.

Gad-el-Hak, M., Ho, C. -M., Blackwelder, R. F. (1984) A visual study of a delta wing in steady and unsteady motion. Workshop on Unsteady Separated Flow, Aug. 10-11, 1983, Air Force Academy, Colorado Springs, pp. 45-51.

Gad-el-Hak, M., Ho, C. -M. (1985) The pitching delta wing, AIAA J. 23, pp. 1660-1665.

Gad-el-Hak, M. (1986) The use of dye-layer technique for unsteady flow visualization ASME J. Fluids Eng. 108, pp. 34-38.

Gad-el-Hak, M., Ho, C. -M (1986) Unsteady flow around three-dimensional lifting surfaces. AIAA J. 24, pp. 713-721.

Gad-el-Hak, M. (1987) Unsteady separation on lifting surfaces. Appl. Mech. Reviews 40, pp. 441-453.

Head, M. R. (1982) Flow visualization in Cambridge University Engineering Department, Flow Visualization II, Hemisphere Publishing Corp., pp. 399-403.

Koochesfahani, M. M. (1987) Vortical patterns in the wake on an oscillating airfoil. Paper AIAA-87-0111, also to appear in the AIAA J.

Lippisch, A (1935) Versuche zur Sichtbarmachung von Stromlinien. Jahrbuch der Vereinigung fur Luftfahrtforschung (VLF), pp. 118-127.

Lugt, H. J. (1983) Vortex flow in nature and technology, John Wiley & Sons Inc.

Maltby, R. L. (1962) Flow visualization in wind tunnels using indicators. AGARD ograph 70.

Merzkirch, W. (1987) Flow visualization, Academic Press.

Modi, V. J., Sun, J. L. C., Akutsu, T., Lake, P., McMillan, L., Swinton, P. G., Mullings, D. (1981) Moving-surface boundary-layer control for aircraft operation at high incidence. J. Aircraft 18, pp. 963-968.

Molloy, M., Lee, M., DePinto, J., Forbes, J., Roberts, B., Somerville, R., Shannon, D. (1987) Investigation of a plunging airfoil as a model of natural propulsion. Senior design internal report, U. of Colorado, Dept. Aerospace Engineering Sciences, May (Available upo request).

Monnet, P., Coutanceau, M., Daube, O., Ta Phuoc Loc (1985) The use of visualization as a guide in numerical determination of the flow around an abruptly accelerated elliptic cylinder or airfoil. Flow Visualization III, Hemisphere, pp. 363-368.

Mueller, T. J. (1983) Flow visualization by direct injection. Fluid Mechanics Measurements, Goldstein, R. E. Editor, Hemisphere, pp. 307-375.

Mueller, T. J. (1987) Visualization of the laminar separation bubble on airfoils at low Reynolds numbers. Flow Visualization IV, Hemisphere Publ. Corp., pp. 359-364.

Mueller, T. J. (1988) Smoke visualization in wind tunnels, Handbook of Flow Visualization, Hemisphere Publishing Corp. To be published.

Pankhurst, R. C., Holder, D. W. (1952) Wind-tunnel technique, Pitman and Sons, p. 141.

Payne, F. M., Ng. T. T., Nelson R. C., Shiff, L. B. (1986) Visualization and flow survey of the leading edge vortex structure on delta wing planforms. Paper AIAA-86-0330.

Peterka, J. A., Cermak, J. E. (1974) Wind engineering study of One Williams Center, Tulsa. Technical Report, Fluid Mechanics Program, Engineering Research Center, Colorado State University, p. 5, December.

Prandtl, L., Tietjens, O. G. (1934) Applied Hydro- and Aeromechanics, Dover Publications, Inc., New York, p. 268.

Settles, G. S. (1986) Modern developments in flow visualization, AIAA J. 24, pp. 1313-1323.

Simmons, L. F. G., Dewey, N. S. (1931) Photographic records of flow in the boundary layer. Reports and Minutes, National Adv. Comm. Aeronautics, No. 134 and 135, London.

Taneda, S. (1977) Visual study of unsteady separated flow around bodies, Prog. Aerospace Sci. 17, pp. 287-348.

Werlé, H. (1971) Visualization hydrodynamique d'ecoulements instationnaires, ONERA Note Technique No. 180.

Werlé, H. (1973) Hydrodynamic flow visualization, Annual Reviews Fluid Mechanics 5, pp. 361-382.

Werlé, H. (1986) Separation structures on cylindrical wings. Rech. Aerospatiale 1986-3, pp. 53-74.

Wille, R. (1963) Growth of velocity fluctuations leading to turbulence in free shear flow, Technical Report, Hermann Foettinger Institute fuer Stroemungstechnik, Technische Universitaet Berlin, June.

Three-Dimensional Quantitative Flow Diagnostics

Richard B. Miles[1]
Daniel M. Nosenchuck[2]

1 INTRODUCTION

Many of the so-called simple flows, including flat-plate boundary layers, two-dimensional compressible flows, and axisymmetric rotating flows, still represent some of the most difficult, ill-understood fundamental problems in fluid mechanics today. Often, such flows contain structures that, though organized, represent a multitude of interacting scales. Inherent nonsteadiness and spacial complexity are rapidly compounded when one extends consideration to full three-dimensional, high Reynolds/Mach number flows. Our success in developing a proper understanding of these complex flows depends largely on our ability to fully characterize them. Such an understanding is a critical step to the design of structures which can operate in unsteady flow regimes and to the control of unsteady phenomena including turbulence and mixing. The challenge facing experimentalists is to augment classical single-point measurements, such as those from hot wires and laser Doppler velocimeters (LDV), with three-dimensional measurements in order to generate a correlated picture of the flow field where all the structure is recorded.

Generally, the first approach to understanding a complex flow field is to invoke some form of flow visualization to convey the lay of the land to the researcher. This qualitative approach is often quite successful in helping put broad boundaries on the parameters of the problem. An example would be the determination of the presence or absence of separation or turbulence. However, even in cases where a question as simple as whether the flow has or has not separated (Settles 1985a; Bogdonoff 1986) the qualitative nature of traditional flow visualization may not be sufficient to provide guidance. Detailed comparisons with both turbulence models and compressible gas models with real-gas effects cannot be made with qualitative information. Furthermore, comparisons with classical point

---
[1]Department of Mechanical & Aerospace Engineering, Princeton University
[2]Department of Mechanical & Aerospace Engineering, Princeton University

measurements are impossible with qualitative data. It is with this in mind that the various *quantitative* aspects of three-dimensional flow visualization are explored in this Chapter. In many instances, only by obtaining direct measurements through observation can even basic determinations be unambiguously made.

At this point the following question must be posed: why use flow visualization for quantitative results when many direct measurement techniques exist? The answer lies in part with the fundamental nature of observation — when faced with a new problem, the knowledge gained from global observations and measurements can efficiently point the direction towards localities where a small number of point measurements can be used to complete the overall understanding of the flow. Not knowing where to initially probe in the flow has given rise to innumerable instances of highly inefficient exploration of what otherwise had proven to be a relatively simple flow field. Thus, it is the field nature of fluid flows that make visualization an attractive and powerful tool for initial observations and follow-up detailed measurements.

It is the intent of this Chapter to discuss the extension of well known flow measurement techniques and to review several recently developed flow visualization techniques, including, but not limited to, volumetric laser-scanning methods, optical tracking of vibrationally excited molecules, and holographic imaging. An overview of the physical phenomena is presented in the following section and some general comments on the range of applications is included with the discussion of each technique. As the data becomes dense, the data presentation also becomes critical. Volumetric, three-dimensional structure must be seen and understood by the researcher. Various display methods will be discussed at the end of this Chapter to give the reader an overview of the display problem and some indication of how that problem is currently being addressed.

Physical phenomena based on the use of light can be divided into four categories: light scattering by small particles and molecules; atomic or molecular fluorescence; phase shifts by variations in the index-of-refraction; and flow marking by molecular tagging. The methods of optically interrogating these phenomena include light sheets, stereoscopic imaging, holography, and tomography. All these methods have the potential of freezing the flow field to generate volumetric images of unsteady phenomena. The use of acoustics and nuclear magnetic resonance for flow imaging represent fundamentally different methods of flow interrogation which are particularly useful for opaque fluids. A brief discussion of these techniques is also included.

The use of light scattering is pervasive in our understanding of fluids, ranging from common everyday flows, to unique situations found only in the laboratory. To a large degree, our intuitive understanding of fluid motion is based on the motion of clouds, fog, smoke, and dust in the air; bubbles in water, cream in coffee, and a myriad of other naturally occurring physical phenomena in which small particles mark the flow. Smoke injection into wind tunnels produced the first quantitative volumetric flow field data (Mueller 1983). Time lines were introduced by A. Lippisch (Borst 1980) using pulsed smoke injection and high-speed movies to produce velocity distributions around models. Thin light sheets generated by high-intensity lamps (McGregor 1961) or lasers (Falco 1978) have extended this technique so that two-dimensional cross sections of three-dimensional flow fields can be visualized. By pulsing the light source or using a short time gate on the camera (Stanislas 1985), the fluid motion can be frozen to give an instantaneous cross section of mixing phenomena or complex structure. If separate particles are distinguishable, two-dimensional velocity fields can be mapped by measuring the length and direction of the particle streak trajectories in the image and relating that to the exposure time (Shiraishi 1985, Dimotakis 1981), or by multiply pulsing the source (Landreth 1987). A review of these techniques is given by Adrian (1986). Numerous examples of particle scattering in both air using smoke or seeded particulates and in water using particulates or bubbles are shown in Van Dyke's, An Album of Fluid Motion (1982). With recently available high-power lasers, scattering from molecules has become a feasible alternative to particle scattering for flow diagnostics. This eliminates problems associated with seeding the flow and gives quantitative information on mixing, density, and temperature. Both Rayleigh scattering, which is direct elastic scattering of light from molecules, and Raman scattering, which is inelastic scattering of light from molecules, have been used to generate cross sectional images (Escoda 1983, Long 1983).

The extension of the particle and molecular scattering techniques into tools for three-dimensional flow field diagnostics has been the goal of several recent research efforts. Four approaches are currently being studied. The first is stereoscopic imaging with high-speed cameras to track individual particles moving in the flow field (Chang 1985a and 1985b; Kent 1982; Brodkey 1986). The second is a holographic recording of the particles using a rapidly pulsed laser and fast camera system (Zarschizky 1983; Weinstein 1986) to, once again, track individual particles in three dimensions as they move in the flow field. The third approach is a rapidly scanned laser sheet where sequential two-dimensional cross sections are imaged in order to generate a volumetric picture of the particle distribution (Utami 1987) or the molecular density (Yip 1988). Here again, sequential volumetric images must be taken in order to track individual particle trajectories. The fourth is simultaneous imaging from different colored light sheets passing through the volume. This has been used to measure density by observing both Rayleigh scattering (Yip 1986a) or scattering from submicron particles (Mantzaras 1988), and could be extended to particle tracking by operating at a high repetition rate.

Those techniques that rely on fluorescence have been possible largely due to the development of lasers. Perhaps the most extensive utilization of this technique has been with fluorescing dyes in water, together with thin laser sheets to generate cross sections of turbulent structure and mixing phenomena. The advantage of this method is that the dye fluoresces in a spectral region redder than the exciting laser beam, so secondary scattering is not a problem (Dewey 1976). Consequently, dye fluorescence can be made very bright and can be easily photographed with a camera or videocamera system. With smoke or particles, secondary scattering may illuminate some of the flow outside of the laser sheet and cause background noise. Sweeping the laser sheet through the region of interest can generate volumetric images of the fluorescing dye (Nosenchuck 1986).

In compressible flows, either a naturally occurring species can be used, such as oxygen (Massey 1984, Lee 1987), or some fluorescing atomic or molecular species can be added to the flow, such as sodium into helium (Zimmermann 1980) or nitrogen (Zimmermann 1981), iodine into helium, nitrogen or air (Cenkner 1982; McDaniel 1983; Zimmermann 1982), or nitric oxide into air (McKenzie 1981). Direct Rayleigh scattering from molecules in the flow, together with simultaneous laser-induced fluorescence, can potentially yield two-dimensional temperature cross sections (Miles 1988b).

The conventional techniques that rely on variations in the index-of- refraction include schlieren, shadowgraph, and interferometry. These methods have contributed enormously to the understanding of fluid motion, but they have the serious limitation of being path integrated. In that sense, they are not truly planar or volumetric, and cannot be used for characterizing complex or unsteady flow fields except in very limited circumstances. For simple structures, three-dimensional information may be obtained by stereographic schlieren images. For more complex fields, optical tomography based on beam deflection (Faris 1987), and interferometry (Sweeney 1974, Snyder 1988) has extended these index-of-refraction-based techniques to two- and three-dimensional temperature and density measurements.

The direct interaction of the light with molecules in the flow has led to the development of new laser tagging methods in which the flow is marked at a particular location and the motion of those marked molecules is followed to generate quantitative velocity information. Time lines similar to those marked by the hydrogen bubble technique can be written into the flow. Early work in this area relied on seeding the flows with molecules which absorbed some of the energy of a high-intensity light and underwent a photochromic chemical change (Popovich 1967 a and b). More recently, photochromic molecules have been activated by UV lasers permitting grid lines to be written directly into the fluid (Falco 1987). The distortion of the grid with time gives velocity and vorticity. Similarly, laser-induced photodissociation of H_2O forms OH which can be used to follow water vapor- bearing flow fields (Shirley 1988). In air, the stimulated Raman effect has been used to vibrationally excite thin lines of oxygen molecules which are subsequently interrogated to give instantaneous velocity profiles (Miles 1987). These techniques show great promise for generating volumetric velocity fields, and, in the latter case, can be coupled with Rayleigh scattering to simultaneously generate density cross sections (Miles 1989).

Virtually all diagnostic techniques have the capability of generating three-dimensional images of steady-state flow fields since long integration times and many steps can be used to accumulate flow field data. The challenge arises when three-dimensional images of time-varying flow fields are required. It is these unsteady phenomena which are of critical interest, and the discussion in the following sections will reflect on the potential of each of these diagnostic techniques for volumetric imaging of unsteady phenomena.

In many cases the flow must be seeded with particulates, dye, or specific molecular or atomic species. If the seed material is to reflect density information, then particular attention must be paid to the uniformity of the seeding. For velocity measurements with particles, seeding too sparsely leads to gaps in the velocity field measurement, and seeding too densely causes secondary scattering, aliasing, and particle ambiguity problems in flow tracking. In high-speed compressible flows, particles cannot follow unsteady phenomena, so particle seeding is of limited utility. In some cases the seed material must be heated before being injected into the flow, so care must be taken to avoid thermally distorting the flow field. Side effects of seeding, including corrosion, noxious odors, and environmental pollution, may also be factors of importance.

2 PHYSICAL PROCESSES

2.1 PARTICLE, RAYLEIGH, AND RAMAN SCATTERING

Perhaps the most easily understood techniques are based on particle imaging where light scattered from particles is recorded so the location and the motion of each particle can be found. The amount of light collected is related to the light scattering cross section, σ. For large particles, σ is simply the projected area of the particle, (πr^2 for a sphere), or the area of the shadow that the particle casts in the illumination beam. Assuming the particle is nonabsorbing, all of the light that falls within the cross sectional area of the particle is scattered.

An additional amount of light is scattered by diffraction around the edge of the particle. This contribution is seen only in the forward direction and corresponds to light diffracted from its otherwise straight trajectory. For very large particles, this light is separable from other scattering and is usually neglected because of its small angle. Babinet's Principle (Van de Hulst 1981) states that the diffraction pattern around an opaque disk is identical to that from a transparent circular hole of the same size, i.e., diffraction is independent of the contrast. Infinitely far from a source of limited size, the light diverges as if it came from a point source, so all the light passing through the hole is scattered. Thus, the true scattering cross section of a large particle is twice the area of the shadow: one contribution from the reflected or refracted light and one contribution from the diffracted light. The diffraction component is only important if the scattering is observed in the far field of the particle, $z > \pi r^2/\lambda$, where λ is the wavelength of the light. In this regime the angular variation of the diffraction pattern becomes the Fourier transform of the shadow. For very small particles the diffraction angle is large and the diffraction component is no longer separable from the reflected or refracted component. Due to diffraction, intermediate-size particles such as dust are easiest to see in forward scattering, and forward scattering is used where larger signal levels are required.

Since the light collection optics occupies only a small portion of the scattering sphere, the differential cross section, $\partial\sigma/\partial\Omega$, is usually used.

$$\sigma = \oint \frac{\partial\sigma}{\partial\Omega} d\Omega \qquad\qquad 1$$

Where $\partial\sigma/\partial\Omega$ is, in general, a function of both θ and ϕ, the two scattering angles (θ is the angle measured from the propagation direction, and ϕ is the angle around the scatterer). For spherical particles, the ϕ-dependence is only due to the polarization of the light. For large particles whose radius, r, is much greater than the wavelength of the light, the dependence of $\partial\sigma/\partial\Omega$ is easily computed by geometric optics. As r decreases to the scale of the wavelength, the scattering becomes very complex due to interference effects at the boundary of the particle (Van de Hulst 1981). In the very small particle regime, the scattering once again becomes simple and is derivable from a forced dipole oscillator model.

$$\left(\frac{\partial\sigma_D}{\partial\Omega}\right)_s = \frac{\omega^4|\alpha(\omega)|^2}{(4\pi)^2 c^4 \epsilon_0^2} \qquad\qquad 2$$

$$\left(\frac{\partial\sigma_D}{\partial\Omega}\right)_p = \frac{\omega^4|\alpha(\omega)|^2}{(4\pi)^2 c^4 \epsilon_0^2} \cos^2\theta \qquad\qquad 3$$

where $\partial\sigma_D/\partial\Omega$ is the differential scattering cross section from a dipole and ω is the radial frequency of the light. The subscript, s, indicates scattering with the polarization of the light perpendicular to the scattering plane (the plane that contains the source, the scatterer, and the observer); and the subscript, p, indicates scattering with the polarization in the scattering plane. c is the speed of light, and $\epsilon_0 = 8.854 \times 10^{-12}$. $\alpha(\omega)$ is the polarizability of the dipole.

$$\alpha(\omega) = \sum_{g,i} \alpha_{g,i}(\omega), \qquad\qquad 4$$

$$\alpha_{g,i}(\omega) = \frac{f_{g,i}e^2\rho_g}{2\omega_{g,i}m_e}\left(\frac{1}{\omega_{g,i}-\omega-\frac{i\gamma_i}{2}} + \frac{1}{\omega_{g,i}+\omega+\frac{i\gamma_i}{2}}\right), \qquad\qquad 5$$

where the subscript g refers to the ground states of the transitions and i refers to the upper states. $f_{g,i}$ is the oscillator strength (between 0 and 1), e is the electron charge, m_e is the electron mass, and ρ_g is the population fraction of the ground state. The temperature dependence is contained in the ρ_g term which is a function of the vibrational state, v, the rotational state, J, and the energy of the states, $E(v, J)$:

$$\rho_g = \frac{(2J+1)\exp[-E(v,J)/kT]}{\sum_{v,j}(2J+1)\exp[-E(v,J)/kT]}. \qquad\qquad 6$$

γ_i is the full width at half-maximum of the resonance and depends predominately on the collision frequency and the lifetime of the upper state:

$$\gamma_i = 2C_i + A_i + D_i + Q_i \qquad\qquad 7$$

where C_i is the collision frequency, A_i is the fluorescence rate, D_i is the predissociation rate, and Q_i is the quenching rate.

If the particles are molecular and constitute a continuous medium, such as air or water, then the ratio $\alpha(\omega)/\epsilon_o$ can be related to the index-of-refraction, n, by the expression:

$$\frac{\alpha(\omega)}{\epsilon_0} = \frac{3}{N}\frac{n^2-1}{n^2+1} \qquad\qquad 8$$

where N is the density of molecules. At STP, $N = 2.69 \times 10^{19}/cm^3$. Usually, scattering is observed far from resonance, so $\alpha(\omega)$ is a real number as is the index-of-refraction. (Close to resonance, the imaginary part of α becomes significant and corresponds to dipoles being driven into an excited state. This leads to absorption of the laser light and, if the excited dipoles relax radiatively, laser-induced fluorescence.)

Scattering is observed because the phases of the dipoles are uncorrelated due to random motion, thus the scattered light can be separately added from each dipole. This is called Rayleigh scattering, and the total power of the Rayleigh scattered light collected from one volume element ΔV is:

$$P = \eta I N \Delta V \int \frac{\partial \sigma}{\partial \Omega} d\Omega \qquad 9$$

where I is the intensity of the illumination ($watts/cm^2$), P is the collected power, and η is an efficiency factor which includes the detector efficiency and optical losses. The integral is over the collection angle of the optics.

Referring back to Eqs. 2 and 3, we note several important features:

1. The scattering increases as ω^4 (this means that blue light is more strongly scattered, i.e., the sky is blue);

2. There is an equal amount of scattering in the forward and backward direction;

3. There is significant polarization for 90° scattering.

Table I gives values of the index-of-refraction at 0.532 μ (doubled Nd:YAG laser) together with calculated and measured values of the differential Rayleigh scattering cross sections for various molecular gases at STP.

Scattering from particles where $r > \lambda$ is usually called Mie scattering. Tyndall scattering refers to light scattering from particles where $r \leq \lambda$, and Rayleigh scattering refers to scattering from atoms or molecules where $r \ll \lambda$. Smoke experiments generally fall into the Tyndall scattering regime, where the scattering from each particle is very small, but large densities of particles are present. Particle velocimetry and holography require that individual particles be tracked, so large particles must be used at lower densities. Density measurements can be made directly using Rayleigh scattering from the molecules in the flow. The different scattering cross sections of such gases as freon and air also make Rayleigh scattering useful for observing mixing phenomena (Yip 1986b). Since the detected light is at the incident laser frequency, scattering from particles in the flow can easily obscure the Rayleigh signal.

Raman scattering occurs when the scattering process causes the molecular state to change. Due to energy conservation, the scattered light is then at a different frequency than the incident light. Since the energy states of each molecule are well defined, Raman scattering may be used for species identification by observing the energy shift, and for temperature measurements by observing the relative intensities of the scattered light from the various rotational states of the molecule. Table I gives the scattering cross sections and vibrational Raman shifts for various molecules of particular interest to gas diagnostics. Molecules may gain energy, thus causing the scattered light to be at a lower frequency than the incident light (Stokes scattering), or they may lose energy, causing the scattered

GAS	Index-of-Refraction $(n-1) \times 10^3$	Rayleigh Cross-section cm^2/sr Calc. 10^{28}	Rayleigh Cross-section cm^2/sr Meas. (f) 10^{28}	Raman Cross-section cm^2/sr Meas. (g) 10^{32}	Raman Shift cm^{-1}
N_2	.2994 (a)	6.12	6.13	37.1	2331
He	.0349 (b)	0.083	0.097	–	–
Air	.2935 (c)	5.88	6.11	–	–
O_2	.2721 (e)	5.05	5.49	45.5	1555
H_2	.1399 (e)	1.34	1.32	80.0	4156
CO_2	.4511 (d)	13.90	19.43	56.1	1388
				37.0	1285
H_2O	.2531 (e)	4.37	–	91.9	3652
(vapor)				0.354	1595

All Values Scaled to 0.532μ and $0°C$, 760 mm

a. (Peck 1966)   e. (Washburn 1930)
b. (Peck 1983)   f. (Shardanand 1977)
c. (Peck 1972)   g. (Schrotter 1979)
d. (Old 1981)

Rayleigh and Raman Scattering Cross-sections for Common Gases
TABLE I

light to be at a higher frequency than the incident light (anti-Stokes scattering). Due to the Boltzmann distribution of molecular populations (Eq. 6), the ratio of the number of excited molecules to unexcited molecules is related to the temperature, so finding the ratio of anti-Stokes to Stokes scattering is another way of determining the temperature.

The use of Raman scattering as a probe for gas diagnostics is well established (Lapp 1974) as are numerous forms of nonlinear Raman spectroscopy (Harvey 1981). Planar images with Raman scattering have been achieved, but require multiple passes of a very high energy laser to generate enough scattered light to record an image. Long et al. (1983) used a 1.5 J laser source to record a two-dimensional single sheet planar image of CH_4 and of D_2 distributed in an air jet. In order to get enough scattering for an image, the laser had to be multiply passed through the jet using a pair of cylindrical mirrors. This resulted in a factor of 30 increase in the effective laser intensity. In a manner similar to Rayleigh scattering, the Raman scattering cross section scales as ω^4 plus the increase in the polarizability as ultraviolet resonances are approached. Bischel et al. (1983) have shown that the cross section of oxygen is $10^{-27} cm^2/sr$ at 0.193 μ, more than 2000 times its value of 0.532 μ; and that of H_2 is 2×10^{-28} cm^2/sr at 0.193 μ, or 250 times the 0.532 μ value. With new high-energy lasers in the ultraviolet, Raman scattering may become a more practical volumetric diagnostic tool. It is particularly attractive since it is species selective and particle scattering can be eliminated if the scattering at the incident laser frequency can be efficiently filtered out.

2.2 FLUORESCENCE

The fluorescence mechanisms are somewhat more complicated than particle scattering, but the fluorescent signal can again be expressed as a differential scattering cross section which is a function of the polarizability. In this case, the light source (laser) must be tuned onto a specific transition of an atom or molecule which is in the flow. The cross section for light absorption depends on the imaginary part of the polarizability, which is strongly dependent on the frequency of the laser as well as on the laser linewidth and on the atomic or molecular linewidth. Following absorption, some portion of the light is re-emitted as fluorescence, depending on the transition strength and competing processes such as quenching and predissociation. In high-intensity fields, saturation limits the strength of absorption.

Taking these factors into consideration, the differential fluorescence scattering cross section $\partial \sigma_f / \partial \Omega$ can be written:

$$\frac{\partial \sigma_f}{\partial \Omega} = \text{Emission Efficiency} \times \text{Absorption Cross Section} \qquad 10$$

or

$$\frac{\partial \sigma_f}{\partial \Omega} = \sum_{g,i} \left(\frac{A_i}{A_i + Q_i + D_i} \right) \times \frac{\omega_{g,i}}{4\pi c \epsilon_o} \Im\{\alpha_{g,i}(\omega)\} \qquad 11$$

where $\Im$ signifies the imaginary part.

The values of the fluorescence rate, A_i, the quenching rate, Q_i, and the predissociation rate, D_i, are strongly dependent on the particular transition excited. The quenching rate also depends on the frequency of collisions and the species with which the collisions occur. In a moving gas, dipoles are Doppler shifted by the motion, so the frequency, $\omega_{i,g}$ becomes $\omega'_{i,g} = \omega_{i,g}(1 - v/c)$, where v is the velocity component in the direction of illumination. This shift can be used to generate velocity and temperature measurements (Miles 1975, McDaniel 1983).

The fluorescence lines from molecules span a broad frequency spectrum, lying predominately to the red of the exciting laser. Atoms have a few well defined spectral lines. In some cases, specific fluorescence lines are selected by optical filters, but, in general, detectors collect the broad bandwidth radiation. The dependence of fluorescence on laser frequency, transition strength, and the local environment significantly complicates efforts to use fluorescence intensity to make quantitative measurements such as temperature and density. Velocity measurements use either the Doppler shift or line tagging, both of which are not dependent on the absolute value of the fluorescence, so they are somewhat easier to accomplish. By scanning the exciting laser across an absorption line and observing the fluorescence, temperature, velocity, and density can be extracted from the line profile since temperature and velocity contribute to the Doppler shift and density contributes to the collision frequency (Miles 1975). Figure 1 is a composite picture taken from a movie which shows the fluorescence from a sodium seeded Mach 3.4 nitrogen jet as a function of laser frequency (Zimmermann 1985). The images begin at the top with the laser tuned to highlight the low-velocity components and sequentially move toward the high-velocity components. The jet is underexpanded so the repeated shock structure is apparent. By quantitatively following the intensity at any point, the velocity, temperature, and density can be determined.

Scanning the laser over a broader range produces a picture of the absorption band, which, in the case of molecules, is made up of numerous transitions from the various rotational states. For heavier molecules, such as iodine, this

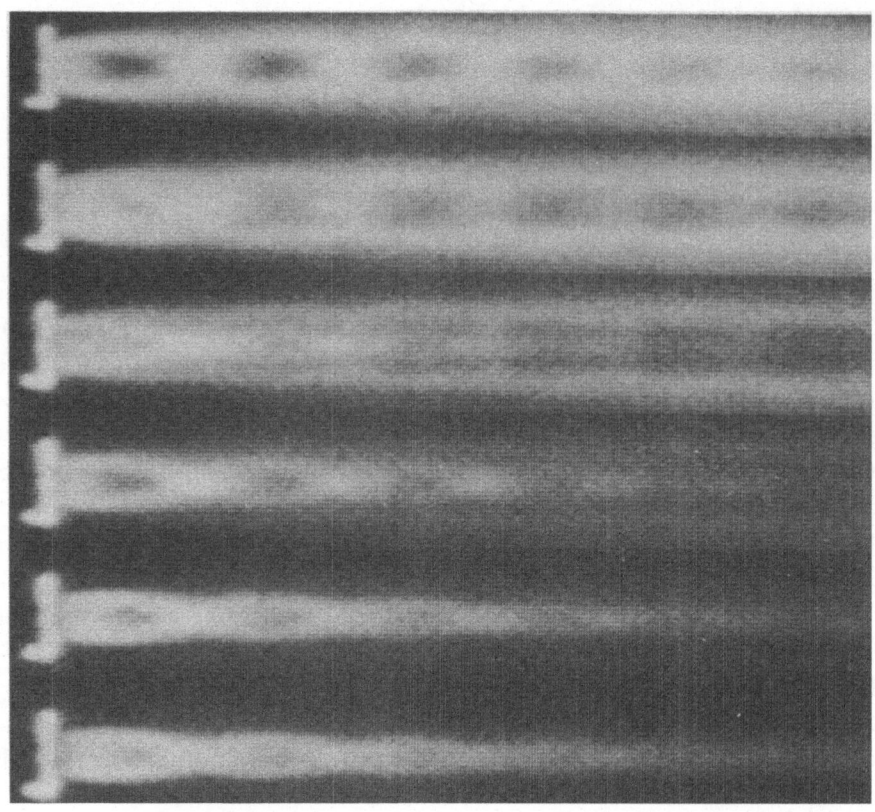

Figure 1. Cross-sectional images of a Mach 3.4 nitrogen jet seeded with trace amounts of sodium vapor. The illuminating laser is scanned in frequency to sequentially highlight the low (top) to high (bottom) axial velocity components. By continuously recording the fluorescence intensity as the laser is tuned, velocity, temperature, and nitrogen density at any point can be determined. (Zimmermann 1985)

structure can be significantly complicated by hyperfine splitting due to interactions with the nucleus. Since the rotational populations are thermally determined (see Eq. 6), recording the fluorescence as the laser is tuned through transitions from different rotational states can give the temperature if the quenching of each state is know or if the quenching is constant for all the states observed.

An indication of the temperature dependence and the complexity of the absorption band structure is shown in Fig. 2 for oxygen observed over the tuning range of an argon-fluoride laser (Miles 1988b). The dashed lines correspond to the gh scattering, which depends on $|\alpha|^2$, and the solid lines correspond to the laser-induced fluorescence, which depends on $\Im\{\alpha\}$. Due to a very large predissociation rate ($D > 10^5 \times A$), quenching and collisions are relatively unimportant and fluorescence cross sections are rather low, but the spectrum of oxygen is practically independent of collisions and quenching. Temperature can be found from a comparison of the relative strength of various absorption lines by observing the different fluorescence intensities (Massey 1984). Alternatively, one could simultaneously observe Rayleigh scattering and fluorescence to measure density and temperature (Miles 1988b). If the pressure is constant, the temperature can be found from the density (Lee 1987). In combusting or seeded flows, OH or NO can be used for temperature and density measurements (Cattolica 1985; McKenzie 1981; Kychakoff 1983), but quenching must be taken into account.

2.3 INDEX-OF-REFRACTION AND INTERFERENCE

The electric field of a single frequency plane wave, propagating through a medium, is expressed:

$$E = E_o \cos(\omega t - kz), \qquad 12$$

where k is the propagation constant which depends on the index-of-refraction of the medium:

$$k = \frac{\omega n}{c} \qquad 13$$

In vacuum, the index-of-refraction is one, but in transparent media such as air and water, the index-of-refraction is directly related to the polarizability and the density as was shown in Eq. 8. In gases at reasonable pressure the index-of-refraction is close to one, so Eq. 8 simplifies to:

$$n = 1 + \frac{N\alpha(\omega)}{2\epsilon_o}. \qquad 14$$

Close to resonance, the index-of-refraction becomes complex, leading to absorption (and fluorescence).

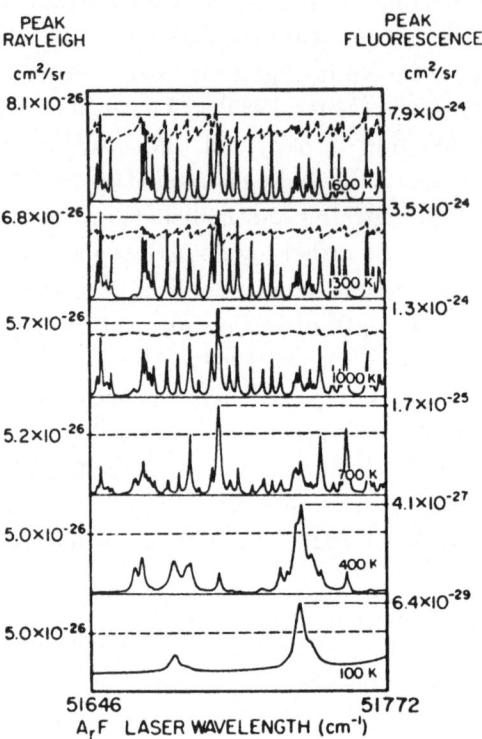

Figure 2. Total fluorescence (solid lines) and Rayleigh scattering (dashed lines) of oxygen as a function of ArF laser wavelength for temperatures between 100K and 1600K. (Miles 1988b)

With a given wavelength and a given molecular species or mixture, $\alpha(\omega)$ is constant. Consequently, variations in the index-of-refraction are directly related to variations in the gas density. If different gases are not well mixed, then a variation of the index-of-refraction will occur due to the different polarizabilities of the gases.

If the index-of-refraction has spatial variations, the phase fronts of the propagating plane wave passing through the medium are distorted, causing an associated refraction of the light and a local deviation of the beam. A simple projection through a flow with strong shock structure generates a shadowgraph since light passing in the vicinity of the shocks has been bent away, casting a shadow behind the shock. In schlieren, the unbent light is focused to a point. The light which has been refracted does not pass through this point so various methods of filtering to highlight the deflected light at the image plane can be done (Holder 1963, Settles 1985b) For example, if the light at the point is blocked, only the deflected light will pass, giving an image with contrast reversal. In both shadowgraphs and schlieren, the light bending is related to the gradient of the index-of-refraction integrated through the sample volume.

If the original light was spatially coherent, as is the case when a laser is used, then one may also construct an interference pattern by combining the light which is passed through the sample volume with a plane wave at the same frequency. Usually this is done by splitting off part of the laser beam and passing it around the test section to act as the reference. The interference occurs because the intensity is proportional to the time-averaged square of the electric field. In this case the electric field is made up of two components with phase differences that add and interfere due to the different effective path lengths traveled. In the absence of any flow structures, straight lines appear due to the intersection of two plane waves. In the presence of flow structure, the lines are distorted giving a quantitative measure of the index-of-refraction variation integrated along a path. With a pulsed laser these interferograms can be recorded virtually instantaneously, freezing unsteady flow structure (Bryanston-Cross 1986).

2.4 FLOW MARKING

Hydrogen Bubbles

Perhaps the most effective marking technique is the use of electrolytically generated hydrogen bubbles in a water channel. This technique has been applied since the mid-1950's (Geller 1955) and has been quite useful in measurements of the structure of turbulent boundary layers. A review of the hydrogen bubble technique is given by Merzkirch (1987) in his book on flow visualization.

The concept relies on the fact that when two electrodes are placed in water and are driven with a voltage greater than the decomposition potential, hydrogen bubbles are formed at the cathode and oxygen bubbles at the anode. If the electrodes are thin wires, these bubbles will act as tracers in the flow field. The smaller the bubble, the slower the rate of rising motion and the more accurately it follows the flow. Since the hydrogen bubbles from the cathode are much smaller than the oxygen bubbles, they are used as tracers and the anode is located away from the test region. Smaller wires generate smaller bubbles and a typical wire diameter is 25 microns (Smith 1983). High-voltage pulses on the order of a few hundred volts for a few microseconds in duration generate time lines in the flow which are then photographed using a synchronized strobe light and a camera. The velocity must be corrected for the wake defect close to the wire, and very close to the wall reflections of the strobe light make bubble line discrimination difficult. Lu and Smith (1985) have developed computer algorithms to track the bubbles and identify time lines. An example is shown in Fig. 3. The computerized data can then be used to generate sequences of instantaneous velocity profiles and time-averaged statistics. Wires can be partially insulated or kinked to develop streak line patterns and several wires can be placed in the flow to generate volumetric flow patterns. By crossing these wires, grid structures can be marked giving both time lines and streak lines (Matsui 1979). At high velocities the bubbles rapidly diffuse so this flow marking technique is limited to low-speed flows.

Photochromism

It has long been known (Brown 1971) that light can induce a reversible color change in certain chemical species. Goldfish et al. (1965) used the 'blueprint' reaction between ferrous and ferricyanide ions illuminated by a high-intensity xenon flashlamp to tag a 1" long by 3/8" in diameter portion of the fluid. The technique was extended by Popovich and Hummel (1967a and b) to a line marked by focusing the light from a flash tube. They used 2-(2,4- dinitro-benzyl)-pyridine dissolved in 95% ethyl alcohol, and were able to observe flow conditions in the viscous sublayer. The marked region of the flow was interrogated by a second flash tube which was delayed between 60 microseconds and a few milliseconds by an electronic time-delay circuit.

More recently, the photochromic flow tagging method has been signif-icantly improved by using an ultraviolet laser and a pair of oversized, blazed reflection gratings to generate a two-dimensional grid of lines in the flow (Falco 1987). In this case the chemical used was Kodak 1,3-Trimethyl-8- nitrospiro[2-H-

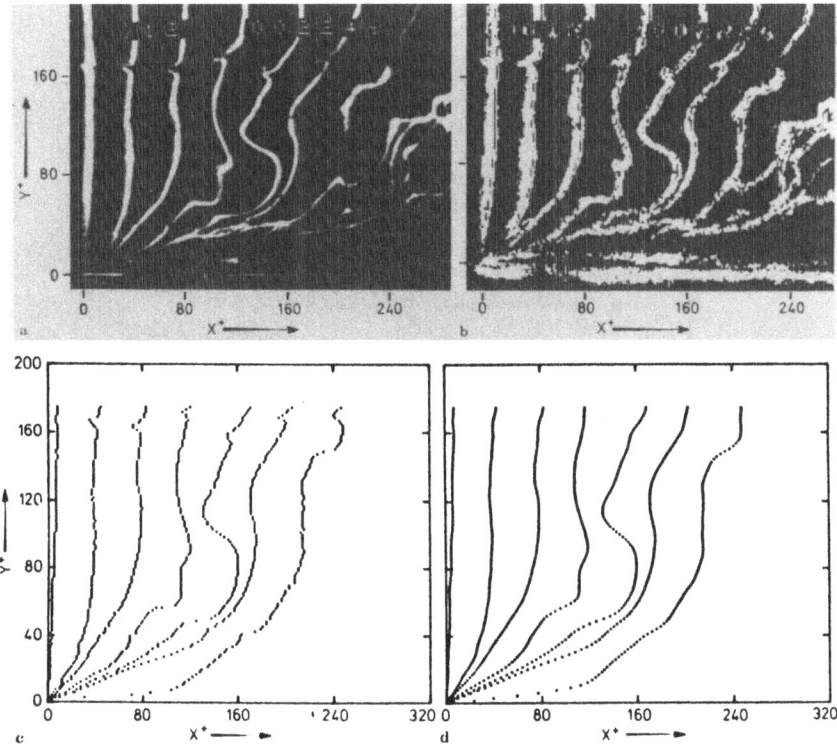

Figure 3. Individual frame of hydrogen bubble time-line pattern in a turbulent boundary layer. Re = 1,120 and the bubble pulse frequency = (1/30) sec. (a) Original video frame; (b) redisplayed video frame; (c) computer-identified time-lines; (d) filtered time-lines. (Lu and Smith 1985)

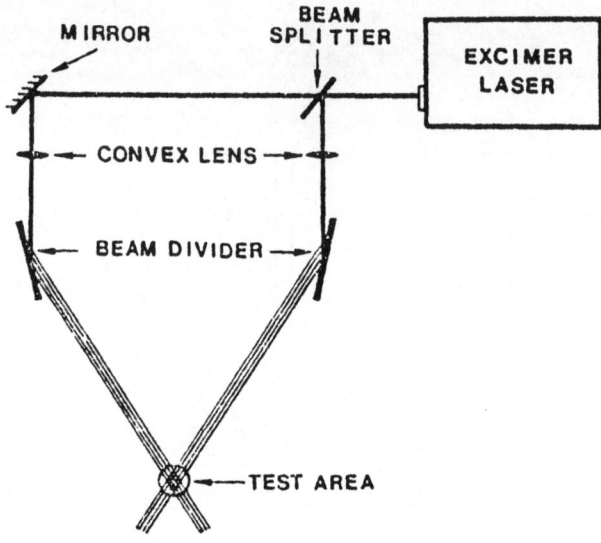

Figure 4. Optical setup to produce a grid of photochromically tagged lines. An XeF laser is scattered off gratings at grazing incidence to generate the multiple lines. (Falco 1987)

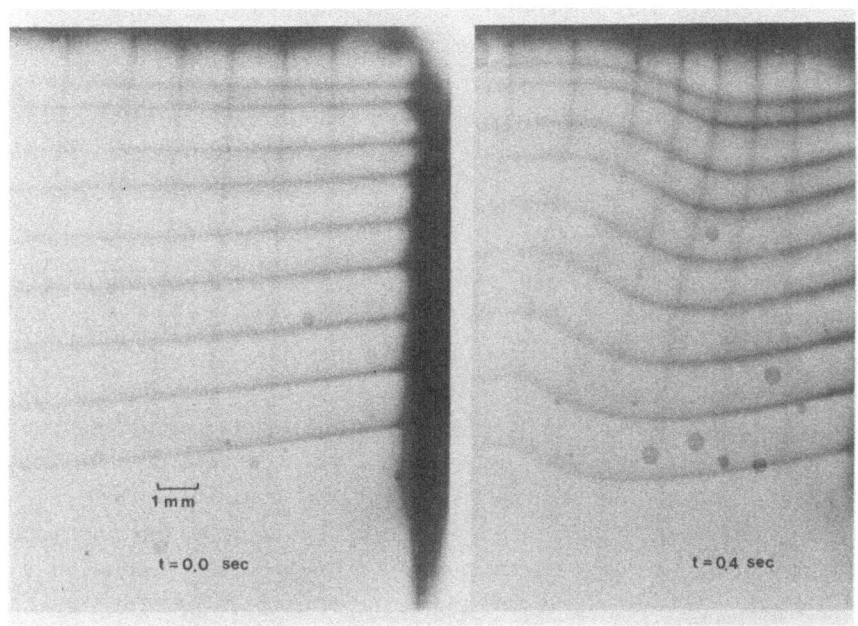

5a   5b

Figures 5a - 5b. Example of a photochromically marked
grid (a) shortly after the laser pulse and (b) 0.4 sec later.
(Falco 1987)

1-benzopyran-2,2'-indoline] excited at 0.351 μ using a XeF excimer laser with a 100 mj output pulse of 20 nsec duration. The working fluid was kerosene, and the photochromic effect was observed to last on the order of tens of seconds. Photoactivated fluid reflects blue light which is photographed by a camera. A diagram of the experimental set-up is shown in Fig. 4. Examples of the grid shortly after the laser pulse and 0.4 seconds later are shown in Figs. 5a and 5b. The accuracy of this technique is limited by the grain size of the film used, the size of the lines written, diffusion, and distortion during the developing process.

Photodissociation

Photodissociation of water molecules has been used in a similar manner to generate time-gated lines of OH molecules in water vapor-bearing flows (Shirley 1988). A KrF laser (0.248 μ) is focused to a line passing through the flow and generates OH molecules by two-photon photodissociation. Some of the OH is produced in the electronically excited state, so fluorescence recorded by a vidicon marks the initial location of the tagged line. A short time later, a high-intensity, pulsed dye laser excites the OH transition at 0.3081 μ which causes the molecules to fluoresce. The image of this displaced line is also recorded on the vidicon. The displacement of the line then gives a measure of the velocity. Resolution is limited by the thermal diffusion of the OH molecules caused, in part, by local heating during the photodissociation step.

Vibrationally Excited Molecules

A method similar to the OH tagging has been reported (Miles 1987) which combines Raman Excitation and Laser-Induced Electronic Fluorescence (RELIEF) to vibrationally tag and subsequently interrogate oxygen molecules in air over a wide range of velocity, temperature, and density. The long vibrational lifetime of oxygen allows the marked pattern to be interrogated at a later time to give quantitative velocity measurements and instantaneous images of the flow structure. A simultaneous cross section of density can be recorded using Rayleigh scattering (Miles 1989) and, by tuning the tagging lasers, the time-averaged temperature can be found (Miles 1988a).

A diagram of the experimental setup is shown in Fig. 6. The tagging lasers raise oxygen from its ground state ($v" = 0$) to the first vibrationally excited state ($v" = 1$). This is done with a pair of high-power lasers which cause a stimulated Raman transition. These lasers are separated in frequency by the vibrational frequency of the oxygen molecule. They are shown entering the test chamber

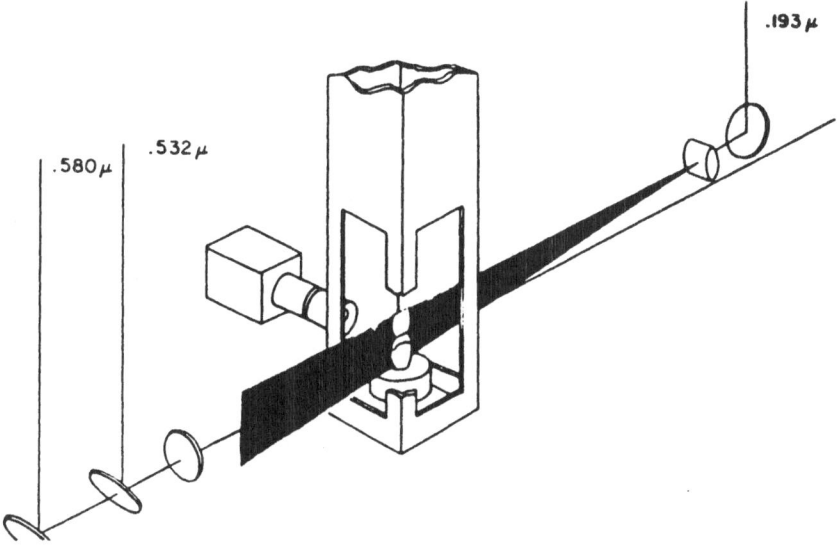

.193 μ

.532 μ

.580 μ

Figure 6. Diagram of oxygen molecular tagging in an underexpanded sonic jet by Raman Excitation followed by Laser-Induced Electronic Fluorescence (RELIEF). The tagging lasers enter from the left and the interrogation laser enters from the right. (Miles 1989)

from the left in Fig. 6.

The rate at which molecules are excited per unit volume can be found from the expression:

$$\dot{N}_{ex} = \frac{4\pi^2}{\hbar c^2} I_1 I_2 \chi'',$$ 15

where $\hbar$ is Planck's constant divided by 2π, I_1 and I_2 are the two laser intensities, and χ'' is the imaginary part of the nonlinear susceptibility.

$$\chi = \chi' + i\chi'' = \sum_{i,j} \frac{c^4}{\hbar \omega_R^4} \left(\frac{\partial \sigma_R}{\partial \Omega}\right)_{i,j} \frac{(N_i - N_j)\left[\Delta\omega_{i,j} - \Delta\omega + i\frac{\gamma_{i,j}}{2}\right]}{\left[(\Delta\omega_{i,j} - \Delta\omega)^2 + \left(\frac{\gamma_{i,j}}{2}\right)^2 \left(1 + \frac{I_1 I_2}{K_{i,j}}\right)\right]}$$ 16

$\partial \sigma_R / \partial \Omega$ is the differential Raman scattering cross section, N_i and N_j are lower and upper state molecular population densities, ω_R is the frequency of the Raman shifted (lower energy) pump laser, $\Delta\omega_{i,j}$ is the frequency associated with the ground to the vibrationally excited state transition, $\Delta\omega$ is the frequency difference between the two lasers, $\gamma_{i,j}$ is the transition linewidth (full width at half maximum), and K_{ij} is a saturation constant. Values of the differential Raman cross section for various gases at 0.532 μ are given in Table I. Throughout the visible, these vary as ω^4, similar to Rayleigh scattering. In the ultraviolet, the polarizability is affected by resonances and the Raman cross section increases more rapidly. For tagging, the laser intensity must be high enough to drive a significant fraction of the available molecules into the excited state in order to get strong signals. Intensities on the order of 50 GW/cm^2 are ideal and can easily be achieved from commercial pulsed lasers. In this intensity regime, strong saturation is observed (Miles 1988a) indicating strong pumping. The population of the various rotational levels is reflected in the $(N_i - N_j)$ term, and tuning the frequency difference, $\Delta\omega$, through numerous rotational states gives a time-averaged measure of the temperature.

Since two lasers are required, they may be made colinear to mark a line in the flow or can be crossed to mark points. Crossing at an angle marks a short line segment which may be used to measure vorticity. The stimulated Raman process does not significantly affect the beam quality or the energy of the laser beams, so these two lasers can be refocused back through the sample region to mark multiple lines or multiple points. Experiments to-date have used a frequency-doubled high-energy Nd:YAG laser, part of whose output is used for one tagging laser beam, and part is used to pump a dye laser to generate the other tagging laser beam. The pulse duration of these lasers is approximately 10 nsec, leading to virtually instantaneous tagging. The oxygen molecules are pumped to the same rotational state that they were originally in, so there is very little heating of the flow due to the pumping process.

The tagged molecules move with the flow for some specified time interval and are interrogated by an ultraviolet argon-fluoride (ArF) laser. This laser is loosely focused into a sheet which intersects the flow longitudinally as shown in Fig. 6. The ArF laser further excites the tagged molecules into the Schumann-Runge electronic band from which they immediately decay, generating UV and visible fluorescence which is recorded with a high-sensitivity camera. The intensity of the ArF laser is high enough to excite virtually all of the vibrationally excited molecules lying in the selected rotational state. Fluorescence is weak because only one out of approximately 10^5 of these molecules fluoresce due to the rapid predissociation rate of electronically excited oxygen. As the temperature of the O_2 increases, the thermal population of the v" $= 1$ state increases, generating a background fluorescence that will obscure the marked line at temperatures greater than approximately 700K.

Figure 7 is a picture showing a single marked line together with the Rayleigh scattering density cross-section near the Mach disk of an underexpanded sonic jet. The line was straight when marked, so the deformation is a quantitative measure of the velocity profile. The Rayleigh scattering is a quantitative measure of the density cross-section, and is automatically recorded simultaneously with the interrogation step. Rayleigh scattering can be eliminated by placing a far-UV blocking filter in front of the camera so only the fluorescence is detected. This is typically done when the line profiles are to be interpreted by computer line identification. Otherwise, the background contributed by the Rayleigh scattering reduces the line contrast causing errors in the line identification procedure. Figure 8 shows a magnified view of four lines taken across the free shear layer of the sonic jet at the Mach 3.6 location, just upstream of the Mach disk. Each of the four images in this picture is a separate experiment, and the time delay was $2.0 \pm 0.005 \mu\mathrm{sec}$. The maximum distance traveled during this time interval is 1.19 mm, corresponding to a flow velocity of approximately 600 m/sec on the far right-hand side of the photographs. Similar images with double line marking have also been made.

The fundamental limit to the accuracy of this flow tagging method is molecular diffusion which causes the width of the line to increase with time according to the relation:

$$w = (4\tau D + w_o^2)^{1/2} \qquad 17$$

where D is the molecular diffusion constant, τ is the time between tagging and interrogation, and w_o is the tagging linewidth at $\tau = 0$. A Gaussian profile has been assumed. Assuming the line center can be found to within 10% of the width,

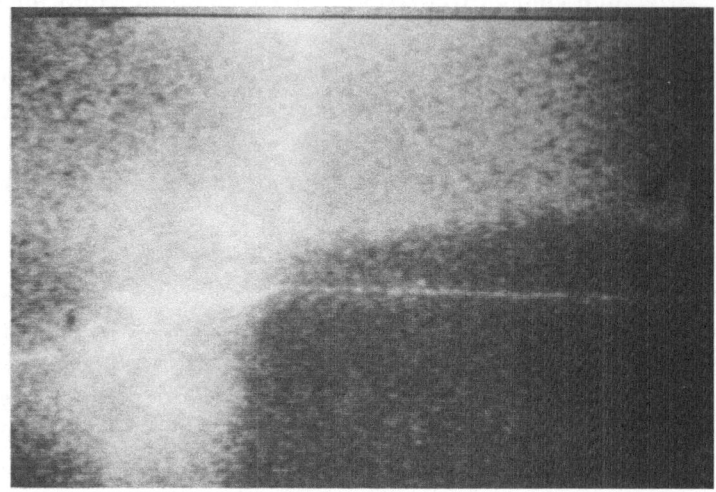

Figure 7. Marked line and instantaneous density cross section of an underexpanded sonic jet near the Mach disk (flow is from bottom to top). The line was written into the flow 2μsec before the image was taken, so the deformation of the line is a measure of the instantaneous velocity profile. The Mach number of the flow at the point where the line is marked is computed to be 3.6.

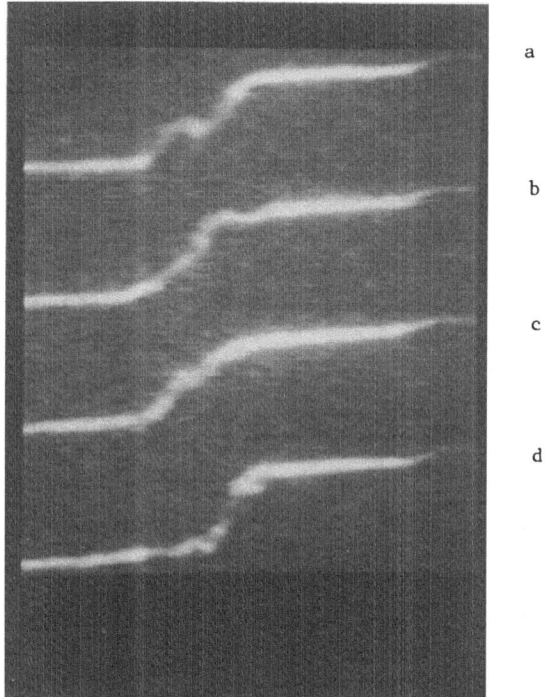

a

b

c

d

Figures 8a - 8d. Composite picture of four separately recorded images of lines across the free shear layer at the Mach 3.6 location in an underexpanded sonic air jet. Flow is from bottom to top and the lines were straight when marked, 2μsec before interrogation. (Miles 1989)

the accuracy of the measurement will be:

$$\xi \cong \frac{0.1w}{D} = \frac{0.1(4\tau D + w_o^2)^{1/2}}{\tau u} \qquad\qquad 18$$

where D is the distance that the tagged region has moved in the time, τ, and u is the flow velocity. For oxygen at room temperature and at 1 atmosphere pressure, $D = 0.21 \mathrm{cm}^2/\mathrm{sec}$.

In the supersonic flows such as that shown in Fig. 8, the time interval required is on the order of microseconds, so diffusion is insignificant. The lines shown in that figure are approximately 50 microns in diameter, so the measurement accuracy limit is 0.4%. In the actual measurement, the accuracy was ±3%, due largely to the uncertainty in the distance the line had moved. This uncertainty is mostly from the limited resolution of the camera. Volumetric grids or line structures have not yet been written into the flow, but doing so is a straightforward extension of current technology. Volumetric patterns could be followed stereoscopically to generate three-dimensional velocity fields.

2.5 SENSITIVITY LIMITS

The fundamental limit to the sensitivity of light scattering diagnostic probes is shot noise. Photons striking the detector surface create photo-induced electrons, or photoelectrons, which are then electronically amplified to produce the measured signal. Due to the Poisson statistics of randomly occurring events, when N photoelectrons are expected, the uncertainty is equal to $\sqrt{N}$. This basic principle can be applied to virtually all light scattering phenomena to determine the signal-to-noise ratio or, equivalently, the sensitivity. In the case of simple scattering mechanisms, including Rayleigh, Raman, fluorescence, and particle scattering, one needs to determine how many detected photoelectrons are generated from each resolvable element during the sampling interval. Equation 9 gives an expression for the collected power from a volume element. This expression is for Rayleigh scattering and can be used for Raman scattering and fluorescence by correcting for the lower photon energy of the Raman or fluorescing light. Equation 9 can be recast in terms of the number of detected photoelectrons by recalling that the energy of a single photon is equal to $\hbar\omega$. The intensity then converts to an incident photon flux per unit area:

$$\frac{\dot{N}}{A} = \frac{I}{\hbar\omega} \qquad\qquad 19$$

and the power becomes the number of photoelectrons per unit time:

$$\dot{N} = \frac{P}{\hbar\omega} \qquad\qquad 20$$

Equation 9 can then be rewritten in terms of the number of photons:

$$N_D = \Delta t\eta \left(\frac{\dot{N}_I}{A}\right) N\Delta V \int \frac{\partial\sigma}{\partial\Omega} d\Omega \qquad\qquad 21$$

where N_D is the detected number of photoelectrons, $\dot{N}_I/A$ is the incident photon flux, and Δt is the collection time interval. Written in this form the equation is also directly applicable to fluorescence and Raman scattering since only the number of photons is important. (The frequency of the incident photons affects the scattering cross-section.)

In order to determine the sensitivity of any scattering measurement, the total number of photons collected from each resolvable element must be calculated. For example, one cubic millimeter of air illuminated for 10 nsec by a 100 mJ, 0.532 micron laser focused into a 1 cm by 1 mm sheet ($10^8 watts/cm^2$, or 2.7×10^{26} $photons/cm^2/sec$) and collected by optics subtending 0.1 steradians with a throughput of 25% and a quantum efficiency of 10% ($\eta = .025$), yields 10^5 detected photoelectrons. (See Table I for the Rayleigh cross-section.) The shot noise is the square root of this number or 300 photoelectrons, which is 0.3%. If better resolution than one cubic millimeter is desired, the laser may be focused into a tighter sheet and the increase in intensity just offsets the decrease in volume in the sheet thickness direction. Increased resolution in the other two dimensions (horizontal and vertical as observed by the collector), reduces the number of photons per resolution element, increasing the noise. For example, a resolution element 100 microns on a side, illuminated with a 10 mJ sheet of light 1 cm by 100 μm , will give only 10^3 photoelectrons, raising the shot noise to 3% of the signal. Similar calculations can be made for Raman scattering and fluorescence measurements. Very low cross sections significantly limit the resolution and sensitivity of Raman scattering (see Table I). The fluorescence scattering cross sections, such as those of oxygen shown in Fig. 2, are strongly affected by quenching and predissociation.

Scattering from a single particle can be found by replacing the density volume product $(N\Delta V)$ by one. The large particle scattering cross sections can produce large signals which may be collected in a single resolution element, so long as the particle image is smaller than the size of the element. If particle sizes are important, or if the position of the particle needs to be measured accurately, then many resolution elements must be illuminated by the scattering from the particle, and the number of photons per resolution element drops accordingly.

Other background noise can also contribute to a reduction of the detection sensitivity. This noise usually arises from other light sources or from undesired scattering or fluorescence. Generally, it can be reduced significantly by using the proper color filters in both the source and the detector so that only the desired wavelength illuminates the region of interest and only the proper wavelength is detected. Time-gating the detector is also an effective way of eliminating light from other sources. Careful masking is important to minimize reflection and undesired scattering. Often, observing with an imaging system gives an immediate picture of where noise is coming from.

3 VOLUME IMAGING

3.1 LIGHTSHEET VOLUME IMAGES

Lightsheet-swept volume images have been made using particle scattering (Utami 1984), fluorescence (Nosenchuck 1986), and Rayleigh scattering (Yip 1988). Since the development of the 'vapor screen' method of flow visualization (McGregor 1961), the advantage of a thin illumination sheet in viewing cross sections of complex flow fields has been evident. Many researchers continue to use lasers and light sheets together with dye, smoke, particulates, and water vapor to see volumetric images of steady- state phenomena and time-varying images of flow cross sections. The challenge has been to extend these techniques to volumetric images of unsteady phenomena.

Utami and Ueno (1984) used light from a high-intensity lamp passing through a thin slit to illuminate a horizontal cross section of a turbulent open channel flow. The light sheet was 3 mm thick and polystyrene beads, ranging in diameter from 0.1 to 0.5 mm were used as tracers in water. The seeding concentration was approximately 10 particles/cc, the mean flow velocity was 9.8 cm/sec, and the Reynolds number was 4000. The light sheet was moved upwards at a speed of 8 mm/sec, and two cameras, each with 0.25 sec exposure times, looked through a rotary shutter which caused the particles, which are moving in each cross sectional view, to appear as dashed lines. The whole photographic system was moved downstream in the flow direction at 7.91 cm/sec, slightly slower than the mean velocity of the flow.

The dotted lines enabled Utami and Ueno to determine which of those tracers moved into or out of the illumination region during the exposure time. The coordinates of both end points of every tracer were digitized so the flow patterns could be presented in a variety of ways. For example, Fig. 9 shows

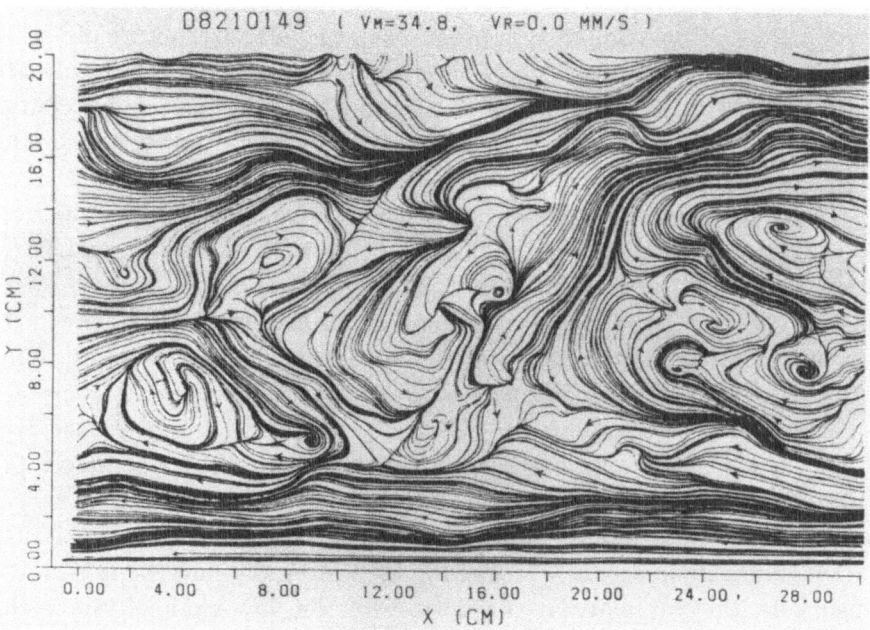

Figure 9. Pattern of streamlines in one horizontal cross section viewed by an observer moving at a speed equivalent to the mean flow velocity. Vector data was collected by observing light scattering from particles in a flow channel and the pattern was generated by drawing lines tangent to each flow vector. (Utami and Ueno 1984)

stream lines which would be seen by an observer moving with the flow. Vertical cross sections were constructed assuming frozen turbulent structure in each of the recorded horizontal cross sections of the flow. The origin of the coordinates for each picture was shifted by the value of the time lag, and the vertical component of the velocity was calculated from the continuity equation. An example of a longitudinal, vertical cross section of the flow is shown in Fig. 10.

The use of rapidly scanned laser-sheets to image a volume has been developed by Nosenchuck (1986) to observe complex transitional and turbulent boundary- layer structures in a dye-seeded flow channel. A full, three-dimensional picture was obtained by rapidly traversing a laser sheet through the flow field in a period short compared to the time scales of interest in the flow. The resultant data consisted of a series of two-dimensional sheets which represented a space-filling three-dimensional volume. The net effect is to record the entire laser-dye field each time the sheet is traversed through the flow. The resultant images provide a qualitative picture of the flow, as well as quantitative information concerning the degree of turbulent mixing which has occurred.

Disodium fluorescence dye was injected in to the flow using a slot which was flush with the flat plate at the bottom of the flow channel. Since the dye is injected at the wall, it 'tags' the low-speed fluid. During the bursting process this low-speed, high-vorticity fluid moves away from the wall and mixes with the outer flow. This process, which produces a large portion of the turbulent energy, can be examined using the laser sheet-scanning technique. Typical experiments were run over a 2.0 cm portion of the boundary layer with the laser sheet rapidly stepped in 32 increments. The intersheet normal separation was, therefore, 0.62 mm, setting a rough maximum on the laser sheet thickness. The laser sheet was produced by appropriately expanding and focusing an argon-ion laser beam (501 nm) with a series of vertically- and horizontally-oriented cylindrical lenses. A rotating mirror was placed in the optical path to deflect the defocused beam in the y direction. The optics ensure that a thin sheet was formed, which, when deflected, remained parallel to itself over a height of up to 10 cm. The mirror was capable of generating up to 2000 steps per second, but the experiments were limited to 30 steps per second by the imaging video recorder. Horizontal and planform views of the optical experimental setup are shown in Figures 11a and b.

Several representative sheets from a single rapid scan of the laser sheet through a turbulent spot are shown in Fig. 12. These digitized images may be processed in multiple ways to yield both qualitative and quantitative insights into the nature of the flow. For example, the degree of mixing (most of which is attributable to turbulence) can be determined by distinguishing between grey

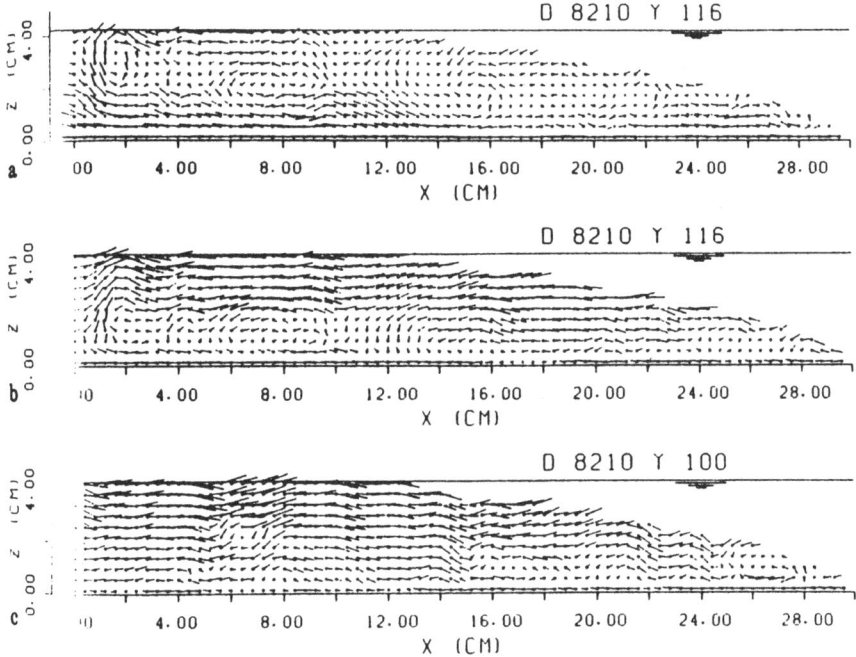

Figures 10a-10c. Distribution of velocity vectors in a longitudinal vertical cross section of the flow. (a) Distribution of velocity vectors in the section y = 11.6 cm (see Fig. 9) viewed by an observer moving at the same speed as the measuring frame. (b) Same as (a) but the observer is moving 2 cm/sec slower. (c) Same as (b) except in the section y = 10.0 cm. (Utami and Ueno 1984)

Planform View

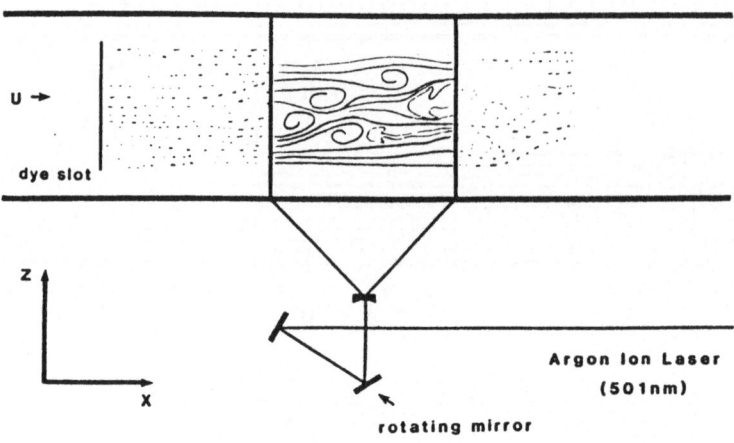

Side View

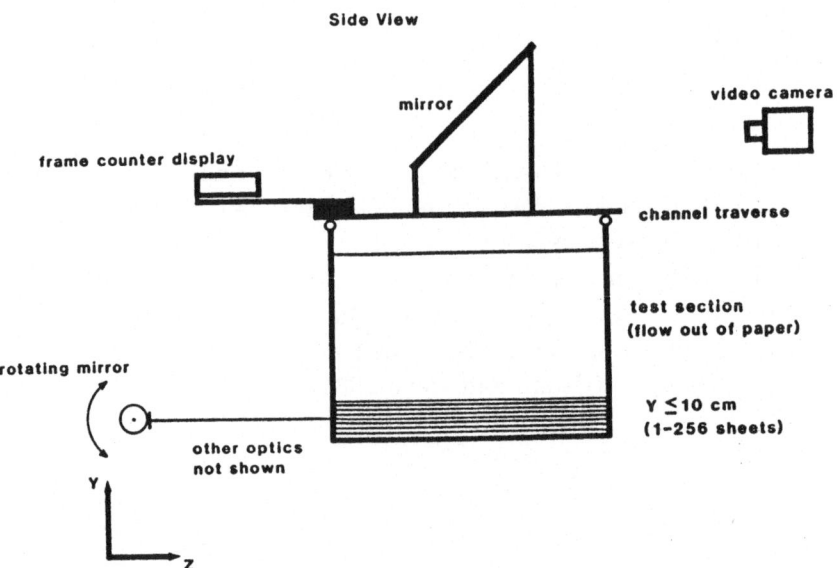

Figure 11. Planform and elevation views of laser-sheet scanning setup. (Nosenchuck 1987)

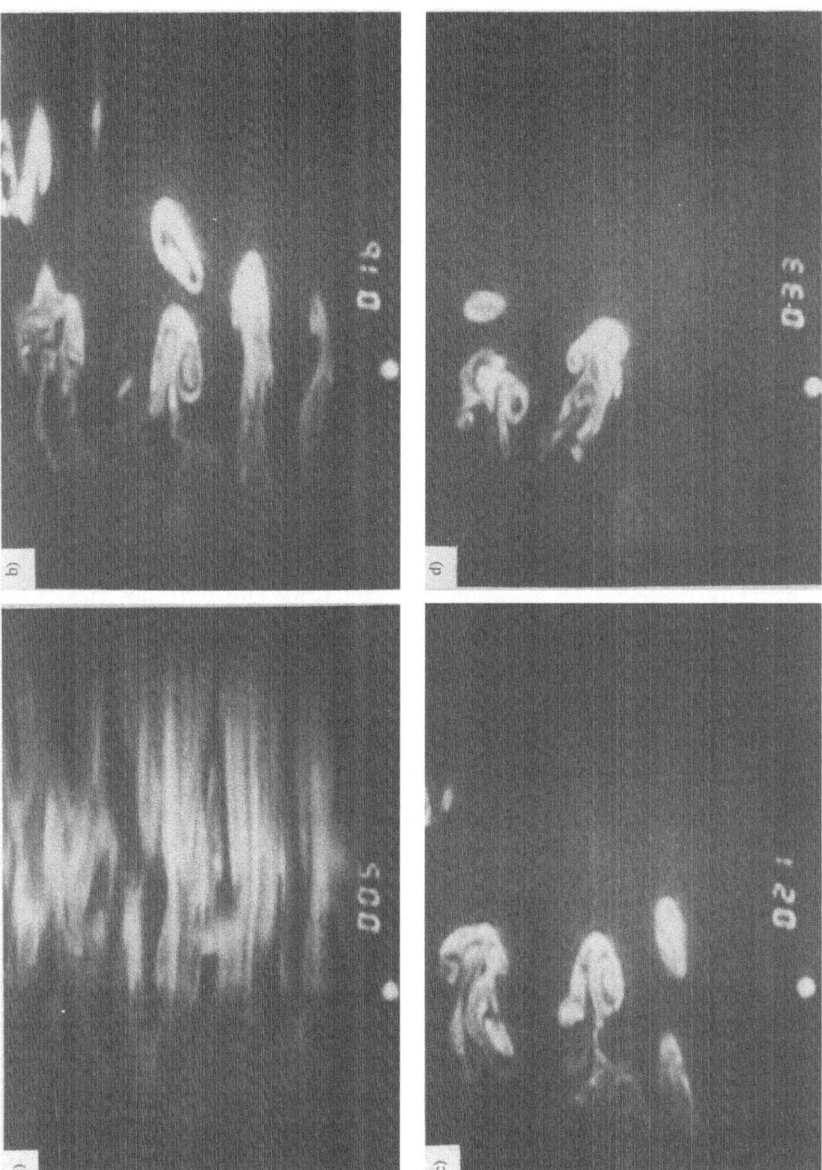

Figure 12. Views of several cross sections of a turbulent water flow marked by a sheet of dye which was injected into the sublayer at the bottom of the channel. (Nosenchuck 1987)

levels (thus, dye concentration) in a single sheet. For complete spatial and temporal diagnostics, three-dimensional image processing routines, such as those discussed later in this Chapter, can be applied to the data in order to generate a full, three-dimensional view. A reconstruction using this technique is shown in Fig. 13. Other examples of this dataset are shown at the end of this Chapter in Figures 25a, 26a, 27 and 28. Nosenchuck is currently applying this technique to view the laminar- turbulent transition process and develop mechanisms for actively controlling turbulent flow instabilities (Nosenchuck 1987).

To capture unsteady phenomena, the period of a scan must be less than that associated with one-half of the period of the highest frequency present in the flow. This is to reduce bias between the upper and lower portions of the scan and, also, to ensure that repetitive scans accurately capture the nonsteady nature of the flow. If the highest frequency is 20 Hz, then a scan must be completed in 25 msec. For 32 steps, this requires that 1300 sheets be generated each second, neglecting the time taken to reposition the starting sheet after each scan. A high framing rate film or video camera, or an electronic framing camera, is then required to record the images. By sequentially sweeping the volume, the three-dimensional time evolution can be tracked. If scattering particles are added to the flow and the sequential sweeps are rapid enough, single particle trajectories can be followed, yielding the three-dimensional velocity field. For flow velocities and seeding concentrations similar to those of Utami and Ueno (10 cm/sec, 10 particles/cm^3), each particle must be viewed every 10 msec to avoid aliasing. For a 32-step scan, this requires a scan rate of 3200 steps per second and an imaging system capable of recording 3200 frames per second.

By using a resonant scanning mirror and an electronic framing camera, Yip, Schmidt, and Long (1988) have achieved sweep rates equivalent to 20,000,000 frames per second. In their experiment (Fig. 14), a 1.4 microsecond duration, high-energy pulse laser was rapidly swept across a turbulent jet. The electronic framing camera imaged light scattered by the gas molecules at 90° producing as many as 12 cross-sectional planes of the flow region. The output of the framing camera was imaged directly onto a two-dimensional charge coupled device (CCD) array camera to permit data digitization as the experiment was performed. The laser was the flashlamp-pumped dye laser (DIANA) at the Sandia Combustion Research Facility in Livermore, California. The pulse energy was 550 mj at 440 nm (blue). The jet was moving at a velocity on the order of 10 m/sec and with a resonant scanning mirror, the laser sheet moved at approximately 4.9 km/sec. The laser was focused to a sheet 250 microns thick and 3 cm tall. The framing camera produced exposures of 10 nsec each, separated by 50 nsec. This timing matched the thickness and speed of the laser so that the entire volume was observed. The

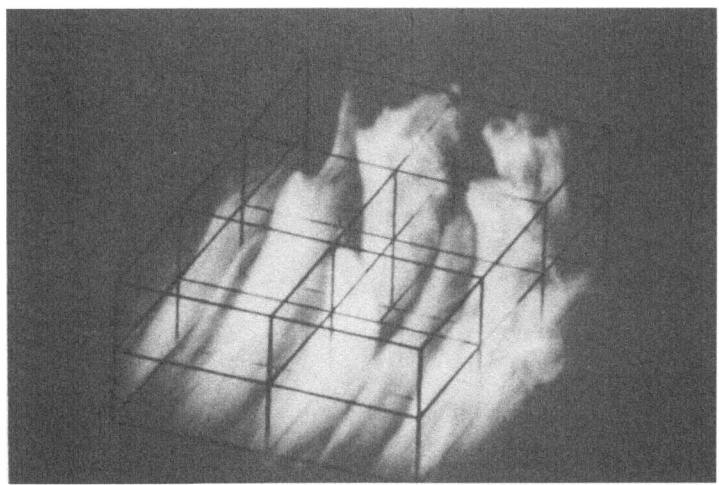

Figure 13. Three-dimensional reconstruction of the trailing edge of a turbulent spot. Data was acquired by a rapidly swept laser sheet passing through a flow which had had dye injected into the sublayer at the bottom of the channel (Nosenchuck 1986). The reprojection method is discussed in Section 4 (Russell 1987).

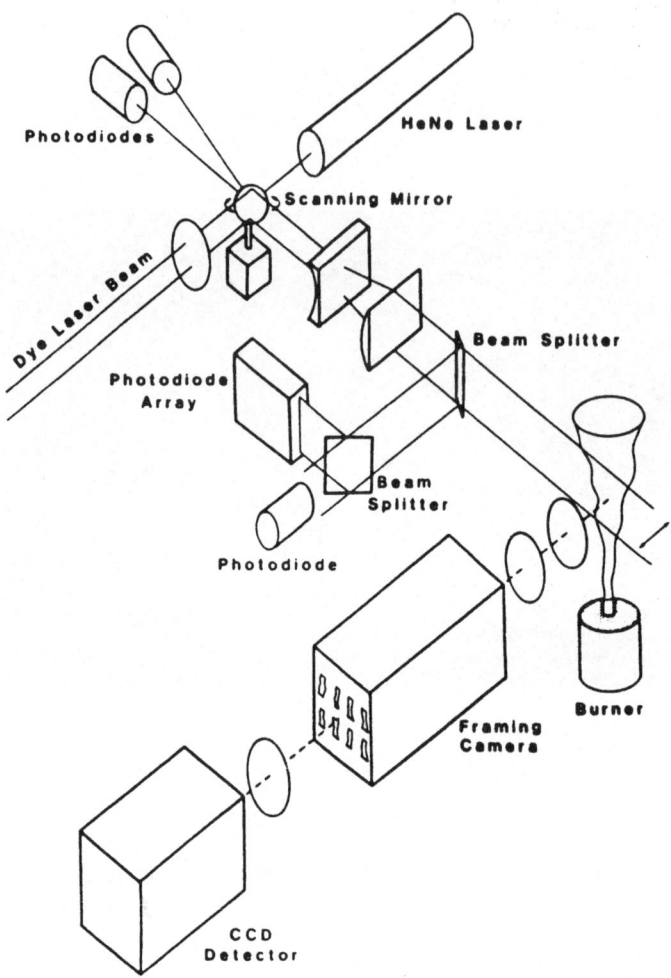

Figure 14. Apparatus for a rapidly swept laser sheet imaging system capable of acquiring a cross sectional image each 50 nsec. Up to 12 such images can be recorded during a single 1.4μsec laser pulse. (Yip 1988)

output of the framing camera was imaged onto a 584 x 390 pixel CCD array camera, so that each of the 12 images occupied 58 x 120 pixels, corresponding to 6 mm x 12 mm in real space. The temporal and spacial profiles of the laser were simultaneously monitored so that nonuniformities in the illumination sheets could be corrected.

Several different light scattering mechanisms were used. Rayleigh scattering from Freon 12, which has a scattering cross section 15 times that of air, produced signals approximately twice the system noise level. Figure 15 shows an instantaneous constant gas concentration surface in a turbulent jet (Reynolds number = 13800) from that experiment. Mie scattering was used from both submicron-sized aerosols seeded into nonreacting and reacting flows, and from soot particles marking flame fronts in low-speed, non- premixed flames. Biacetyl vapor was also used as a flow marker and Fig. 16 shows a constant fuel air concentration surface in a premixed methane-air- biacetyl flame.

Extending this technology to measure three-dimensional velocity fields will require at least two sequential sweeps of the laser through the volume to observe particle motion. The two sweeps must be delayed on the order of 50 μsec to allow the particles to move far enough to get reasonable measurements of the velocity. This could be done by using a pair of lasers, or, possibly, multiply pulsing a single laser.

An alternative to sweeping the beam is simultaneously collecting scattering from multiple laser sheets. Early work in this area was done by Yip and Long (1986a) who used a dye laser at 0.563 microns and a doubled Nd:YAG laser at 0.532 microns to generate two simultaneous cross sections of a turbulent gas jet. They relied on Rayleigh scattering from freon 12 and two separate cameras which observed the flow through color filters so the light scattered from each plane could separately be recorded. More recently, this technique has been extended to four simultaneously pulsed sheets of light which are imaged by a single camera which viewed the flow through a quadrupling prism and appropriate filters (Mantzaras 1988). The light sheets were passed through a transparent cylinder head and the structure of turbulent flames was recorded by observing scattering from submicron titanium dioxide particles. Figure 18 is a sketch showing the optics for creating the laser sheets. A high-intensity frequency-doubled YAG laser is passed through a hydrogen Raman shifter to generate Stokes and anti-Stokes beams at 0.436 microns and 0.683 microns, respectively. Those plus the original laser beam, which is at 0.532 microns, are passed through a prism and form three of the sheets used. A fourth sheet at 0.355 microns was generated by frequency tripling the Nd:YAG laser. Figure 18 shows three-dimensional images taken with equivalence ratios of 1 and 0.59 at 300 rpm and 1200 rpm. The time resolution of

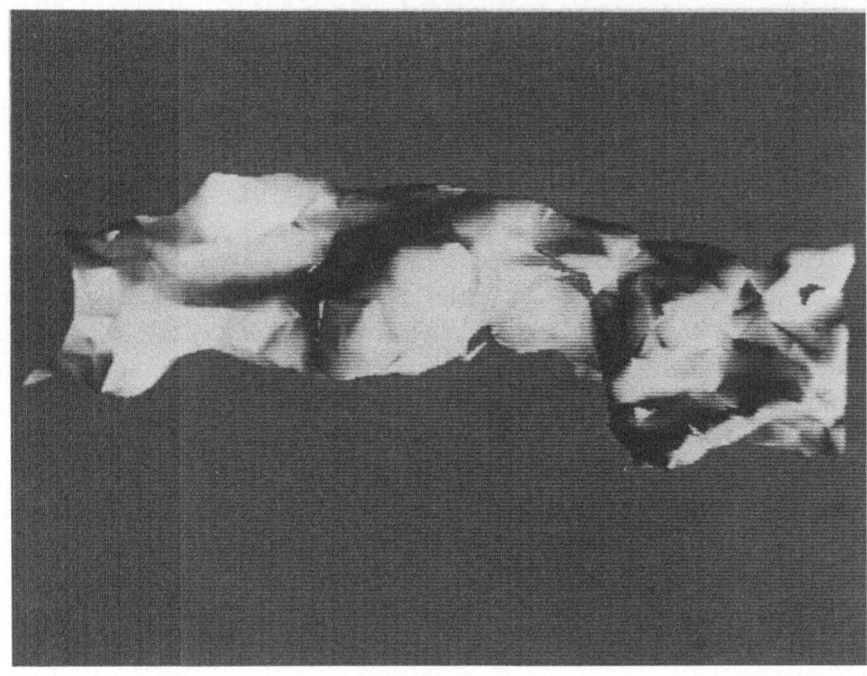

Figure 15. Instantaneous constant-gas concentration surface contour in a turbulent jet (Reynolds number 13,800)
measured by imaging Rayleigh scattering from Freon 12.
Flow is from left to right. (Yip 1988)

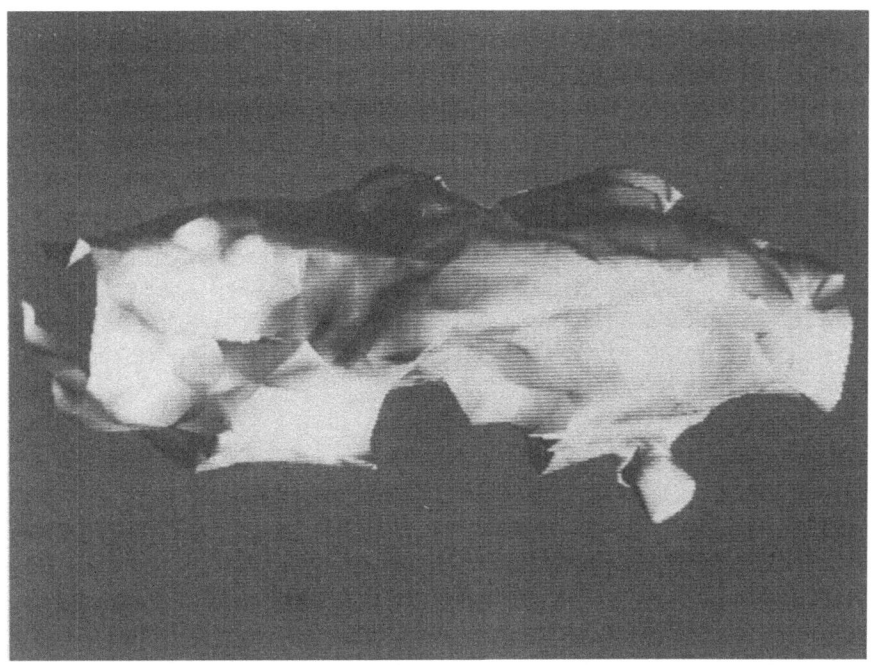

Figure 16. Instantaneous constant-fuel-gas-concentration surface contour in a turbulent premixed flame. The flame was seeded with biacetyl vapor which acted as a fluorescent marker. Flow is from left to right. (Yip 1988)

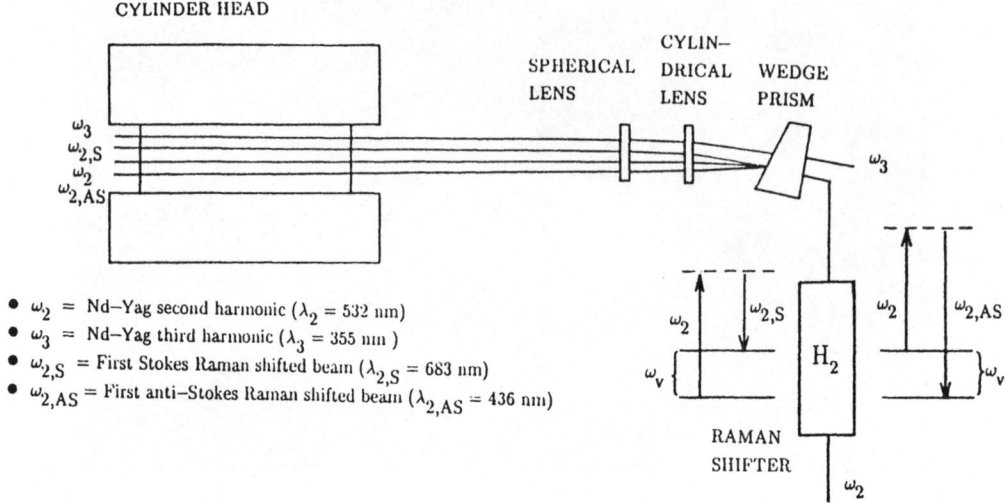

Figure 17. Optics for creating four simultaneous laser light sheets (Mantzaras 1988).

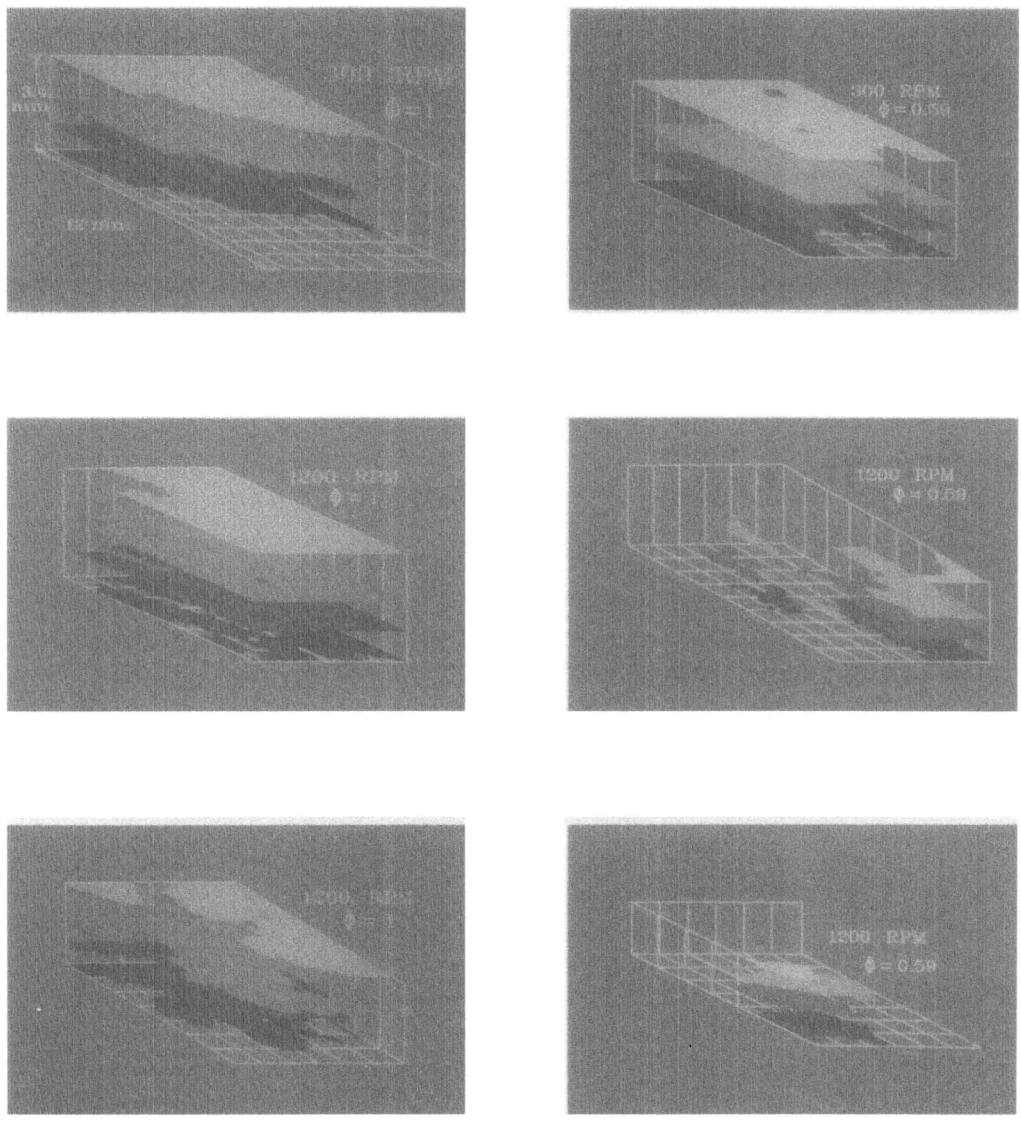

Figure 18. Three-dimensional representation of flame fronts in a cylinder head at equivalence ratios of 1 (left) and .59 (right) for 300 and 1200 rpm. The four slices are simultaneously recorded in each image using four laser sheets (Mantzaras 1988).

the experiment was set by the 10 nsec pulse length of the Nd:YAG laser, causing motion to be frozen. The strong variation in intensity indicates the location of the flame front during ignition. The dark represents cool gases where the density is high, and the light represents hot gases where the density is low. The authors have used these images to study the structure of premixed turbulent flames in internal combustion engines. This imaging technique can, in principle, be extended to many more simultaneous cross sections by using additional light sheets and multiple cameras with narrow linewidth filters. The major constraint is that the scattering mechanism must be inelastic. The color of the scattered light must be the same as that of the illuminating laser, so that the separate regions can be identified. This is true for Rayleigh scattering, Tyndall scattering, and Mie scattering, but is not the case for laser- induced fluorescence. For tracking particles, a multiple pulse laser must be used.

3.2 STEREO IMAGING

An obvious alternative to light sheets is direct stereo imaging of particles or other material moving with the flow. This technique dates back to the 19th century (McKay 1951) and essentially duplicates the way our eyes observe motion in three-dimensions. The scattering centers can be particles or bubbles which are randomly seeded into the flow, or they can be hydrogen bubbles, photoactivated dyes, or vibrationally excited molecules which are written into the flow at well defined locations and times using thin wires or laser beams. In the case of particles and bubbles, the test region is illuminated with the bright white light source. To observe photochromic features, a blue light source is used, and to see vibrationally excited molecules, an ultraviolet light source tuned to an appropriate molecular transition generates laser-induced fluorescence. An example of lines produced by hydrogen bubble wire are shown in Fig. 3. A grid produced by laser-induced photochromism is shown in Figs. 5a and 5b, and lines written across the supersonic portion of a free air jet by molecular tagging of oxygen are shown in Fig. 8.

Stereoscopic images are generated by observing the scattering simultaneously from two different angles. If an individual particle or flow feature can be identified in each of the images, then its position in three-dimensional space can be determined. Generally, attempting to identify three-dimensional streaks from long exposures is avoided because the image becomes too cluttered to analyze unless the density of particles is very low. Thus, one must turn to a high-speed camera using either high repetition rate pulsed illumination or fast shutters to permit proper tracking of particles from frame-to-frame. In the case where patterns are written into the flow at well known locations, only a single pair of images needs to be taken at some well defined time after the pattern is written to generate a three-dimensional velocity field. Sequential images may be desired to follow the evolution of the velocity field in time.

An example of work done on stereo imaging of particles is that by Chang, Watson, and Tatterson (1985 a and b), who observed turbulent flow in a 0.914 meter diameter plastic tank equipped with an impeller and baffles. Neutrally buoyant tracer particles, ranging in size from 400 to 500 microns, were seeded into the flow at a density of approximately 30 particles/liter. The flow was photographed with a 16 mm cine camera operating at 400 frames/sec. Stereo imaging was accomplished with a Bolex stereoscopic lens mounted on the camera. Particles were identified and tracked on the film using a software package which was developed by the research team. The highest particle velocity measured was 3.20 mm/sec. Estimated errors in particle location were 25.0 microns in the x and y coordinates, and 1,046 microns in the z, or depth, coordinate. This large uncertainty in the depth dimension was due to the small distance (64 mm) between the two collection lenses compared to the distance from the scatterers (approximately 0.5 meters–an angle of approximately 0.13 radians). Better depth measurements were achieved by Kent and Eaton (1982) who used an effective collector separation of 0.24 radians to get accuracies of 60 microns in the x and y directions, and 460 microns in the z direction for 2 mm diameter helium bubbles seeded into a transparent engine cylinder and observed from a distance of 77 cm. In their experiment the camera was operated at 1000 frames/sec. and velocities up to 7.4 m/sec. were observed. Two bubbles were tracked by hand analysis of the data.

An obvious limitation of the above two experiments is the very low concentration of scatterers necessary for successful data processing. A more recent effort by Brodkey (1986) has concentrated on increasing the density by color coding particles so they are easier to track in three dimensions. The fundamental density limit of stereo imaging occurs when the identification of selected particles as seen by the two different imaging systems is ambiguous. This is related to the 'uncertainty volume' occupied by the particle. Assuming that the alignment of the two imaging systems is accurately known, perhaps by using reticules scribed on a glass plate in the field of view, then a single particle seen in one image defines a line in the other. The thickness of this line is the recorded diameter of the particle, and the extent of the line is through the entire sample volume or the focal depth of the camera. If the two imaging systems are orthogonal, then each particle viewed in one image produces a line all the way across the other image. The volume swept out by this line is the 'uncertainty volume' associated with a single image. The point where this line intersects the other projected view of the particle identifies the true location of the particle in three dimensions. The particle is best located using orthogonal projections. As the angle between the two projections becomes smaller, the intersection point in the depth dimension becomes less certain. When more than one particle appears to lie on the line, the true location of the particle is ambiguous. This means that, from at least one of the two observation angles, one particle has been hidden by another causing confusion in the particle identification procedure. This forms a fundamental limit on the density of the particles in the fluid and depends on the particle diameters and the film resolution. A second limit on the density of the particles is set by the framing rate of the optical system. In order to unambiguously track the motion of a single particle, its displacement from frame-to-frame must be less than one-half the interparticle separation or a confusion in determining which particle has moved to which location results.

This aliasing problem is always found in sampling systems but is somewhat more serious here since turbulent motion and randomly distributed particles are involved. The aliasing problem is significantly reduced if the scattering centers can be precisely written in the flow by thin wires generating hydrogen bubbles, or by lasers inducing photochromism or vibrationally exciting molecules. In these cases, the spacings can be optimally selected to match the sampling rate and the local flow speed. By having the capacity to write specific grid structures, regions of particular interest can be monitored without being subject to arrival statistics of random particles or bubbles. Grids with varying spacing can be written into a flow field so that regions of small-scale and large-scale structure can be simultaneously observed.

3.3 STEREO SCHLIEREN TECHNIQUES

While many visualization techniques currently in use were adapted from existing technologies in fields not directly related to fluid mechanics, schlieren flow imaging represents one of the first 'modern' methods to be extensively investigated and extended by the fluid mechanics community. Much has been written regarding the general nature and techniques of schlieren photography, therefore only those aspects pertinent to three-dimensional imaging will be discussed here.

Schlieren techniques are generally sensitive to gradients in the optical index-of-refraction field that come about due to density and/or temperature changes. Moreover, direct use of this technique presents the researcher with a two-dimensional representation of the integrated index changes along the viewing axis. Thus, traditionally the use of schlieren systems were limited to primarily two-dimensional flows of a compressible fluid (to provide density changes) or of a nonisothermal, generally incompressible flow. If any three-dimensionality was present, the single recorded two-dimensional image of the integrated changes in the first-derivative of the index field provided insufficient information to allow a deconvolution of the field. In an effort to resolve this difficulty and open the use of this method to provide more general field observations, multiple schlieren imaging paths have been combined with computer data processing. These methods include computerized tomography and stereo-schlieren photography.

Computerized tomography (CT) schlieren involves rotating a schlieren system through 360° to provide a finite number of imaging paths through the flow field. Although each path represents an integrated view, through appropriate deconvolution of the multiple images, a three-dimensional index-of-refraction field can be established with some accuracy. This process is roughly analogous to the use of a circumferential acoustic transducer array to obtain time-of-flight data from which flow velocities could be determined. One drawback of this technique is that it is quite cumbersome, given the need to either rotate the test section or a portion of the highly aligned schlieren optics. Another less desirable feature is that in the absence of multiple parallel optical systems, the time associated with the CT scan restricts the technique to primarily steady flows.

The use of a stereo schlieren imaging (Chankaya 1986) permits a limited amount of spacial reconstruction of the index field to occur due to the simultaneous imaging of two displaced field views. The individual schlieren systems are identical to conventional systems set-up with a moderate included angle, typically of the order of $30^o - 45^o$. In the work by Chankaya, a limited amount of depth reconstruction was performed. This technique has a primary drawback in that for the most part, the reconstructions are geometrical in nature, and rely heavily on flow feature identification. This technique is useful, however, in flows where the primary interest is to identify coherent structure, such as the present of organized turbulence or a three-dimensional separation surface in compressible flows.

3.4 HOLOGRAPHIC PARTICLE IMAGING

Early work using holography to size and count particles was done by Thompson et al. (1966), using an in-line holographic technique. This was the original holographic configuration developed by Gabor (1948) in which the unperturbed beam passing through the medium serves as a reference and the scattering from the particles generates the object beam (Fig. 19). The interference pattern of these two beams is recorded on film creating a hologram. Royer (1977) extended the in-line configuration by passing a second beam through the flow field to brighten the scattering so submicron- size particles could be recorded.

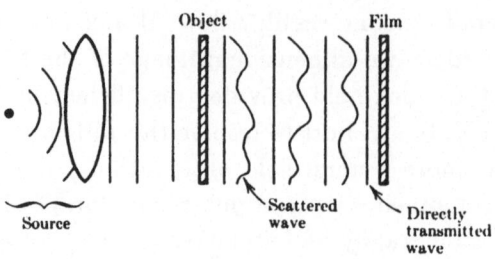

Figure 19. Layout for recording a Gabor hologram. (Goodman 1968)

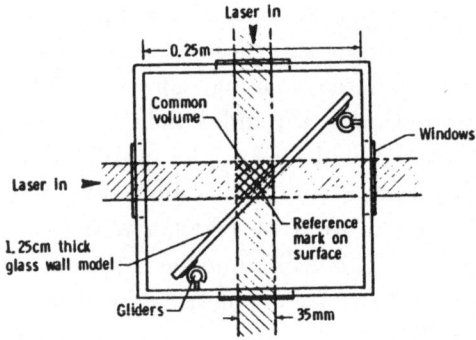

Figure 20. Cross-sectional view of the vertically transported water tunnel showing the laser beam paths for the two simultaneous in-line holograms. (Weinstein 1985)

Numerous papers by Trolinger and others (see Trolinger 1974) examined various configurations for particle holography; and the use of holography to generate three-dimensional images of fast moving gas bubbles to determine their size and position has been developed by Lauterborn and his associates (Haussmann 1980). They point out that the in-line holographic process is limited to particles small enough so that the film plane is in the far field:

$$r < \sqrt{d\lambda} \qquad\qquad 22$$

and large enough so the interference pattern structure is detectable against the grain noise of the photographic plate:

$$r > \sqrt{\frac{d\lambda}{A}} \qquad\qquad 23$$

where r is the particle radius, d is the distance to the film plane and λ is the light wavelength. A is the material constant, and was approximated to be equal to 60 for the Agfa 8E75 plate used. For a distance of 1 meter and a ruby laser (0.6943 μ), the particle diameters must be between 0.2 mm and 1.6 mm. Since Haussmann et al. were looking at a broad range of particle diameters, they used an off-axis holographic set-up which is not subject to the same limitation. The hologram was played back with a second laser, and an image dissector camera was mounted on a computer-controlled translation table to locate the projected real images of the recorded particles. The playback laser was a krypton-ion laser operating at 0.6497 microns, a wavelength close to the ruby laser wavelength, to produce minimal distortion of the reconstructed image. An intensity-weighted gradient operator was used to locate the focal depth of the particles in the real image. In order to resolve the dynamics of cavitation, the images were recorded with a high-speed camera operating at a framing rate of 22.35 kHz (Zarschizky 1983). The main difficulty in image interpretation was background noise due to laser speckle.

The limit on the three-dimensional position accuracy of the holographic method is related to the 'uncertainty volume' occupied by the particles much as it is for stereoscopic particle tracking. In the reconstruction, the x and y coordinates of the particle are relatively well delineated, but the depth location is poor. One can understand this problem by imagining a single particle generating an in-line hologram. The holographic recording corresponds to a series of Fresnel rings, which, when illuminated by a playback laser, act as a lens. The focal point of that lens is the reconstructed image of the particle. The depth of focus is determined by the effective f-number of the lens, i.e., the focal length divided by the lens diameter. With reasonable size holograms, this depth of field is on the order of a few millimeters. The particle location in the other dimension can be found to an accuracy much better than the diameter of the particle. For three-dimensional velocimetry, this depth uncertainty is a critical limitation: it severely limits the accuracy of one velocity component.

Shofner et al. (1970) pointed out that two simultaneous views would remove this uncertainty. At NASA Langley, a dual-image, high-speed holocinematographic velocimeter is being developed (Weinstein 1985) to generate a real-time recording of the three-dimensional motion of a particle seeded flow. In this case, two orthogonal in-line holograms are taken (Fig. 20) through the windows of a vertically transported rectangular water tunnel. Small, hollow glass spheres 40 microns in diameter are being used. These particles are buoyant and rise at a rate of 0.1 cm/sec. The tunnel moves vertically, however, so this rate of motion of the particles can be subtracted out. The maximum flow speed is 0.1 m/sec and a chopped argon-ion laser is used to record the images. The planned facility calls for a 7,000 pulses/sec copper-vapor laser and a 1 m/sec flow with a film exposure rate of 700 frames/sec. Particle seeding is expected to be a maximum of 1/mm, or 10^3/cm^3. This concentration is limited by the camera framing rate and, more fundamentally, by the fact that when two particles occupy the same 'uncertainty volume', the data is ambiguous and must be discarded. Playback is with a helium-neon laser (0.633 μ), and the image is analyzed with a television camera using a microscope objective on a three-axis positioner. The current effort is to develop noise filtering algorithms and automated data reduction schemes to handle the enormous quantity of data that such a system will generate.

3.5 OPTICAL TOMOGRAPHY

The problem with interpreting such line-of-sight optical probes as schlieren, shadowgraph, interferometry, and absorption, is that the three-dimensional features are integrated along the optical path. Consequently, the three-dimensional information is lost, and only dominant flow features, such as shock waves, are observable. However, by passing multiple beams through the same flow field from all angles, the full three-dimensional properties can be reconstructed since all the projectional ambiguities are resolved. The reconstruction is done by expressing the projected images in terms of integrals through the sample volume and performing an inversion (Santoro 1981; Herman 1980). The inversion process is equivalent to a Fourier transform. The resolution of the sample field is limited by the number of projection angles used, with full resolution requiring sampling at all angles from 0 to 180 degrees. Various different algorithms can be applied to achieve this inversion (Hesselink 1988a) depending upon the physical phenomenon being sampled and the structure of the three-dimensional field.

The bulk of the activity in the area of tomography has been directed toward reconstructing medical images (Hesselink 1988b). Work in fluids has used light-absorption in methane-air mixtures (Santoro 1981) and has been proposed for pollution detection using absorption (Stuck 1977) and extinction from Rayleigh and particle scattering (Byer 1979). It is also possible to observe a fluorescing volume from multiple angles and from a similar inversion to generate a three-dimensional volume projection of the fluorescing species.

Interferometry can be combined with tomography for measurements involving the index-of-refraction. Both interferometry and holography have proved to be powerful tools for generating quantitative density measurements in complex flows. A review of these techniques is given by Bryanston-Cross (1986). Recently, Snyder and Hesselink (1988) have reported the instantaneous three-dimensional optical tomographic reconstruction of a nonstationary fluid flow. A schematic of their apparatus is shown in Fig. 21. An argon-ion laser was used with a shutter speed of 320 μsec. Eighteen projections were simultaneously made through a co-flowing jet with helium in the core and air around the perimeter. A similar optical arrangement with a second laser beam was used to generate reference beams for the film and the interference patterns were recorded. The large difference in the index-of-refraction between helium and air generated an interferogram which was a measure of the path-integrated index-of-refraction variations due to mixing of the gases as seen from that particular angle. A holographic interferogram was created by exposing two of these interferograms on the same film, one without the flow and one with the flow. The reference beam was slightly tilted between exposures to give a reference fringe pattern. The developed holographic interferogram, which recorded projections from all angles, was then illuminated with a playback laser beam to produce a second interferogram that was recorded on film for data processing. This second interferogram was simply a series of parallel lines in the absence of the flow. The flow created a contoured fringe pattern as shown in Fig. 22. The flow velocity was approximately 6 m/sec, leading to a resolution limit of 2 mm from the exposure time. The data processing was done by digitizing the fringe pattern on a 512×512 grid of points and using and inverse Fourier transform inversion method. A two-dimensional slice of the reconstructed jet is shown in Fig. 23, where the maximum value represents 100% pure helium. The full three-dimensional map of the helium concentration is created by stacking sequential two-dimensional reconstructions, all of which were obtained during the same 320 microseconds.

This single-pulse tomographic set-up can be extended to measure concen-

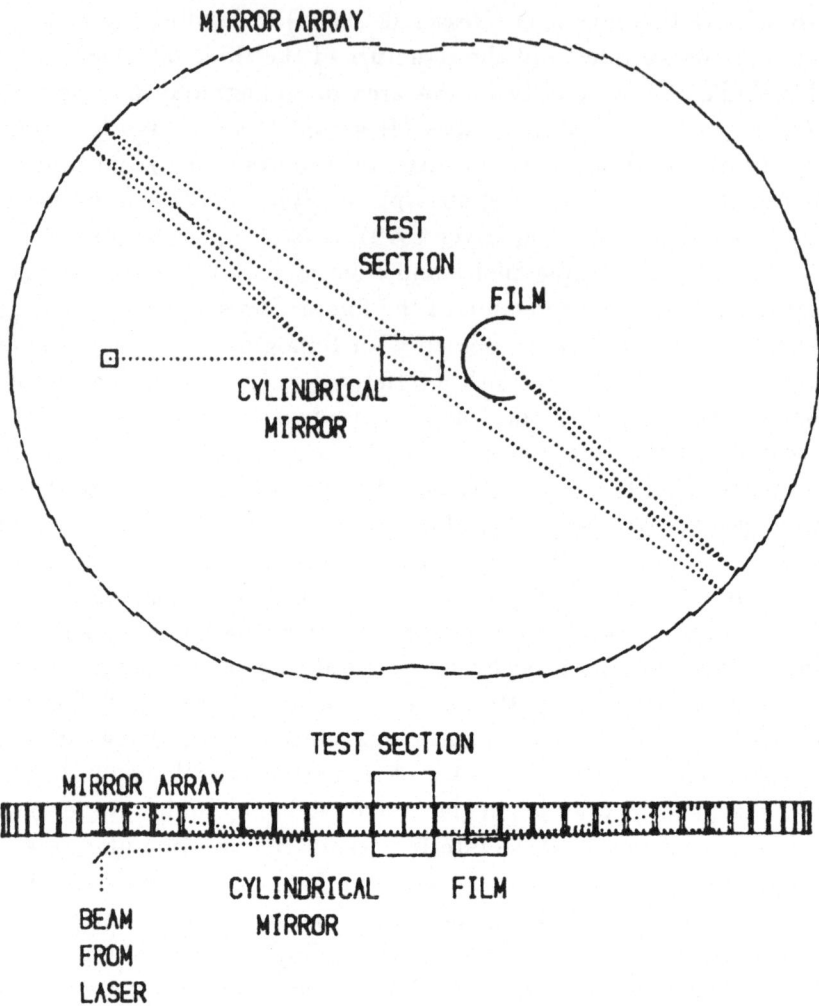

Figure 21. Schematic top and side views of experimental apparatus for instantaneous tomographic recording of a nonstationary fluid. Dotted lines indicate the path of the object beam for a single projection. (Snyder and Hesselink 1988)

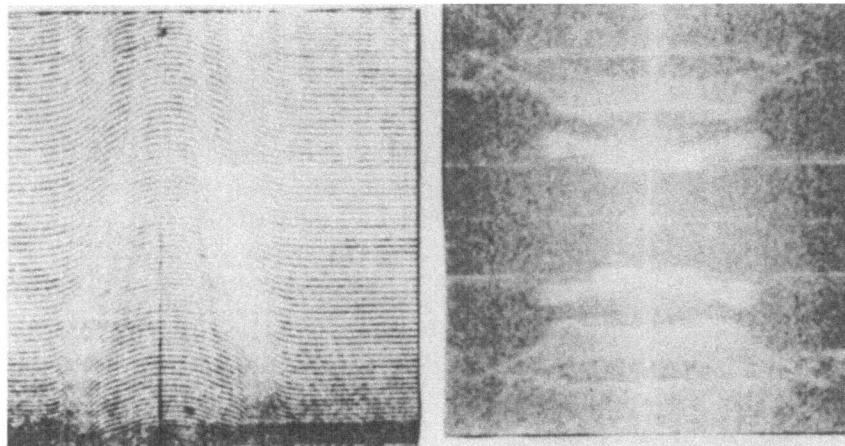

Figure 22. (Left) Raw interferogram data for one projection through a co-flowing helium-air jet. (Right) Two-dimensional Fourier transform of a digitized interferogram showing spatial frequency components grouped around the bias frequency and harmonies. The origin of the coordinate system is in the center of the image. (Snyder and Hesselink 1988)

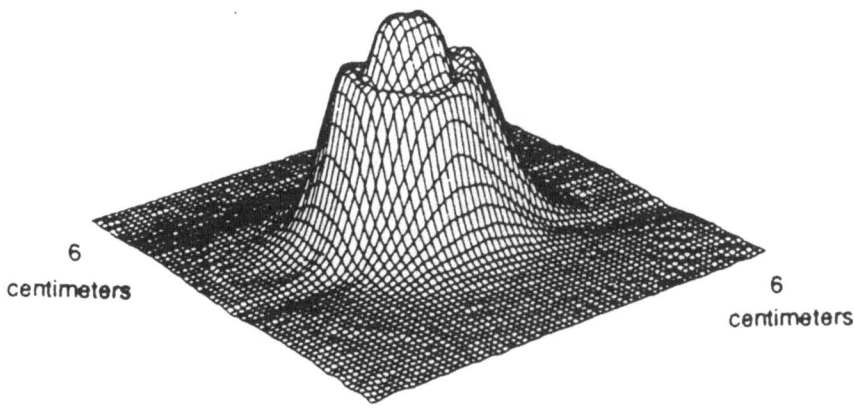

Figure 23. Helium concentration cross section in the co-flowing helium-air jet. The slice is taken 5.4 diameters downstream of the nozzle and the maximum value represents 100% pure helium; the background is air (Snyder and Hesselink 1988).

trations of absorbing species by selecting a laser whose wavelength is coincident with an absorption line of the species of interest. The time evolution of the density distribution or species concentration can be followed using rapidly pulsed lasers and a high-speed camera at the image point. With the apparatus as it is described, the 18 views yield a lateral resolution in a 4 cm diameter jet of 3 mm. Higher resolution will require a larger number of views. This implies that such a measuring device will not be capable of following small particles in the flow, and, consequently, cannot be used for velocity measurements.

An alternative approach uses the index-of-refraction induced deflection of a laser beam as it is swept across a supersonic jet (Faris 1987). Once again, the laser must be incident from many angles in order to generate the tomographic inversion. Using a swept laser sheet, volumetric projections can be done, but simultaneous rapidly swept sheets from multiple angles will be required to freeze unsteady phenomena.

3.6 Acoustic Imaging

There are a number of instances when the use of light to 'visualize' the objects in a flow field is impractical or undesirable. Such instances involve the flow of optically dense materials, including various petrochemicals such as gas/oil/tar mixtures, as well as fluids comprised of optically sensitive materials that would degrade when illuminated. To study and monitor these flows, the use of acoustic imaging has made inroads over the past several decades. Much of the interest arose after the rapid successes of the introduction of acoustic methods in nondestructive testing. However, acoustic imaging for flow diagnostics involves relatively low-contrast, unsteady changes in the acoustic index-of-refraction, in comparison to the stationary, often high-contrast fields encountered in solids testing. Acoustic imaging is generally sensitive to changes in the acoustic index

$$n = \frac{a_0}{a},$$

24

where a_0 is the unperturbed speed-of-sound, and a is the local sound speed dependent on the local temperature, and variable due to the local velocity field with respect to a fixed receiver.

Early work in acoustic imaging was performed in a pipe flow in water (Johnson et. al. 1976) through the use of a circumferential transducer array that was used to gather time-of-flight information about discrete paths through the velocity fields. Comparable work in air (Schmidt 1975) was performed using a receiver-transmitter that was mechanically traversed through the flow field. Both of these methods rely on relatively long temporal scales in the flow due to the rather long periods associated with data acquisition. Through the use of individual planar acoustic waves and an array of acoustic receivers (Trebitz 1982), the period during which flow was assumed to be steady was markedly decreased. By launching successive waves it was demonstrated that acoustic imaging could be used to provide some limited information relative to a slowly changing velocity field. Trebitz specifically demonstrated that, for a rotationally symmetric velocity field with a finite tangential velocity field and zero-velocity radial field (found in potential line vortices, for example), the tangential velocity $v(r)$ could be found from the measured phase deviations $\delta\phi$ as a function of array position y:

$$\frac{v(r)}{r} = \frac{a_0}{\pi k_0} \int_{r^2}^{R^2} \frac{d}{dy^2} \left(\frac{\delta\phi(y)}{y} \right) \frac{dy^2}{\sqrt{y^2 - r^2}}. \qquad 25$$

where a_0 is the sound speed in the vortex core and k_0 the corresponding wavenumber, y is the coordinate along the incoming planar acoustic wave, r is the radius from the axis of symmetry, and R is the radius of the velocity disturbance. This approach is roughly analogous to solving Abel's equation. Thus, with only phase-change information, a three-dimensional (albeit rotationally symmetric) flow field was successfully reconstructed using acoustic imaging.

To summarize the relative merits of acoustic imaging to optical imaging, it is imperative that one remember that most of the analogies between the behavior of light and sound are qualitative, and that many of the attractive features of optical imaging found, for example, in holography, are not capable of being reproduced at present using acoustic means. Two of the primary drawbacks of acoustic imaging are the relatively slow propagation speeds of the waves, limiting use mainly to incompressible flows, and the high ultrasonic frequencies needed to provide adequate spacial resolution. However, the optical properties of the fluid being observed are of no consequence; coherent acoustic wave generation is quite easy, and, with suitable receiver arrays, direct measurement of phase and amplitude is straight-forward in comparison to similar determinations using light where interferometric means must be invoked. Film and current digital arrays do enjoy the advantage of being capable of recording high-resolution images, while current acoustic array technology is such that, typically, 1,000 sensors per array is a practical limit.

3.7 MAGNETIC RESONANCE IMAGING

When an atom with a non-zero nuclear spin is placed in a strong magnetic field, it acts much like a spinning top in a gravitational field. By using a radio frequency pulse, the atomic spins can be aligned and their interaction with the surroundings can be observed by watching both the spin dephasing time and the relaxation time. Magnetic resonance imaging interrogates hydrogen atoms in a magnetic field gradient to generate sequential two-dimensional slices across the three-dimensional volume. The different dephasing and relaxation times of hydrogen in its various molecular environments produce exceptionally good images of three- dimensional structure. This imaging technique is widely used for medical diagnostics. Moving fluids, such as water and blood, can be observed due to the fact that the spinning nuclei are moving.

Early work on quantitative flow measurements with nuclear magnetic resonance used time-of-flight methods to follow this motion (Singer 1959). More recently, imaging using the phase shift induced by the spin motion along the magnetic field gradient has produced high-quality angiograph-type images of blood flow (Dumoulin 1987). By careful selection of magnetic field gradient pulses and averaging over the cardiac cycle, a three-dimensional image giving the average flow velocity of both venus and arterial flow can be generated. By observing only the difference of rapidly acquired data, nonmoving materials and patient motion can be suppressed. An example of such a three-dimensional image is shown in Fig. 24 (Dumoulin 1987). An extension of this method to the observation of flow profiles and shear stress inside pipes and other opaque objects looks promising (Menditto 1988). Due to the rather low data acquisition rate, magnetic resonance imaging is best suited for either laminar or repetitively pulsed flows, but no optical access is needed.

4 VISUALIZATION OF SPACE-FILLING DATA

The experimental methods discussed above have either already generated, or have the potential of generating, space-filling three-dimensional data. Similar space-filling data sets are being generated computationally, and medical diagnostic technologies such as computer-assisted tomography, magnetic resonance imaging, and ultrasound, produce very complex space-filling data sets which are becoming critical for medical diagnostics. Developing methods to display such data sets in an understandable format is increasingly important, particularly in the fluid dynamics area where embedded structures and gradual boundaries make standard three-dimensional projection methods inappropriate.

Currently, three projection techniques are being used. The first is a graphic rendering technique in which the volumetric field is represented by a three-dimensional contour, creating a complex solid surface (Yip 1986b). This surface is then rendered using a conventional three-dimensional graphic display workstation

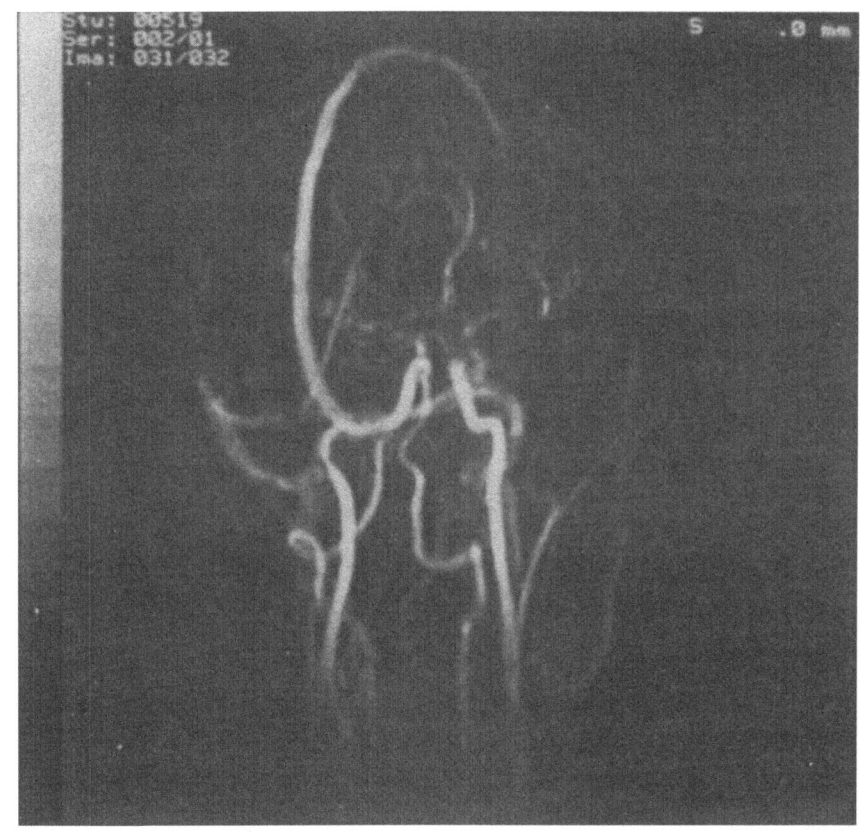

Figure 24. Magnetic resonance image of an oblique head projection showing moving blood. The image was acquired in 4.5 minutes. (Dumoulin 1987)

to produce an object whose lighting, shading, and perspective gives the viewer an excellent perception of the surface contour (see for example Figs. 15 & 16). This object may be dynamically rotated to give an even stronger sense of the structure of the three-dimensional surface. New techniques of 'volume rendering' are capable of rapidly projecting such three-dimensional surfaces.

The second method also relies on three-dimensional surface contours, but the projection is done by a holographic reconstruction (Agui 1985) to give the viewer an immediate sense of the three-dimensionality of the object. An apparent light source illuminating the object from the side gives further definition to the surface characteristics by casting shadows.

The third method involves a projection of all the data simultaneously. A simple transparent projection of the full three-dimensional data set generates a confused or muddy appearance from which volumetric features are difficult to extract. One solution to this is the varifocal mirror (Mills 1984) which uses the combination of a mirror with a variable focus and a rapidly updated video display to step through sequential cross sections of the data so that they are projected in their proper spatial order and location. This generates a full three-dimensional object which is perceived with 100% transparency, since structure which is closer to the viewer cannot obscure that which is farther away.

Total transparency is one extreme of three-dimensional projection technology, whereas a totally solid object is the other. Both the solid projection and the totally transparent projection severely limit one's ability to interpret the data. In the case of the solid projection, if the data does not contain distinct surfaces, these surfaces must be imposed where none actually exist, creating visual artifacts. Gradual variations in the data are often important and cannot be visualized within the solid model environment. Furthermore, embedded structure is obscured by the solid surface and can only be seen in cut-away views. Such views indicate some apriori knowledge of the flow structure, so important unanticipated features may be obscured. The 100% transparent additive projections generate a highly confusing image in which there is so much information presented that it cannot be clearly understood.

Several groups have reported work which lies between these extremes. Jaffey et al. (1984) have described a source-attenuation model which defines the continuum of transparent through solid projections. However, they have not looked in detail into applying this model in the intermediate domain. Lenz et al. (1984) have experimented with additive projections including obscuration using a dynamic two-dimensional display. Obscuration, which is the property of objects to cover or hide regions behind them, is an important visual cue. It serves to simplify a picture (by hiding some of the data content) and to yield relative depth cues to aid in comprehension. Pictures with complete obscuration (solid) hide too much, and those with no obscuration become too confusing. Selective use of partial obscuration can produce very informative, comprehensible pictures (Russell 1987). By varying the obscuration factor, the data ranges from a solid object to totally transparent. The proper value of obscuration is that which leads to the best comprehension. Examples of high transparency projections and solid projections are shown in Figs. 25a and 25b, and 26a and 26b, which show the trailing edge of a three-dimensional turbulent spot (taken with a swept laser beam by Nosenchuck as described earlier in this chapter) and X-ray tomography data of a head. A projection with intermediate transparency and a computer generated three-dimensional grid is shown in Figure 13. Stereo pairs showing projections of the turbulent spot with partial obscuration are shown in Figs. 27 and 28. In these pictures, the features which are close are easily distinguishable from those which are far away and the partial transparency permits structures in the back to be faintly seen through those in the front.

When confronted with a true, three-dimensional complex image, our first instinct is to either move the object or move our heads so that we can see the three-dimensional structure. Our eyes are able to convert the time- evolving motion into a three-dimensional perception. In a sense, our eyes do a tomographic inversion, but they need to be presented with sequential angular views updated at a rate of approximately ten views per second. When partial obscuration is coupled with dynamic motion (Russell 1987), the relative location of features becomes clear and the full three-dimensional data set becomes understandable.

The algorithm which is used to generate these partially obscured projections is equivalent to the source-attenuation model described by Jaffey (1984). Each volume element is assigned an obscuration, a, and a brightness, b. Projections are then made by beginning with the plane farthest from the observer and sequentially stacking the data, with the volume elements of each new plane adding their individual brightness values to the image and attenuating those volume elements which they obscure by the factor a. This algorithm has been implemented by Russell using a Cartesian scan order. An intermediate projection is used to minimize aliasing due to the mapping of the volume elements onto the picture elements of the particular projection chosen. The imaging algorithm has also been extended to use color to indicate areas of particular interest, including marking organs in medical images and designating levels of vorticity in three-dimensional fluid projection.

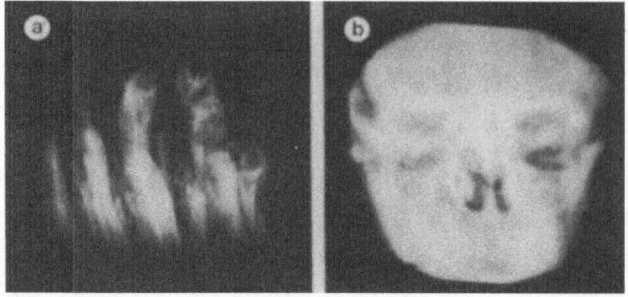

Figure 25. High transparency projections. (a) The trailing edge of a three-dimensional turbulent spot looking downstream from above. (b) Computed X-ray tomographic image of a head. (Russell 1987)

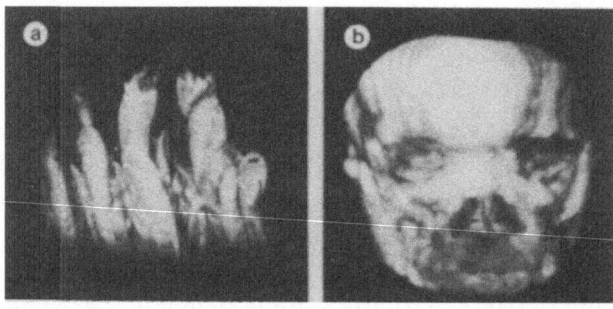

Figure 26. Solid projections. (a) The trailing edge of a three-dimensional turbulent spot looking downstream from above. (b) Computed X-ray tomographic image of a head. (Russell 1987)

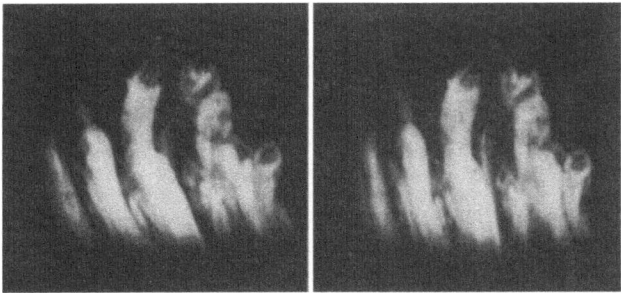

Figure 27. Turbulent spot–view looking downstream (stereo pair). (Russell 1987)

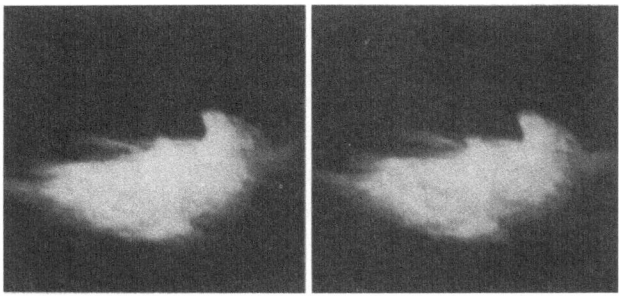

Figure 28. Turbulent spot–view across the spot looking slightly upstream (stereo pair). (Russell 1987)

5 SUMMARY AND CONCLUSIONS

The selection of a particular diagnostic technique for quantitative three-dimensional measurements depends critically on the geometry and character of the flow field to be studied. Tables II and III give an overview of the various techniques discussed and the quantities that they are able to measure for both incompressible and compressible flows, respectively. The numbers in the tables indicate the current and short-term projected capabilities of each of the techniques. As was previously mentioned, virtually all of these techniques can be extended to three-dimensional measurements. In some cases, such as fluorescing dyes and particle and Rayleigh scattering, this has already been done. In others, including laser-induced fluorescence and flow tagging, volumetric scans are a relatively straightforward extension of the current technology.

5.1 INCOMPRESSIBLE FLOWS

Particle scattering and fluorescing dyes have already been used for volumetric imaging of incompressible flows. Due to the complexity of photographing and tracking a large number of particles, particle scattering techniques are currently limited to Reynolds numbers below 1500 or so. This, of course, is strongly dependent on the volume of the flow field observed and the accuracy with which the velocities are measured. Four methods are currently under investigation for volumetric imaging. These include stereoscopic photography, holography, swept laser sheets, and simultaneous multi-color laser sheets. Holography and stereoscopic imaging are only useful with a relatively small number of scatterers in the flow, so that each particle can be separately imaged and tracked. Swept laser sheets can be used in conjunction with fluorescing dyes or particle scattering to yield volumetric scans of the flow field. New technologies for rapidly sweeping laser sheets, coupled with fast-framing cameras, suggest that this diagnostic method has the potential of time-freezing high Reynolds number flow fields. Simultaneous scattering from multiply colored laser sheets can be used for particle-laden, incompressible flows to generate an instantaneous view of the particle density, or, with multiple pulses, of the particle velocities. Both the swept laser and the multi-colored laser sheets can be high enough power to generate Rayleigh scattering for direct measurements of density and mixing in gases.

Flow marking with hydrogen bubbles is limited to low Reynolds number due to bubble break-up problems. Marking fluids with photochromic dyes is applicable to high Reynolds numbers flows and can be configured to give a direct measurement of vorticity (Falco 1988). Measurements to date have been done

	Velocity	*Mixing*	*Comments*
Particle or Bubble Scattering	3 (low Reynolds number)	3	Requires particle seeding. Mixing is by selective seeding.
Fluorescing Dyes		3	Requires selective seeding with dye. Rapid volumetric scans possible. Compatible with particle scattering.
H$_2$ Bubble	1 (low Reynolds number)		Bubble lines show velocity profiles and structure.
Photochromic Dyes	2	2	So far done in kerosene, may be possible in water. Mixing is by selective seeding.
Acoustic Imaging	$\bar{2}$		So far needs many angles or a symmetric flow. No optical access required. Currrently for repetitive or slowly varying flows.
Magnetic Resonance Imaging	$\bar{3}$	$\bar{3}$	No optical access required. Only for repetitive or slowly varing flows.

1	Line Instantaneous	$\bar{1}$	Line averaged
2	Plane Instantaneous	$\bar{2}$	Plane Averaged
3	Volume Instantaneous	$\bar{3}$	Volume Averaged

Incompressible Flows Current Diagnostic Capabilities
TABLE II

	Velocity	Temp.	Den.	Mixing	Comments
Particle Scattering	3 (low Re)			3	Requires particle seeding. Mixing is by selective seeding
Rayleigh Scattering			3	3	Subject to interference from particle scattering.
Raman Scattering		2	2	2	Very low signal levels.
Fluorescence					
O_2		(2)	2	2	Low signals at low temperature.
I_2	$\bar{2}$		2	2	Requires approx. 1 part/10^3 seeding. Low signal levels.
NO	(2)	2	2	2	Requires approx. 1 part/10^3 seeding.
OH	2	2	2	2	Useful in combusting environments.
Na	$\bar{2}$	$\bar{2}$	$\bar{2}$	2	Requires seeding approx. 1 part/10^8. Not useful in low temp. air. Strong signal levels.

TABLE III (Continued on next page)

TABLE III (Continued from preceding page)

	Velocity	Temp.	Den.	Mixing	Comments
Optical Tomography			2, $\bar{3}$	2	Requires extensive optical access.
Photochromism	2			(2)	Flow must be seeded with liquid droplets
Oxygen Tagging (RELIEF)	2	($\bar{2}$)	2	2	Limited to < 700 K. Density and mixing are by simultaneous Rayleigh scattering

1 Line Instantaneous $\bar{1}$ Line Averaged
2 Plane Instantaneous $\bar{2}$ Plane Averaged
3 Volume Instantaneous $\bar{3}$ Volume Averaged

( ) Currently Under Development

Compressible Flows Current Diagnostic Capabilities
TABLE III

in kerosene, but recent work indicates that new water-soluble dyes may become available.

In flows with no optical access, acoustic and magnetic resonance imaging can be used. Both of these are rather slow and can only be applied to steady or slowly varying flows, or synchronously applied to flows with repetitive fluctuations such as the pulsatile nature of blood flow.

5.2 COMPRESSIBLE FLOWS

The diagnostic tools which can be applied in compressible flows are shown in Table III. Particle scattering has virtually the same limitations as it does with incompressible flows with the additional caveat that particles will not follow the flow where steep velocity gradients occur in high-speed environments. Rayleigh scattering is an attractive method of measuring density and can be used to see mixing if the Rayleigh scattering cross sections of the two mixing species are significantly different. For example, hydrogen or helium mixed into air show up as dark regions due to their low Rayleigh scattering cross sections. The Rayleigh scattered light is at the same frequency as the illumination and is, therefore, subject to strong interference from background scattering. Of particular concern is the presence of particles in the flow, since scattering from them can easily obscure the Rayleigh signal. Raman scattering may solve the problem of interference from particles and has the additional capability of yielding temperature and species, but low signal levels currently limit its utility.

Fluorescence from atomic or molecular species can be used in a variety of ways. Oxygen fluorescence can potentially be used to measure instantaneous temperatures and densities and observe mixing, but suffers from very low signals below about 500K. This is especially true if the oxygen is excited by an ArF laser. Iodine, seeded into the flow, can be used to measure velocity and can also be used to measure mixing if only one component is seeded. Due to its small absorption cross section and low fluorescence, iodine must be seeded at densities on the order of one part per thousand, and may cause serious corrosion problems. NO can also be used for temperature measurements and has been proposed for velocity measurements (Hanson 1988). Again, it requires seeding at relatively high concentrations in low temperature air. In combusting flows OH can be used for temperature measurements and, by photodissociating water, can become a line marker for velocity measurements. Sodium can be used for time-averaged velocity, temperature, and density measurements, as well as mixing, but cannot be used in air except at high temperatures, since it forms molecular compounds.

Very low concentrations of sodium are needed (approximately one part in 10^8), and the fluorescence is very strong, so high-speed fluctuations can be monitored.

Optical tomography has been used to measure density and species concentrations, but requires extensive optical access for non-symmetric flow fields. Molecular tagging by photochromism can be done in aerosols, which can then be used to mark the flow. With very small droplets, particle density can be made high so continuous lines can be written. The deviation of these lines then gives an instantaneous map of the velocity and, as in the liquid case, can be used to measure vorticity. Oxygen flow tagging is an alternative method of marking the flow field which gives instantaneous velocity profiles and can simultaneously be used to measure density and mixing by observing Rayleigh scattering. By tuning the flow marking lasers, the time-averaged temperature profile can also be recorded. Oxygen flow tagging has been demonstrated from low-speed flows up to approximately Mach 4, but is limited to temperatures below 700K due to background fluorescence from the thermally excited oxygen molecules.

The preparation of this paper was supported, in part, by AFOSR under Grant 86-0191 and by NASA Grant NAG-1-494.

6 REFERENCES and BIBLIOGRAPHY

ADRIAN, R.J. 1986 Multi-point optical measurements of simultaneous vectors in unsteady flow–A review. International J. of Heat and Fluid Flow 7, 127.

AGUI, J.C. & HESSELINK, L. 1985 Three-dimensional image processing for flow visualization. American Physical Society, Division of Fluid Dynamics, Tucson, AZ, Paper CO3.

BEELER, G.B. & WEINSTEIN, L.M. 1987 Holocinematographic velocimeter for measuring time-dependent, three-dimensional flows. 12th International Congress on Instrumentation in Aerospace Simulation Facilities, Williamsburg, VA.

BISCHEL, W.K. & BLACK, G. 1983 Wavelength dependence of Raman scattering cross sections from 200-600 nm. Excimer Lasers. AIP Conference Proceedings #100, Subseries on Optical Science and Engineering.

BOGDONOFF, S.M. & WANG, S. 1986 Comment on "Conical Similarity of Shock/Boundary Layer Interactions Generated by Swept and Unswept Wings". AIAA Journal 24, 540.

BORST, H.V. 1980 The aerodynamics of the unconventional air vehicles of A. Lippisch. Henry V. Borst and Associates, Wayne, PA.

BRODKEY, R.S. 1986 Image processing and analysis for turbulence research. Chemical Engineering Education 20, 202.

BROWN, G.H. 1971 Photochromism. Wiley-Interscience, New York.

BRYANSTON-CROSS, P.J. 1986 High speed flow visualization. Prog. Aerospace Sci. 23, 85.

BYER, R.L. & SHEPP, L.A. 1979 Two-dimensional remote air-pollution monitoring via tomography. Optics Letters 4, 75.

CATTOLICA, R.J. & STEPHENSON, D.A. 1985 Two-Dimesnional Imaging of Flame Temperature Using Laser-Induced Fluorescence. Dynamics of Flames and Reactive Systems, ed. J.R. Bowen, N. Manson, A.K. Oppenheim, and R.J. Soloukin, Vol. 95 of Progress in Astronautics and Aeronautics, 714.

CENKNER, A.A., Jr. & DRISCOLL, R.J. 1982 Laser-indusced fluorescence visualization on supersonic mixing nozzles that employ gas-trips. AIAA Journal 20, 812.

CHANG, T.P.K., WATSON, A.T. & TATTERSON, G.B. 1985a Image processing of tracer particle motions as applied to mixing and turbulent flow – I. The technique. Chemical Engineering Science 40, 269.

CHANG, T.P.K., WATSON, A.T. & TATTERSON, G.B. 1985b Image processing of tracer particle motions as applied to mixing and turbulent flow – II. Results and discussion. Chemical Engineering Science 40, 277.

CHANKAYA, K.M. 1986 Development of a Stereo Schlieren Technique for Visualizing Three-Dimensional Unsteady Flows. Masters Thesis, T-1728, Princeton University.

DAVIS, J., BOUTTCHER, J. JOHNSON, G. & MARSCHALL, E. 1985 Photochromic flow visualization in single-phase and two-phase flows. International Symposium on Physical and Numerical Flow Visualization, SME/FEd, 22, 75.

DEWEY, C.F., Jr. 1976 Qualitative and quantitative flow field visualization utilizing laser-induced fluorescence. AGARD Conference Proceedings No. 193 on Applications of Non-Intrusive Instrumentation in Fluid Flow Research, AGARD-CP-193, 17-1.

DIMOTAKIS, P.E., DeBUSSY, F.D., & KOOCHESFAHANI, M.M. 1981 Particle streak velocity field measurements in a two-dimensional mixing layer. Physics of Fluids 24, 995.

DUMOULIN, C.L., SOUZA, S.P. & HART, H.R. 1987 Rapid scan magnetic resonance angiography. Magnetic Resonance in Medicine 5, 238.

ELKINS III, R.E., JACKMAN, R.R., JOHNSON, R.R. & LINDGREN, E.R. 1977 Evaluation of stereoscopic trace particle records of turbulent flow fields. Rev. Sci. Instrum. 48, 738.

ESCODA, M.C. & LONG, M.B. 1983 Rayleigh scattering measurements of the gas concentration field in turbulent jets. AIAA Journal 21, 81.

FALCO, R.E., 1978 Combined simultaneous flow visualization/hot-wire anemometry for the study of turbulent flows. Nonsteady Fluid Dynamics, Eds. D.E. Crow and J.A. Miller, ASME, New York, p. 73.

FALCO, R.E. & CHU, C.C. 1987 Measurement of two-dimensional fluid dynamics quantitites using a photochromic grid tracing technique. SPIE Vol. 814 Photomechanics and Speckle Metrology, 706.

FALCO, R.E., CHU, C.C., HEATHERINGTON, M.H. & GENDRICH, C.P. 1988 The circulation of an air foil starting vortex obtained from instantaneous vorticity measurements over area. AIAA Paper No. 88-3620.

FARIS, G.W. & BYER, R.L. 1987 Quantitative three-dimensional optical tomographic imaging of supersonic flows. Science 238, 1700.

GABOR, D. 1948 A new microscope principle. Nature 161, 777.

GELLER, E.W. 1955 An electrochemical method of visualizing the boundary layer. J. Aero. Sci. 22, 869.

GOLDFISH, L.H., KOUTSKY, J.A., & ADLER, R.J. 1965 Tracer Introduction by Flash Photolysis. Chem. Engineering Sci., 20, 1011.

HANSON, R.K. 1988 Private Communication.

HARVEY, A.B. 1981 Chemical Applications of Nonlinear Raman Spectroscopy. Academic Press, New York.

HAUSSMANN, G. & LAUTERBORN, W. 1980 Determination of size and position of fast moving gas bubbles in liquids by digital 3-D image processing of hologram reconstructions. Applied Optics 19, 3529.

HERMAN, G.T. 1980 Image reconstruction from projections. Academic Press, New York.

HESSELINK, L. 1988a Digital image processing in a flow visualizaton. Annual Reviews of Fluid Mechanics, Palo Alto, Annual Reviews, Inc.

HESSELINK, L. 1988b Optical tomography. Handbook of Flow Visualization, Hemisphere Publishing.

HOLDER, D.W. & NORTH R.J. 1963 Schlieren Methods. British National Physical Laboratory Notes on Applied Science, 31.

HUMPHREY, J.A. 1977 Effect of turbulence on the precision of velocity and velocity fluctuation data obtained by photochromic visualization. Can. J. of Chem. Eng. 55, 126.

JAFFEY, S., DUTTA, K. & HESSELINK, L. 1984 Digital reconstruction methods for three-dimensional image visualization. Proceedings of the Society of Photo-Optics Instrumentation Engineering, Vol. 507, 155.

JOHNSON, S.A., GREENLEAF, J.F., HANSEN, C.R., SAMAYOA, W.F., TA-NAKA, M., LENT, A., CHRISTENSEN, D.A., & WOOLLEY, R.L. 1976 Reconstructing three-dimensional fluid-velocity vector fields from acoustic transmission measurements. Acoustical Holography, 7, 307-326.

KENT, J.C. & EATON, A.R. 1982 Stereo photography of neutral density He-filled bubbles for 3-D fluid motion studies in an engine cylinder. Applied Optics 21, 904.

KYCHAKOFF, G., HOWE, R.D., & HANSON, R.K. 1983 Flow visualization in combustion gases. AIAA-83-0405, AIAA 21st Aerospace Sciences Meeting, January 10- 13, 1983, Reno, Nevada.

LANDRETH, C.C., ADRIAN, R.J., & YAO, C.S. 1987 Double pulsed particle image velocimeter with directional resolution for complex flows. Experiments in Fluids, Springer Verlag 5, 1.

LAPP, M. & PENNEY, C.M. 1974 Laser Raman Gas Diagnostics, Plenum Press, New York.

LAUTERBORN, W. & VOGEL, A. 1984 Modern optical techniques in fluid mechanics. Annual Review of Fluid Mechanics, 16, 223.

LEE, M.P., PAUL, P.H. & HANSON, R.K. 1987 Quantitative imaging of temperature fields in air using planar laser-induced fluorescence of O_2. Optics Letters, 12, 75.

LEE, Y.G., FOURNEY, M.E. & MOULTON, R.W. 1974 Determination of Slip Ratios in Air-Water Two-Phase Critical Flow at High Quality Levels using Holographic Techniques. AIChE J. 20, 209.

LENZ, R., DANIELSSON, P-E. & GUDMUNDSON, B. Interactive display of 3-D images in picap II. Proceedings of the Society of Photo-Optics Instrumentation Engineering, Vol. 507, 19.

LIBURDY, J.A. 1987 Holocinematographic velocimetry: resolution limitation for flow measurement. Applied Optics 26, 349.

LONG, M.B., FOURGUETTE, D.C., ESCODA, M.C. & LAYNE, C.B. 1983 Instantaneous Ramanography of a turbulent diffusion flame. Optics Letters 8, 244.

LU, L.J. & SMITH, C.R. 1985 Image processing of hydrogen bubble flow visualization for determination of turbulence statistics and bursting characteristics. Experiments in Fluids 3, 349

MANTZARAS, J., FELTON, P.G., and BRACCO, F.V. 1988 Three-dimensional visualization of premixed-charge engine flames: Islands of reactants and Products; fractal dimensions; and homogeneity. SAE/SP-88/759 Proceedings of the International Fuels and Lubricants Meeting and Exposition, Portland, Oregon, October 10-13, 1988.

MASSEY, G.A. & LEMON, C.J. 1984 Feasibility of measuring temperature and density fluctuations in air using laser-induced O_2 fluorescence. IEEE J. of Quantum Electronics, Vol. QE-20, 454.

MATSUI, T., NAGATA, H. & YA-SUDA, H. 1979 Some remarks on hydrogen bubble technique for low speed water flows. Flow Visualization, (ed. T. Asanuma), Hempshire, Washington, D.C., 215.

McDANIEL, J.C. 1983 Quantitative measurement of density and velocity in compressible flows using laser-induced iodine fluorescence. AIAA 21st Aerospace Sciences Meeting, Reno, Nevada.

McGREGOR, I. 1961 The vapor-screen method of flow visualization. Fluid Mech. 11, 481.

McKAY, H.C. 1951 Three-dimensional photography. Jones Press, Minneapolis, MN.

McKENZIE, R.L. & GROSS, K.P. 1981 Two-photon excitation of nitric oxide fluorescence as a temperature indicator in unsteady gas dynamic processes. Applied Optics 20, 2153.

MENDITTO, S.A., STREAK, L.M., SEBOK, D.A. & MEZRICH, R.S. 1988 Magnetic resonance imaging of shear stress. Presented at the Socity of Magnetic Resonance in Medicine, 7th Annual Meeting, San Francisco, August 22, 1988.

MERZKIRCH, W. 1987 Flow Visualization. 2nd Ed., Harcourt-Brace-Jovanovich.

MILES, R.B. 1975 Resonant Doppler velocimeter. Physics of Fluids 18, 751.

MILES, R., COHEN, C., CONNORS, J., HOWARD, P., HUANG, S., MARKOVITZ, E., & RUSSELL, G. 1987 Velocity measurements by vibrational tagging and fluorescent probing of oxygen. Optics Letters 12, 861.

MILES, R.B., CONNORS, J.J., MARKOVITZ, E.C. & ROTH, G.J. 1988a Coherent anti-Stokes Raman scattering (CARS) and Raman pumping lineshapes in high fields. SPIE Meeting, Los Angeles, CA, January 11-15, 1988 (to be published in SPIE Conference Proceedings).

MILES, R.B., CONNORS, J.J., HOWARD, P.J., MARKOVITZ, E.C. & ROTH, G.J. 1988b Proposed single-pulse two-dimensional temperature and density measurements of oxygen and air. Optics Letters 13, 195.

MILES, R.B., CONNORS, J.J., MARKOVITZ, E.C., HOWARD, P.J., and ROTH, G.J. 1989 Instantaneous Supersonic Velocity Profiles in an Underexpanded Jet by Oxygen Flow Tagging. (To be published in Physics of Fluids, February 1989).

MILLS, P.H., FUCHS, H. & PIZER, S. 1984 High-speed interaction on a vibrating-mirror 3-D display. Proceedings of the Society of Photo- Optics Instrumentation Engineering, Vol. 507, 93.

MUELLER, T.J. 1983 Smoke visualization in wind tunnels. Astronautics & Aeronautics, January 1983, 50.

NOSENCHUCK, D.M. & LYNCH, M.K. 1986 Three-dimensional flow visualization using laser-sheet scanning. AGARD-CPP-413, Paper #18.

NOSENCHUCK, D.M., LYNCH, M.K. & STRATTON, J.P. 1987 Active control of sublayer disturbances using an array of heating-elements. 1987 ASME/ JSME Thermal Engineering Joint Conference, Vol. 2, Book No. 10219B.

OLD, J.G., GENTILI, K.L., & PECK, E.R. 1981 Dispersion of CO_2. JOSA 61, 89.

PECK, E.R. & KHANNA, B.M. 1966 Dispersion of nitrogen. JOSA 56, 1059.

PECK, E.R. & REEDER, K. 1972 Dispersion of air. JOSA 62, 958.

PECK, E.R. 1983 Sellmeier fits with linear regression; multiple data sets; dispersion formulas for helium. Applied Optics 22, 2906.

POPOVICH, A.T. & HUMMEL, R.L. 1967a A new method for non-disturbing turbulent flow measurements very close to a wall. Chemical Eng. Sci 22, 21.

POPOVICH, A.T. & HUMMEL, R.L. 1967b Experimental study of the viscous sublayer in turbulent pipe flow. AIChE Journal 13, 854.

PRENEL, J.P., PORCAR, R., REINICHE, S. & DIEMUNSCH, G. 1986 Optical oscilloscope for three-dimensional flow visualization. Optics and Laser Technology, August 1986, 208.

PRENEL, J.P., PORCAR, R., REINICHE, S. & DIEMUNSCH, G. 1986 Visualistions tridimensionnelles d'ecoulements non axisymetriques par balayage programme d'un faisceau laser. Optics Comm. 59, 92.

ROYER, H. 1977 Holographic velocimetry of submicron particles. Optics Communications 20, 73.

RUSSELL, G. & MILES, R.B. 1987 Display and perception of 3-D space-filling data. Applied Optics 26, 973.

SANTORO, R.J., SEMERJIAN, H.G., EMMERMAN, P.J. & GOULARD, R. 1981 Optical tomography for flow field diagnostics. Heat and Mass Transfer 24, 1139.

SCHMIDT, D.W. 1975 Akustische Messung der Zirkulation von Wirblen und von Airkulatinsverteilungen bei Modelluntersuchungen in Windkanaelen. Mitteilungen aus dem Max-Planck Insitut fuer Stroemungsforschung und der Aerodynamischen Versuchsanstalt, Nr. 61.

SCHROTTER, H.W. & KOCKNER, H.W. 1979 Raman scattering cross sections in gases and liquids in Raman spectroscopy of gases and liquids. Topics in Current Physics, Vol. II, Ed. A. Weber (Springer-Verlag, Berlin), p. 123.

SETTLES, G.S. & Lu, F.K. 1985a Conical Similarity of Shock/Boundary Layer Interactions Generated by Swept and Unswept Wings. AIAA Journal, 23, 1021.

SETTLES, G.S. 1985b Colour-coding Schlieren Techniques for the Optical Study of Heat and Fluid Flow. International Journal of Heat and Fluid Flow, 6, 3.

SETTLES, G.S. 1985c Flow visualization techniques for practical aerodynamic testing. Flow Visualization 3, (ed. by W.J. Yang) Hemisphere Publishing Corp., 306.

SHARDANAND & PRASAD RAO, A.D. 1977 Absolute Rayleigh scattering cross sections of gases and freons of stratospheric interest in the visible and ultraviolet regions. NASA TN D-8442.

SHEU, Y.-H.E., CHANG, T.P.K., TATTERSON, G.B. & DICKEY, D.S. 1982 A three- dimensional measurement technique for turbulent flows. Chemical

Engineering Communications 17, 67.

SHIRAISHI, T., MUNEZANE, S. & MAKINO, M. 1985 Measurement of low velocity field in a plenum using the flow visualization technique. Proceedings of the Int'l Symposium on Fluid Control and Measurement, Tokyo, 1985, Pergamon Press, 655.

SHIRLEY, J.A. & BOEDEKER, L.R. 1988 Nonintrusive space shuttle main engine nozzle exit diagnostics. AIAA/ASME/SAE/ASEE, 24th Joint Propulsion Conference, July 11-13, 1988, Boston, MA, AIAA #88-3038.

SHOFNER, F.M., WEBB, R.O., MENZEL, R.W. & HEIFNER, R.L. 1970 Optical processors for holographic velocmetry data. Air Force Flight Dynamics Laboratory Report #AFFDL-TR-69-100.

SINGER, J.R. 1959 Blood Flow Rates by Nuclear Magnetic Resonance Measurements. Science 130, 1652.

SMITH, C.R. & METZLER, (1983) S.P. The Characteristics of Low-speed Streaks in the Near-Wall Region of a Turbulent Boundary Layer. Journal of Fluid Mechanics, 129, 27.

SNYDER, R. & HESSELINK, L. 1984 Optical tomography for flow visualization of the density field around a revolving helicopter rotor blade. Applied Optics 23, 3650.

SNYDER, R. & HESSELINK, L. 1985 High speed optical tomography for flow visualization. Applied Optics 24, 4046.

SNYDER, R. & HESSELINK, L. 1988 Measurement of mixing fluid flows with optical tomography. Optics Letters. 13, 87.

STANISLAS, M. 1985 Ultra-high speed smoke visualization of unsteady flows using a pulsed ruby laser. Flow Visualization 3, Proceedings of the Third International Symposium on Flow Visualization, Sept. 6-9, 1983, University of Michigan, Ann Arbor, MI, U.S.A. (ed. by W.J. Yang), 41.

STUCK, B.W. 1977 A new proposal for estimating the spatial concentration of certain types of air pollutants. J. of the Optical Society of America, 67, 668.

SWEENEY, D.W. & VEST, C.M. 1974 Measurement of three-dimensional temperature fields above heated surfaces by holographic interferometry. Int. J. Heat Mass Transfer 17, 1443.

THOMPSON, B.J. & WARD, J.H. 1966 Particle Holography. Science Research 37.

THOMPSON, B.J., WARD, J.H., & ZINKY, W.R. 1967 Application of Hologram Techniques for Particle Size Analysis. Applied Optics, 6, 519.

THOMPSON, B.J. 1984 Holographic particle characterization size, position, and velocity. Image Technology Research and Development, 12-14.

TREBITZ, B.O. 1982 Acoustic Transmission Imaging for Flow Diagnostics. Ph.D. Thesis, California Institute of Technology.

TROLINGER, J.P. 1969 Holographic Techniques for the Study of Dynamic Particle Fields Appl. Opt. 8, 957.

TROLINGER, J.D. 1974 Laser instrumentation for flow field diagnostics. AGARD AG-186 (ed. S.M. Bogdonoff).

TROLINGER, J.D. 1977 Diagnostics of turbulence by holography. SPIE 125, 105.

UTAMI, T. & UENO, T. 1984 Visualization and picture processing of turbulent flow. Experiments in Fluids 2, 25.

UTAMI, T. & UENO, T. 1987 Experimental study on the coherent structure of turbulent open-channel flow using visualization and picture processing. J. Fluid Mech. 174, 399.

VAN DeHULST, H.C. 1981 Light scattering by small particles. Dover Publications, Inc., New York.

VAN DYKE, M. 1982 An album of fluid motion. Parabolic Press, Stanford, CA.

WASHBURN, E.W., ed. 1930 International Critical Tables of Numerical Data: Physics, Chemistry, and Technology, Vol. VII, McGraw-Hill, NY, pp. 2-11.

WEINSTEIN, L.M., BEELER, G.B. & Lindemann, A.M. 1985 High-speed holocinematographic velocimeter for studying turbulent flow control physics. AIAA Shear Flow Control Conference, Boulder, CO.

WEINSTEIN, L.M. & BEELER, G.B. 1986 Flow measurements in a water tunnel using a holocinematographic velocimeter. AGARD, Fluid Dynamics Panel Symposium on Aerodynamics and Related Hydrodynamic Studies Using Water Facilities, Monterey, CA.

YIP, B. and LONG, M. 1986a Instantaneous planar measurement of the complete three-dimensional scalar gradient in a turbulent jet. Optics Letters 11, 64-66.

YIP, B., FOURGUETTE, D.C. & LONG, M.B. 1986b Three-dimensional gas concentration and gradient measurements in a photoacoustically perturbed jet. Applied Optics 25, 3919.

YIP, B., SCHMITT, R.L. & LONG, M.B. 1988 Instantaneous three-dimensional concentration measurements in turbulent jets and flames. Optics Letters 13, 96.

ZARSCHIZKY, H. & LAUTERBORN, W. 1983 Digital picture processing on high speed holograms. ICIASF'83 Record, 49.

ZIMMERMANN, M. & MILES, R.B. 1980 Hypersonic-helium-flow-field measuremetns with the Resonant Doppler Velocimeter. Appl. Phys. Lett. 37, 885.

ZIMMERMANN, M., CHENG, S. & MILES, R.B. 1981 Measuremetns of supersonic nitrogen flow properties with the resonant Doppler velocimeter. IEEE/OSA Conference on Lasers and Electro-Optics, Washington, DC, June 1981. CLEO'81 Technical Digiest, Paper WO2, Washington, DC, Optical Society of America, 62.

ZIMMERMANN, M., CHENG, S. & MILES, R.B. 1982 Flow visualization in supersonic air with the resonant Doppler velocimeter. 1982 IEEE/OSA Conference on Lasers and Electro-Optics, Phoenix, AZ, April 1982. CLEO'82 Technical Digest, Paper THS4, Washington, DC, Optical Society of America.

ZIMMERMANN, M., CHENG, S. & MILES, R.B. 1985 Velocity Selective Flow visualization in a Free Supersonic Nitrogen Jet with the resonant Doppler velocimeter. Flow Visualization III, W.J. Yang, ed., Springer Verlag.

PARTICLE TRACING: REVISITED

M. Gharib and C. Willert
Department of Applied Mechanics and Engineering Science
University of California, San Diego
La Jolla, California 92093

SUMMARY

Particle tracing is a simple technique used to infer global quantitative information regarding velocity and vorticity fields from flow images. In this paper, recent advances in solving major drawbacks associated with conventional particle tracing are discussed. A quantitative automatic scheme which eliminates the photographic step and is capable of constructing the velocity field directly from the experimental set up is presented.

INTRODUCTION

Flow visualization is a specialized tool used to better understand complex flow structures. Usually, the nature of this understanding is qualitative, therefore a subsequent quantitative measurement is essential for further progress. Quantitative single point measurement techniques such as hot wire anamometry or laser doppler velocimetry are often used to obtain Eulerian velocity fields. These two techniques are typically limited to simultaneous sampling at a few spatial locations. In this respect, the global nature of the information contained in a single flow image invites one to develop quantitative methods to obtain multi-point spatial and temporal velocity or vorticity fields from such images.

Flow images produced by the pathlines of small distinct particles suspended in fluid are good candidates for quantitative as well as qualitative flow observations. The patterns generated by the short pathlines are instantaneous and do not carry time history of past events which are often misleading in the continuous dye visualization techniques (Taneda 1985). The velocity field can be obtained from trace photographs by measuring lengths and location of streaks. Modern history of fluid mechanics credits Ahlborne (1902) and Prandtl (1934) for development of the particle tracing technique. In this regard particle tracing is different from other techniques that rely similarly on the presence of particles to extract velocity information such as Laser Doppler velocimetry or speckle techniques. This paper will only review the advances in the particle tracing method and will avoid discussion of the other techniques.

A major drawback in using particle tracing techniques has been the unacceptable amount of manual work which is required to obtain the velocity field from a large number of traces on each image. In recent years there have been renewed interests in making particle tracing techniques less laborious by incorporating image digitizers and computer image processing into the process of information extraction from the flow pictures. The degree of successfulness of these recent attempts should be judged based on the level of advancement that each method introduces to different steps of the particle tracing method. These steps include the type of particles, illumination techniques, photo or video graphic technique, digitization, trace recognition, sense of direction, error detection provisions, flow computation of randomly distributed data and overall reduction of processing time by automation of the different steps.

In this paper, the parameters that are important in the practical use of the technique will be discussed, and the state of the art as well as some new ideas and methods that have been developed and implemented by the present authors and others will be reviewed.

CRITERIA FOR THE PARTICLE SIZE

The basic assumption in the particle tracing method is that the motion of the particles accurately represents the fluid motions. For photographic reasons, large particle sizes are desirable. However, large particle sizes and density mismatch between the particle and fluid might result in large differences between fluid and particle velocity.

For the case of the particle with density matching that of the fluid and small values of the particle Reynolds number, i.e.

$$R_p = \frac{(U_f - U_p)D}{\nu}$$

where U_f and U_p are the fluid and particle velocity respectively, ν is the kinematic viscosity, and D is the particle diameter, Merzkirch (1974) gives the particle response time to a step change in the fluid velocity as

$$\tau_p = \frac{D^2}{18\nu}$$

Therefore for flows varying slower than τ_p, particles can accurately respond to the drag forces generated by the relative fluid motion. Large particles can also experience lift in flows with strong velocity shear. For particles smaller than the local Kolmogorov scale, this lift force is negligible (Saffman, 1965). Agui and Jimenez (1987) give the following relationship for the relative error of tracking for small density differences:

$$\frac{|U_f - U_p|^2}{|U_f|^2} \approx \frac{r^2 D^2}{10 \, \tau_{op} \, \nu}$$

where

$$r = \frac{\rho_p}{\rho_f} - 1$$

(ρ_f is the fluid density and ρ_p is the particle density and τ_{op} is a typical large eddy turn-over time.)

In general, it is important that particles be neutrally buoyant. However, since inertial and gravitational forces decrease rapidly with the particle size, particles with a size in the range of the local Kolmogorov scale should follow the flow correctly.

ILLUMINATION AND OPTICAL MODULATION OF THE PARTICLES

The original experiments of Ahlborne and Prandtl, which involved free surface visualization, did not require sophisticated illumination methods. However, in order to visualize two components of the velocity in a three-dimensional flow field, illumination of a thin but finite cross-section of the flow is required. A sheet of light can be generated by passing a white light beam through a slit (Wiese Nielson 1970, Imaichi and Ohmi, 1983). The practical difficulties of collimating the light sheet for uniform thickness or generating a thin light sheet from an incoherent light source can be solved using lasers. Laser beams can be thinned, collimated and expanded to a sheet of light through the use of cylindrical lenses, mirrors or rapid light scanners. Dimotakis et.al. (1981) pioneered the laser sheet method for particle tracing and obtained improved pictures due to the small depth of field in their images.

Since the laser sources have a narrow band of wave lengths, particles coated with fluorescent or phosphorescent materials can be selected to absorb and emit light at different wavelengths. Therefore, with proper photographic filtration the background noise due to the scattering of the light by the unwanted particles can be eliminated. Another advantage of using luminescent particles is the capability of effectively adding information regarding the flow direction on the trace. Gharib and Hernan (1985) used the after glowing properties of optically activated phosphorescent particles to generate traces with variable intensity. The intensity decay on each trace was used to infer the flow direction.

With fluorescent particles, which are easier to make, one can obtain a similar effect by chopping the laser beam with a variable density filter. These filters, which are used as intensity modulators, can be designed to generate reference points to mark the beginning and ending of each trace. Gharib et.al. (1986) used two small round openings at the two ends of the chopping filter's window to generate the reference points. A similar technique has been adopted by Agui and Jimenez (1987).

In the chopping filter method a desired flow time scale could be resolved by controlling the illumination period through the rotation rate of the filter. The shape and intensity of the light pulse could only be changed by modifying the window's geometry and filter. In the latest version of the particle tracing technique developed by the authors, the chopping filter has been replaced by an acousto-optic device (Bragg cell). A single line laser beam diffracts to several frequency shifted beams as it goes through the Bragg cell. The frequency shifted beams are harmonics of the main beam which is also available after passing through the Bragg cell. The intensity distribution among the main beam and its harmonics is controlled by the RF voltage that is provided by a driver device. The intensity of the main beam can be modulated by an arbitrary wave function which is provided by an IBM AT to the driver device (Fig.1).

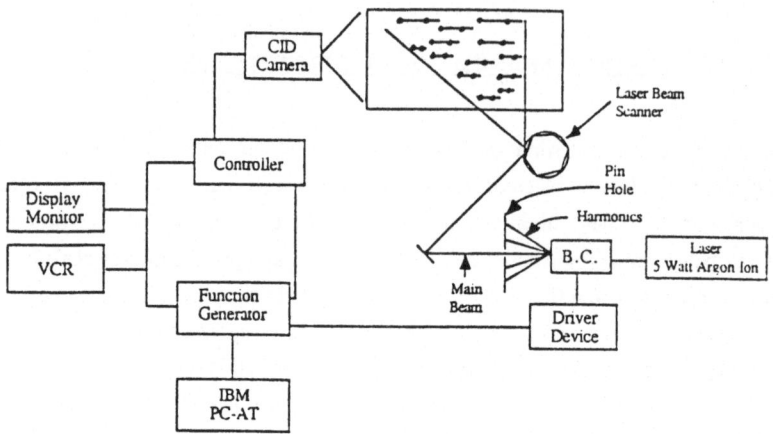

Figure 1: Schematic of the illumination and imaging system for the reference point technique.

IMAGE ACQUISITION AND DIGITIZATION

Photographic recording of tracer images is still popular when high spatial resolution is required. The photographic recording time is normally controlled by the illumination period. In conventional modern particle tracing methods, either the photographic negative or positive is digitized by digitizing tables (Imaichi and Ohmi, 1983, Dimotakis et.al., 1981) or video frame grabbers (Gharib and Hernan, 1985). The digitizing process is controlled by a host computer. Commercial frame grabbers available for the PC type computers can digitize an image into a 512 X 512 pixel matrix with a 256 (8 bit) gray level resolution.

A real-time video recording system is desirable in order to eliminate the required development time associated with photographic negatives and to eliminate non-uniformity in the image signal-to-noise ratio, which is a function of the processing method. One major problem with commercial tube type or CCD (Charged Coupled Devices) based video systems is that their recording time is fixed at 1/30 second (or 30 frames per second). Therefore, generating a streak which requires a continuous recording of the particle motion within a single frame is difficult if not impossible. To eliminate the photographic step, one needs to use CID (Charge Injection Device) type cameras. The CID camera is the digital counterpart of photographic cameras. Its exposure time is controllable. Even though the exposure time may be varied from a few milliseconds to several minutes, the updating framing rate is usually fixed at 1/30 of a second. In the remainder of this section an integrated illumination and recording system based on the CID cameras will be described.

A CID camera (GE TN2250) with a 512 X 512 pixel resolution is synchronized with the Bragg cell through the IBM AT. At a command from the operator, the controller for the CID camera starts integrating the light signal received from the flow for a period of ΔT_1. At a later time, a specified light pulse similar to the one generated by the chopping filter starts illuminating the flow field with a period of ΔT_2 where $\Delta T_2 <$ ΔT_1. The reference points are separated by a period of ΔT which can be controlled independently. At the end of each ΔT_1 period, an image is sent to the IBM AT for digitization or to a video recorder (Sony 3/4" VCR) for a later digitizing process. The updating speed of the system is limited by the video rate which is 1/30 of a second per image. See Figure 2 for the timing of different steps of synchronization. Figure 3 shows a sample streak image which was recorded using the CID camera in the configuration described above. A typical trace generated by the reference point method is shown in Figure 4. The trace ends with a tail which indicates the flow direction.

By removing the photographic steps, the quality of the digitized images has improved drastically. Therefore, any post-processing needed to remove noise has been eliminated.

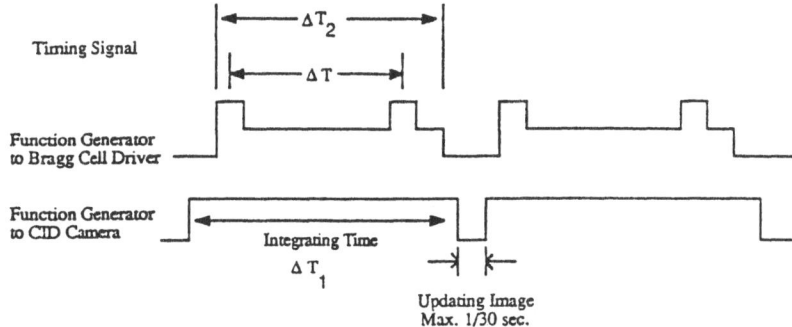

Figure 2: Timing diagram for different steps of the synchronization.

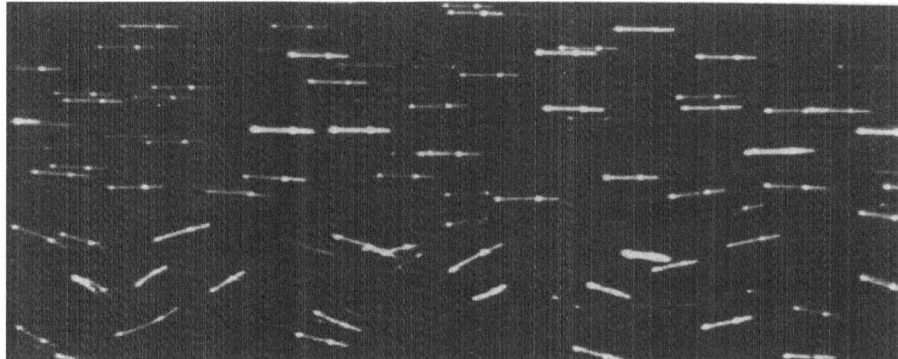

Figure 3: A typical streak image recorded with the charge injection device (CID) camera.

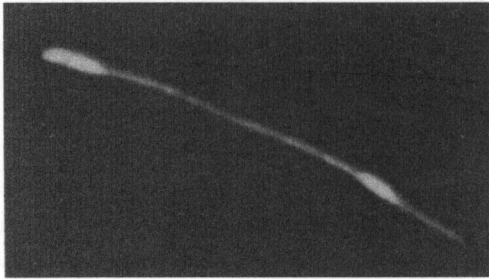

Figure 4:   A typical trace generated by the reference point technique.

AUTOMATIC PARTICLE TRACING

The next step in obtaining the velocity field is to educate the computer to recognize particle traces in the digital image. Automating this process has been the major challenge since the introduction of computers to the particle tracing technique. The methods that use digitizing tablets and human operators to recognize traces and transform them to the digital form do not fall in the category of automatic particle tracing methods. These efforts suffer from the fact that recognizing process depends on the operator's judgement, who should bear in mind the flow direction and use some undefined criterion to identify the position and the length of each trace (Wiese Nielson, 1970, Imaichi & Ohmi, 1983, among others).

The first attempt to automatically identify a contiguous group of data, representing traces from individual particles, was made by Dimotakis et.al. (1981). In their scheme a cluster of connected dots that had a predefined number of dots was recognized as a trace. A least squares parabolic fit was then computed for each separate group. The endpoints and length of each trace were saved for a subsequent interactive step to assign the local velocity vector and eliminate crossed or overlapping traces. This final stage of checking and restoring, which is interactive and time consuming in its nature, is common to many recent works that claim a certain degree of automation in their techniques (Altman, 1985, Jonas & Kent, 1979, Elkins et.al,1977).

The rest of this section is devoted to the description of the modified version of a fully automatic particle tracing technique that was originally developed and used by Gharib et.al. (1986). The essence of the technique is based on the assumption that the trace images generated by the particles contain the reference points and a tail to indicate the direction. In the present arrangement the digital images can be obtained from recorded images on VCR or directly from the camera. The software was designed to operate either on the displayed image on the monitor or on the stored information on the hard disk. The first step in the analysis of the digital images is to define the boundaries of each trace. By pre-examination of a sample image, a gray level threshold is input to determine the minimum pixel value of what will be considered as a trace. The quality of the images make it possible to identify the trace regions with a single threshold value. The intensity distribution over a sample trace is presented in Figure 5. Once the threshold value is determined, it is applied to the remainder of the sequence of images. Using this threshold value, the processing program then sets any pixel value below the threshold level equal to zero (black) and saves the thresholded image.

In the next step, the program scans the image line by line to detect the first pixel greater than zero (or grey) of any potential trace region. To define a region, its perimeter of black pixels is found first by clockwise tracing around the region. The area on the interior of the perimeter is the trace region whose grayscale information and pixel coordinates stored in an array for further analysis (Figure 6). To eliminate any strongly curved traces, the program proceeds to calculate the centroid of the trace region. A centroid that does not lie on the perimeter or in the interior of the trace region indicates a curved trace leading to elimination of the entire trace region.

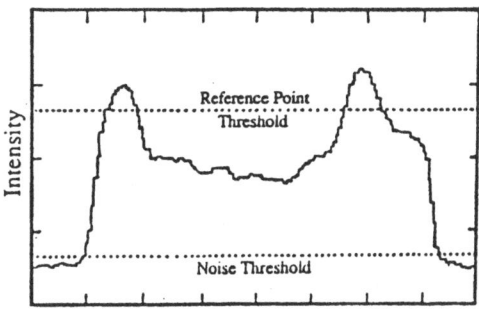

Figure 5: The intensity distribution for the trace in Figure 4.

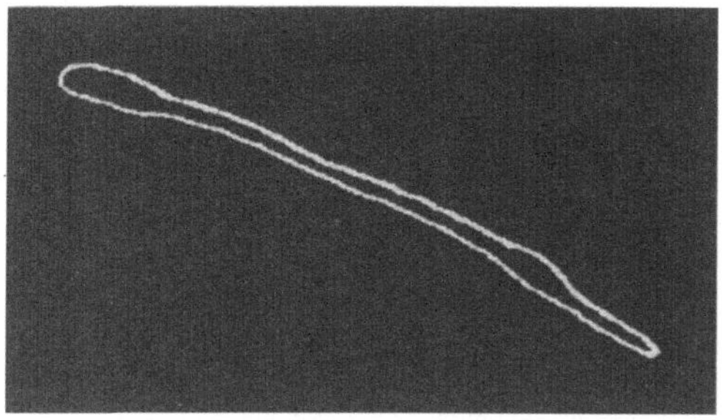

Figure 6: Boundary of the trace region.

The next step is an automatic search inside the trace region for the reference points by scanning for pixel values greater than a computed peak-threshold. This peak-threshold is a function of the average gray level within the region and the standard deviation of this gray level average. The threshold value typically is greater than the average pixel value. Once the outline boundary of a reference region is defined in the same fashion that a trace region is defined, the centroid of it will be determined and assigned to the corresponding region (Figure 7). By counting the number of centroids in the region the program disqualifies any region with other than two reference points. Therefore, any region which was generated by an incomplete trace or by two crossed or overlapping traces is eliminated from further processing (Figure 8).

The length of the line that connects the reference points will be measured as the magnitude of the velocity vector and will be assigned to the halfway point between the two reference points (Figure 9). The flow direction is found by summing the distances from each reference point to all pixels of the trace region.

The greater sum indicates the trailing end of the trace. Finally the velocity scalar, direction and mid-location of the trace is sorted in a file and the gray level value of the completed trace region is set to zero. Therefore, when the program continues the search for the next trace, it will not define the already detected trace region.

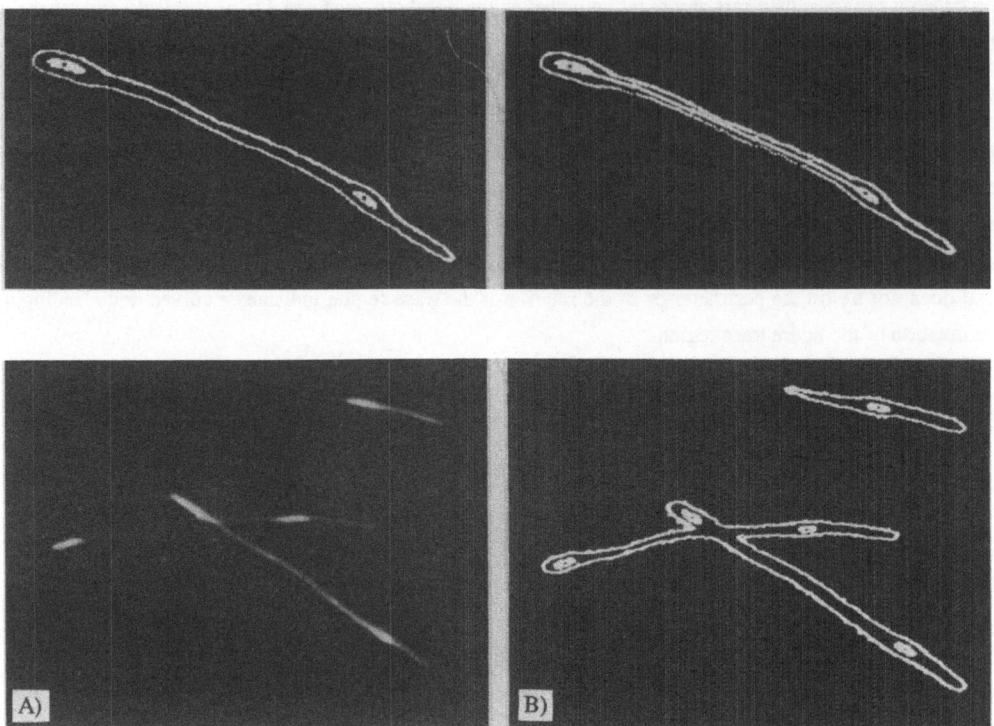

Figure 8:   Samples of incomplete and crossed traces.   A) Original image
            B) Processed image

It is important to mention that the net time for the image recording from experiment, the digitization, the trace detection and the velocity vector field computation is approximately two minutes per image. We expect to optimize our software to reduce this time to less than a minute per image. A typical reported time for obtaining similar information by other investigators (time for photographic processing not included) is on the order of one hour (Imaichi & Ohmi, 1983, Dimotakis et.al., 1981, and Altman, 1985 to mention a few).

In the next step a grid system is superimposed on the entire digitized field. The data for each trace is then read and assigned to the mesh square into which the mid-point of the trace falls. A vector summation of all the traces associated within a certain mesh square results in the mean velocity vector for the mesh square.

Gharib et.al. (1986) applied the described technique to a recirculating flow inside a cavity with a width-to-depth ratio of .8. At this width-to-depth ratio, the flow inside the cavity is essentially a steady 2-dimensional recirculating stream. For a larger cavity width-to-depth ratio (b/h > 1), flow becomes unstable on a large scale and shows strong three-dimensional behavior. Figure 10 presents the trace images and vector field inside the cavity with b/h = .8 obtained from averaging 60 consecutive images.

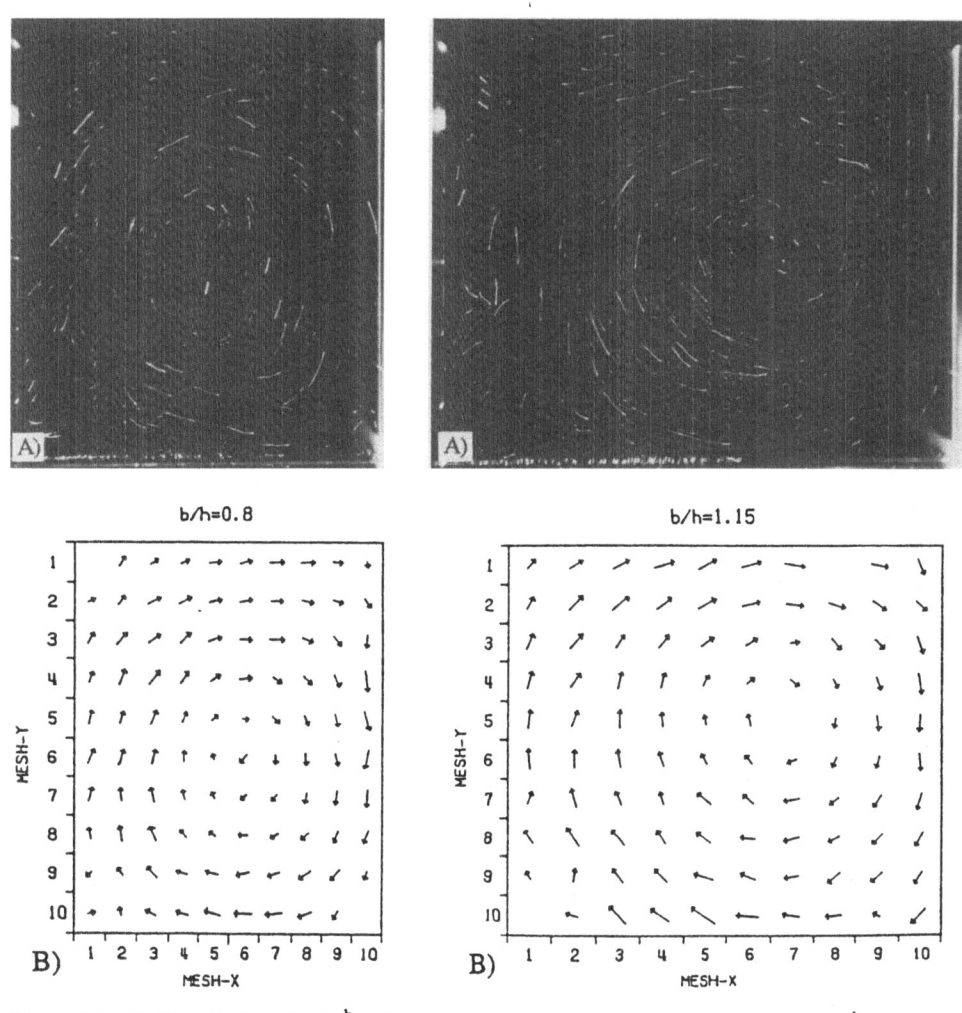

Figure 10: A) Sample flow field, $\frac{b}{h}$ =.8.
B) Averaged velocity vector field.

Figure 11: A) Sample flow field, $\frac{b}{h}$ =1.15.
B) Averaged velocity vector field.

Figure 11 shows a similar sequence for the b/h = 1.15. To compare the Lagrangian velocity information to the information obtainable from single point Eulerian measurement techniques such as laser Doppler velocimetry, Gharib, et.al. obtained one traverse of the velocity field at the middle of the cavity (b/h = .8). Figure 12 shows an excellent agreement between this velocity profile and one obtained from the particle tracing method. Such agreement can be seen only on the upper half of the velocity profile for the large cavity configuration with b/h = 1.15 (Figure 13). The disagreement between the two velocity profiles can be attributed to the strongly three-dimensional flow near the bottom of the cavity which drastically reduced the number of valid traces. Therefore, it was rarely possible to retrieve a sufficient number of particle streaks from even 60 realizations.

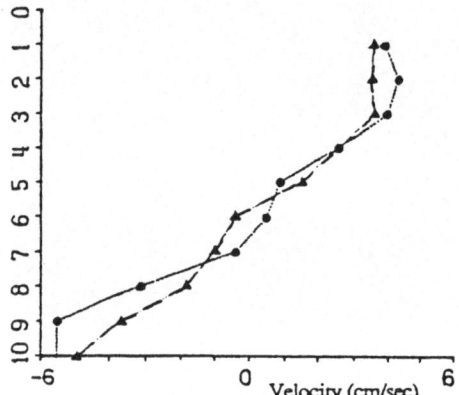

Figure 12: Comparison of particle tracing results (▲) to LDV measurements (●) for $\frac{b}{h}$ =.8.

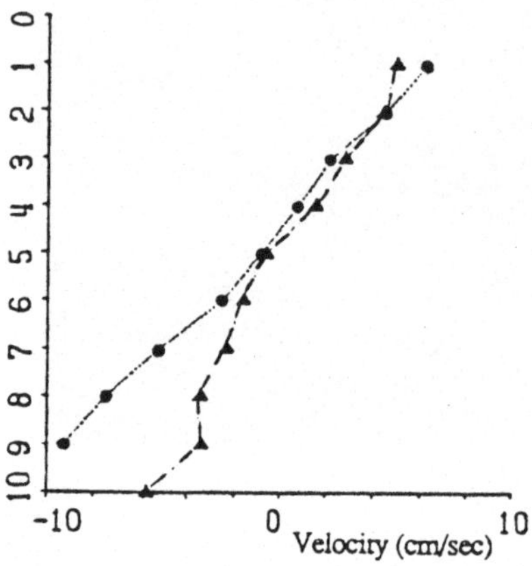

Figure 13: Comparison of particle tracing results (▲) to LDV measurements (●) for $\frac{b}{h}$ =1.15.

In another application of this technique we mapped the flow inside of a 7° angle diffuser. Figure 14 presents the velocity profiles at three stations inside the diffuser. Each station presents average velocities at 20 to 30 grid points. Each grid point velocity presents an average of 30 velocity traces. It is interesting to mention that a mass conservation check at each station revealed an error of less than 5%.

The reference point method was also applied to the turbulent wake of a cylinder at a Reynlods number of 440 at x/d ≅ 10. Both the free stream and the mean wake velocity profiles are plotted in Figure 15. Each data point on the profiles represents the average of 30 - 50 traces from a total of 50 images. The drag coefficient calculated using a control volume analysis on the mean profile is 1.18 which is in good agreement with the direct force measurement value. The Reynolds stress $\overline{(u'v')}$ and the longitudinal $\overline{(u'^2)}$ and transverse $\overline{(v'^2)}$ turbulence intensities were computed from the original data. The normalized distribution of the turbulent quantities which are presented in Figure 16 compare well with the available data from single point measurement techniques.

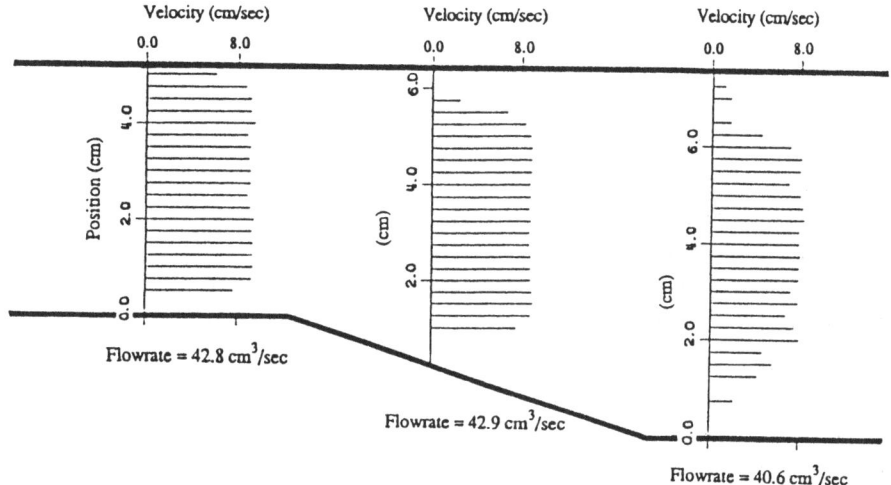

Figure 14: Velocity profiles at three stations inside a diffuser obtained by the reference point method.

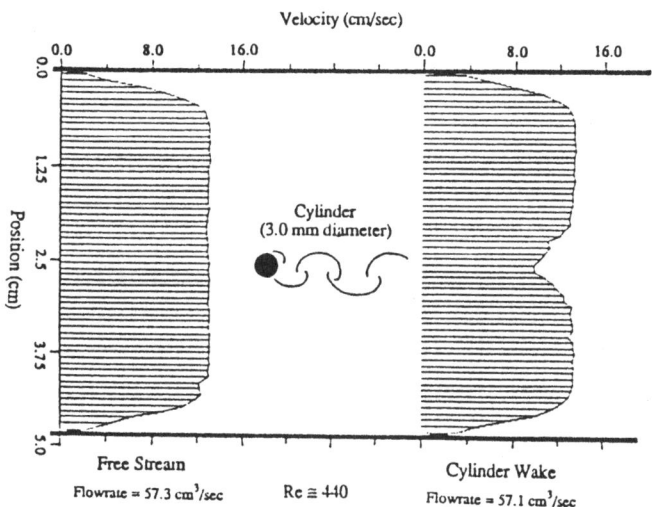

Figure 15: Mean velocity profiles for a cylinder in a channel flow.

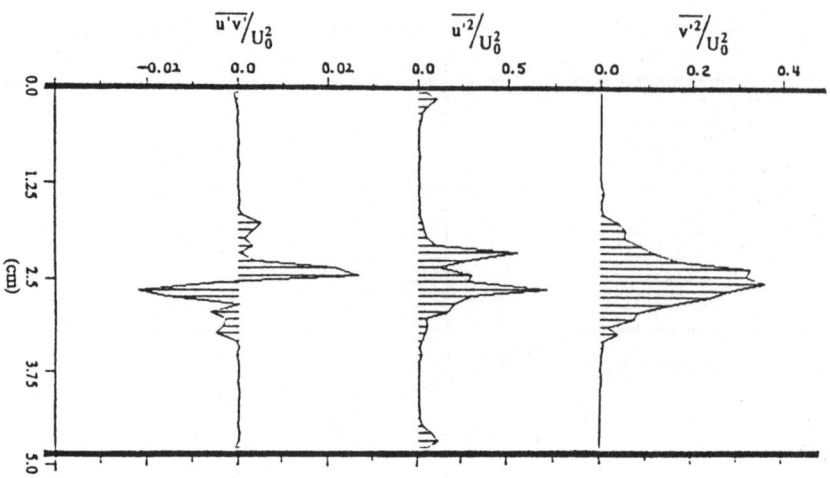

Figure 16: Normalized Reynolds stress $\overline{u'v'}$ and turbulence intensities $\left(\overline{u'^2}, \overline{v'^2}\right)$ for the cylinder wake in Figure 15.

SOURCES OF MAJOR ERRORS

The accuracy of the measurements and flow calculations in the described technique can be greatly influenced by several parameters. Optical and imaging errors which are common in most of the techniques that involve optical imaging have been discussed by several investigators (Wolf, 1983, Imaichi & Ohmi, 1983) and in some cases several techniques have been suggested to reduce the error (Browand & Plocher, 1985). In this section we will discuss some fundamental sources of errors that directly influence the flow computation and arise from imperfect conditions of the measurement technique. The magnitude of these errors, under certain circumstances, is by far larger than optical and imaging errors. In some cases, techniques that are developed to reduce these errors of measurement will be discussed.

ERRORS OF COMPUTING EULERIAN VELOCITY FROM LAGRANGIAN MEASUREMENTS

In the particle tracing technique we assume that the traces generated by the Lagrangian velocity of the particles are representing the Eulerian instantaneous velocity vector. However one needs to carefully consider the conditions that might violate the aforementioned assumption.

The trace generated by a particle is a result of the particle's Lagrangian motion. Therefore its length can be obtained by integrating its Lagrangian velocity

$$\ell = \int_0^\tau \frac{ds}{dt}\, dt$$

where $\underline{V} = \frac{ds}{dt}\,\underline{e_t}$,

where $\underline{e_t}$ is the tangential unit vector and s is the curvilinear coordinate. Following Altman (1985), one can Taylor expand the Lagrangian velocity and integrate to obtain

$$\ell = \int_0^\tau V(s(t),\, t)\, dt$$

$$= \int_0^\tau \left(V_E(0,\, 0) + \left.\frac{\partial V}{\partial s}\right|_{(0,0)} \tau^2 + \cdots \right) dt$$

$$\doteq V_E(0,0)\,\tau \; + \; \left.\frac{\partial V}{\partial s}\right|_{(0,0)} \tau\ell \; + \; \left.\frac{\partial V}{\partial t}\right|_{(0,0)} \tau^2 \; + \; H.O.T.$$

where V_E is the Eulerian velocity and H.O.T. are higher order terms.

The measured velocity is usually obtained by dividing the trace length by exposure time,

$$V = \frac{\ell}{\tau} = V_E|_{(0,0)} + \left.\frac{\partial V}{\partial s}\right|_{(0,0)} \ell \; + \; \left.\frac{\partial V}{\partial t}\right|_{(0,0)} \tau \; + \; H.O.T.$$

Therefore the trace measurement will be a reasonable approximation of the Eulerian velocity if the velocity field changes slowly in space and time. If we assume that

$$V = \frac{ds}{dt}\,\underline{e_t} = \dot{s}\,\underline{e_t},$$

then

$$\frac{dV}{dt} = \dot{s}\,\underline{e_t} - \frac{\dot{s}^2}{\rho}\,\underline{e_n},$$

where ρ is the principal radius of curvature and $\underline{e_n}$ is the normal unit vector. It is immediately apparent that trace curvature or an acceleration of the particle will have a substantial contribution to the higher order terms. It is interesting to mention that large particle acceleration will cause a non-uniform intensity variation along the trace. Therefore one should avoid curved traces with non-uniform intensity for the purpose of velocity measurement.

There is also a tradeoff between the length of the trace and the accuracy of the velocity measurement, even for flows varying very little spatially. The measurement of the separation between the two bright spots of each trace shows a smaller relative error for longer traces. But for spatially varying flows, acceleration and curvature effects become increasingly significant for longer traces.

ERRORS DUE TO THE THREE-DIMENSIONAL MOTION OF THE PARTICLES

As has been addressed by Gharib et.al. (1986) and Agui & Jimenez (1987), the most fundamental source of error in particle tracing methods is the erroneous trace length reading due to the lack of information regarding the entrance or exit of the particles from the light sheet. The nature of this error reveals no indication of its frequency of occurence.

To understand the problem, let us define the traveling time of particle $\Delta\tau$ as the total time that the particle has traveled through the laser sheet during the illumination time period, ΔT. In conventional particle tracing methods, ΔT corresponds to the camera's exposure time and the basic assumption is that $\Delta T = \Delta\tau$ and particle velocity is usually defined as

$$U_p = \frac{\Delta L}{\Delta T}$$

where ΔL is the photographically registered trace length. However, in a real situation, particles that enter the sheet after starting of the illumination or exit earlier than ending of the illumination period will have a traveling time shorter than ΔT. Usage of an improper ΔT causes erroneous velocity readings as large as one hundred percent of the local mean velocity. This problem can be easily solved by the reference point technique because any particle that has remained in the sheet during the illumination period should have two bright reference points. Lack of one or both points on a trace disqualifies the trace for the velocity calculations.

In the cavity experiment with b/h = 1.15 the disqualified particle traces were more than 60% of the total traces on each image. Other flow measurement techniques such as laser speckle or particle image displacement techniques (Adrian, 1984; Lourenco et.al., 1986) also suffer from the three-dimensional out-of-plane motion of the particles. However, only the methods that implement the reference point technique can accurately solve the problem.

SAMPLING ERRORS

The Nyquist sampling criterion suggests that the shortest resolvable wavelength in a flow image is twice as long as the average distance between traces. To avoid trace overlap problems, this average distance is usually set close to the average trace length in the image. Therefore features of the flow with a wavelength shorter than twice the average trace length will be undersampled. The resulting error in the velocity will be proportional to the integrated amplitude of the fluctuations that are filtered out by the undersampling. Agui and Jimenez (1987) show that while this error might not affect the mean velocity measurements, it certainly contributes a major error to turbulent intensity measurements.

ERROR IN FLOW COMPUTATION

The trace detection schemes such as those described in this paper that can accurately identify and reject traces due to overlapping and three-dimensional effects will severely reduce the number of traces per image. This results in a random distribution of traces over the image, and an interpolating scheme is required to reduce the irregularly spaced data to a regular rectangular grid. Several interpolation schemes are available where velocities at each mesh point are reconstructed by convolution with an adaptive Gaussian window (Agui & Jimenez, 1987) or a spline-thin-shell (Paihua de Montes, 1987) method over the entire velocity field.

To estimate the errors of interpolation Agui & Jimenez (1987) suggested an ingenious bootstrapping method where a data field composed of N velocity measurements interpolated onto a grid is successively resampled with N random samples. In this new sample some data from the original set are repeated while others are omitted. By comparing the velocities obtained through interpolating the new sample at the positions where trace velocities are known from an original data set, one can estimate the interpolation errors. The error estimates using the same bootstrapping technique is less reliable (mostly conservative) in the case of vorticity calculations. This is due to the fact that no *a priori* knowledge of the vorticity field is available.

The Adaptive Gaussian Window (AGW) and spline-thin-shell (STS) techniques in combination with the bootstrapping method have been comprehensively implemented by Rignot and Spedding (1987) and Spedding et.al. (1987) to study the flow behind an unsteady flap. The velocity vector field which was

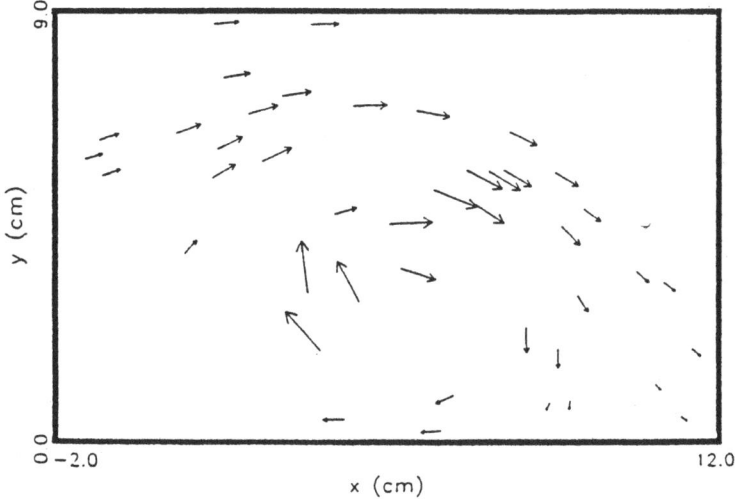

Figure 17: Velocity vector field behind an unsteady flap obtained by
the reference point method. (Rignot and Spedding,
1987)

obtained by the reference point technique is shown in Figure 17. Figures 18 and 19 show the interpolation velocity and vorticity fields, respectively, using the AGW and STS interpolating methods. Even though the reconstructed velocity fields are similar, isometric surfaces of the vorticity on the two grids clearly demonstrate how the vorticity distributions may be distorted by interpolation errors. Rignot & Spedding (1987) concluded that one should make careful and informed decisions concerning the choice and implementation of interpolating techniques: the STS method would be the algorithm of choice where N, the number of particles, is small. If original data has large cumulative errors, then AGW, being fast and easy to implement, may be a reasonable alternative.

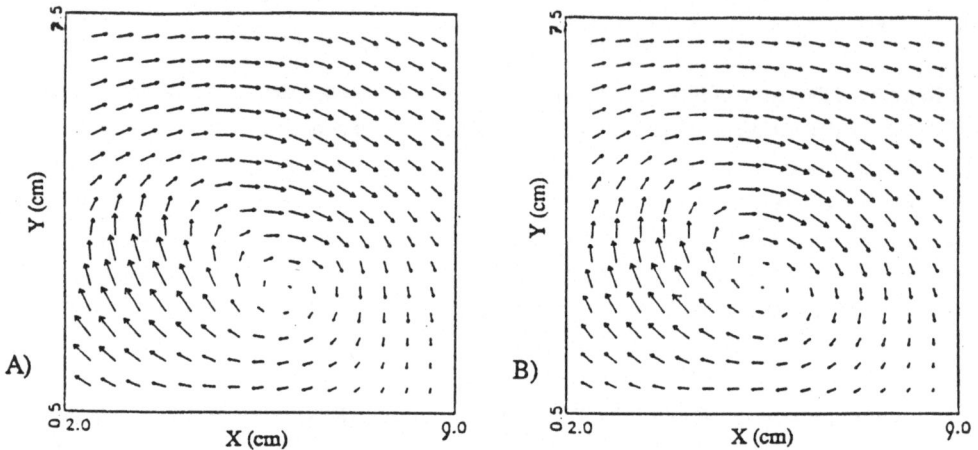

Figure 18: Interpolated velocity vector field for the flow in Figure 17.
A) STS method
B) AGW method

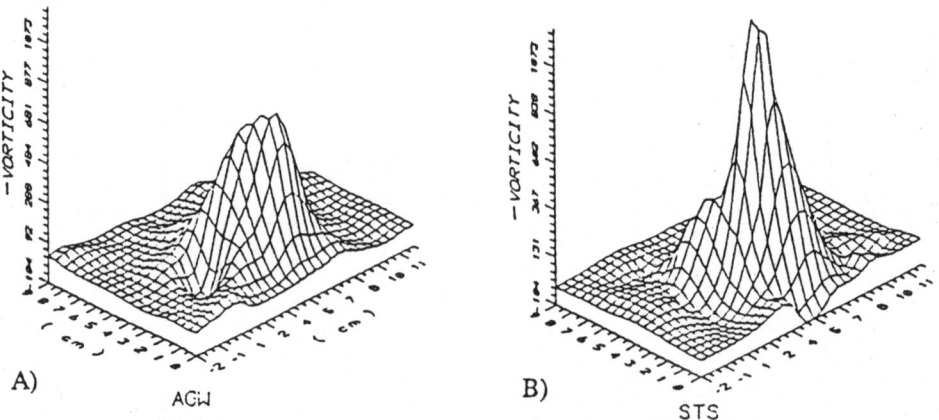

Figure 19: Interpolated vorticity vector field for the flow in Figure 17.
A) STS method
B) AGW method

CONCLUSION

A combination of the automatic particle tracing and one of the described interpolation methods can provide accurate and reliable velocity and vorticity measurements with an unprecedented speed. Currently, the speed of processing, which includes image acquisition and vector field construction, is one minute per image. With the use of more powerful computers this time can be accelerated to the video rate (30 vector fields per second). The fact that fancier measurement methods such as laser speckle photography and speckle image displacement methods depend on the processing of photographic negatives makes the method described in this paper a candidate for becoming a real-time particle tracing technique. This is an important point when one considers the future uses of global velocity or vorticity information in real time active control of the various flow fields. A fundamental challenge of the future is to deal with the fast automatic construction of three-dimensional flow field with preferably a single camera. Methods that can simultaneously provide velocity vector and scalar fields such as temperature and concentration are in great demand.

ACKNOWLEDGEMENTS

An initiation grant from the Engineering Foundation (GR Engng Fdn RI-A-86-7) has been instrumental in the development of the automatic particle tracing technique described in this paper. Currently a grant from DARPA/ACMP is supporting us in efforts to add the three-dimensional and multifield capability to the technique. Many colleagues and former students have helped us in bringing the technique to its present stage. We would like to thank Dr. Miguel Hernan for helping with the initial development of the idea, Barry Dyne and Oran Thomas for helping in the development of software, Thomas Phillips and Chee Yap for different stages of hardware development and Dr. Kathie Stuber for her careful laser doppler measurements.

REFERENCES

Adrian, R.J. (1984), "Scattering Particle Characteristics and their Effect on Pulsed Laser Measurements of Fluid Flow: Speckle Velocimetry vs. Particle Image Volumetry.", App. Opt. 23, 1690-1691.

Agui, J.C. and J. Jimenez (1987), "On the Performance of Particle Tracking." J. Fluid Mech. 1985, 447-468.

Ahlborne, F. (1902), "On the Mechanism of Hydrodynamic Drag." (German) Abhandl Gebiete Naturwiss 17, Hamburg

Altman, D.B. (1985), "Laboratory Studies of Internal Gravity Wave Critical Layers." Ph.D. Thesis, University of California, San Diego.

Browand, F.K. and D.A. Plocher (1985), "Image Processing for Sediment Transport." Proceedings of the International Association for Hydraulic Research, 21 Congress, 1985.

Dimotakis, P.E., F.D. Debussy, and M.M. Koochesfahani (1981), "Particle Streak Velocity Field Measurements in a Two-Dimensional Mixing Layer." Phys. Fluids 24, 995-999.

Elkins, R.E. III, G.R. Jackman, R.R. Johnson, E.R. Lindgren and J.K. Yoo (1977), "Evaluation of Stereoscopic Trace Particle Record of Turbulent Flow Fields." Review of Scientific Instrumentation 48, 738-746.

Gharib, M., B. Dyne, O. Thomas, and C. Yap (1986), "Flow Velocity Measurements by Image Processing of Optically Modulated Traces." Paper #22, Proceeding of the AGARD Conference on Aerodynamic and Related Hydrodynamic Studies Using Water Facilities, Preprint # 413.

Gharib, M. and M. Hernan (1985), "Flow Velocity Measurement by Image Processing of Optically Activated Tracers." AIAA 85-0172-AIAA 23rd Aerospace Sciences Meeting, Reno, Nevada 1985.

Imaichi, K. and K. Ohmi (1983), "Numerical Processing of Flow-visualization Pictures-Measurement of Two-dimensional Vortex Flow." J. Fluid Mech. 129, 283-311.

Jonas, P.R. and P.M. Kent (1979), "Two-dimensional Velocity Measurement by Automatic Analysis if Trace Particles." Journal of Physics E. Scientific Instrumentation 12, 604-609.

Lourenco, L., A. Krothapalli, J.M. Buchlin and M.L. Reithmuller (1986), "Noninvasive Experimental Technique for the Measurement of Unsteady Velocity Fields.", AIAA J. 24, 1715-1717.

Merzkirch, W. (1974), "Flow Visualization." Academic Press, New York, 250 pp.

Paihua de Montes, L. (1978), "Quelques Methodes Numeriques pour le Calcul de Fonctions Splines á une et Plusieurs Variables.", These de 3e Cycle, Universite de Grenoble.

Prandtl, L. and O.G. Tietjens (1934), "Applied Hydro- and Aeromechanics." McGraw-Hill, New York, 311 pp.

Rignot, E. and G.R. Spedding (1987), "Performance Analysis of Automated Image Processing and Grid Interpolation Techniques for Fluid Flows.", USC Aerospace Engineering Internal Report, USCAE 143, March 1988.

Saffman, P.G. (1965), "The Lift on a Small Sphere in a Slow Shear Flow." J. Fluid Mech. 22, 385-400.

Spedding, G.R., T. Maxworthy and E. Rignot (1987), "Unsteady Vortex Flows over Delta Wings." Proc. 2nd AFOSR Workshop on Unsteady and Separated Flows, Colorado Springs, July 1987.

Taneda, S. (1985), "Flow Visualization." Recent statistics on Turbulent Phenomena.

Wiese Nielson, K. Th. (1970), "Vortex Formation in Two-dimensional Periodic Wakes." Ph.D. Thesis, Oxford.

Wolf, P.R. (1983), "Elements of Photogrammetry, 2nd edition." McGraw-Hill, New York, 628 pp.

Particle Image Velocimetry

Outline

1. Introduction

One of the most challenging and time-consuming problems in experimental fluid mechanics is the measurement of the overall flow field properties, such as the velocity, vorticity, and pressure fields. Local measurements of the velocity field (i.e., at individual points) are now done routinely in many experiments using hot-wire (HW) or laser velocimetry (LV). However, many of the flow fields of current interest, such as coherent structures in shear flows or wake flows, are highly unsteady. HW or LV data of such flows are difficult to interpret, as both spatial and temporal information of the entire flow field are required and these methods are commonly limited to simultaneous measurements at only a few spatial locations.

Interpretation of these flow fields would be easier if a quantitative flow visualization technique was used in conjunction with the flow field measurements. Such a technique would provide both spatial and temporal information. One such method is termed particle tracing (Gharib, Dyne, Thomas, and Yap, 1987) and consists of measuring the streak lengths and orientation generated by injected particles. However, this method only provides partial results, because of its limitations in accuracy and spatial resolution (Lourenco, 1986).

Although the vorticity field is an essential property of most flows of current interest, measurements of this quantity have exceeded experimental capability. This difficulty arises from the fact that vorticity is a quantity defined in terms of local velocity gradients. In contrast, the currently available flow measurement techniques, such as hot-wire anemometry or laser velocimetry, are sensitive only to the local velocity. Hence, measurements must be made over several points and the resulting velocity components are then analyzed by finite difference schemes. However, the errors produced by the necessary differentiations limit the accuracy and spectral range. In addition, the spatial resolution of this method is often not sufficient to measure small-scale fluid motions of rapidly changing velocity gradients. As a consequence, the measured vorticity field is a type of spatially averaged estimate of the actual vorticity field. Finally, this method provides information at only a single point. If information on the entire flow field is required, measurements must be carried out sequentially one point at a time. This sequential method, although laborious, is straightforward in applications involving steady flows. However, the method becomes very difficult, if not impossible, when studying unsteady flows. Direct measurement of vorticity has been tried, for instance by the injection of spherical particles which rotate in the flow with an angular velocity proportional to the local vorticity (Frish and Webb, 1981). Such methods suffer the same drawback of insufficient spatial resolution just mentioned and also can be quite complex.

Recently, a novel velocity measurement technique, commonly known as Laser Speckle Velocimetry (LSV) or Particle Image Velocimetry (PIV), has become available. This technique provides the simultaneous visualization of the two-dimensional streamline pattern in unsteady flows as well as the quantification of the velocity field over an entire plane. The advantage of this technique is that the velocity field can be measured over an entire plane of the flow field simultaneously, with both accuracy and spatial resolution. From this the instantaneous vorticity field can be easily obtained. This constitutes a great asset for the study of a variety of flows that evolve stochastically in both space and time, such as unsteady flow separation or vortex-surface interaction. This article describes the principle of this technique, various methods of data acquisition and reduction, some parameters that affect its utilization, and some examples of its use.

2. Principle of The Technique

The application of PIV to the measurement of the velocity in a fluid involves two steps. First it is necessary to "create" a selected plane or surface within the flow field. The orientation of this plane should be such that it contains the dominant flow direction, if one exists. For instance, if the technique is used to measure the velocity field over a model in a wind tunnel, the plane will be parallel to the wind tunnel flow axis. The plane itself is created by seeding the flow with small tracer particles, such as those used in LV applications, and illuminating them with a thin sheet of coherent light, as depicted in Figure 1. A pulsed laser, such as a Ruby or a Nd-Yag laser, or a CW laser with a shutter, is normally used as the light source. The laser sheet itself is formed, for example, by focusing the laser beam first with a long focal length spherical lens (to obtain minimum thickness), and then diverging the beam in one dimension with a cylindrical lens. The light scattered by the tracer particles in the illuminated plane provides a moving pattern. When the seeding concentration is low, the instantaneous pattern consists of resolved, diffraction-limited images of the particles. When the concentration increases, the images overlap and interfere to produce a random speckle pattern. A multiple-exposure photograph, of two or more of the instantaneous patterns and taken in quick succession, is used to record the data. When the time interval between exposures is appropriately chosen, the tracer particles will have moved only a few diameters, far enough to resolve their motion but less than the smallest length scale of the flow. Thus, information on the local fluid velocity is stored on the photographic image and can be retrieved by subsequent analysis.

In a second step the local fluid velocity is derived from the ratio of the measured spacing between the images of the same particle, or speckle grain, and the time between exposures. The recorded image, whether formed by isolated disks, in the case of low particle concentration, or speckle grains for high particle concentration, is a complicated random pattern. Several methods exist to convert the information contained in the multiple-exposed photograph, or specklegram, to flow field data such as velocity, streamlines, or vorticity. These methods can be grouped into two broad categories. In the first category the distance between particle pairs is evaluated directly. That is, the absolute locations of the particles' corresponding images in the photograph are measured, for instance using a digitized version of the photograph, and the velocity is determined by computing the relative position of the corresponding images. The second category covers those techniques that evaluate the particles' image spacings indirectly. They exploit the property that all particles in a small region (small relative to the length

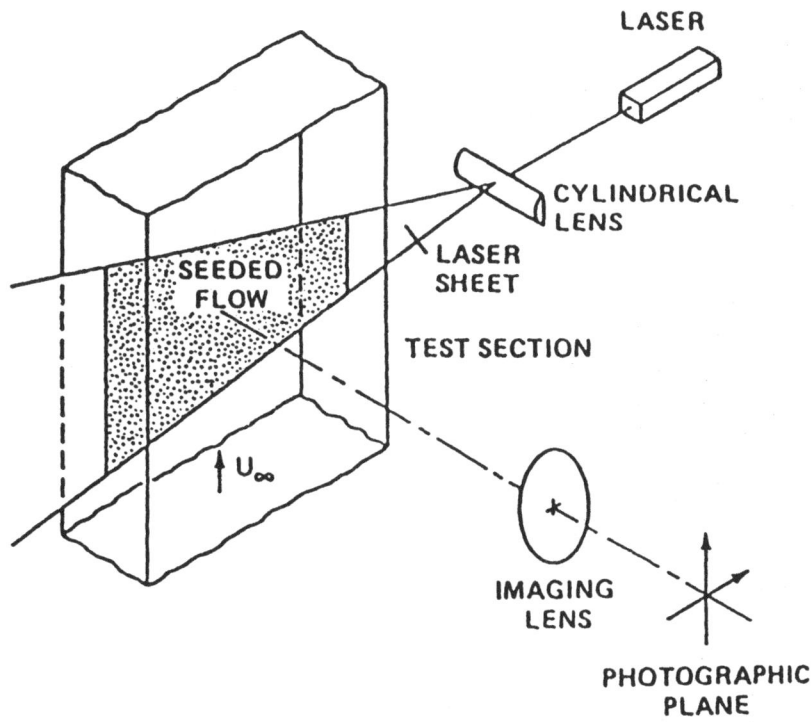

Figure 1. Schematic of Experimental Setup for Particle Image Velocimetry

scales of the fluid under study) are displaced roughly the same amount between exposures. This property is exploited in several ways in both hardware and software recently developed in digital image processing to determine the velocity. One specific method in this second category, evaluation by Young's fringes, is described in detail in Section 4.2, following a brief overview of other techniques in Section 4.1. However, a brief background of how this analysis technique first originated is described first.

For instance, it is conceivable to measure the local displacements by visual or computer-aided inspection. The data reduction systems that have been proposed are based on the digitization of the entire photograph into a very large number of pixels and the development of algorithms to permit computer identification of individual streaks or pairs of particles. Methods for this direct analysis of these images have been developed, but with limited success. Gharib et al, (1987), Elkins et al (1977), and Dimotakis et al (1981) are recent studies where this technique has been used. Photographs based on these techniques are difficult to interpret when the mean distance between two independent particles is the same order of magnitude as the particle displacement. This difficulty is usually circumvented by using low particle concentrations. However, flow field information is then restricted to those isolated locations where particles are present. This results in velocity measurements with low spatial density. Spatial derivatives of the velocity (e.g., vorticity) are then difficult to estimate and must be inferred by indirect arguments, such as described by Dimotakis, et al, (1981).

It is important to realize that the multiple-exposure photograph produces a locally periodic random image. This periodicity is proportional to the local velocity and can be determined using Fourier or autocorrelation techniques. To obtain the velocity field, the photograph can be scanned on a point-by-point basis, which yields measurements of the local displacement (i.e., velocity), or with a whole field filtering technique, which yields isovelocity contours. An example of this latter method is given by Meynart, (1980). Recently, an anamorphic optical system has been proposed by Collicott & Hesselink (1985). This method performs a 1-D Fourier transform in the x-direction for measuring the x-velocity component, and images the speckle pattern in the y-direction. This results in curved fringes which have a local spacing inversely proportional to the x-velocity at that point. Simultaneous multiple point measurements are obtained by imaging in the y-direction. Thus, it is possible to measure a velocity component along a selected line in the flow.

3. History of Particle Image Velocimetry

The terms Laser Speckle Velocimetry and Particle Image Velocimetry (or sometimes Particle Image Displacement Velocimetry) are often used interchangeably. However, these terms reflect an important distinction, related to the particle density in the flow field. To understand these differences, it is first necessary to describe how the application of this technique to the measurement of fluid flows developed.

The term "speckle", or the speckle phenomenon refers to the granular appearance that diffusely reflecting and transmitting surfaces take on when illuminated by a laser beam. This grainy appearance is caused by constructive and destructive interference of coherent light (i.e., the laser beam) scattered from a surface element whose roughness is large compared with the wavelength of the laser. For example, when a sheet of white paper is placed in the path of a laser beam, the reflected light contains information on the roughness of the paper. In the field of holography this is sometimes referred to as speckle noise (e.g., Collier, Burckhardt, and Lin, 1971). Actually, it is not noise but rather unwanted information in the context of holography. It is this information that is utilized in the laser speckle context. The first applications that made use of the speckle phenomena were in the field of solid mechanics. It was originally used to measure in-plane displacement and strain of solids with diffusely scattering surfaces and has also been applied to surface roughness measurement, vibration, and deformation analysis. Several early applications of the laser speckle concept are described in Erf (1980) and Stetson (1975).

The technique as used for the measurement of displacements in solid mechanics is essentially as follows. A surface is first illuminated by a laser beam, as shown in Figure 2. When this surface is imaged through a lens onto a photographic plate, the interference of the scattered light wavelets gives rise to a speckle pattern. The speckle size is a statistical average of the distance between adjacent regions of maximum and minimum brightness and can be estimated (Erf, 1980) by the Rayleigh resolution criterion

$$d_s = (1.2)\lambda f_N(1+M) \tag{1}$$

where d_s is the size of a speckle grain, λ is the wavelength of the illuminating laser, f_N is the f-number of the recording optics, and M is the

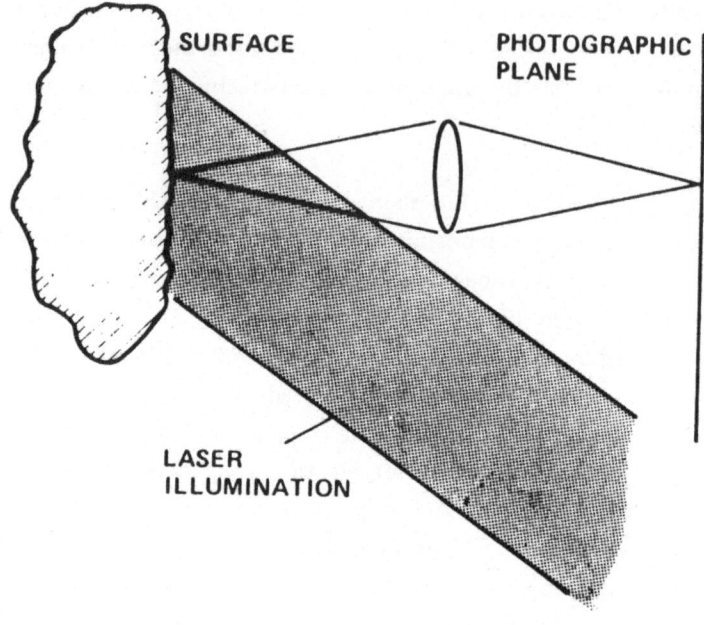

Figure 2. Laser Speckle Concept in Solid Mechanics

magnification. A doubly exposed photography of the speckle pattern is then recorded, once before and once after a lateral displacement is introduced. This photograph, or specklegram, contains two correlated grids which can be analyzed as a non-uniform diffraction grating. The technique can only be used when the displacement between exposures is greater than the speckle size, d_s, but not so great as to destroy correlation. Thus, the individual speckle size sets the lower measurable limit. Analysis of this specklegram was performed by generating Young's fringes from the specklegram transparency, as described in Section 4.2.

Although speckle photographs obtained in solid mechanics and fluid mechanics are similar, there are two fundamental differences. The first is that the fluid is illuminated by a sheet of laser light whose thickness is Δz. Therefore, scattering occurs from a volume distribution of particles rather than a surface distribution. Secondly, the number density of particles per unit volume (seeding concentration) can vary over a wide range of values. Strictly speaking, for a true speckle pattern to exist, the number of scattering particles must be so high that the images overlap and interfere to produce a random speckle pattern in the image plane. When the particle concentration is lower than this level (and reasons why this can often be an advantage will be discussed in a later section) discrete images of the particles will be photographed instead. This low particle concentration is referred to as Particle Image Velocimetry (PIV), reserving the term Laser Speckle Velocimetry (LSV) for the high particle concentration levels where a random speckle pattern is usually formed. Most of the early applications of this technique to fluid flows (e.g., Simpkins and Dudderar, 1978) used the LSV method, whereas for reasons to be described later, most of the more recent studies use the lower concentration PIV method.

4. Particle Image Analysis

4.1 Overview of Analysis Techniques

The most common methods of analysis are point-by-point techniques. In this type of analysis a small portion of the multiple-exposed photograph is examined, over which the velocity field is assumed constant. Several techniques have been developed to extract the flow field information. One approach consists of analyzing the position of the particles in the image plane and measuring directly the image pair spacings in the photographs. In this method the local particle displacements are measured, for example, by determining the two-dimensional correlation of the image field within the interrogation region. The spot is digitized in a NxN format (where N is the number of pixel rows or columns) and a two-dimensional correlation is performed. This results in a digital autocorrelation function with a maxima at the coordinates corresponding to the average displacement of the tracer particles. The major drawback of this method is that the computation of the autocorrelation function requires large data arrays and becomes extremely slow when N is large. A new processing method, developed and used by Yao and Adrian, (1984), reduces the general NxN element of a two-dimensional problem into two N element one-dimensional problems, by compressing the information in two orthogonal direction using integration techniques. In this method, called "orthogonal image compression", the 2-D image of an interrogation region is split, and optically compressed onto two orthogonally-aligned linear detector arrays. Particle images in the 2-D region appear as peaks in the 1-D distributions of each of the two array signals. The optimal method for determining the separation of the peaks, and thus the velocity, depends on the image density, defined as the mean number of particle image pairs in the interrogation region. If the image density is less than one, the peak separation is measured directly in each orthogonal direction. If the image density is greater than one, the peak separation is evaluated using 1-D spatial correlation. A recent study by Landreth, Adrian, and Yao, (1988) has indicated, however, that the correlation distributions given by this technique sometimes included random peaks in addition to the peaks created by the particle image pairs, resulting in incorrect measurements. These extraneous peaks seem to be due to random image pairings. Modifications to this method to prevent this possibility are currently being investigated.

4.2 Analysis of Young's Fringes

An alternate method for the measurement of the local displacement between the two images of the particle pair is by the use of Young's fringes. These fringes are obtained by illuminating a small portion of the specklegram, or multiple-exposed photograph, with a focused laser beam. The diffraction produced by coherent illumination of the multiple images in the negative generates a fringe pattern in the Fourier plane of a lens, provided that the particle images correlate. This is shown schematically in Figure 3. These fringes have an orientation which is perpendicular to the direction of the local displacement and a spacing which is inversely proportional to the displacement. If Δ is the real translation of the object, then this is related to the distance between the particle images, s, by

$$\Delta = s/M \qquad (2)$$

The spacing between Young's fringes can be shown to be (e.g., Born and Wolf, 1980),

$$d_f = \lambda f_L / s \qquad (3)$$

where d_f is the fringe spacing and f_L is the focal length of the converging lens. Thus, the displacement of the images is given by

$$\Delta = \lambda f_L / M \, d_f \qquad (4)$$

This technique offers an important advantage over those that directly analyze the particle images in that it eliminates the difficulties associated with finding the individual image pairs on the photograph.

The basis of the Young's fringe method can be described as in the following. Consider the function $D(r)$ describing the light intensity in the image plane of a photographic camera, where $r(x,y)$ are the plane coordinates. Considering that there is an in-the-plane displacement dy of the scatterers, the image will be translated by Mdy between exposures, where M is the magnification of the camera lens, and the resulting intensity distribution of the specklegram is

$$D(x,y) + D(x,y + Mdy) = D(x,y) \otimes [\delta(x,y) + \delta(x,y + Mdy)] \qquad (5)$$

where $\delta(x,y)$ is the Dirac delta function centered on $r(x,y)$, and considering that a translation can be represented as a convolution with a delta function. This total intensity is recorded on the photographic plate. After development, the transmittance, τ, of the negative is given by

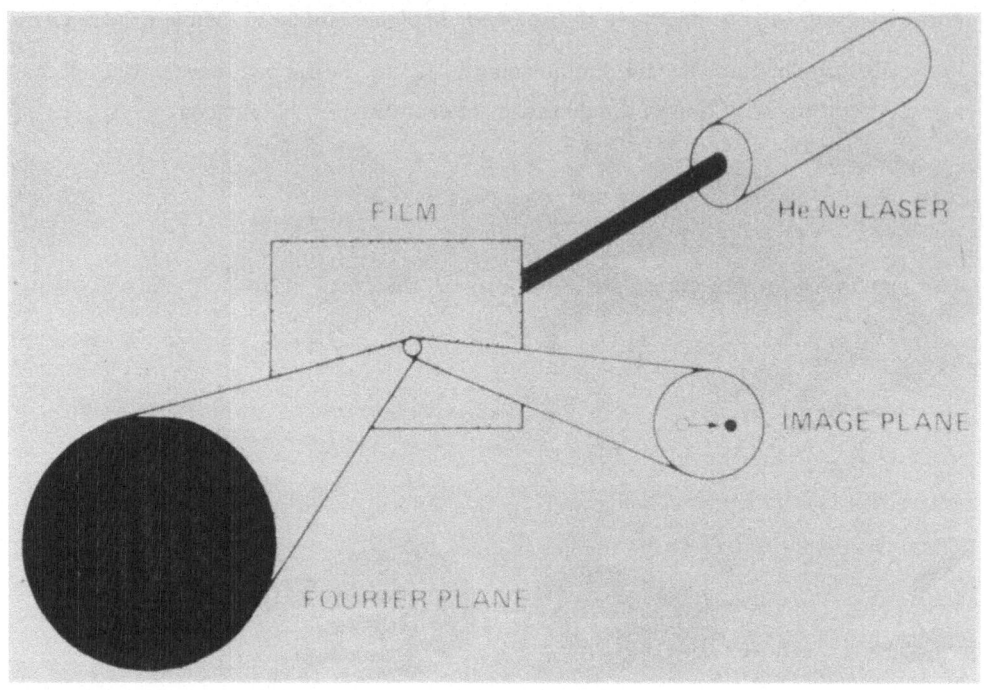

Figure 3. Schematic Diagram of Young's Fringe Formation for Specklegram Analysis

$$\mathcal{T}(r) = a + bD(x,y) \otimes [\delta(x,y) + \delta(x,y+Mdy)] \tag{6}$$

where a and b are characteristic constants of the photographic emulsion. Local analysis of a small portion of the film negative with a probe laser beam will produce in the far field an optical two-dimensional Fourier transform of the transmittance distribution, with an intensity distribution given by

$$\mathcal{T}(u,v) = a\,\delta(u,v) + b\,D(u,v)[1+\exp(i2\pi v\,Mdy)/\lambda_a] \tag{7}$$

where τ represents the Fourier transform of τ, u, and v are the angular coordinates of a point in the Fourier plane, and λ_a is the wavelength of the interrogating laser beam. The first term, $a\,\delta(u,v)$, on the right hand side of equation (7) represents the image of a point source, (i.e., the interrogating beam), when diffraction effects are neglected. This image is seen as a small bright spot in the center of the Fourier plane. The second term is composed of a fine speckle structure D, modulated by the expression

$$[1+\exp(i2\pi v\,Mdy)/\lambda_a] \tag{8}$$

The intensity distribution for the second term is obtained by multiplication with its complex conjugate, resulting in

$$|D(u,v)|^2[4\cos^2(2\pi v\,Mdy)/\lambda_a] \tag{9}$$

The diffuse background, given by $|D(u,v)|^2$ and called the "diffraction halo", is modulated by a set of Young's fringes whose spacing is given by equation (3). Knowing M, f_L, λ_a, and measuring d_f the displacement dy is easily found from equation (3). The direction of motion is perpendicular to the orientation of the fringes. There is a 180 deg. direction ambiguity in that the motion is known only as perpendicular to the fringes. A method to resolve this ambiguity is described in Section 5.8.

4.3 Limitations of the Young's Fringes Method

The formation of the fringe pattern that occurs when a local region of the specklegram is illuminated with a coherent laser beam requires that the displacement of the particle pairs within the interrogation region be correlated.

Identical shifting of all particle pairs will result in perfect correlation. Two factors reduce this correlation and can eliminate the fringe pattern. The first is when there is a slight out-of-plane motion of the particles, due to three-dimensional motions in the flow. The tolerance to out-of-plane motion is basically equivalent to the width of the illumination sheet and the depth of field of the recording optics. The time between exposures and the width of the illuminating sheet need to be carefully selected to avoid too many particles entering or leaving the sheet between exposures. The second factor is when the velocity varies across the interrogation region. This will cause the various particle image pairs to be displaced by different amounts. This will not occur when the diameter of the interrogation region is smaller than the smallest length scale of the flow being studied.

4.4 Data Processing

The Young's fringe pattern, produced by the techniques described in Section 4.2, are analyzed using a digital image analysis system, which typically consists of a host computer, a digital image processor, a frame digitizer, pipeline processor, and a video output controller to convert digital to analog information for display on a monitor. The system also usually includes a two-dimensional traversing mechanism and a controller for the purpose of automatically scanning the film transparencies. Analysis of the fringes can occur in either an interactive mode, which requires the assistance of an operator, or in an automated procedure.

The interactive method consists of first obtaining a 1-D periodic signal from the straight fringes. This is performed by determining the fringe angle relative to a predetermined reference line, followed by an averaging over the lines of the digitized picture as given by the following relation.

$$f(m) = \Sigma I[m+(n-255)\tan\alpha,n], \quad 0 \le m \le 511 \tag{10}$$

where $f(m)$ is the resulting periodic signal, $I(m,n)$ represents the digitized picture, and α, is the angle of the fringes with the reference n axis. In this equation it is assumed that the image is digitized with a 512x512 format with 256 shades of gray. The extraction of the frequency from this signal is straightforward. The Fourier transform of $f(m)$ displays a peak at the frequency proportional to the velocity component parallel to the m axis. However, due to low frequency modulation of the fringes, which is a consequence of the non-uniform light intensity distribution in the diffraction halo, it is sometimes difficult to identify this peak, especially if the fringes have a low frequency (i.e., few fringes). To remove this modulation, the fringe signal can be passed through a high pass filter prior to processing.

The advantage of this one-dimensional averaging technique is rapidity, in that only one line of the fringe pattern needs to be digitized. The computation, which includes the determination of the fringe angle by the operator and position updating of the film transparency scanning mechanism, can be completed in just a few seconds. The disadvantage of this method is the need for an external adjustment of the angle of the fringes by an operator. This inconvenience can be corrected by using the automated method.

The second method, which does not require an operator, consists of computing the velocity components along independent directions. The basis for this method is that each line of the fringe frame can be considered as a noisy periodic signal with variable phase. Then the automatic determination of a velocity component can be performed simply by averaging over a quantity independent of this phase. The autocorrelation for each line, or its Fourier transform for the power spectrum, satisfies this requirement. Using the autocorrelation, the m velocity component can be computed from the relation

$$g(u) = \Sigma[(\Sigma[I(m,n)I(m+u,n)]/\Sigma[I(m,n)]^2], \quad -511<u<511 \tag{11}$$

where g(u) is the resulting periodic signal, based upon the autocorrelation of the intensity distribution. This algorithm has been implemented on a pipeline processor by Lourenco and Krothapalli, (1988) to compute the autocorrelation for all lines of a frame simultaneously. They found that, for an accurate estimate of both the velocity magnitude and direction, four such full image operations, giving four autocorrelation functions, were required. From these autocorrelation functions the velocity vector can be determined by selecting the values of the components which have been computed from autocorrelations having the highest signal-to-noise ratio, and visibility. The determination of the velocity vector typically takes on the order of two to five seconds.

The obvious advantage of this technique is that no external operator is required. However, a shortcoming of this technique is the difficulty in measuring the velocity when the fringe density is too low (typically less than three bright fringes). In this case, the velocity can still often be evaluated by the interactive, one-dimensional averaging method. Hence, these two methods are actually complementary.

5. Detailed Considerations

This section describes some of the parameters that can affect the use of particle image velocimetry. The impact of these parameters is discussed and, where possible, recommended values are given.

The technique relies on the ability to detect and record on a photographic plate the images of the tracer particles. This image is a function of the scattering power of the particles within the fluid, the amount of light in the illuminating sheet, the length of time the film is exposed, magnification of the recording optics, and film sensitivity at the wavelength of the illuminating laser light. The specific parameters playing a role in PIV include the following. Additional details are contained in Lourenco and Krothapalli, (1987) and Smith, Lourenco, and Krothapalli, (1986).

Light source;	strength and duration of pulses
Tracer particle;	type, dimension, and concentration
Exposure parameters;	duration, time between exposures, and number of exposures
Film parameters;	sensitivity, grain size, and resolution
Recording optics;	magnification and lens aperture

These parameters are strongly interrelated and depend upon such factors as the type of fluid, velocity range and length scales of the flow being studied, and the required spatial resolution in the results.

5.1 Light Source

The light recorded on the specklegram is that which has been scattered 90 deg. to the incoming laser light sheet. Extremely bright light sources are usually required because of the low efficiency of this scattering process. Although the particle detection increases proportionally with increasing power of the illuminating laser, it is important to keep the laser power requirement to a minimum, primarily because of the expense. The specific amount of laser light energy required is a function of tracer size and concentration, recording lens aperture and magnification, and film sensitivity. For a successful photographic recording of a particle image, the mean exposure of an individual particle image must be greater than the film sensitivity at the wavelength of the illuminating laser. This minimum sensitivity is sometimes referred to as the "gross fog" level (Adrian and Yao, 1985). In analytical terms this is expressed as

$$E = \int_{\Delta t} \tau dt \; > \; CE_0 \qquad\qquad (12)$$

where E is the mean exposure of an individual particle image, τ is the average intensity of light scattered by a particle, E_0 is the film fog level, and C is a constant between 1 and 10. The fog level is defined as the exposure level below which the transmissivity of the film is independent of the incident intensity, as shown in Figure 4.

The mean intensity of the light, τ, of the light scattered by a single particle can be expressed as

$$\tau = (4/\pi\kappa^2 d_i^2) \; I_0 \int_\omega \sigma^2 \, d\omega \qquad\qquad (13)$$

where κ is the wavenumber of the illuminating laser light, d_i is the nominal diameter of the particle, including diffraction, I_0 is the intensity of the illuminating sheet, σ is the Mie parameter, and ω is the solid angle subtended by the camera lens. The effective dimension of the particle image, d_i, can in turn be expressed as

$$d_i = \{ M^2 \, d_p^2 + d_e^2 \}^{1/2} \qquad\qquad (14)$$

where d_p is the actual particle diameter, and d_e is the diffraction-limited spot diameter of the particle image, given by

$$d_e = 2.44 \, (1 + M) \, f_N \lambda \qquad\qquad (15)$$

Equation (14) is an approximate relation representing the combined effects of magnification and image blurring in determining the final image diameter. When a pulsed laser is used, and assuming that the particle is stationary during exposure, the laser power required is determined from equation (12).

$$I_0 > \{ CE_0 \, (\pi\kappa^2 d_i^2) \} \, / \, 4 \int \sigma^2 \, d\omega \qquad\qquad (16)$$

The recommended value for the constant C in equation (16) is between three and

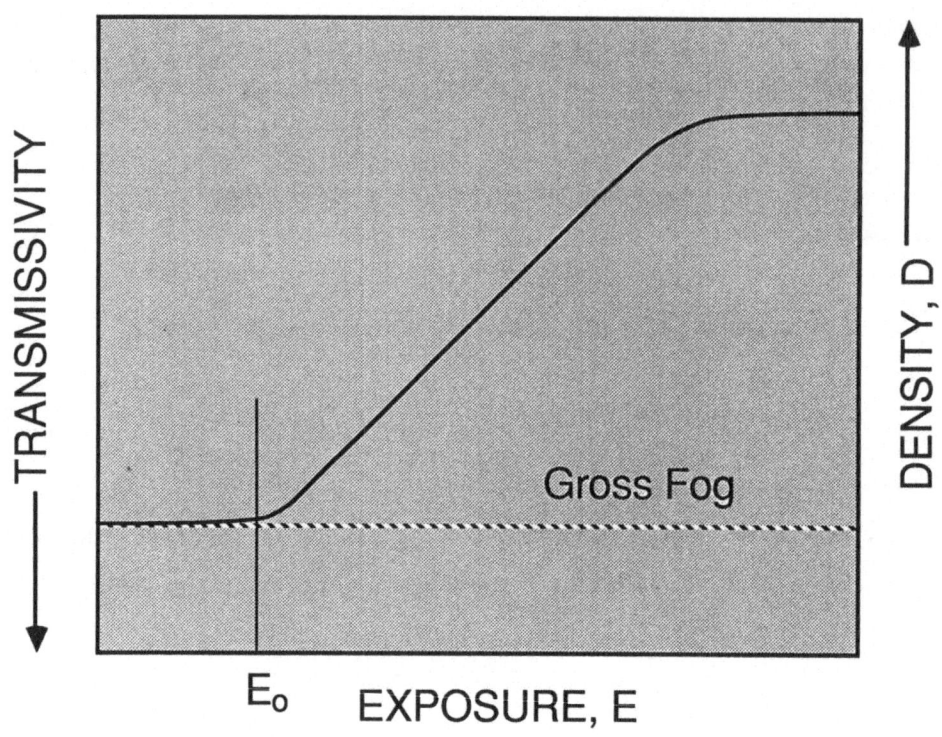

Figure 4. Density-Exposure Curve for a Film

five to account for the film reciprocity effects caused by very short exposures. Similar calculations for a CW laser are discussed by Lourenco, (1986).

5.2 Tracer Particles

The basic assumption in all techniques that use tracer particles added to the flow, whether it's LV, PIV, or some other technique, is that the motion of the particle accurately follows the motion of the fluid. These requirements are usually met by tracers used in LV applications. In air flows for instance, oil smoke produces a relatively uniform seeding. The minimum detectable particle diameter is a function of the recording optics and the laser input energy. Particles ranging from 0.5-10 μm are fairly typical. It should be noted that, for particles of this size, the diameter of the recorded image is relatively insensitive to the actual particle diameter. In this range of particle sizes, the image size is dominated by diffraction effects from the photographic lens.

5.3 Particle Concentration

The Laser Speckle mode of operation relies upon identical, laterally shifted speckle patterns. For these speckle patterns to exist, the number of scattering sites per unit volume must be high enough that many images overlap with random phase in the image plane. With lower concentrations the mode of operation changes from the speckle mode to the particle image mode (PIV) where the pattern consists of discrete images of particles. Velocities determined from the LSV mode suffer inaccuracies associated with the randomness of the speckles, but the velocity can be determined at any point in the flow. In the PIV mode, regions of the flow field may be left out due to poor seeding (sometimes referred to as signal drop-out) but the velocity can be measured without the inaccuracies of the speckle mode.

Slight out-of-plane motion of the particles, due to three-dimensional motions in the flow, will result in speckle patterns that are not entirely similar. As a consequence, the correlation between patterns decreases and the fringe pattern is suppressed of eliminated. This poses a severe limitation in the use of the laser speckle mode for the study of turbulent flows, or flows with a significant velocity component in the direction perpendicular to the laser sheet. However, the fringe quality is less dependent on out-of-plane motion in the PIV mode of operation. In this case the tolerance to out-of-plane motion is roughly equivalent to the width of the illuminating sheet and the depth of field of the recording optics.

There are practical bounds to the particle concentration in both modes of operation. In the PIV mode the upper boundary is set simply at that value above which a speckle pattern is formed, as shown in Figure 5. If C_p is the particle concentration and Δz the width of the laser sheet, the maximum concentration is given by

$$\{1/(\Delta z C_p)\}^{1/2} \gg d_i/M \qquad (17)$$

where d_i is the image diameter, given in equation (14).

The lower end in the PIV mode can be determined by the criterion that, in order to have a valid experiment, a minimum number of particle image pairs must be present in the area being scanned by the interrogation beam. The case of a single particle image pair is ideal because it yields fringes with optimum signal-to-noise ratio. However, this situation can only be achieved by lightly seeding the flow, which gives rise to signal drop-out. An interesting case occurs when two particle pairs are present in the interrogation area. The resulting diffraction pattern includes multiple equally intense fringe patterns due to cross interference of non-corresponding image pairs. In this situation, shown in Figure 6, the local displacement cannot be resolved. As the number of particle image pairs in the interrogation area increases, the cross interference fringes become weaker in comparison with the main fringe pattern, which reflects the local displacement. These cross-interference fringes are sometimes called "background speckle noise". Experience shows that, for reasonable fringe quality, at least four particle image pairs should be present in the interrogation area.

At the high end of the particle concentration scale, the LSV mode, the particle concentration is governed by convenience, economics, and flow distortion. Attempting to obtain these high particle concentrations in large scale flows or in high speed flows, such as in a wind tunnel, can become exceedingly difficult, as well as expensive as the actual number of particles increases. Finally, the high concentration of particles required by the LSV mode may influence or distort the flow field being studied. For these reasons, the PIV mode is normally used.

5.4 Exposure Parameters

The exposure parameters are chosen in accordance with the maximum expected velocity in the flow field and the required spatial resolution. The spatial resolution, which in turn is equal to the cross-sectional area of the interrogating laser beam, is dictated by the scales associated with the fluid motion. So as not to

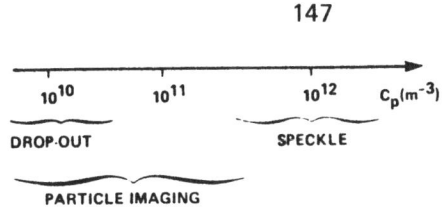

Figure 5. Mode of Operation vs. Particle Concentration

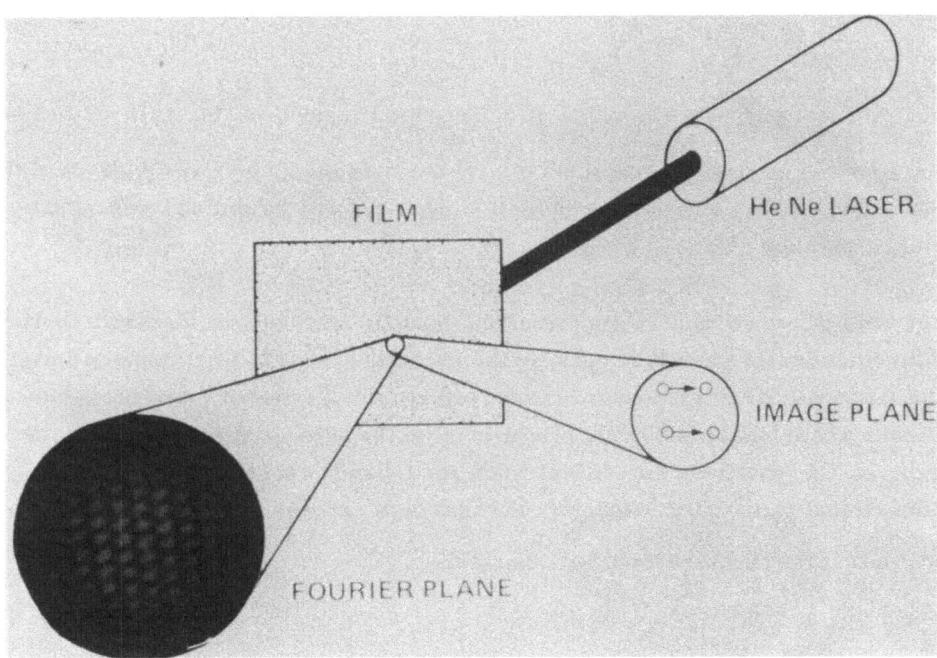

Figure 6. Cross-interference Fringe Pattern Generated by Two Particle Pairs

lose any information, the spatial resolution should be less than the smallest scale in the flow being studied.

The time between exposures, T, is determined by the maximum permissible displacement of a particle such that a correlation is obtained when the negative is analyzed locally with a probe laser beam. A necessary condition to obtain a fringe pattern is that the distance between adjacent particle images be less than a fraction of the analyzing beam diameter. In practice, the maximum permissible displacement that can be detected corresponds to the case when the fringe spacing, d_f, is larger than the diffraction limited spot diameter, d_l, of the interrogating optics. In analytical terms,

$$d_f = \lambda_a f_L / MTV_{max} > d_l = 4 \lambda_a f_L / \pi D \qquad (18)$$

The time between exposures can then be expressed as

$$T = (0.5)D/MV_{max} \qquad (19)$$

For practical purposes the value of (0.5) is used instead of the mathematically correct ($\pi/4$), as given in equation (18). This points up another advantage of this technique, in that the velocity sensitivity range can easily be shifted by altering the pulse separation, T.

For very short exposures the recorded particle images are identical to the diffraction limited particle images, as the particles appear to be stationary during the exposure. When the exposure time is increased, the recorded images becomes streaks whose length is directly proportional to the exposure time. The diffracted light in the spectrum is concentrated in a band whose width is inversely proportional to the streak length. For optimum exposure, Lourenco (1984) has shown that the exposure time, Δt, should be

$$\Delta t = d_i / MV_{max} \qquad (20)$$

where d_i is the particle image diameter.

5.5 Film Parameters
The technique relies on the ability to detect and record on photographic media the

images of the seeding particles. This is a function of the scattering power of the particles within the fluid, the amount of light in the illuminating sheet, camera lens and film sensitivity at the wavelength of the illuminating laser light. Although the particle detection increases proportionally with increasing power of the illuminating laser, it is of great importance to keep the laser power requirement to its minimum for economy.

To compensate for the many cases when limited laser power is available for illumination, the film used to record the particle images should have good sensitivity, but without sacrificing film resolution. Unfortunately, for commercially available photographic films, speed and resolving power are inversely related. Hence, choices range between high speed, low resolving power and low speed, high resolving power. For most practical applications good precision is necessary, so the advantage lies with the high resolution films (with about 300 line-pairs/mm). Of probably more importance, however, is the film grain. When illuminating the film negative to produce Young's fringes, the film's unexposed grains can introduce amplitude and/or phase changes into the wavefront of the analyzing laser beam, thus creating additional noise. This noise has a frequency content which is in the same general range as the fringes. Therefore, its elimination is difficult either by optical or digital filtering. Lourenco and Whiffen (1984) reduced this noise source by producing a positive copy through contact printing, on a very high resolution fine grain film. The positive copy was analyzed in the same manner using the probe laser beam. Problems arising from film grain will always be present, especially in applications where fast film is used to cope with low power density of the illuminating sheet. However, in general, the grain size should be much smaller than the diameter of the particle image to avoid this unwanted noise.

5.6 Recording Optics

As mentioned previously, the minimum detectable particle diameter is a function of the f-number of the recording optics. In fact, the f-number can have a significant effect on the mean exposure of an individual particle image. When the particle diameter is small compared with the diffraction-limited spot diameter, d_e, then d_i, the dimension of the particle image including diffraction, is independent of the actual particle diameter and instead is proportional to the f-number. The minimum detectable particle diameter can be shown to increase sharply for apertures smaller than f-11. Of course, any size particle can be detected provided the concentration is high enough. However, there are many reasons, described in other sections, for not increasing the concentration.

5.7 Dynamic Range

The dynamic range of the technique refers to its ability to resolve large velocity gradients in the flow field. It is defined as the largest velocity difference that can be detected. The low end of the dynamic range is determined by the requirement that the spacing between successive particle images be well resolved. That is, simply that they do not overlap. In analytical form,

$$l_s = d_i + V_{min} \, \Delta t \; < \; TV_{min} \qquad (21)$$

where l_s is the spacing between successive particle images. It was shown in the previous section that T, the time between exposures, is related to the maximum velocity V_{max} as

$$T = (0.5) \, D \, / \, MV_{max} \qquad (19)$$

For a pulsed laser and low speed flows it is sufficient to assume that $\Delta t = 0$. Thus, combining these Equations (21) and (19),

$$V_{min} = 2M \, d_i \, V_{max} \, / \, D \qquad (22)$$

The velocity dynamic range is then defined as the normalized velocity difference,

$$\Delta V = (V_{max} - V_{min}) \, / \, V_{min} \qquad (23)$$

and is written as

$$\Delta V = (D \, / \, 2M \, d_i) \, - 1 \qquad (24)$$

for a pulsed laser. Considering typical values of $d_i = 0.3$ mm, $D = 5$ mm, and $M = 1$, a dynamic range of 7.5 can be expected. A relation similar to equation (24) can be obtained for a CW laser, using $\Delta t = d_i \, / \, V_{max}$. Following the same line of reasoning,

$$\Delta V = (D - 2 \, d_i) \, / \, (2 \, M \, d_i) \, - 1 \qquad (25)$$

for a CW laser. A powerful advantage of this technique is evident in these equations. That is, by adjusting the magnification M in recording the specklegrams, the system sensitivity can be altered to accommodate the amount of motion anticipated in the experiment.

5.8 Direction Ambiguity

One disadvantage of using a multiple-exposure photograph to extract velocity data is the 180 deg. ambiguity in determining the direction of the velocity vector. That is, given identical conditions in recording each of the two exposures, there is no property identifying the order in which the two images of the particle were recorded. Measuring the separation of the particle image pair provides the magnitude of the velocity at that point, but is insufficient to give the direction of the velocity vector field. Thus, a given displacement will indicate a velocity of +/- U, with the sign being ambiguous. Many flow fields of interest (such as wakes and separated flows) contain regions of reversed flow and the direction may not be known a priori. Thus, a means to remove this ambiguity would be very useful.

A method to resolve both this ambiguity of the velocity vector, as well as to improve the dynamic range of the measurements, has been developed by Adrian, (1986) and Lourenco, et al (1986). This method is termed "spatial image shifting" or "velocity bias technique". The method consists conceptually of recording the flow field in a moving reference frame, and thus superposing a known velocity bias to the actual flow velocity. This is accomplished by shifting the image by a known displacement between the two exposures. The image can be shifted physically using a moving camera, or by optical means using rotating or scanning mirrors. This effect is demonstrated in Figure 7. Consider a flow field with regions of reversed flow, and with velocities ranging up to a value V. Four such velocities, one in each of the four quadrants, are shown in the figure. Using standard PIV recording techniques, it would not be possible to resolve between velocities lying in the first (or second) quadrant from those lying in the third (or fourth) quadrant. Now impose a velocity V*, much larger than V, as in Figure 7. The four velocity vectors are now transformed into four distinctly different vectors, depending upon their direction. The correct velocity, with its direction, can now be easily obtained upon removal of the velocity bias.

The method currently employed uses a scanning mirror to displace the image during the exposure with a predetermined velocity. A schematic of the scanning mirror arrangement is shown in Figure 8. Consider two particle pairs A B and C D, having equal displacements in opposite directions in the object plane. By introducing a mirror placed at 45 deg. between the camera lens and the object

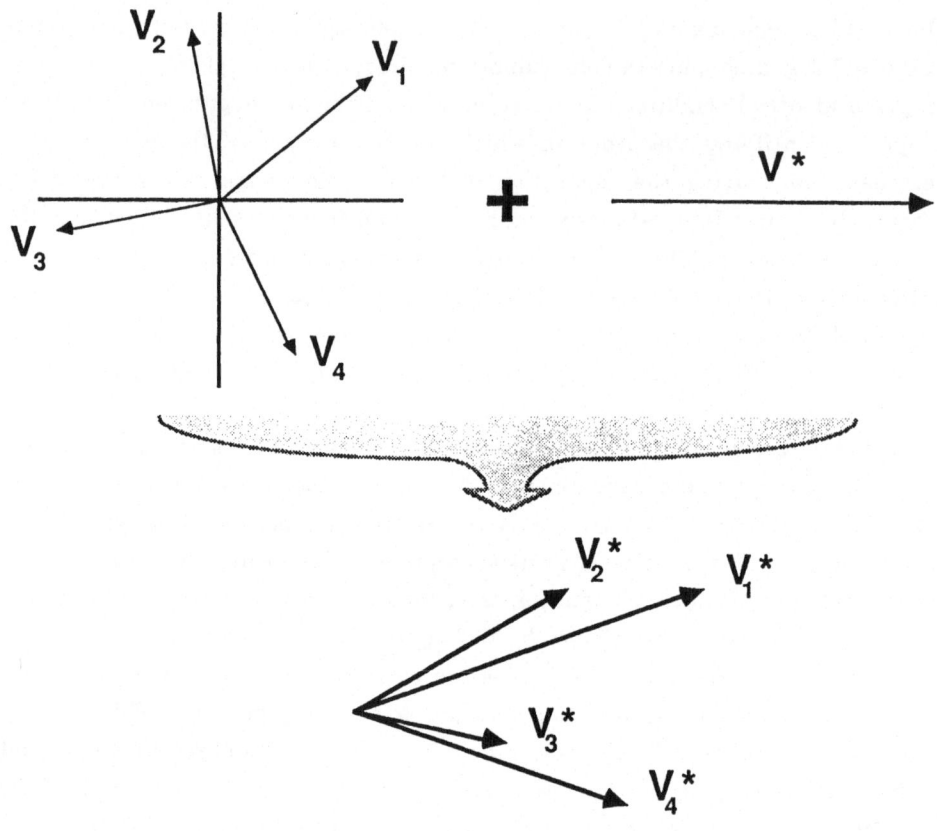

Figure 7. Removal of Direction Ambiguity Using Velocity Bias

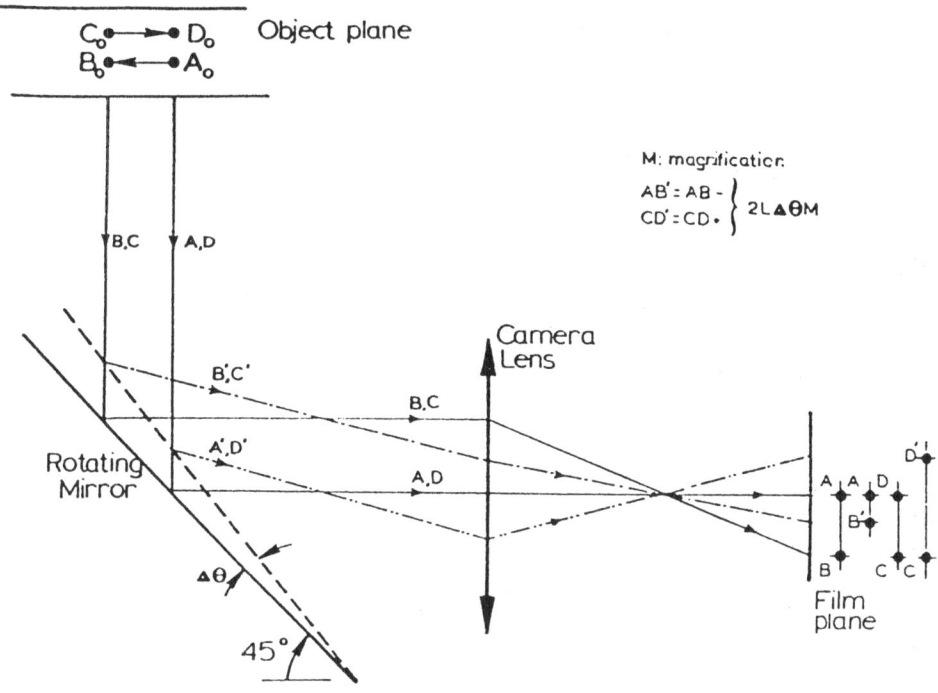

Figure 8. Scanning Mirror Arrangement for Imposing Velocity Bias

plane, the corresponding displacements appear in the film plane as AB and CD with equal magnitudes. When the mirror is rotated by an angle of $\Delta\theta$ between exposures, the displacements corresponding to A B and C D appear in the film plane as AB' and CD', with different magnitudes, resolving the direction ambiguity.

Figure 9a is a double-exposure photograph of the flow past in impulsively started airfoil captured at a stage of its development corresponding to a non-dimensional time of $t^* = tU/c$, where t is the time from the start of the motion, U is the free-stream velocity, and c is the airfoil chord. This figure depicts a complex flow field and exhibits large regions of flow reversal. Analysis of this photograph would yield velocity vector information only within the restriction of the 180 deg. direction ambiguity. In addition, there would also be regions of drop-out where the flow velocity is less than the lower velocity range limit of this technique. Using the velocity bias technique, with a bias velocity equal to two times the free stream velocity, gives the biased images shown in Figure 9b. The velocity field obtained by analyzing Figure 9b is shown in Figure 10a. The actual velocity field, in the reference frame of the airfoil, is given in Figure 10b, upon removal of the velocity bias.

5.9 Overall Accuracy of the Technique

It has been pointed out that out-of-plane motion (i.e., three-dimensional motion) is a severe limitation to this technique in the application to fluid flow. The reason for this limitation is that out-of-plane motion by the tracer particles results in patterns that are poorly correlated. Consider the imaging system shown in Figure 11, with the particle in position P_0 within the laser sheet. The particle will more to a new position R_0 between exposures due to the fluid motion, including an out-of-plane motion dz. In the image plane the corresponding positions are P_L and R_L. The coordinates of these latter two locations are, neglecting second-order and higher terms for simplicity,

$$P_L; \{-M_x, -M_y, (d_0 + d_L)\}$$

$$R_L; \{-M(x + dx)(1 + dz/d_L), -M(y + dy)(1 + dz/d_L), (d_0 + d_L)\} \qquad (26)$$

The displacement between these two locations, $P_L R_L$, determined by the method of Young's fringes, is given by

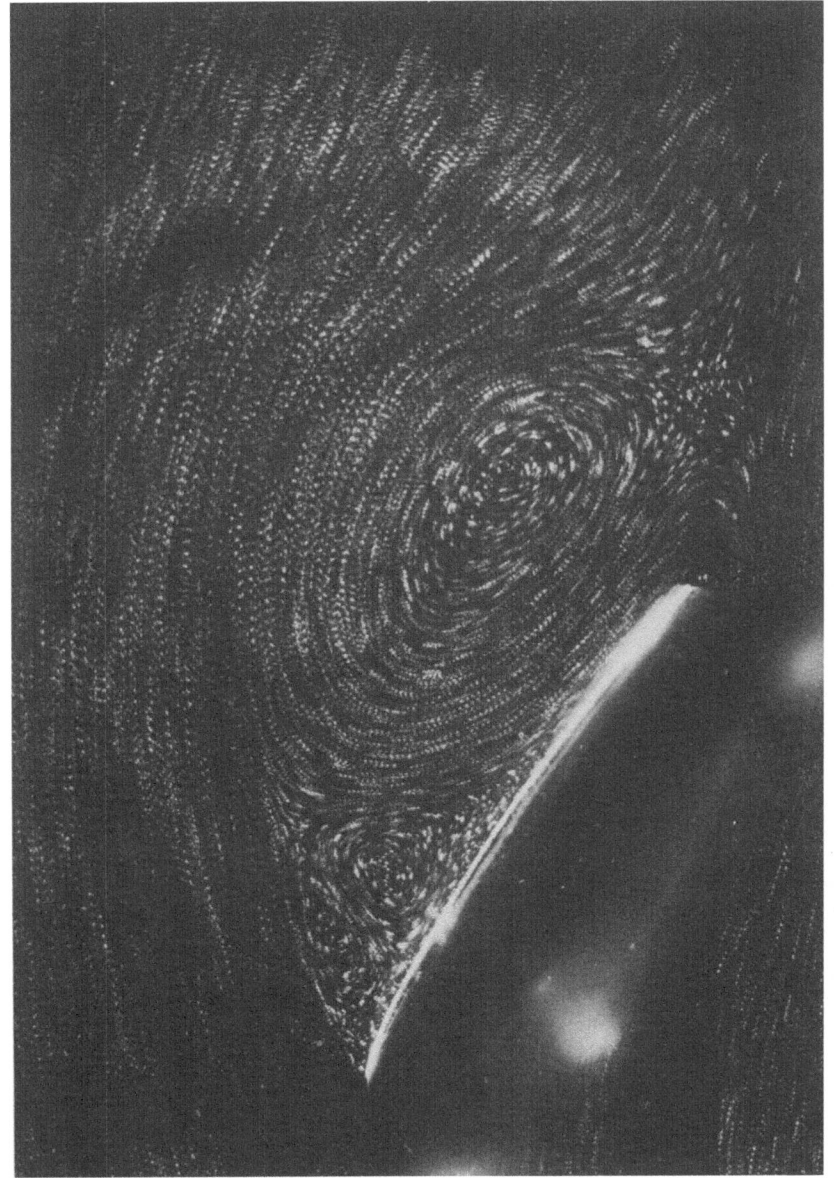

Figure 9a. Double-Exposed Photograph of Airfoil Impulsively Started from Rest; Unbiased Image

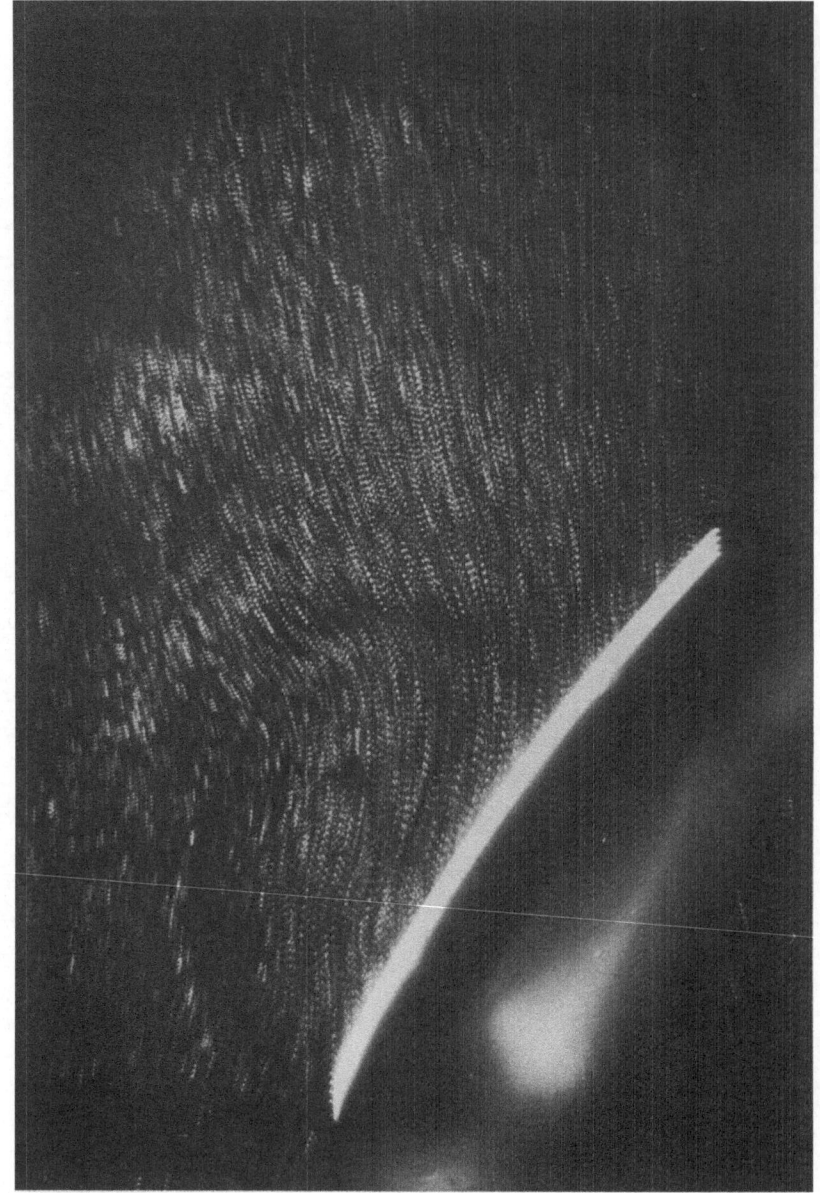

Figure 9b. Double-Exposed Photograph of Airfoil Impulsively Started from Rest; Biased Image

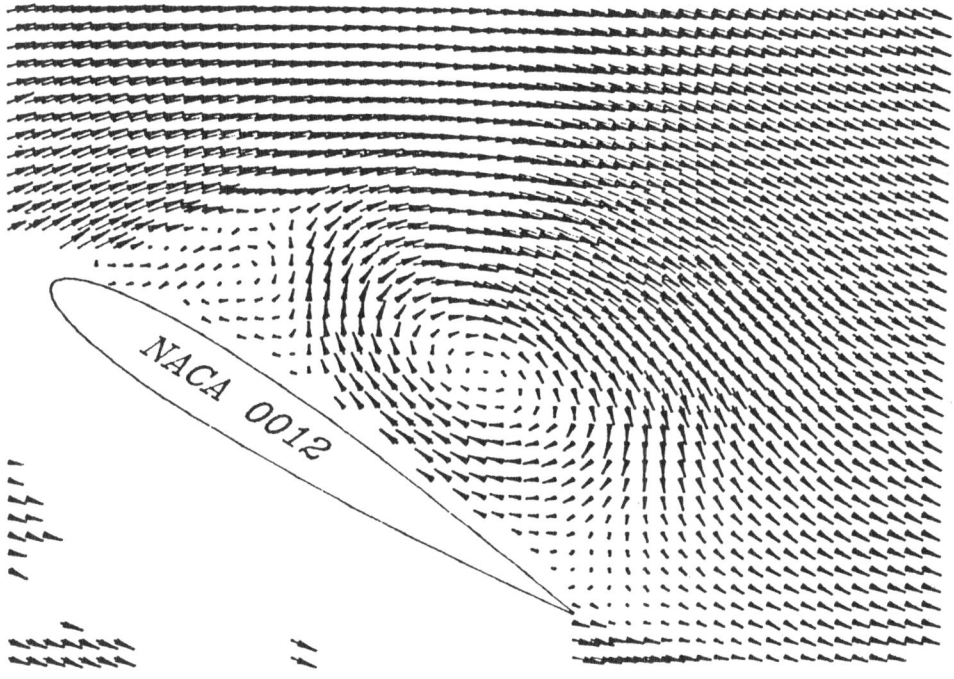

Figure 10a. Instantaneous Velocity Field of Airfoil Impulsively Started from Rest; Before Removal of Velocity Bias

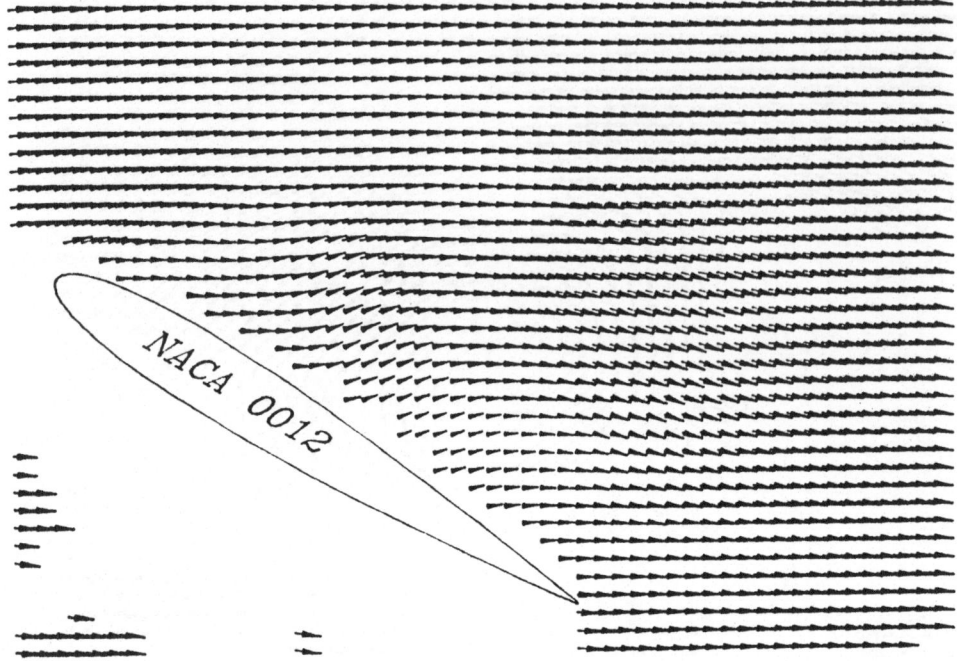

Figure 10b. Instantaneous Velocity Field of Airfoil Impulsively Started from Rest; After Removal of the Velocity Bias

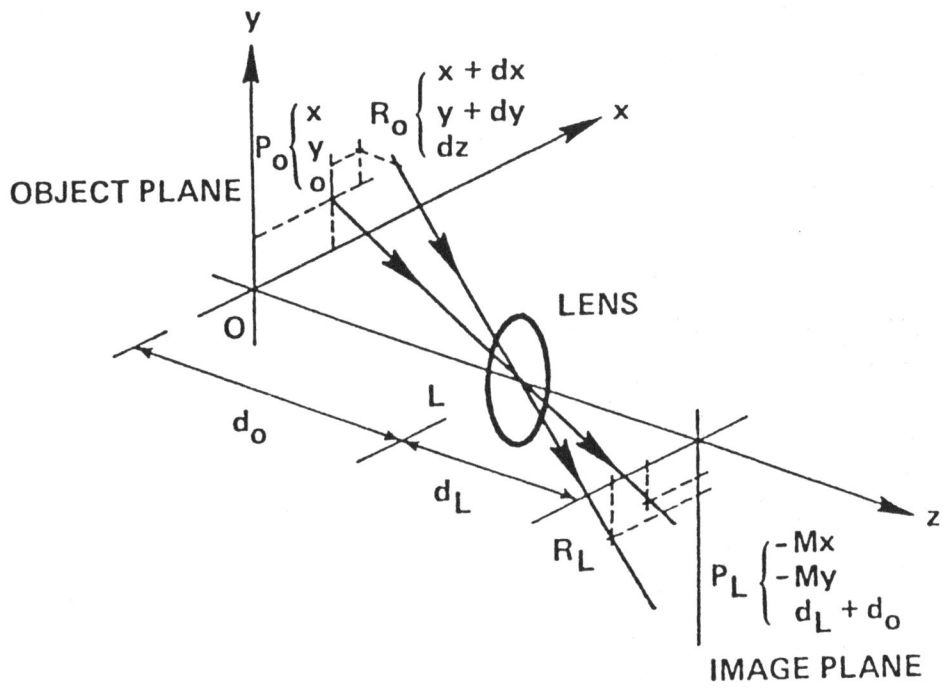

Figure 11. Schematic of Imaging System Estimating the Effect of Out-Of-Plane Motion

$$dx_m = M \, dx \, (1 + x \, dz/dx \, d_L), \quad dy_m = M \, dy \, (1 + y \, dz/dy \, d_L) \tag{27}$$

Thus, the measured displacement components, referred to the object plane, become

$$dx_0 = dx + x \, dz/ \, d_L, \quad dy_0 = dy + y \, dz/ \, d_L \tag{28}$$

The contribution of the out-of-plane displacement to the measured displacement is given by the two parasite terms, $x \, dz/ \, d_L$ and $y \, dz/ \, d_L$. The error produced by three-dimensional motion is a function of the distance from the optical axis. While negligible near the optical axis, it increases linearly, and may become important, farther away. The importance of the out-of-plane motion therefore becomes particularly important when using short focal lengths and wide angle objectives.

The overall accuracy of the technique can be evaluated by studying the uniform flow field created by towing a camera at constant speed past a quiescent flow. Several multiple-exposure photographs were taken, with differing times between exposures, thus resulting in photographs with particle pairs at different known distances. The range of time between exposures, as well as the distance between corresponding particle images in the film plane are presented in Table 1. A large number of points (100) of these five photographs were analyzed using the methods described previously. Uncertainties in the experiment include errors introduced during the recording of the multiple-exposure photograph, such as the ones introduced by distortion of the scene being recorded by the camera lens, limited film resolution, and inaccuracies due to the processing algorithms.

In the absence of a systematic bias, the standard deviation of the measured velocity distribution is an estimate of the mean measurement error. Analysis of the film transparencies using the two techniques (the interactive, one-dimensional averaging method and the automatic autocorrelation method) yields the same mean value with a nearly equal standard deviation (Table 1). The values in Table 1 indicate that using these methods, inaccuracies of the order of 1-2 % are expected. It is believed that these inaccuracies are due to a combination of the limited resolving power of film used for recording (only about 100 lines/mm) and the limited response of the camera lens. Another source of error which is not accounted for in this analysis is the one due to the spurious contributions on the in-plane displacement recording by the out-of-plane motions (Lourenco & Whiffen, 1984, Lourenco, 1986).

Table 1. Overall Accuracy of the Technique

Time Between Exposures, msec	Fringe Frequency	RMS Fringe Frequency	Measured Distance, μM
22.2	33.447	0.257	194
25.0	37.045	0.208	218
28.6	42.950	0.381	249
33.3	49.037	0.710	291
40.0	60.693	1.190	350

6. Examples

6.1 Flow Behind a Circular Cylinder

The time-space development of the near wake flow behind a circular cylinder impulsively accelerated from rest to a constant velocity is studied in this first example. This flow is excellent as a first example because it contains large-scale vortical motions and has extreme velocity gradients. Also it is a well-studied flow and there are several theoretical analyses.

A classic flow visualization study of this flow was performed by Prandtl (1927) and reveals several interesting flow features. Soon after the motion begins, the boundary layer separates and vorticity is convected away from the rear of the cylinder. Two symmetric eddies are formed behind the cylinder, each containing vorticity of opposite sign. The two separating streamlines that surround these eddies join downstream of the eddies and form a closed vortex region. The size of this region grows with time and eventually becomes larger than the cylinder itself. As time increases still further, perturbations cause the standing vortices to develop asymmetric oscillations. Eventually, some of the vorticity in the larger eddy breaks away and moves downstream. The process repeats itself with the other eddy and the flow develops into the familiar Karman vortex street.

The experiment was conducted by towing circular cylinder, 25.4 mm in diameter through a towing tank measuring 300x200x600 mm. The towing carriage is driven by a variable DC motor, and the towing velocity was 22 mm/sec. The Reynolds number, based on cylinder diameter, was 550. The fluid used in the experiment was water seeded with 4 micron metallic-coated particles. For the illumination, a laser beam from a 5 Watt Argon-Ion laser is steered and focused to a diameter of 3 mm using an inverse telescope lens arrangement. A cylindrical lens, with a focal length of -6.34 mm, is used to diverge the focused beam in one dimension, creating a light sheet. The laser sheet was 70 mm wide and the illuminated the mid-span section of the cylinder. For the multiple exposure, the CW laser beam was modulated using a Bragg cell. In this experiment the laser power density, I_0, of the sheet was 0.27 W/mm^2. A 35 mm camera, attached to the towing carriage, was used to record the flow field. The frequency at which the multiple exposures were taken was 1.7 Hz. The aperture of the lens, with a focal length of 50 mm and a space of 12 mm, was set at F#5.6 and the resulting magnification factor was 0.40. The exposure time, t, and the time between exposure, T, were chosen by the criteria described in Section 5 and are 3 msec and 30 msec, respectively. These two parameters, along with the diameter of the analyzing beam and the particle image diameter, determine the dynamic range of the velocity (see Section 5). For this experiment, the dynamic range was roughly 6.

The flow was captured at several stages of its development, corresponding to t*, where t* = tU/D, the non-dimensional time, t is the time from the start of the motion, U is the free stream velocity, and D is the cylinder diameter. Figure 12a-d show typical multiple-exposure photographs of this flow field. In Figure 12a, at t* = 2.2, the two symmetric eddies are clearly seen in the wake of the cylinder and the closed vortex region is roughly the same size as the cylinder diameter. At a later time, t* = 3.2, (Figure 12b) the eddies are still symmetrical but has grown much larger. At a still later time, t* = 4.2, (Figure 12c) the asymmetry is just beginning. Finally, at t* = 5.2, (Figure 12d) the flow field is completely asymmetric and vorticity from the upper eddy is about to break away and move downstream.

The velocity data are acquired in a square mesh by digital processing of the Young's fringes, produced by point-by-point scanning of the positive contact copy of the photograph (Lourenco, 1986). The scanning step size and the dimension of the interrogating laser beam are both 0.5 mm, which, with the magnification of 0.40 corresponds to a spatial resolution of about 1.25 mm in the object plane. This is about 1/20 of the diameter of the cylinder. The fringes were processed using the methods described in Section 4. The resultant two-dimensional velocity fields, corresponding to Figures 12a-d, are shown in Figures 13a-d are a good representation of the expected flow pattern. The length of each vector in the Figure 13 is proportional to the local velocity at that point.

Because of the high spatial resolution of these data, vorticity contours can be derived by taking spatial derivatives of the velocity data. Letting each grid location be labeled with indices (i,j), the vorticity component at location (i,j) is given by

$$\Omega_{i,j} = 1/2\{(V_{i+1,j} - V_{i-1,j})/2\Delta X - (U_{i,j+1} - U_{i,j-1})/2\Delta Y\} \qquad (29)$$

where $\Omega_{i,j}$ is the vorticity at point (i,j), U and V are the longitudinal and lateral velocities, and Δx and Δy are the mesh intervals in the streamwise and cross-stream directions, respectively. Figure 14a-d show the smoothed vorticity contours, normalized with respect to the free steam velocity and the cylinder diameter. To aid in the understanding of this flow, the value of vorticity can be

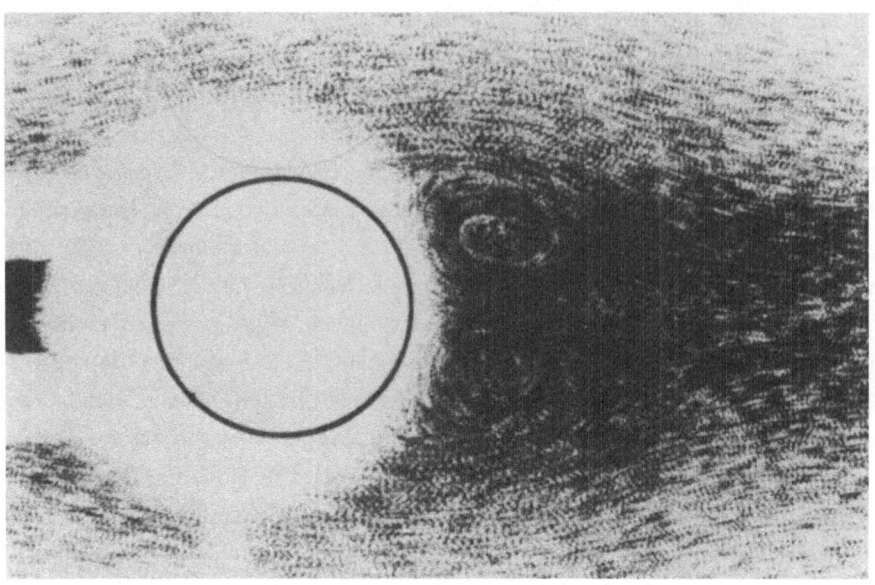

(a)

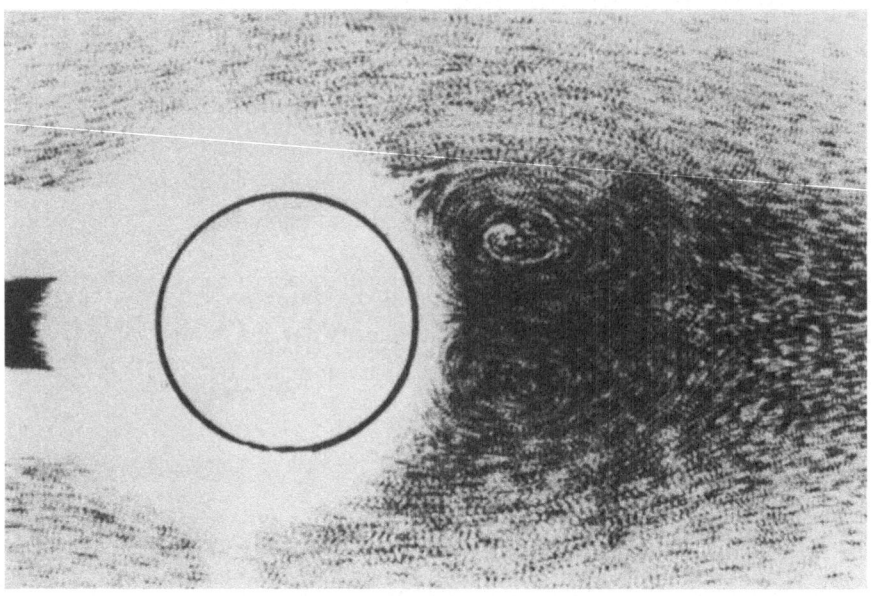

(b)

Figure 12. Multiple-Exposed Photographs of the Wake Flow Field Behind a
Circular Cylinder; a) $t^* = 2.2$; b) $t^* = 3.2$; c) $t^* = 4.2$; d) $t^* = 5.2$

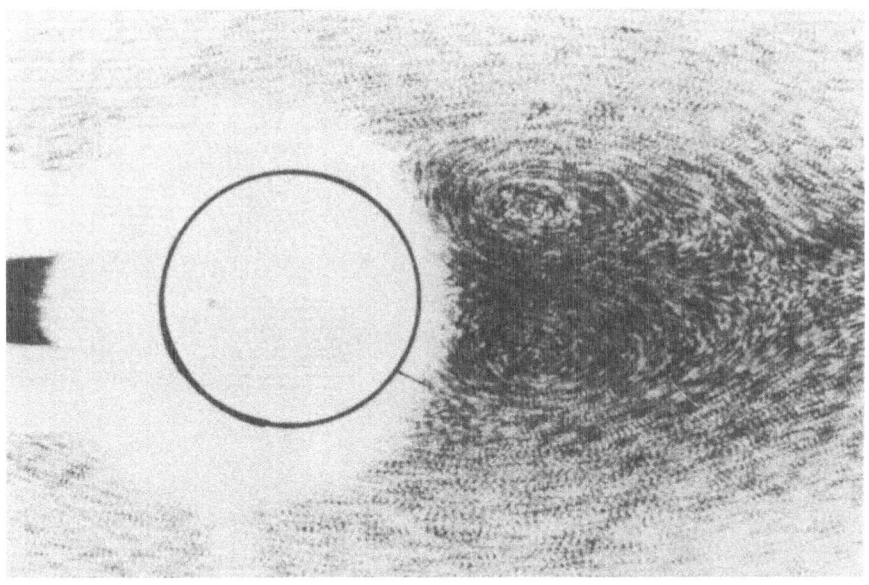

(c)

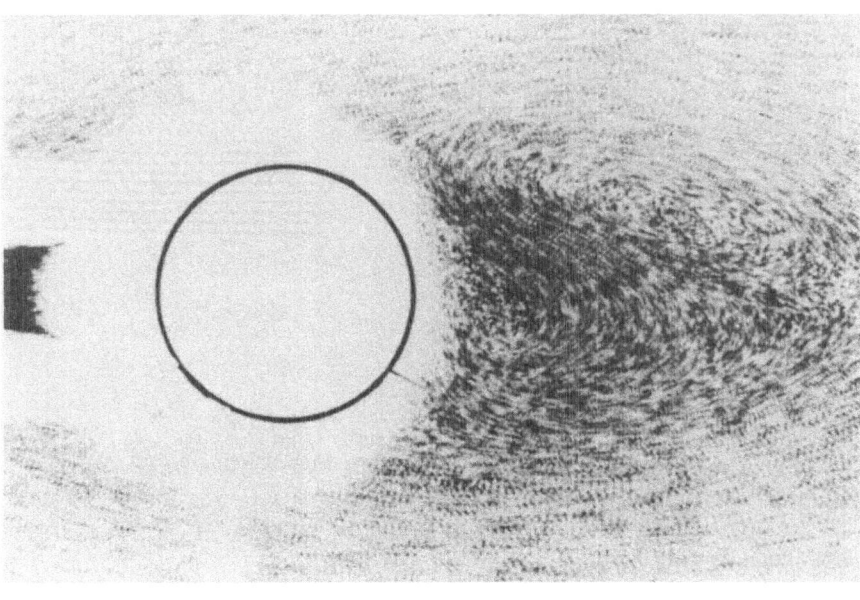

(d)

Figure 12. Multiple-Exposed Photographs of the Wake Flow Field Behind a
Circular Cylinder (Concluded); a) $t^* = 2.2$; b) $t^* = 3.2$; c) $t^* = 4.2$; d) $t^* = 5.2$

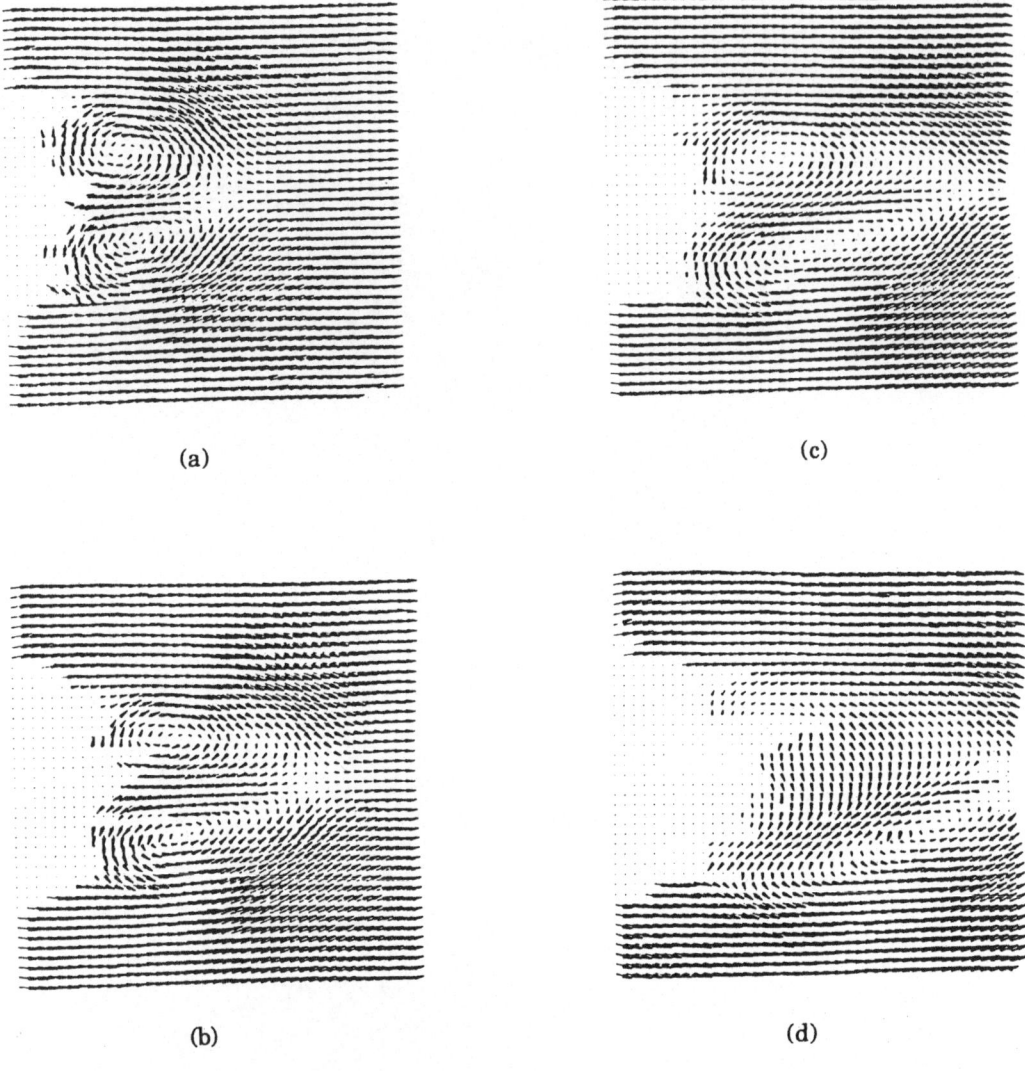

(a)

(c)

(b)

(d)

Figure 13. Instantaneous Velocity Field of the Wake Flow Field Behind a Circular Cylinder; a) $t^* = 2.2$; b) $t^* = 3.2$; c) $t^* = 4.2$; d) $t^* = 5.2$

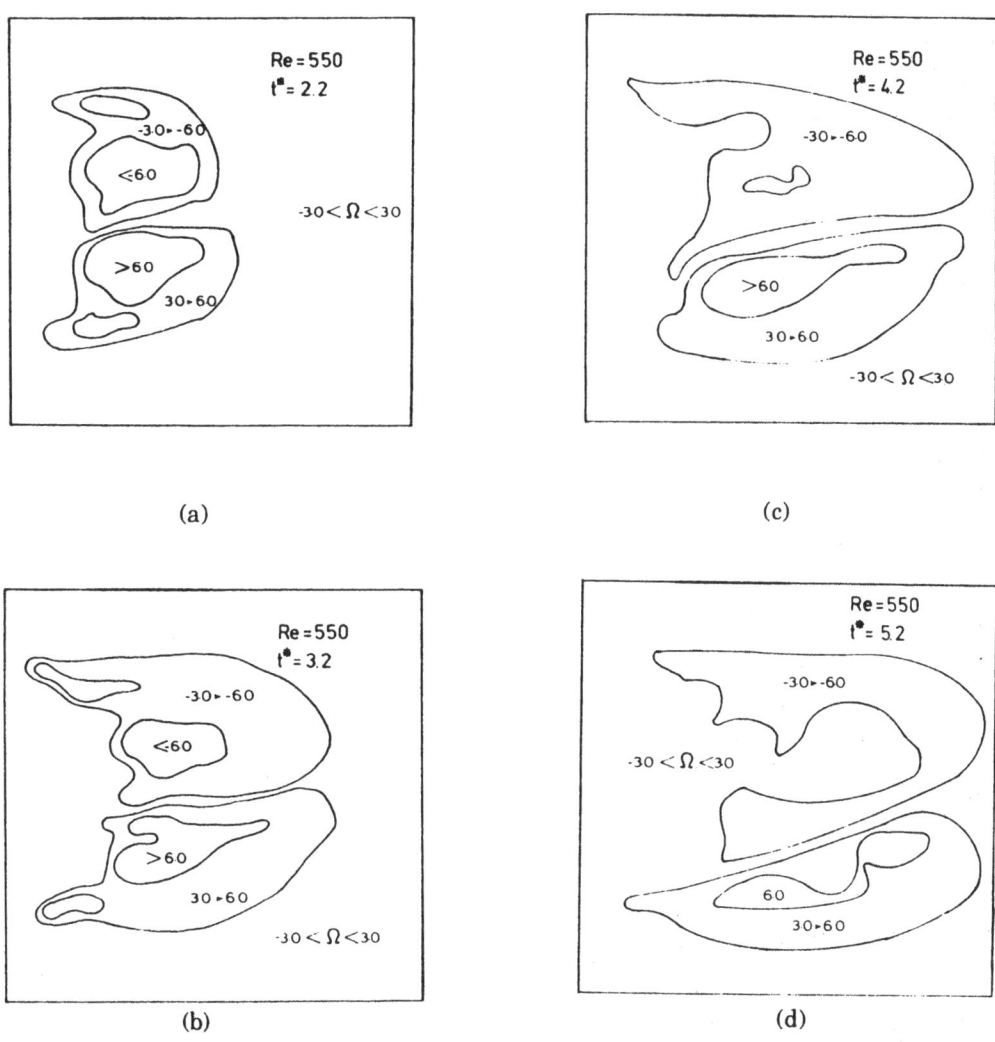

(a)

(b)

(c)

(d)

Figure 14. Constant Vorticity Contours of the Wake Flow Field Behind a Circular Cylinder; a) $t^* = 2.2$; b) $t^* = 3.2$; c) $t^* = 4.2$; d) $t^* = 5.2$

superposed on the velocity data of Figure 13. This is shown in Figure 15a-d, where the vorticity is displayed by color coding each velocity vector. The color code represents the vorticity level, the magnitude of which is given by the color bar on the top of each figure. The red and blue colors represent the peak positive and negative vorticity regions, respectively.

Analysis of these figures reveals some interesting features. Two primary regions of high vorticity form at the rear of the cylinder, corresponding to the startup vortices, while two secondary high vorticity regions are observed further outward. This is especially clear in the vorticity contours shown in Figure 14. The primary vorticity regions may possibly correspond to the "vorticity peak" reported by Bouard and Coutanceau (1980), whereas the secondary regions may be related to the breakup of the feeding sheet as suggested by the flow visualization of Tietjens (1970). Also, it is interesting to observe that the vorticity field (Figure 14) displays earlier evidence of asymmetry than the velocity field (Figure 13).

Using the velocity data of Figure 13, global wake characteristics can also be determined. One example is the growth in the size of the closed vortex region, or wake bubble, with time. Figure 16 displays the development of the wake length, measured in terms of the distance between the cylinder surface and the saddle point (zero velocity) where the two counter-rotating wake vortices join. These values, which are plotted in terms of the non-dimensional time, t^*, compare well with available experimental data by Honji & Taneda (1969) and numerical predictions by Loc (1980) and van Dommelen (1981). It was observed that the distance between the two twin vortices remained constant at a value of about 0.55D, where D is the cylinder diameter, throughout the experiment. This is also in good agreement with the observations reported by Honji & Taneda (1969).

6.2 Flow Past an Airfoil at Angle of Attack

The time-space development of the unsteady separated flow generated by an NACA 0012 airfoil at an angle of attack of 30 deg. and started impulsively from rest is studied in this section. The flow is created by towing the airfoil in the same towing tank as described in the previous section. The airfoil chord is 60 mm and was towed with a velocity of 22 mm/sec. The corresponding Reynolds number was 1400. In order to record the time development of the flow field, the camera was attached to the towing carriage and the frequency which the multiple exposures were taken was set at 2 Hz. Typical multiple exposure photographs of this flow are shown in Figure 17. The photographic arrangement was purposely adjusted to enhance the view of the flow field on the upper surface of the airfoil rather than to show the entire flow around the airfoil. Consequently, the details of the flow

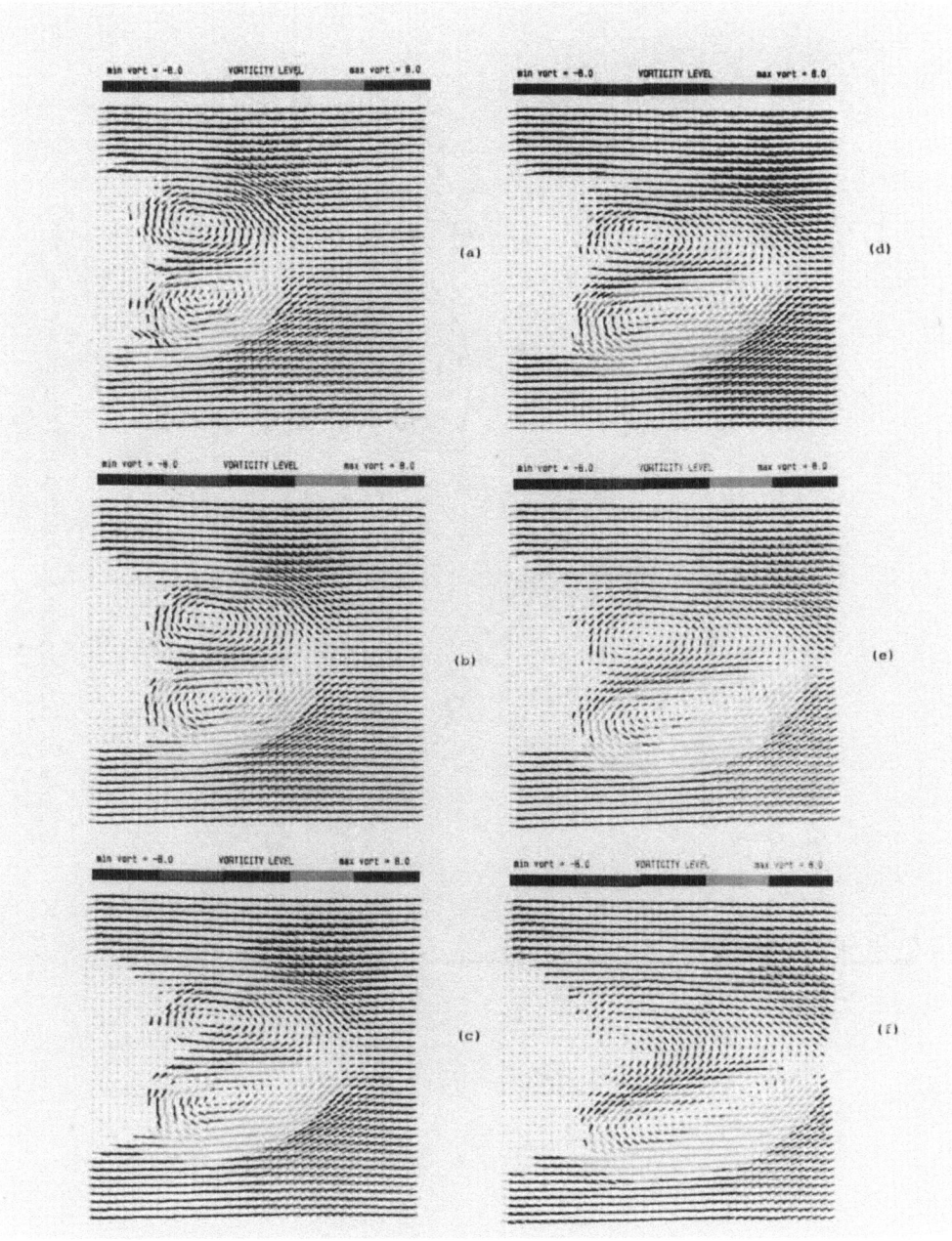

Figure 15. Superposition of Velocity and Vorticity Fields for the Circular Cylinder Wake Flow; a) $t^* = 2.2$; b) $t^* = 2.7$; c) $t^* = 3.2$; d) $t^* = 3.7$; e) $t^* = 4.2$; f) $t^* = 4.7$

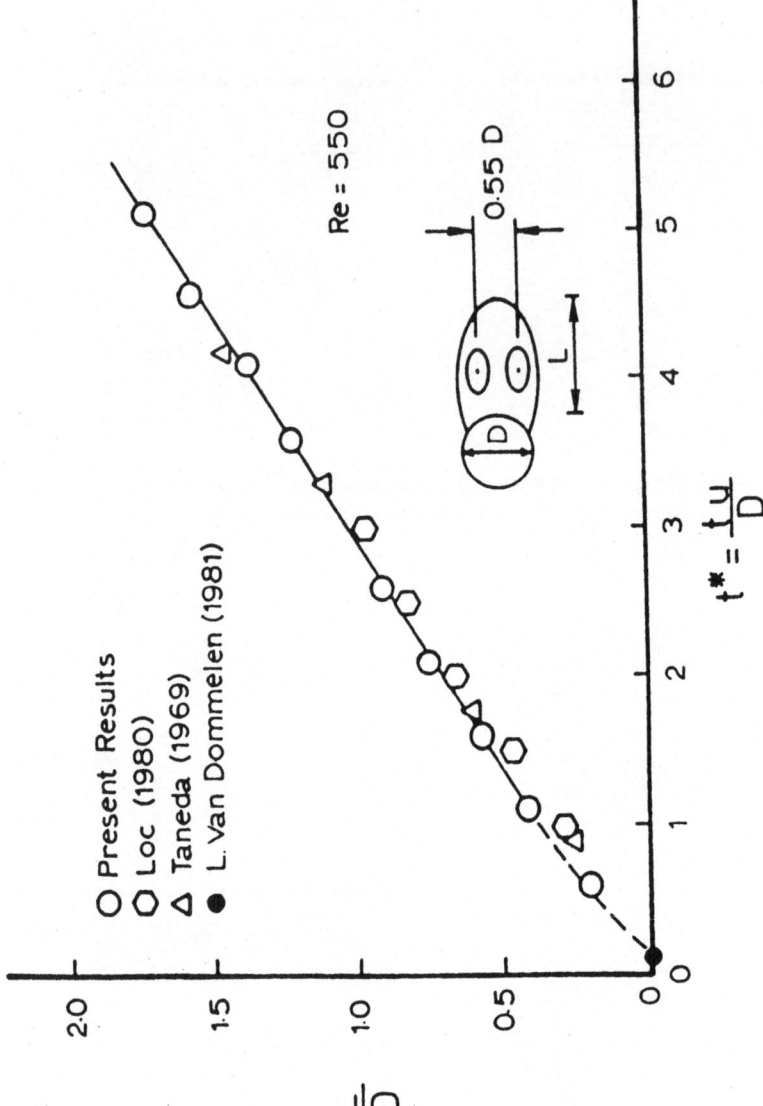

Figure 16. Development of Cylinder Wake Size with Time

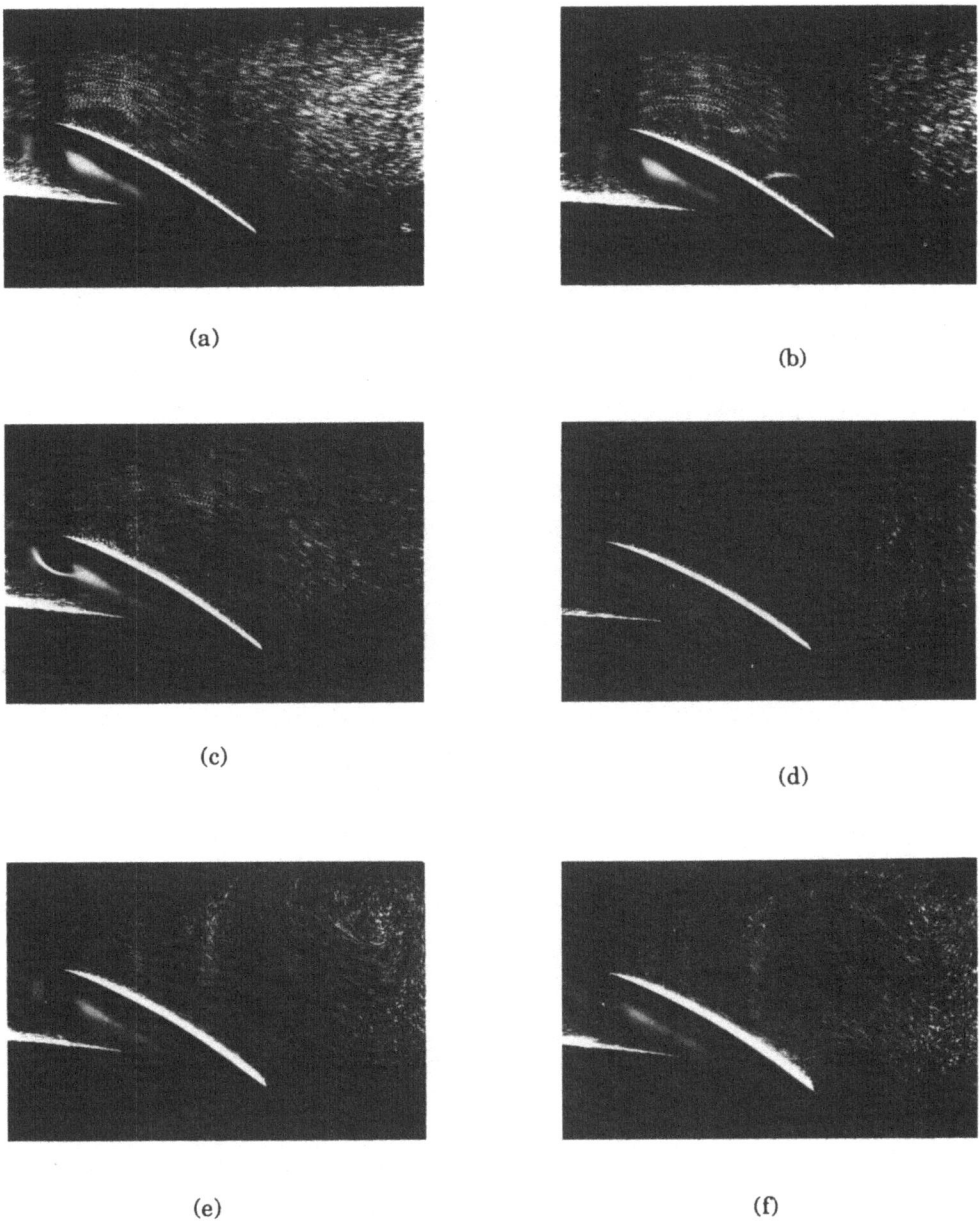

Figure 17. Multiple Exposure Photographs of the Flow Over an Airfoil at Angle of Attack; a) $t^* = 0.68$; b) $t^* = 1.02$; c) $t^* = 2.02$; d) $t^* = 3.02$; e) $t^* = 4.02$; f) $t^* = 4.85$

under the airfoil cannot be seen clearly in these photographs. These photographs display the flow field from the leading edge of the airfoil to a downstream location of about 1-1/2 chords. The quadruple exposures shown here increase the SNR (signal-to-noise ratio) as well as the fringe visibility and provide an excellent flow visualization.

When the airfoil is at angles of attack of ten deg. or less, the flow is well behaved and attached over the entire impulsive process. However, at larger angles of attack ($\alpha \geq$ 20 deg.), the flow separates on the upper surface of the airfoil and generates large scale vortices. The photographs shown in Figure 17 reveal that when the airfoil is first started, a vortex at the trailing edge, commonly referred to as the "starting" vortex, is generated and is carried away from the body. Concomitant with this is the generation of a separation bubble at the leading edge of the airfoil. At a later time, the separation bubble grows into an isolated primary vortex with secondary vortices following behind it. A similar type of vortex structure was also observed in the flow behind a circular cylinder. This multiple vortex structure continues to grow together and move along the upper surface until it reaches the trailing edge. At this point the primary vortex induces a vortex at the trailing edge. At a later time the primary vortex abruptly moves away from the surface of the airfoil leaving behind a vortex-sheet-like structure. This vortex sheet rolls up into distinct vortices and they grow in size with time. During this process the trailing edge vortex also grows, creating a very complex flow field. Close to the surface of the airfoil a small vortex remains present for $t^* > 3.0$. This vortex has the same sign as the trailing edge vortex. A similar vortex structure was observed by Ho (1986), who call it an "induced vortex" and associates it with the unsteady separation phenomenon.

Typical measurements of the instantaneous velocity field are shown in Figure 18. The data are presented in a body-fixed reference frame. The starting vortex and the initial separation bubble at the leading edge can be seen clearly at $t^* = 0.68$. At $t^* = 2.02$ the primary vortex with the secondary vortices behind it can be seen. The trailing edge vortex has just formed and is starting to move downstream at $t^* = 3.02$. Also, at $t^* > 3$, the vortex sheet structure described in the previous paragraph can be seen. This structure may be attributed to the interference of tip vortices generated at the tips of the wing.

Two-dimensional computational results from random-walk vortex simulations of the full Navier-Stokes equations are shown in Figure 19. The angle of attack and the Reynolds number are the same as those in the experiment. The streamline pattern, along with vorticity, which is represented in bit-mapped graphics as half

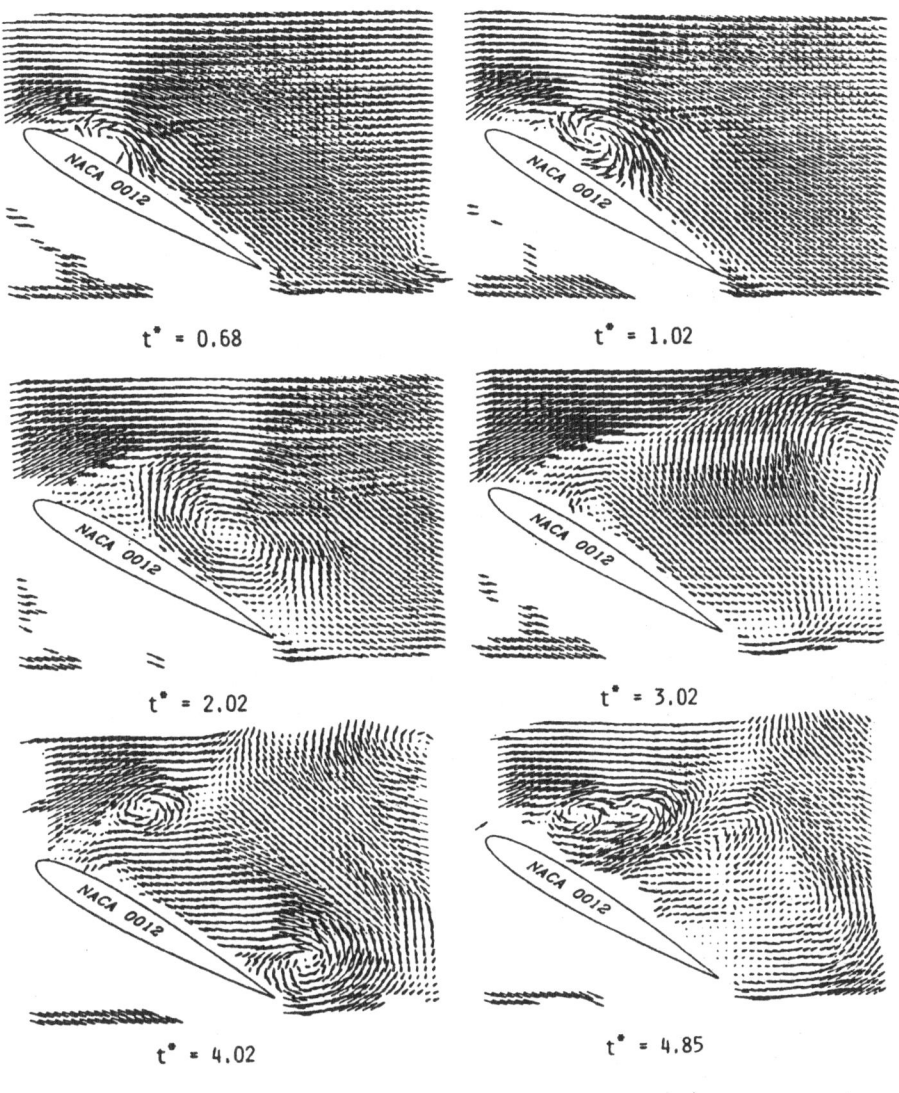

t* = 0.68

t* = 1.02

t* = 2.02

t* = 3.02

t* = 4.02

t* = 4.85

Figure 18. Instantaneous Velocity Field of the Flow Over an Airfoil at Angle of Attack; a) t* = 0.68; b) t* = 1.02; c) t* = 2.02; d) t* = 3.02; e) t* = 4.02; f) t* = 4.85

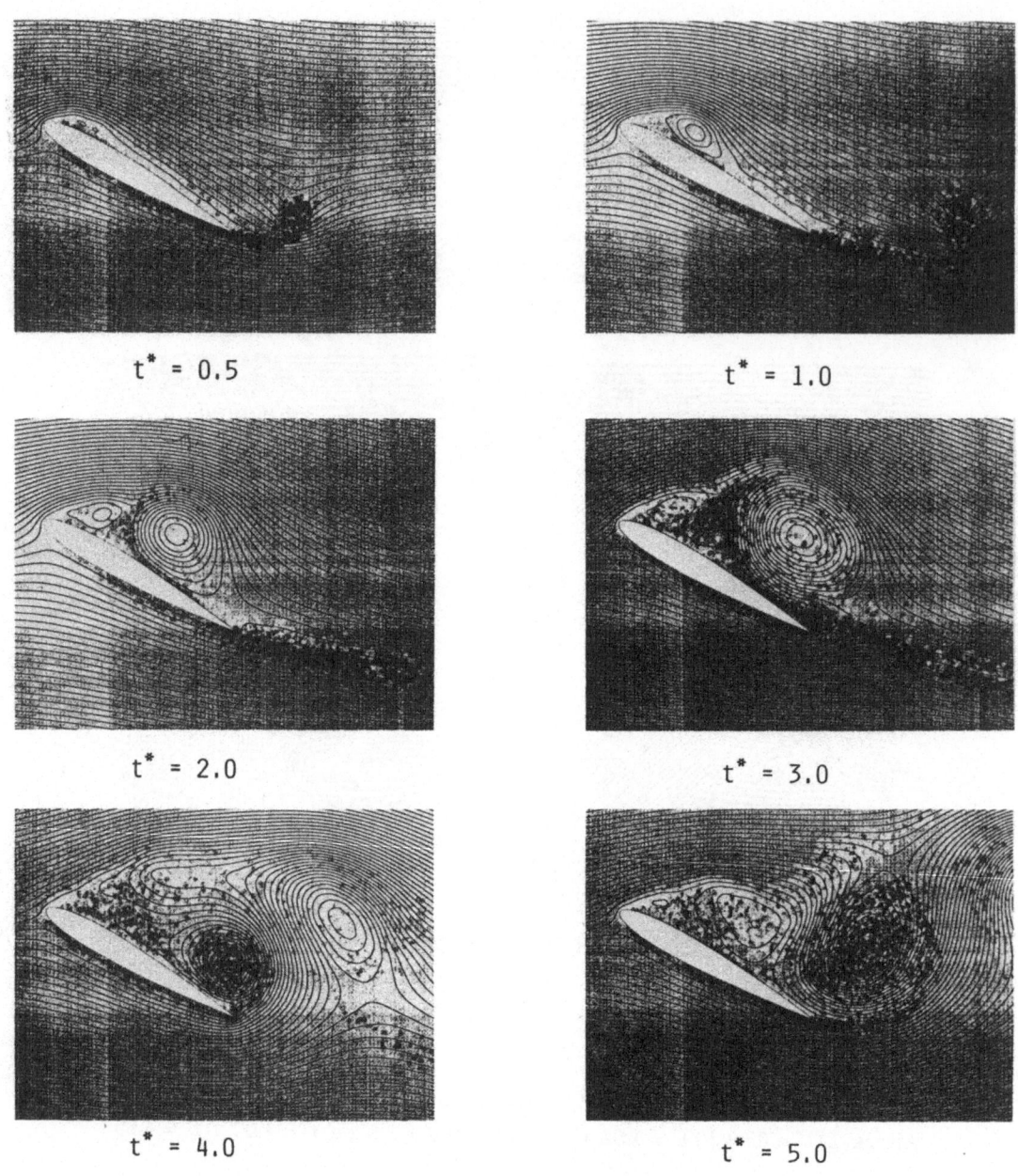

Figure 19. Two-Dimensional Computational Results of the Flow Field Over an Airfoil at Angle of Attack; a) $t^* = 0.5$; b) $t^* = 1.0$; c) $t^* = 2.0$; d) $t^* = 3.0$; e) $t^* = 4.0$; f) $t^* = 5.0$

tones are shown in the figure. Except for the effect of the finite aspect ratio of the airfoil, the streamline patterns look very similar to the patterns observed in Figure 17. To further evaluate these results, the locus of the primary vortex as it develops in time is shown in Figure 20. The computational results agree well with the experiment for $t^* \leq 2$. Beyond this time, it is expected that the experimental flow field was influenced by the tip vortices, making the flow three dimensional. The coefficients of lift and drag as obtained from the computations are shown in Figure 21. As expected, the coefficient of lift increases with t^* up to a point where the primary vortex is attached to the upper surface. For later times, where the primary vortex leaves the upper surface, the coefficient of lift drops significantly.

6.3 Three-Dimensional Turbulent Jet Flow

The flow field considered is a three-dimensional incompressible jet of air issuing from a rectangular nozzle of aspect ratio 4. The structure and development of such a jet is markedly different from those issuing from two-dimensional or axisymmetric nozzles, the focus of most previous investigations on turbulent jets (e.g., Krothapalli, Baganoff, and Karamcheti, 1981). With renewed interest on thrust vectoring and mixing devices, emphasis is now shifting to the study of three-dimensional nozzles. The structure and development of these jets contain many interesting features and are yet to be thoroughly understood. One such feature is the "cross-over" phenomenon, which is generally characterized by the switching of the major and minor axes downstream of the nozzle exit. Recent experiments, conducted by Ho and Gutmark, (1987) on low aspect ratio elliptic jets suggest that an initial instability process, which is accompanied by large vortices, may influence the position of the cross-over point, and thus the development of the jet. The example described here examined the structure and growth of the mixing layer region of the jet. Additional details may be found in Lourenco and Krothapalli, (1988) and Lourenco, Krothapalli, and Smith (1988).

A simple low speed air supply system was used to provide the airflow to a cylindrical settling chamber 27 cm in length and 10 cm in diameter. A honeycomb and a series of screens at the inlet of the nozzle are used to further reduce flow disturbances. The cross section of the contraction changes from a circular cross section, 10 cm in diameter, to a rectangular cross section, 3 cm by 0.75 cm. The contraction contours in the two central planes were fifth-order polynomials. Seeding of the jet was accomplished by using a theatrical-type smoke generator, which produces smoke particles in the sub-micron range. Smoke and ambient air were mixed in a large settling tank and then supplied to

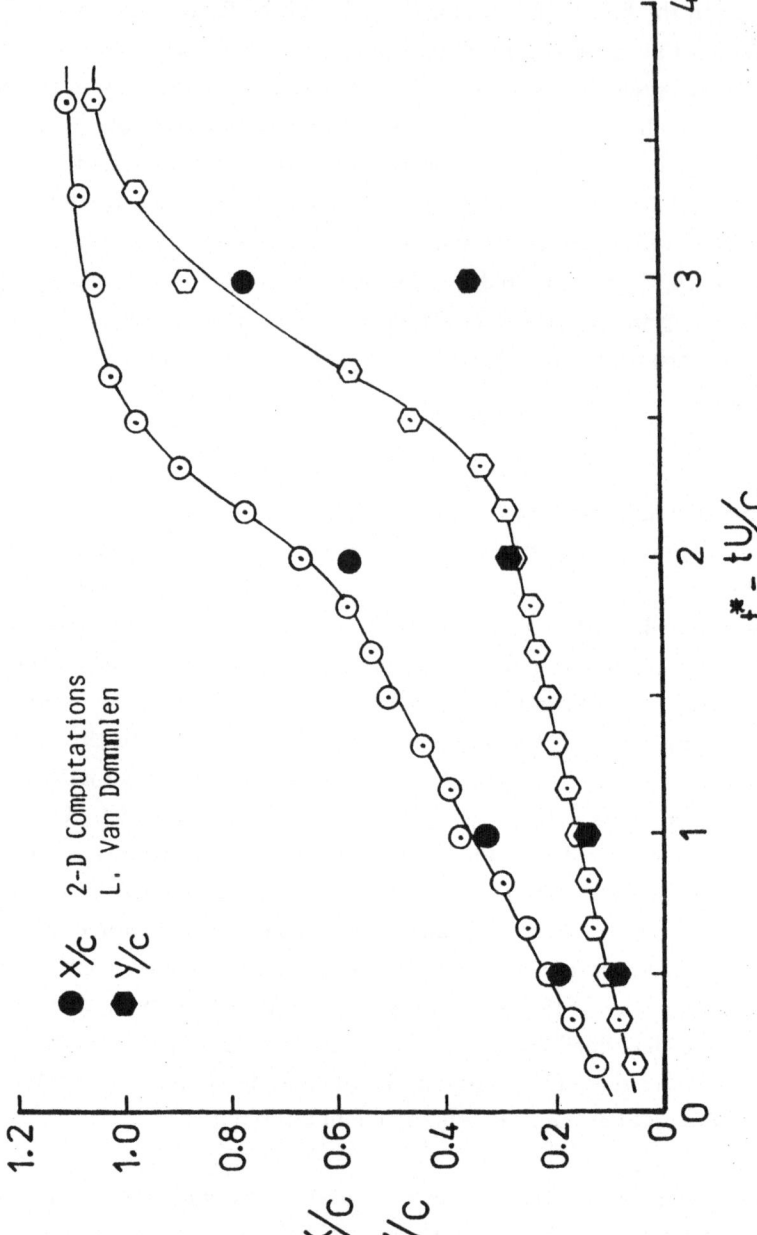

Figure 20. The Variation of the Primary Vortex Location with Time

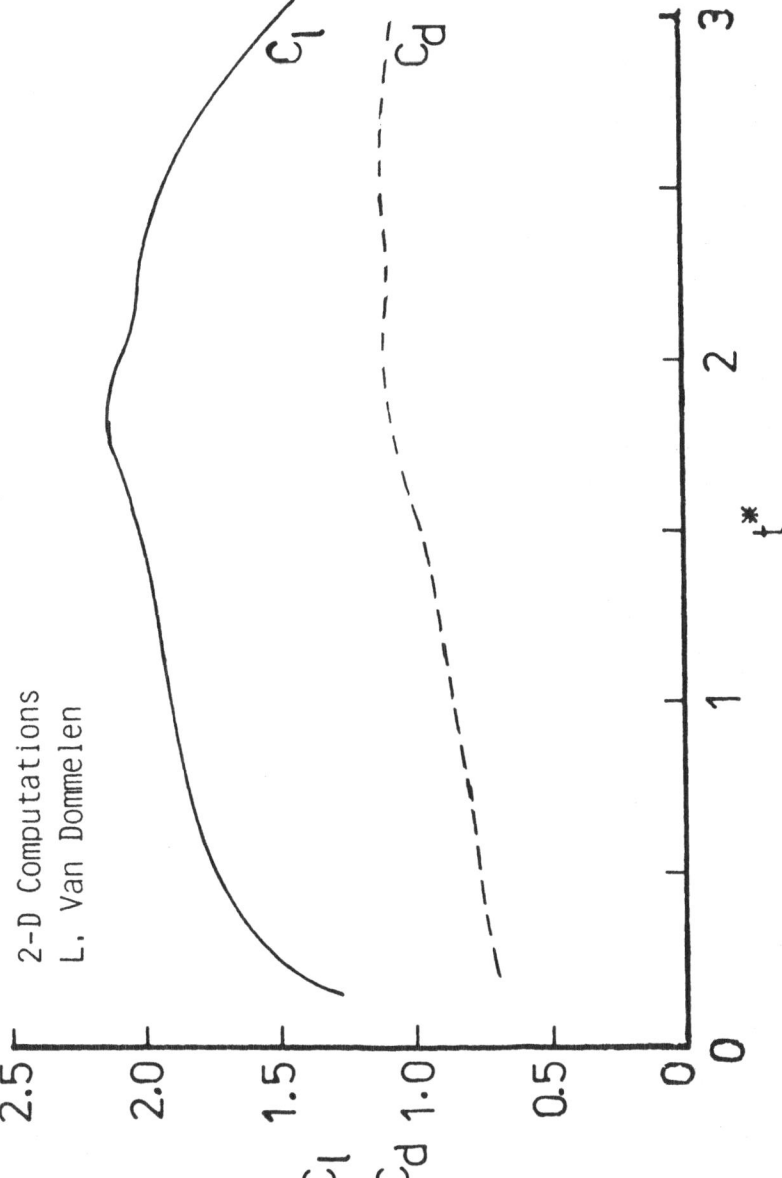

Figure 21. Computed Lift and Drag Variation with Time

the settling chamber of the jet using a small axial fan. A second smoke generator of the same type was used to seed the outside ambient flow surrounding the jet. A schematic of the arrangement is shown in Figure 22. The mean velocity at the exit of the jet was 4.5 m/sec. This resulted in a Reynolds number of 3600, based on the hydraulic diameter (four times the nozzle area divided by the perimeter). Basing the Reynolds number on the jet width (0.75 cm) gives a value of 2250, which may be more appropriate when discussing the stability of the jet. The mean velocity profile at the exit plane of the nozzle was flat with a laminar boundary layer at the walls.

The laser light sheet was created with a frequency doubled, double-pulse Nd:Yag laser, with a similar inverse telescope lens/cylindrical lens arrangement as described in Section 6.1. The laser sheet was 60 mm wide and illuminated the central plane through the small dimension of the nozzle (i.e., the X-Y plane in Figure 22). Two laser pulses with a duration of 10 nsec and a separation of 50 microsec were used to create the specklegram. The pulse separation of 50 microsec is much smaller than any relevant time scale of the flow field and thus the double exposure photograph truly represents a flow field at a given instant of time. The velocity bias technique, described in Section 5, was used to resolve the ambiguity of the velocity vector. A 35 mm camera was used to record the specklegram, using Kodak TMAX 400 ASA film, which has good sensitivity at the frequency of the illuminating laser light. The magnification was 0.50.

Typical double-exposure photographs of the jet, taken at two different times, are shown in Figure 23. These pictures display the flow field from the nozzle exit to about eight jet widths downstream. The photographs were taken using the velocity shift technique described in Section 5 and with external seeding of the ambient medium. From these results, along with other pictures, several observations can be made. The jet consists of three regions: the region in which the initial shear layer is unstable and rolls up into discrete vortices; an interaction region in which the vortices pair with each other; and a region in which the vortices break up into random, three-dimensional motion. In spite of the relatively large aspect ratio of the nozzle exit (AR = 4), the rectangular jet organizes itself into a structure similar to that of an axisymmetric jet, as shown by Bouchard and Reynolds, (1982). The pairing process is also quite similar. In this process, the trailing vortex catches up with the leading vortex, decreases in size and passes through the leading vortex, which has slowed down and grown in size. The vortex cores rotate around each other and ultimately merge, producing a single vortex. A number of vortex pairings can occur before vortex breakdown occurs. Figure 23b shows such a vortex pairing in progress. The physical

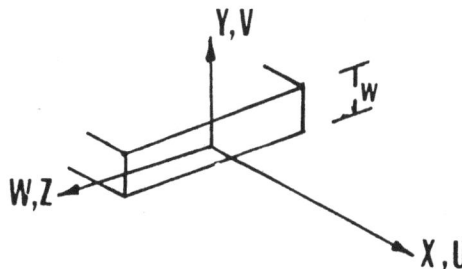

Figure 22. Schematic of Three-Dimensional Jet Experiment

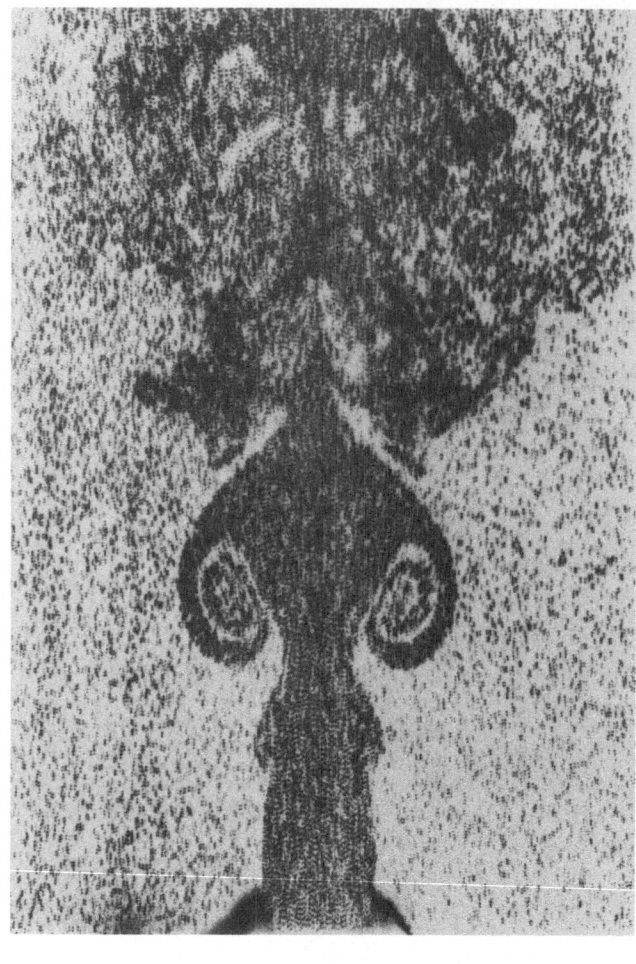

(a)

Figure 23. Instantaneous Double-Exposed Photographs of the Central Plane of the Jet

(b)

Figure 23. Instantaneous Double-Exposed Photographs of the Central Plane of
the Jet (Concluded)

regions where these phenomena takes place overlap and depends on the phase of the development of the jet. The jet Strouhal number, St (non-dimensional passage frequency of the vortices prior to pairing), is given by St = fw/U, where f is the passage frequency, w the nozzle width, and U is the mean exit velocity of the jet. For this experiment, St is estimated to be about 0.7. This Strouhal number is close to that measured for an axisymmetric jet by Becker and Massaro, (1968), at a comparable Reynolds number. Examination of several photographs suggests that the vortex breakdown enhances the mixing in the plane of the small dimension of the nozzle. No increase in mixing was observed in the central plane containing the long dimension of the nozzle.

The instantaneous velocity field, for two typical phases of the development of the jet, are shown in Figure 24. In this figure the velocity is given in the laboratory reference frame; that is, the velocity bias has been removed. The length of each vector is directly proportional to the magnitude of the velocity. Because the velocity gradients are largest in the transverse direction, the velocity data were acquired using a rectangular mesh with a mesh spacing of 2 mm in the jet axial direction and 0.5 mm in the jet transverse direction. The velocity fields displayed in Figure 24 describe in great detail all of the aforementioned regions of the jet flow field, from the initial shear layers to regions with highly three-dimensional motion. Such an accurate and detailed representation of the flow field was a consequence of the use of the velocity bias technique together with judicious management of the seeding.

Examination of these velocity fields further reinforces the previous analysis made on the basis of the flow visualization pictures. The jet structure is further enhanced by presenting the velocity field in a reference frame with a convection velocity of the vortical structure, estimated at about 70 percent of the jet exit velocity, as shown in Figure 25. In this reference frame, the large scale vortical structures are clearly observed which shows the nature of the symmetric instability. The instantaneous velocity profiles provide a unique means to quantify the extent of the jet unsteadiness, the existence of the coherent motions, their interactions, and subsequent generation of the random three-dimensional motions. As an example, Figure 26 shows the instantaneous distribution of the axial centerline velocity along the jet axis obtained from the data of Figure 24. As expected, the centerline velocity distribution is phase dependent and does not display a monotonic behavior as commonly observed in mean velocity distributions. The peaks and valleys in this phase-dependent distribution are a consequence of the vortex dynamics. Thus, a complete understanding of the structure and development of the jet must include a detailed study of the time evolution of the whole flow field.

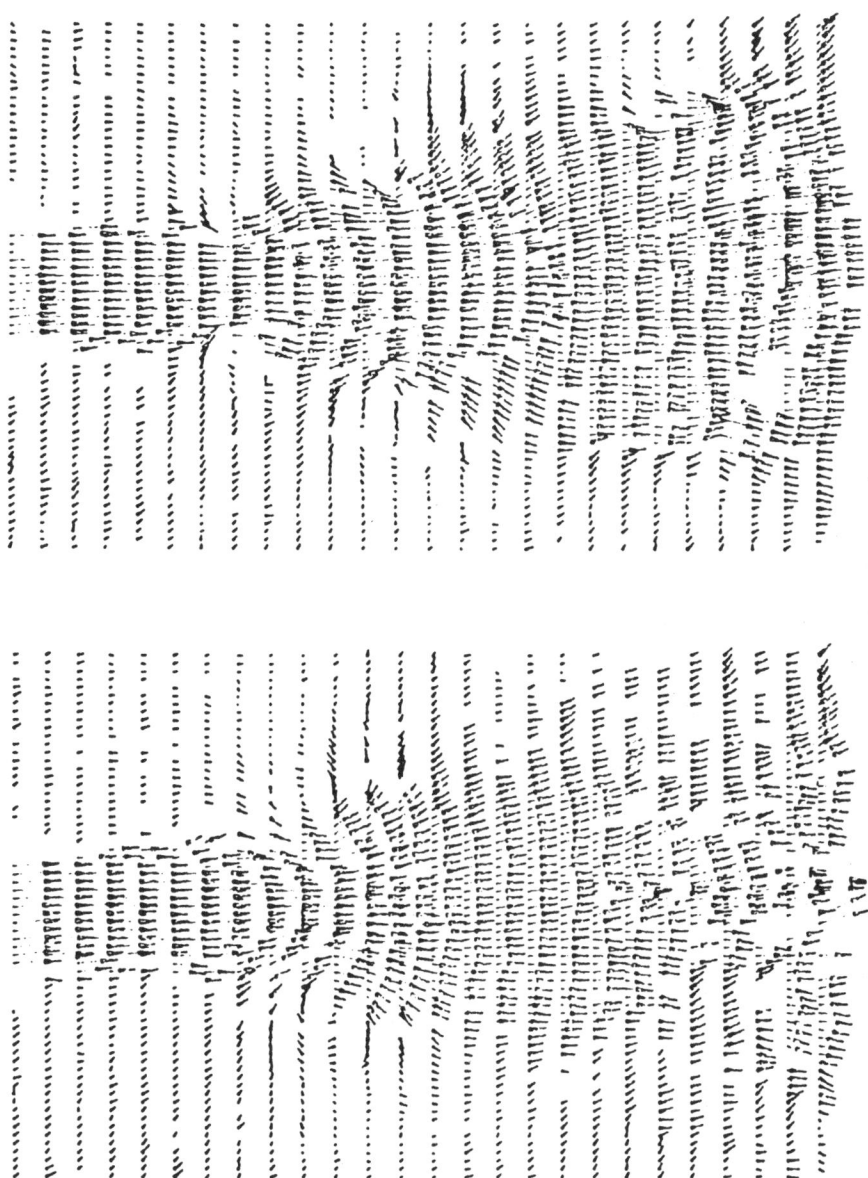

Figure 24. Instantaneous Two-Dimensional Velocity Field in the Central Plane of the Jet

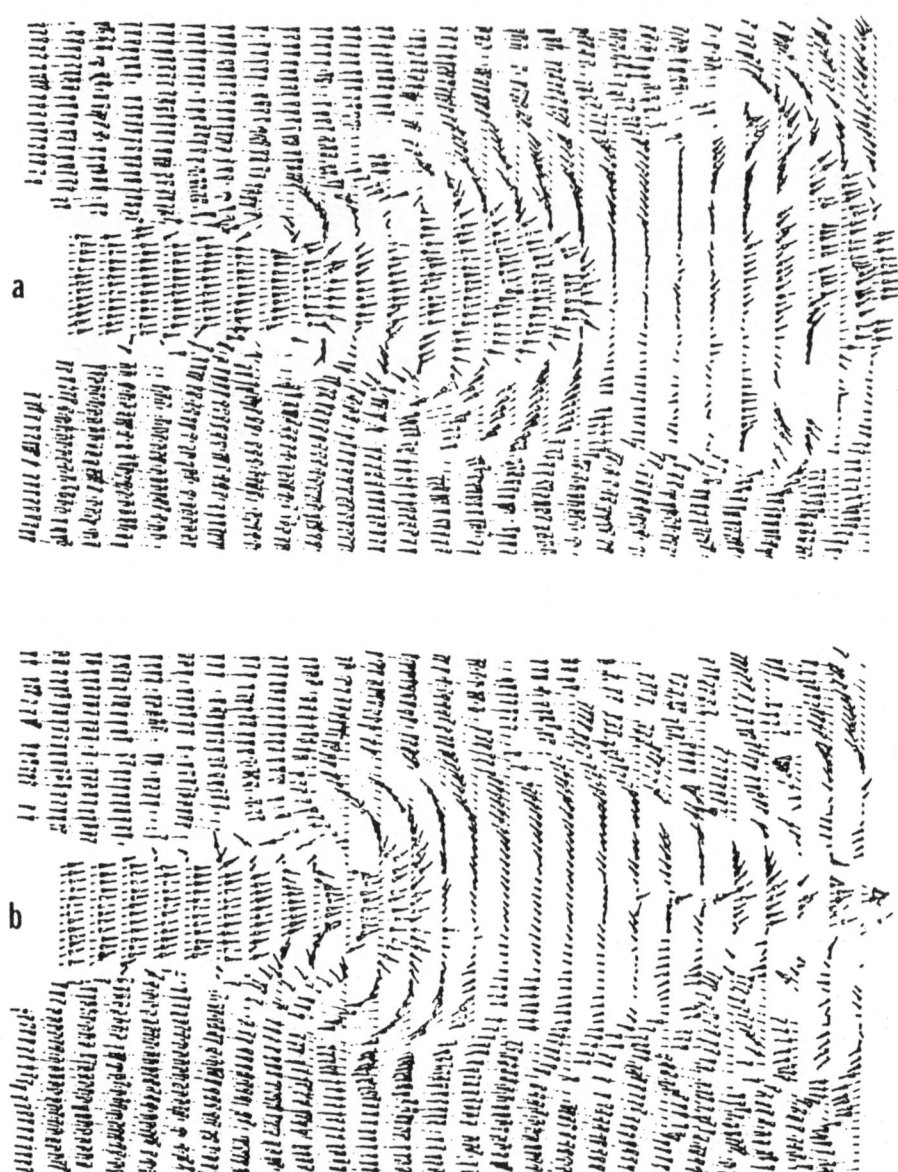

Figure 25. Instantaneous Two-Dimensional Velocity Field Shown in the Reference Frame with a Convection Velocity of the Vortical Structures.

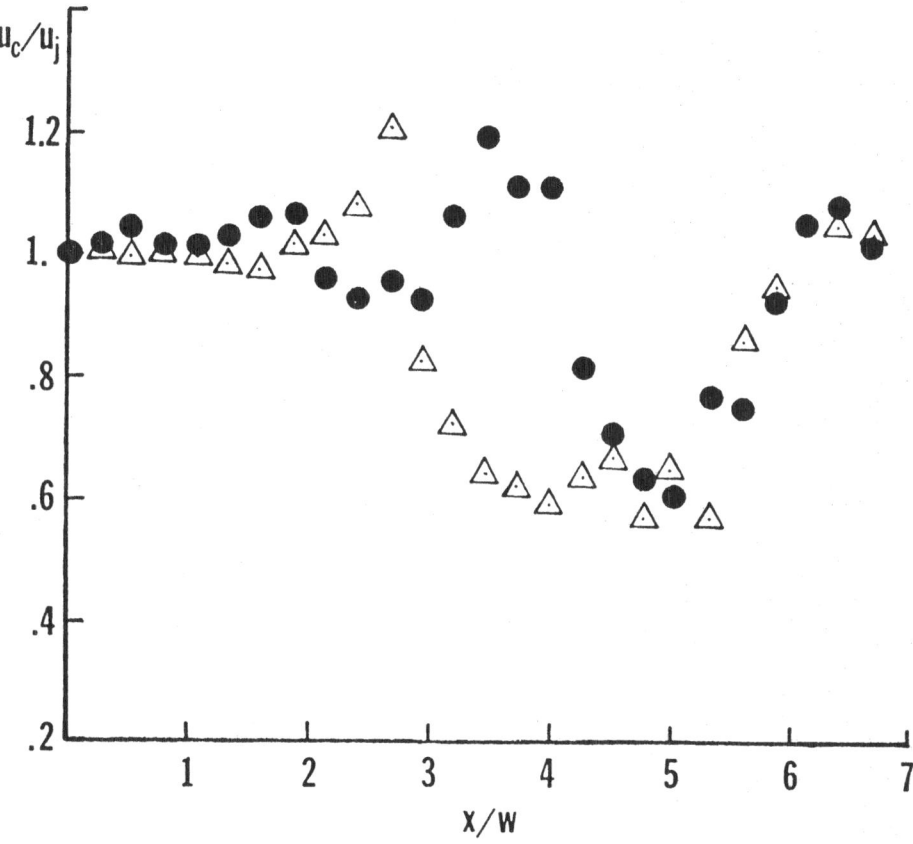

Figure 26. Variation of the Instantaneous Centerline Axial Velocity with Downstream Distance for the Two Cases Shown in Figure 25; •) Figure 25a, Δ) Figure 25b.

7. Recent Developments

7.1 High-Speed Measurements

A recent study by Kompenhans and Hocker, (1988) has demonstrated conclusively that PIV can be successfully applied to the study of high-speed flows. In their experiment, a pair of Nd:Yag lasers were combined by means of polarizing prisms. The schematic of this system is shown in Figure 27. Utilizing a digital clock to control the exact time delay between the two pulses, they studied the flow of a circular jet at exit Mach numbers ranging from 0.1 to almost 1. The seeding particles, created by injecting pressurized air into olive oil, were of the order of 1

μm in diameter. The flow facility consisted of a 15 mm diameter circular air jet, at ambient temperature. The Reynolds number, based on jet diameter, ranged from 4×10^4 to 3×10^5. Both the jet flow and the ambient air outside the jet were seeded in order that the particle concentration in the shear layer region, where ambient air is entrained into the jet, was sufficiently high to yield good results. The multiple-exposure photographs were analyzed using the Young's fringes technique.

7.2 Three-Dimensional Measurements

Until recently, applications of PIV have been limited to two-dimensional flows. The physical limitation of the system is that the particle must be in the illuminated sheet during both exposures. Any out-of-plane motion of particles into or out of this sheet reduces the particle correlation and can result in a loss of the signal. However, many flows of interest are three-dimensional and ways of extending PIV to the study of such flows are of obvious interest. One such method, pioneered by Riethmuller and his colleagues at the von Karman Institute (VKI), is described in this section, which summarizes the work presented by Gauthier and Riethmuller (1988).

The VKI method is a stereoscopic scheme in which the flow in the illuminated plane is viewed from two different directions simultaneously. Thus, with this technique measurements over a single plane are still made, but all three components of velocity are obtained. The constraint is that the time delay between exposures must be chosen such that the maximum particle motion normal to the illuminated plane is much less than the thickness of the sheet. With this scheme it is often necessary to add a velocity bias to the particle images to keep them in the range of the analysis system.

In the stereoscopic method the flow is the flow is viewed from two different

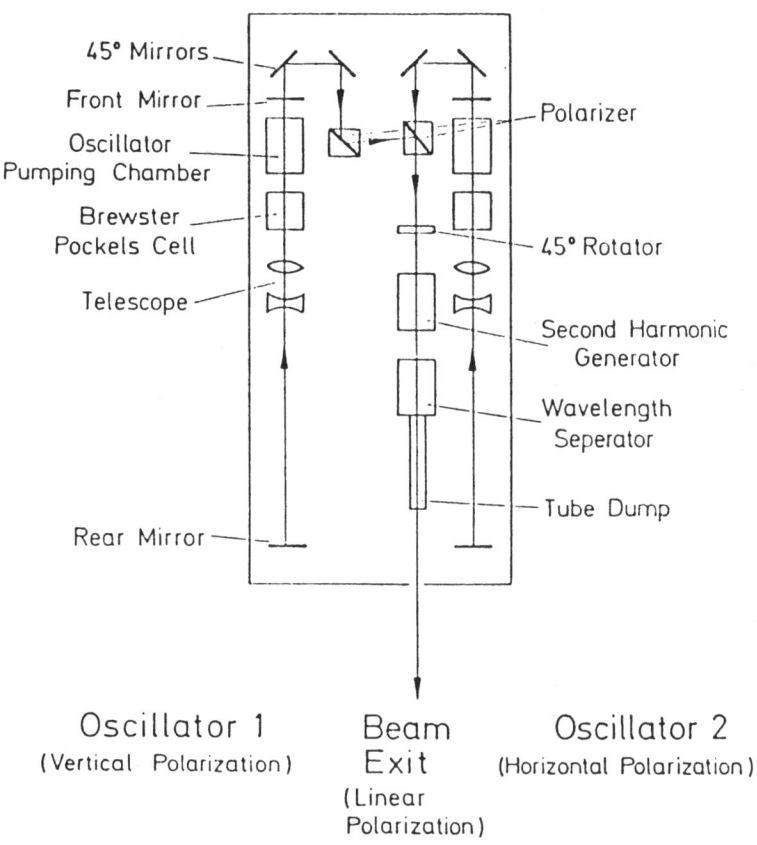

45° Mirrors

Front Mirror

Oscillator
Pumping Chamber

Brewster
Pockels Cell

Telescope

Polarizer

45° Rotator

Second Harmonic
Generator

Wavelength
Seperator

Tube Dump

Rear Mirror

Oscillator 1
(Vertical Polarization)

Beam
Exit
(Linear
Polarization)

Oscillator 2
(Horizontal Polarization)

Figure 27. General Layout of the Double Pulse Laser System used by Kompenhans and Hocker, (1988)

directions and the three velocity components are obtained from their projections and the geometrical characteristics of the optical system. Two different stereoscopic systems have been tried thus far by the VKI group. The first is called an angular displacement method and is one in which the optical axes are not perpendicular to the illuminated sheet but are inclined at an angle β to the normal to the sheet, as shown in Figure 28. To estimate the error in measuring the displacement, Gauthier and Riethmuller considered a displacement of 250 μm at an angle of θ to the illuminated plane, a magnification of 0.4 and a resolution of 2.5 μm. Results are shown in Figure 29. The error is minimized when the optical axes are at an angle of 45 deg. to the illuminated plane, and for small values of θ (which can be obtained by superposing a velocity bias to the flow if necessary).

The second stereoscopic scheme is referred to as the translation method and is shown schematically in Figure 30. In this technique the optical axis of each camera is perpendicular to the illuminated sheet and the distance between the two axes provides the stereoscopic effects. The error associated with this method is shown in Figure 31 for the same conditions as in Figure 29. Although the error decreases with increasing distance between the two optical axes, so does the overlap or common area recorded by the two cameras.

Gauthier and Riethmuller applied each of these methods to a simulated 3-D flow by measuring the uniform flow in a rectangular duct (25x40 mm) with a laser sheet at an angle of 20 deg. to the duct axis, as shown in Figure 32. The velocity was 5 m/s (no velocity bias was used), the magnification was 0.5, and the time between exposures was 50 μs. The scatter in the out-of-plane displacement for the two methods is shown in Figure 33. There is a large scatter of 32 deg. in the measurements using the translation method, and a scatter of 8 deg. with the angular displacement method. These results confirm the predicted errors in Figures 29 and 31, which indicated better accuracy for the angular displacement method. These basic experiments demonstrate the applicability of PIV for the instantaneous measurements of the three components of velocity in a plane of a fluid flow.

7.3 Suggestions for Future Work

Although some success in demonstrating the applicability of this technique on high-speed flows and three-dimensional flows has been demonstrated, much work remains to be done. For example, application of PIV to the study of

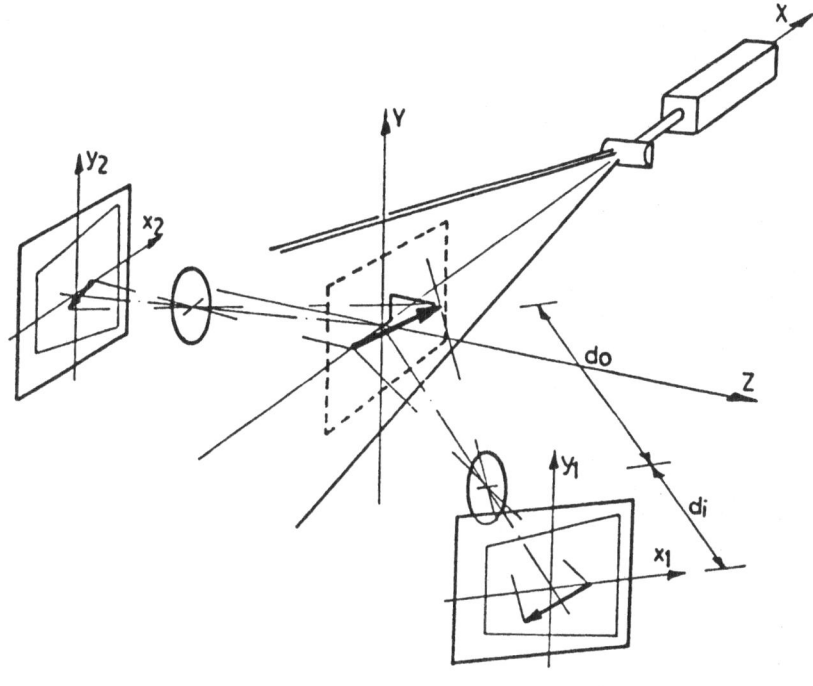

Figure 28. Stereoscopic Angular Displacement Method, from Gauthier and Riethmuller, (1988)

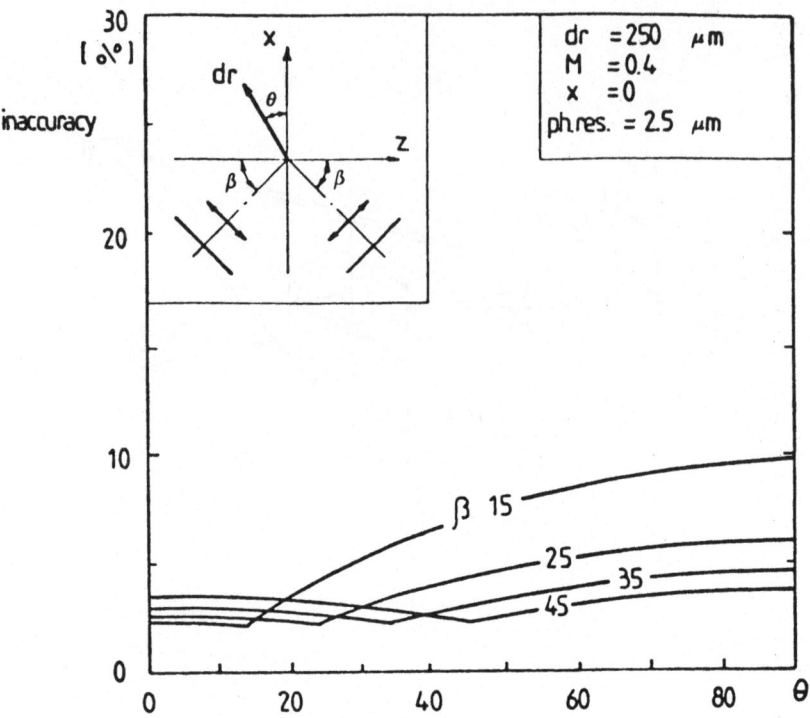

Figure 29. Inaccuracy of the Angular Displacement Method, from Gauthier and Riethmuller, (1988)

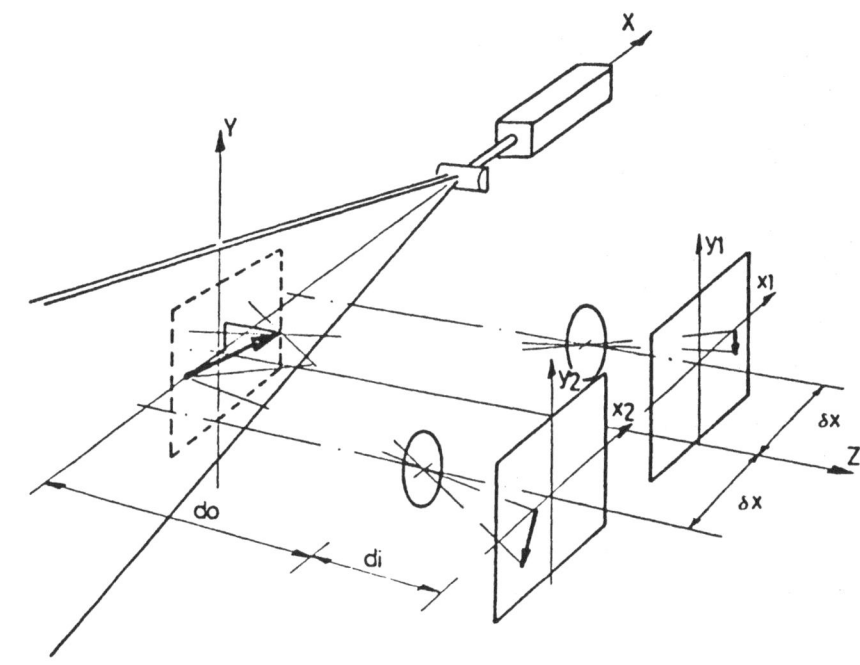

Figure 30. Stereoscopic Translation Method, from Gauthier and Riethmuller, (1988)

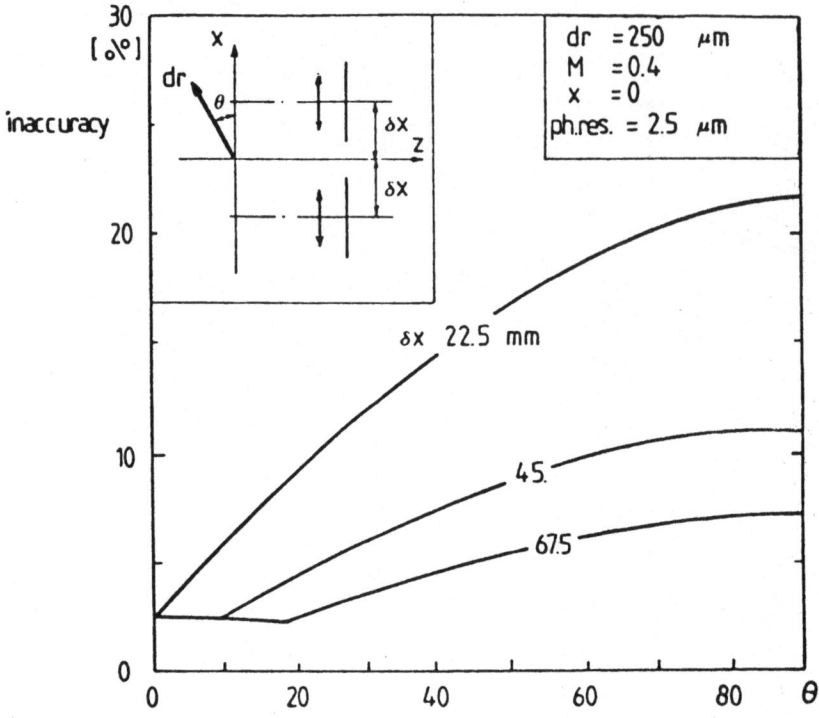

Figure 31. Inaccuracy of the Translation Method, from Gauthier and Riethmuller, (1988)

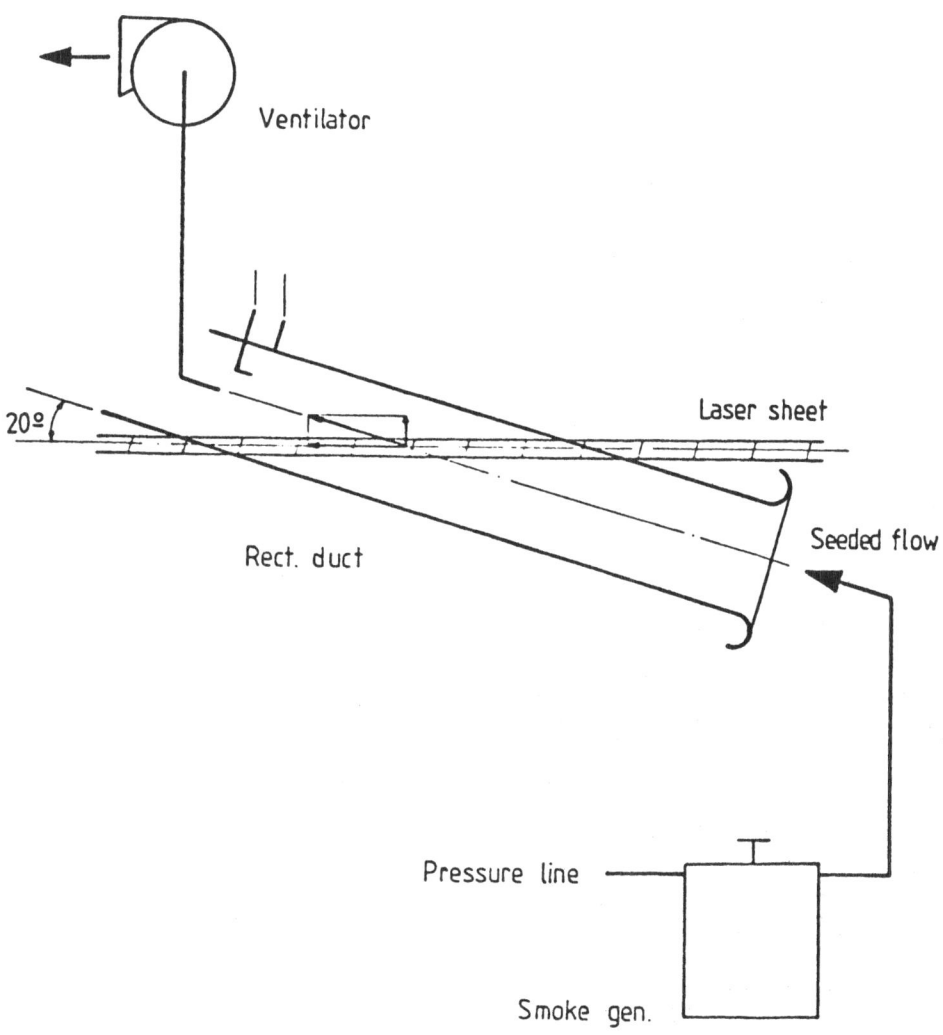

Figure 32. Experimental Setup for Uniform Out-of-Plane Motion, from Gauthier and Riethmuller, (1988)

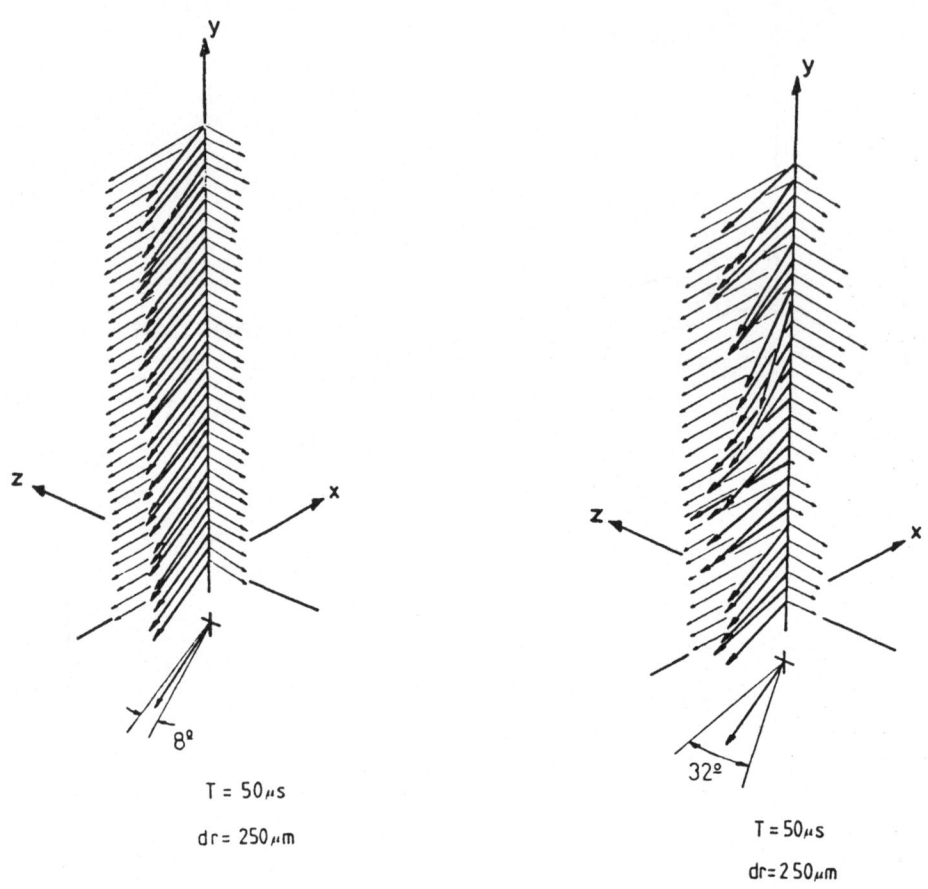

Figure 33. Scatter in Out-of-Plane Displacement Measurements, from Gauthier and Riethmuller, (1988); a) Angular Displacement Method, b) Translation Method

three-dimensional flows with large velocity gradients, such as the tip vortex wake behind an aircraft wing, will require much study. Also, the relationship between the laser sheet thickness, the flow quantities (such as velocity, spatial resolution, time and length scales), and the processing techniques needs additional study. Finally, the use of PIV as a means to validate CFD (computational fluid dynamics) codes is a fruitful area for further work.

8. Conclusions

The concept of Particle Image Velocimetry has been described and shown to be applicable to the study of flows with vortical motions. At present, most of the work in this field has been towards developing the technique itself, and very little has been done in applying this technique to fluid flows of research interest. Most of the work to date has been limited to low speed, primarily two-dimensional flows, although some research has been done to extend the technique to higher speeds and all three dimensions. The use of dual systems (i.e., two lasers to study high speed flows and two recording systems to study three-dimensional flows) looks especially promising. These activities should be continued. One area that has received very little attention to date is an effort to increase the size of the viewing area. Most research has been done on small, laboratory-type experiments and a field in which PIV would prove especially useful is the study of wind tunnel flows. A means to measure three-dimensional, unsteady, vortical flow fields in wind tunnels at high speed and high Reynolds number and on a regular basis would increase the productivity of these facilities by several orders of magnitude.

9. References

Adrian, R.J.;"An Image Shifting Technique to Resolve Directional Ambiguity in Double-Pulsed Laser Velocimetry", Appl. Opt., vol 25, pp 3855-3858, 1986.

Becker,H.A. and Massaro,T.A.;"Vortex Evolution in a Round Jet", J. Fluid Mech., vol 31, pp 435-448, 1968.

Born,M. and Wolf,E.;Principles of Optics, Pergamon Press, 1980.

Bouard,R. and Coutanceau,M.;"The Early State of Development of the Wake Behind an Impulsively Started Cylinder for $40<Re<10^4$",J. Fluid Mech., vol. 101, pp 583-607, 1980.

Bouchard,E.E. and Reynolds,W.C.;"The Structure and Growth of the Mixing Layer Region of a Round Jet", Rept. TF-17, Thermosciences Division, Dept. of Mech. Engrg., Stanford Univ., 1982.

Collicott, S.H. and Hesselink, L.;"Anamorphic Optical Processing of Laser Speckle Anemometer Data", Bull. Amer. Phys. Soc., vol 30, pg 1728, 1985.

Collier, R.J., Burckhardt, C.B., and Lin, L.H.;Optical Holography, Academic Press, New York, 1971.

Dimotakis, P.E., Debussy, F.D., and Koochesfahani, M.M.;"Particle Streak Velocity Field Measurements in a Two-Dimensional Mixing Layer", Phys. Fluids, vol 24, pp 995-999, 1981.

Elkins, R.E.,III, Jackman, G.R., Johnson, R.R., Lindgren, E.R., and Yoo, J.K.;"Evaluation of Stereoscopic Trace Particle Records of Turbulent Flow Fields", Rev. Sci. Instrum., vol 48, pp 738-746, 1977.

Erf, R.K.;"Application of Laser Speckle to Measurement", in Laser Applications, vol 4, ed. by Goodman, J.W. and Ross, M., Adademic Press, New York, 1980.

Frish, M.B. and Webb, W.W.;"Direct Measurement of Vorticity by Optical Probe", J. Fluid Mech., vol 197, pp 173-200, 1981.

Gauthier,V. and Riethmuller,M.L.;"Application of PIDV to Complex Flow:Measurement of the Third Component", VKI Lecture Series on Particle Image Displacement Velocimetry, Brussels, Mar. 1988.

Gharib, M., Dyne, B., Thomas, O., and Yap, C.;"Flow Velocity Measurements by Image Processing of Optically Modulated Traces", AGARD CP-413, 1987.

Ho,C.M.;"An Alternative Look at the Unsteady Separation Phenomenon", Recent Advances in Aerodynamics, ed. by A.Krothapalli and C.A.Smith, Springer-Verlag, pp 165-178, 1986.

Ho,C.M. and Gutmark,E.;"Vortex Induction and Mass Entrainment in a Small Aspect-Ratio Elliptic Jet", J. Fluid Mech., vol 170, pg. 383, 1987.

Honji,H. and Taneda,S.;"Unsteady Flow Past a Circular Cylinder", J. Phy. Soc. of Japan, vol 27, pp 1668-1677, 1969.

Kompenhans, J. and Hocker, R.;"Application of Particle Image Velocimetry to High Speed Flows", VKI Lecture Series on Particle Image Displacement Velocimetry, Brussels, Mar. 1988.

Krothapalli,A. Baganoff,D. and Karamcheti,K.;"On the Mixing of a Rectangular Jet", J. Fluid Mech., vol. 107, pg 201, 1981.

Landreth, C.C., Adrian, R.J., and Yao, C.-S.;"Double Pulsed Particle Image Velocimeter with Directional Resolution for Complex Flows", Experiments in Fluids, vol 6, pp 119-128, 1988.

Loc,T.P.;"Numerical Analysis of Unsteady Secondary Vortices Generated by an Impulsively Started Circular Cylinder", J. Fluid Mech., vol. 100, pp 111-128, 1980.

Lourenco, L.;"The Fundamentals and Application of Particle Image Displacement Velocimetry", von Karman Institute Lecture Series, Belgium, 1986.

Lourenco, L.;"Application of Laser Speckle and Particle Image Velocimetry in Flows with Velocity Reversal", Bull. Amer. Phy. Soc., vol 31, no 10, 1986.

Lourenco, L.M., Krothapalli, A., Buchlin, J.M., and Riethmuller, M.L.;"A Non-Invasive Experimental Technique for the Measurement of Unsteady Velocity and Vorticity Fields", Aerodynamic and Related Hydrodynamic Studies Using Water Facilities, AGARD CP-413, paper 23, Monterey, CA, 1986.

Lourenco, L.M. and Krothapalli,A.;"The Role of Photographic Parameters in Laser Speckle or Particle Image Displacement Velocimetry", Experiments in Fluids, vol 5, pg 29-32, 1987.

Lourenco, L.M. and Krothapalli,A.;"Instantaneous Velocity Field Measurements of a Turbulent Rectangular Jet (AR = 4) Using Particle Image Displacement Velocimetry", AIAA paper 88-0498, 26th Aerospace Sciences Meeting, Reno, Jan. 1988.

Lourenco, L.M. and Krothapalli,A.;"Application of PIDV to the Study of the Temporal Evolution of the Flow Past a Circular Cylinder", Laser Anemometry in Fluid Mechanics-III, Ladoan-Institute Superior Tecnico, Lisbon, Portugal, pg 161, 1988.

Lourenco,L., Krothapalli,A., and Smith, C.A.;"On the Instability of a Rectangular Jet", Intl. Symp. on Laser Velocimetry, Lisbon, Portugal, July 1988.

Lourenco, L. and Whiffen, M.C.;"Laser Speckle Methods in Fluid Dynamics Applications", Proc. Intl. Symp. on Appl. of Laser Anemometry of Fluid Mechanics, Lisbon, 1984.

Meynart, R.;"Equal Velocity Fringes in a Rayleigh-Benard Flow by the Speckle Method", Appl. Opt., vol 19, pg 1385, 1980.

Prandtl, L., J. Roy. Aero. Soc., vol. 31, pg. 730, 1927.

Simpkins, P.G. and Dudderar, T.D.;"Laser Speckle Measurement of Transient Benard Convection", J. Fluid Mech., vol 89, pp 665-671, 1978.

Smith, C.A., Lourenco, L.M.M., and Krothapalli, A.;"The Development of Laser Speckle Velocimetry for the Measurement of Vortical Flow Fields", AIAA paper 86-0768-CP, 14th Aerodynamic Testing Conf., West Palm Beach, Mar. 1986.

Stetson, K.A.;"A Review of Speckle Photography and Interferometry", Opt. Engr., vol 14, pp 482-489, 1975.

Tietjens,O.;Stromungslehre, 1st Edition, vol. 2, Springer-Verlag, Berlin, pp 105-109, 1970.

Van Dommelen,L.L.;"Unsteady Boundary Layer Separation", Ph.D. Thesis, Cornell University, 1981.

Yao, C.-S. and Adrian, R.J.;"Orthogonal Compression and 1-D Analysis Technique for Measurement of 2-D Particle Displacements in Pulsed Laser Velocimetry", Appl. Opt., vol 23, pp 1687-1689, 1984.

DYNAMIC WALL PRESSURE MEASUREMENTS

Patrick Leehey
Department of Mechanical Engineering, Room 3-264
Massachusetts Institute of Technology
Cambridge, MA   02139   USA

INTRODUCTION

The quest for information on the dynamics of wall pressure fluctuations has been motivated by two principle objectives. The first is to gain more fundamental understanding of the turbulent boundary layer, in particular, the mechanism by which it continually regenerates itself. The wall pressure measurement has the advantage of being, for the most part, non-invasive. On the other hand, it is a weighted integral of the velocity fluctuations in the boundary layer, hence, its picture of boundary layer activity must of necessity be somewhat diffuse. Pressure measurements have been combined with other measurements, in particular, fluctuating velocity and fluctuating wall shear stress. These have taken the form of long-time averages, such as in cross-correlation and cross-spectral density measurements. Short time conditional average measurements have also been taken wherein one or the other of the physical quantities has served as the trigger. A blend of the two techniques has been employed in measuring wall pressure fluctuations in the transition zone between laminar and turbulent flow. The pressure measurement is today an integral part of any serious experimental study of boundary layer dynamics.

The second objective in studying wall pressure fluctuations is structural excitation. To this end we have seen investigators turn away from cross-correlation and cross-spectral density measurements to the direct measurement of wavenumber frequency pressure spectra as being the quantity of predominant interest, particularly for underwater applications.

There have been three recent surveys of wall pressure dynamics. Two with particular emphasis on measurements are those of Willmarth (1975) and Blake (1983). The recent two-volume compendium on flow noise by Blake (1986) contains an extensive discussion of both theoretical and experimental results on wall pressure in the second volume. It would be pointless to duplicate these efforts here; they are too recent to warrant another general survey at this time. We therefore shall concentrate on a number of specific questions related to the resolution of high wavenumber components of spectra, the interpretation of cross-spectral density measurements, and the scaling of low wavenumber components of spectra. Our emphasis shall be on outstanding problems of experiment and theory and their relationship to one another. We shall refer only to those papers that specifically deal with the questions at hand and make apologies at this time to the many substantial contributions that are somewhat apart from the central themes of this paper. We also shall resist the temptation to expand our study to a number of interesting related phenomena, such as wall pressure fluctuations in the transition zone, acoustic radiation from the transition zone, wall pressure fluctuations behind backward-facing steps, and the effects of mean pressure gradients and surface roughness on wall pressure. These matters are of substantial practical interest in their own right, but to deal with them would lead us too far astray from our major objective.

MEASUREMENTS AT HIGH WAVENUMBERS

The high wavenumber portion of the wavenumber-frequency wall pressure spectrum $\phi_p(k_1, k_3, \omega)$ is of primary interest in the determination of the mechanism of turbulence generation. The low wavenumber portion on the other hand, is more responsible for structural excitation. It has not been customary to attempt to measure this high wavenumber portion directly. Since the high wavenumber components are believed to be generated by eddies convecting in the equilibrium layer, it has been customary to predict their behavior either directly from the frequency spectrum using a measured convection velocity and Taylor's hypothesis or by Fourier transforming in space the measured cross-spectral density $\phi_p(r_1, r_3, \omega)$. Generally, only the streamwise quantities k_1 and r_1 are so treated.

It has long been known that the finite size of a pressure transducer limits our ability to resolve the high wavenumber portion of the spectrum. In essence, a flush-mounted sensor cannot resolve pressure scales that are smaller than its effective diameter. A number of corrections have been devised to account for this lack of resolution, e.g., Corcos (1963) or Willmarth and Roos (1965). These resolution analyses have been based upon assumed forms of the wavenumber-frequency spectrum, especially for wavenumbers near the convective ridge, $k_1 = \omega/U_c$. Since these assumptions are based upon measurements of cross-spectra which themselves have been measured by pairs of transducers of limited rersolving capability, the process is akin to lifting oneself by one's own bootstraps. Because of this difficulty experimentalists have continued to develop and utilize transducers with smaller and smaller effective sensing diameters. To a certain extent, this effort has been somewhat self-defeating. For as sensitivity decreases with size, it is necessary to test at higher and higher freestream velocities, i.e. Reynolds numbers, which resulted in a further decrease in the scale size of eddies. Generally these transducers were specially developed, one-of-a-kind type, either piezoelectric, Bull and Thomas (1976), or of the solid dielectric (Sell) type, Schewe (1983). Other investigators, notably, Blake (1970), Emmerling (1973), Burton (1974) and Farabee and Geib (1975), endeavored to maintain sensitivity while at the same time reducing the effective transducer diameter. These latter investigators used the same type of transducer, a Bruel & Kjaer 1/8" condenser microphone. This transducer has excellent pressure sensitivity and very low acceleration sensitivity, qualities prized for wind tunnel testing. For wall pressure measurements the conventional slotted cap on this transducer is replaced by a solid cap with a single pinhole drilled in the center. This pinhole has a diameter of 1/32". There is also a small cavity between this cap and the metal diaphragm of the condenser microphone. As a result, a Helmholtz resonator is set up between the pinhole and the cavity. The measured frequency of this resonance is approximately 17kHz, Blake (1970). Generally this frequency is much too high for the Helmholtz resonator to have any direct effect upon measured wall pressure spectra. There is one weakness to the arrangement, however. During humid weather moisture tends to collect under the cap, causing intermittent breakdown in the air dielectric with resultant intermittent high frequency bursts of noise.

The use of the pinhole microphones revealed for the first time the very substantial high frequency content of the wall pressure spectrum, see Willmarth (1975) for a detailed discussion. These results, however, were brought into question by Bull and Thomas (1976). Briefly, they measured the wall pressure spectrum using four transducer configurations of the same effective diameter: 1/32". The transducers were

a) a Bruel & Kjaer 1/8" microphone with 1/32" pinhole cap;

b) a specially designed flush-mounted piezoelectric transducer without a cap;

c) the piezoelectric transducer with a 1/32" cap; and

d) the piezoelectric transducer with a cap with the the pinhole filled with a silicon grease.

They found that cases (a) and (c) gave the same frequency spectra but that cases (b) and (d) gave frequency spectra that were suppressed in the high frequency ranges by as much as 5dB. They noted further that the presence of a cavity under a pinhole cap had no effect upon response. From this they concluded that the use of a pinhole cap caused an interaction with the turbulent boundary layer that was not related to a Helmholtz resonance but was presumably related to the local removal of the no-slip boundary condition.

   One cannot take issue with the actual experiments performed by Bull and Thomas (1976) as more fully described by Thomas (1977). They appear to have been done with a meticulous attention to calibration, both in a shock tube and in a comparative acoustic calibrator. Two troublesome matters, however, remain. The first is

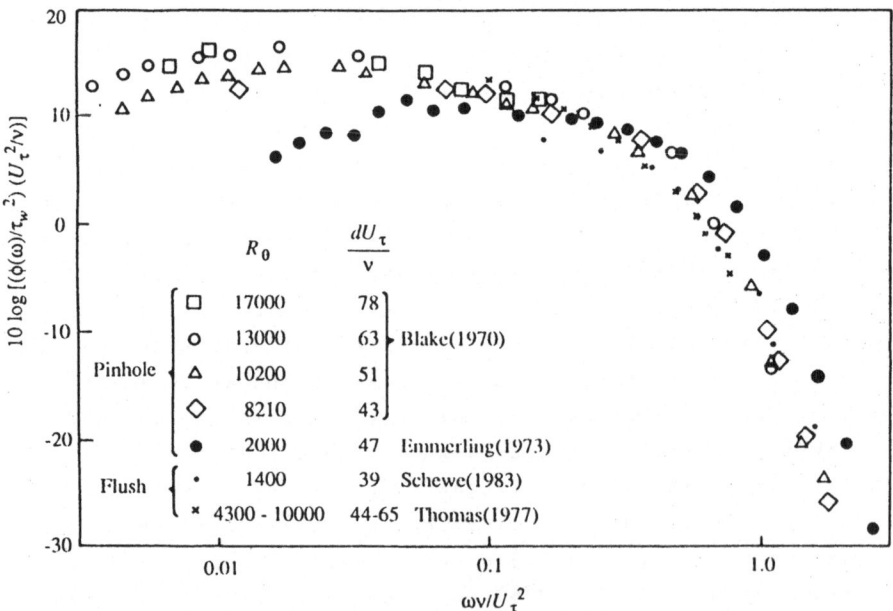

Figure 1. Wall pressure spectra measured with two types of transducers.

that a comparison of spectra measured with pinhole capped and flush
mounted transducers does not show a clear distinction between the
results for the two classes of transducers. Figure 1 shows a
comparison of measurements of four investigators, two using the same
type of B & K 1/8" microphone with a 1/32" pinhole cap and two using
flush mounted transducers. All measurements were made in
essentially zero pressure gradient turbulent boundary layers over a
range of Reynolds number $U\Theta/\nu$ where Θ is the momentum thickness.
The spectra are non-dimensionalized using wall variables u_τ and
ν. Such variables are generally considered preferable for high
wavenumber components of the spectrum generated in the equilibrium
portion of the boundary layer. Blake's and Emmerling's measurements
were done with the pinhole microphone. Schewe's measurements, were
carried out with a specially designed Sell microphone. Bull and
Thomas used a piezoelectric transducer. The effective diameters,
d^+, covered nearly the same range. It is interesting to note that
Blake's spectra are actually somewhat lower than that of Schewe at
high frequency in spite of the fact that Blake used a pinhole
transducer and Schewe used a flush transducer. Clearly factors
other than the pinhole are involved.

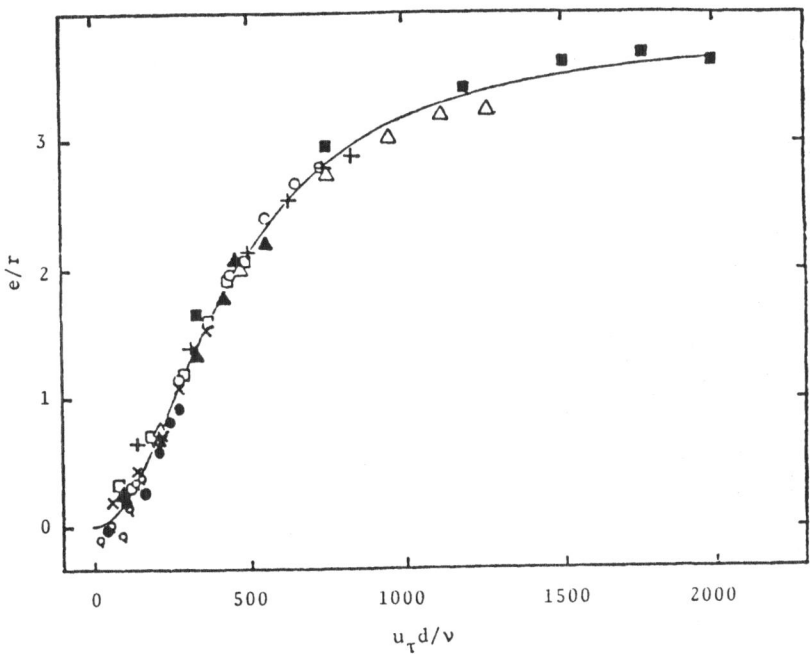

Figure 2. Apparent pressure increase e as a function of orifice
diameter d, from Franklin and Wallace (1970).

The second matter is the difficulty in establishing a physical basis for the interaction caused by the pinhole microphone with the turbulent boundary layer. It is rather well known that an orifice in a wall beneath a turbulent boundary layer measures a mean pressure higher than the true mean pressure at the wall. Figure 2 taken from Franklin and Wallace (1970) shows the apparent pressure increase e divided by the mean wall shear stress τ_w as a function of the orifice diameter d^+ in viscous units. Franklin and Wallace do not give a physical explanation for this effect. It can be demonstrated qualitatively by utilizing the results of triple deck analysis of laminar flow over a trailing edge. Messiter (1971) shows that the pressure distribution in the outer (potential) deck of the flow downstream of a trailing edge is given by the expression

$$\frac{\overline{p}}{q} \sim \frac{A_1 R_L^{-1/2}}{3\sqrt{3}} x^{-2/3}, \qquad A_1 = 1.2881$$

where the Reynolds number, R_L is based upon the plate length L and q is the dynamic head. This positive pressure is impressed across the decks to the wake centerline.

Upstream of the trailing edge, but not immediately adjacent to it, the wall shear stress τ_w is that of Blasius flow:

$$\frac{\tau_w}{q} = \frac{0.664}{\sqrt{R_L}} \quad .$$

From these equations, we find that the mean pressure averaged over a distance d downstream of the trailing edge can be written in the form

$$\frac{\overline{p}}{\tau_w} = \text{const.} \frac{1}{d} \int_0^d x^{-2/3} dx$$

and is independent of the Reynolds number R_L. The fact that the integral diverges is of no consequence; it is well known the triple deck solution fails in the immediate vicinity of the trailing edge where it must be matched to a full solution of the Navier-Stokes equations in a domain of radius $R_L^{-3/4}$. Although the pressure singularity at the trailing edge is removed by the analysis of Hakkinen and O'Neil (1967), the task of obtaining a full solution for a three-dimensional orifice remains formidable. Nevertheless, the major results are clearly correct; that is, there is a pressure increase behind the trailing edge and that it scales on the upstream wall shear stress. Since this pressure increase is associated with the convergence of the mean flow streamlines behind the trailing edge, it might be conjectured that in the case of the turbulent boundary layer, this process could carry the active equilibrium portion of the inner boundary layer in closer to the transducer, thus producing greater apparent components of the high frequency spectrum. The difficulty, however, is evident in Figure 2, wherein it is clear that the effect on mean pressure is negligible for the effective transducer diameters under consideration ($19 < d^+ < 45$) and hence would likely be unimportant for fluctuating pressures as well.

It would seem reasonable that the behavior of pin-holed microphone might be explained by experiments wherein the impedance of an orifice subject to grazing flow was measured, Ronneberger (1972), Kompenhans (1976) and Kompenhans and Ronneberger (1980). In these experiments the orifice impedance was measured by two different techniques. One involved the measurement of the pressure drop across the orifice and the velocity in the orifice. In order to measure the orifice velocity it was necessary to supply a steady flow through the orifice in order to utilize the hot wire anemometer. This was avoided in the second technique where an impedance tube was used on the side of the orifice away from the grazing flow. The results were reported in terms of the difference in orifice impedance from the grazing flow to the no-flow case. In all cases the total orifice resistance was positive, although grazing flow did reduce resistance slightly at high reduced frequencies. Since the studies were done for turbulent as well as for laminar boundary layers, it is difficult to reconcile these results with the findings of Bull and Thomas. A parenthetical remark is in order here, not because it sheds light on the performance of the pinhole microphone, but because it illustrates the subtleties in the experiments which must be considered. Kompenhans and Ronneberger measured the dc orifice resistance as a function of grazing flow velocity and, in fact, found that it agreed remarkably well over a wide reduced frequency range with the impedance tube measurements, indicating a simple quasi-steady law for acoustical resistance of the form

$$R \sim 0.17 \rho_0 U_\infty$$

would be valid for reduced frequencies $\omega a/U_\infty < 0.1$ and for ratios of displacement boundary layer thickness δ^* to orifice radius a greater than 0.4. Oddly enough they did not report the pressure drop across the orifice as a function of grazing flow velocity with no mean flow through the orifice. They simply inferred this as being zero at the limit of zero flow through the orifice, see Figure 3. Yet, for their test conditions where d^+ varied from about 160 to 2000, the pressure increase found by Franklin and Wallace would have completely swamped the pressure differences reported. The only possible conclusion that one can reach is that their static pressure tap by chance or design had the same diameter in viscous lengths as the orifice under study.

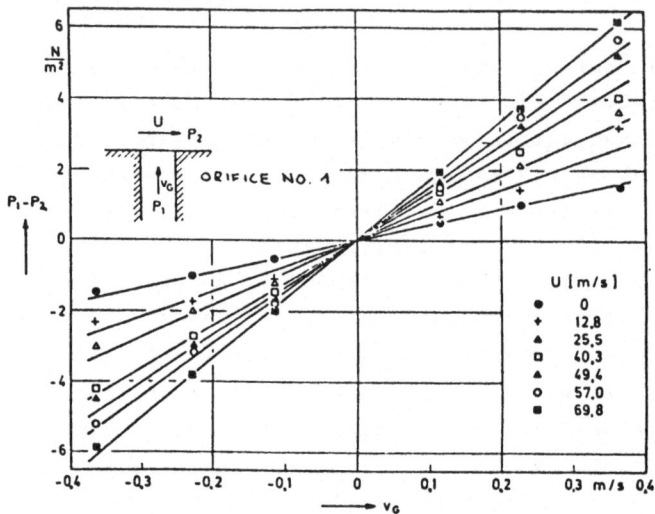

Figure 3. Pressure difference between both sides of the orifice (δ^*~1.9mm). Steady state flow resistance as a function of the flow velocity, from Kompenhans and Ronneberger (1980).

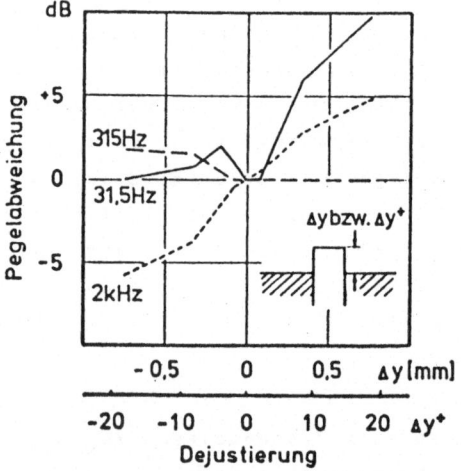

Figure 4. Effect of transducer proudness on one-third octave wall pressure levels, d^+~52, from Langeheineken and Dinkelacker (1977).

We turn our attention now to other effects which may influence the accuracy of measurement of the high wavenumber components of wall pressure spectra. The first is a question of "proudness" of any flush mounted transducer. Figure 4 is taken from Langeheineken and Dinkelacker (1978). The 1/8" Bruel & Kjaer microphone without any cap was mounted in the wall beneath a turbulent boundary layer and the protrusion or proudness of the microphone varied through positive to negative values. The influence of these variations upon the wall pressure spectrum is shown in three 1/3 octave bands as a function of proudness measured in both millimeters and in viscous lengths. Clearly a flush mounted transducer must be extremely fair in the wall, especially in high velocity turbulent boundary layers.

Since the high wavenumber portion of the wall pressure spectrum scales best on inner viscous parameters, it is important that one obtains a very accurate measure of the mean wall shear stress in the neighborhood of the pressure measurement point. Most shear stress measuring devices are non-linear. These include the surface fence, the Preston tube, the Stanton tube and the hot film anemometer. Since the dynamic wall shear stress under a turbulent boundary layer is approximately 35% of the mean (with an error of $\pm$ 20% depending on the measurement technique) it is clear that any non-linear device will bias the mean severely. One must therefore calibrate in a turbulent flow environment such as in fully developed pipe flow and hope that the test environment has similar dynamics. Measurement of the mean wall shear stress by determining the terms of the momentum integral seem to be recommended primarily by those who have never attempted to do this. Finally, there is the simple technique and still one of the best, of fitting a semi-log plot of the mean velocity profile to the Karman slope. This works reasonably well if you have an excruciatingly accurate measure of the distance of your hot wire from the wall. Even then, one must bear in mind that the Karman slope itself is known to at best two significant figures. The scatter in the data in Figure 1 could easily be attributed to a 5% error in the measurement of mean wall shear stress.

Finally, there is a question as to whether or not the above scaling of wall pressure spectra on inner variables remains valid at very high Reynolds numbers. Efimtsov (1984) has assessed a very large body of Soviet measurements of wall pressure spectra. These cover a Mach number range from 0.015 to 4.0. He finds no Mach number effect that is independent of its influence upon wall quantities which go to make up the Reynolds number $Re = u_\tau \delta/\nu$. As illustrated in Figure 5 he finds that for $Re \gg 3 \times 10^3$, the spectra

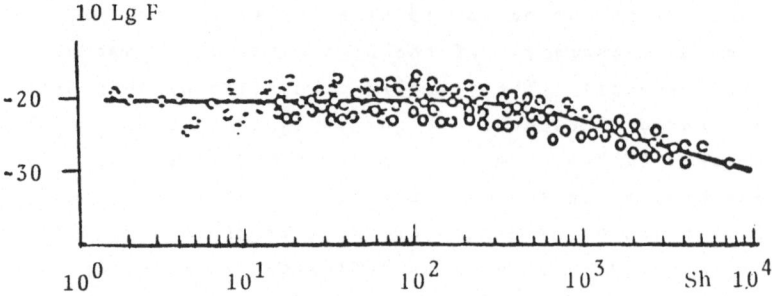

Figure 5. Wall pressure spectral density at high Reynolds number
$Re=u_\tau\delta/\nu$; $F=\phi(\omega)/u_\tau^3\rho^2\delta$, $Sh=\omega\delta/u_\tau$, from Efimtsov (1984).

collapse on mixed parameters $\Phi_p(\omega)/u_\tau^3\rho\delta$ and $\omega\delta/u_\tau$. These spectra
have been corrected for the effect of transducer size,
unfortunately, by methods reported in publications presently
unavailable to the author. For lower Reynolds numbers the high
frequency spectra collapse on the usual inner parameters. The low
frequency portions of the spectra do not, these being presumably
influenced by the large eddies in the outer flow.

INTERPRETATION OF CROSS-SPECTRAL
DENSITY MEASUREMENTS OF WALL PRESSURE

Longitudinal and lateral cross-spectral density measurements of
wall pressure fluctuations serve two useful purposes. First, they
are essential to any procedure for determining the resolution of a
transducer of finite size. Second, they are useful for determining
structural response to turbulent boundary layer excitation when the
principal mechanism governing the response is that of hydrodynamic
coincidence (see the next section for a more complete discussion).
In this respect they are markedly superior to the older space-time
correlation measurements. With questions of structural response and
reradiation, it is what is going on in a given frequency band that
is of practical interest. All too often correlation measurements
are contaminated by extraneous frequency components.

There are two aspects of the cross-spectral density
measurements which to this day are still only partially understood.
The first has to do with the determination of streamwise convection
velocity. This is determined by setting the phase of the
longitudinal cross-spectrum equal to $\omega r_1/U_c$ where r_1 is the
longitudinal separation of the transducers. Hopefully, the

convection velocity U_c is principally a function of frequency ω only and not of r_1, for otherwise the process would be a pointless exercise. Figure 6 shows the result of such measurements by Bull (1963) and Blake (1969). Bull's measurements were done in 1/3 octave frequency bands with transducers of an effective diameter d^+ ~ 170. Blake's measurements on the other hand, were carried out using a 5Hz bandwidth filter up to a frequency of 250Hz and a 50Hz bandwidth filter for higher frequencies. His transducer had a much smaller effective diameter, d^+ ~ 45. Although there is a good correspondence between the results of the two sets of experiments for $\omega\delta^*/U_\infty > 1.5$, below this frequency the results diverge dramatically. Blake (1970) has argued that the reason for this discrepency lies in the inherent dispersiveness of the wall pressure spectrum and its influence upon the differences in measurement technique. In essence, Bull measured a group velocity U_{cg} ~ $\Delta\omega/\Delta k$ where $\Delta\omega$ is the 1/3 octave bandwidth and Δk is proportional to the reciprocal of d^+. Blake, on the other hand, measured something much nearer to a phase velocity U_{cp} ~ ω/k_1. Further, it is evident that the phase velocity is increasing with frequency for $\omega\delta^*/U_\infty < 0.5$ and hence the group velocity must be greater than the phase velocity in this range. One should also note that Blake's phase velocities increase with streamwise transducer separation r_1/δ^* at low frequencies.

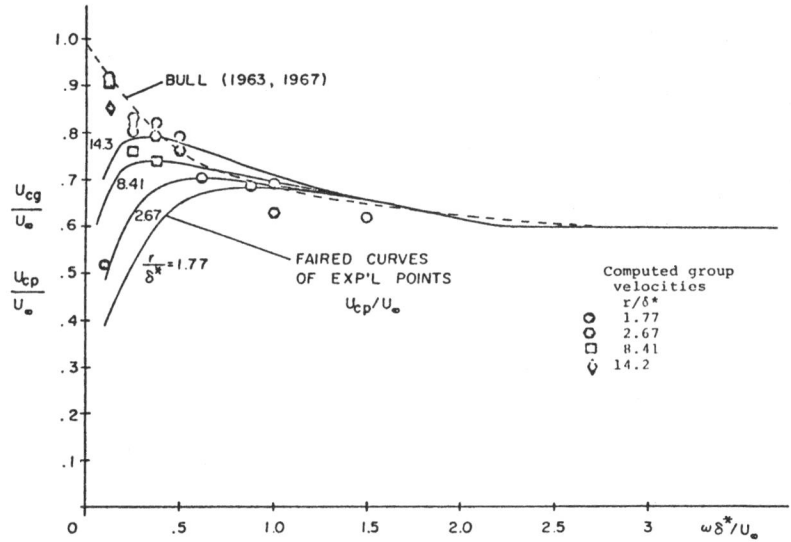

Figure 6. Comparison of phase and group convection velocities, smooth wall, from Blake (1969).

Another question has to do with the measurement of the normalized amplitude functions for the longitudinal and the lateral cross spectral density, see Figures 7 and 8 for the results of Bull (1963). These have been plotted in terms of the so-called Corcos

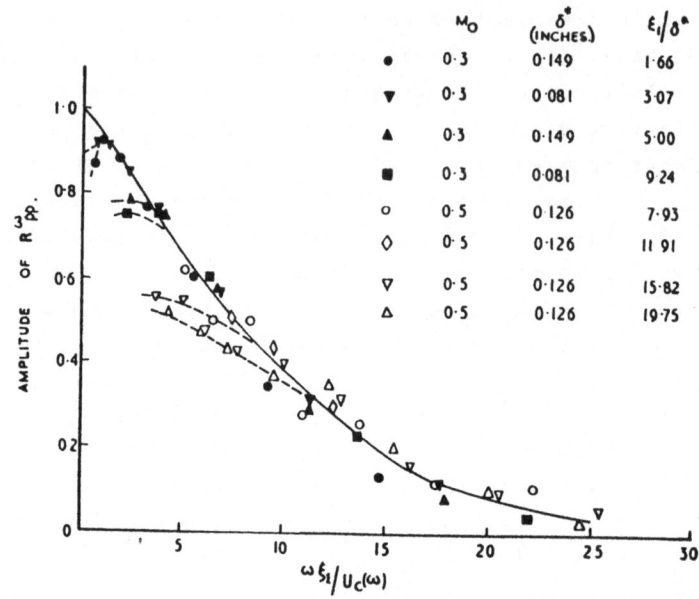

Figure 7. Amplitude of narrow-band longitudinal space-time correlations of the wall-pressure field, from Bull (1963).

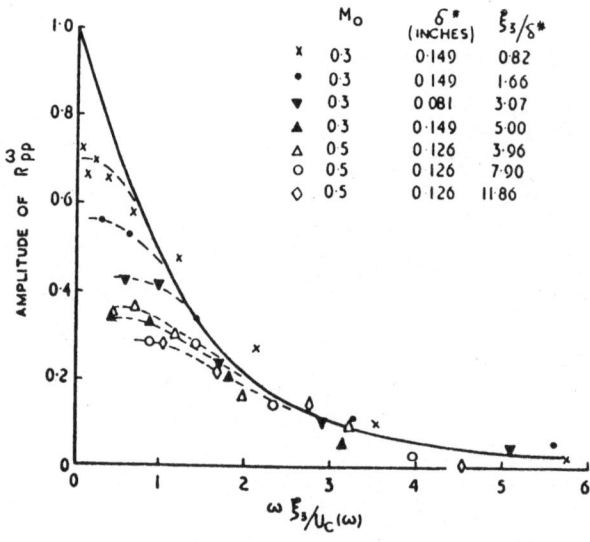

Figure 8. Amplitude of narrow-band lateral space-time correlation of the wall-pressure field, from Bull (1963).

(1963, 1967) similarity parameters $\omega\xi_1/U_c$ and $\omega\xi_3/U_c$. In both cases at low frequencies the amplitude functions decrease with increasing transducer separations. Blake's data on the other hand, showed no such dependencies on separation distance. Of course, one would expect the amplitude function curves to shift left at low frequencies if the group convection velocity rather than the smaller phase convention velocity is used to make the frequency non-dimensional. However, the phase velocity is closer to the group velocity for large separation distance r_1/δ^*, an effect opposite to that which would be required to explain Bull's results in Figures 7 and 8. Recently, Moller (1987) made additional cross-spectral density measurements using methods similar to those of Blake but for significantly greater separations. He found a tendency for the the amplitude functions to fall off at low frequencies in a fashion similar to that seen by Bull. It is clear then that the Corcos similarity representation

$$\phi_p(r_1,r_3,\omega) = \phi_0(\omega)\ A(\frac{\omega r_1}{U_c})\ B(\frac{\omega r_3}{U_c})\ \exp(\iota\omega r_1/U_c)$$

is inadequate. The A and B functions must also be dependent on r_1/δ^* and r_3/δ^*, respectively. Clearly, as r_1 approaches zero, the A amplitude function must approach one. However, this is not necessarily so if the frequency ω alone approaches zero for finite r_1. This difference in limiting processes is masked when only the similarity parameter $\omega r_1/U_c$ is considered. A corresponding remark applies to the lateral amplitude function. This dependency on spatial separation is probably not of too great consequence when one uses Corcos' form in determining the resolution of small transducers, Corcos (1963), but it clearly would not be appropriate to Fourier transform the same form in space in order to obtain a wavenumber-frequency spectrum. Attempts to do this have generally yielded wavenumber-frequency spectra that are too high even fairly close to the convective ridge.

LOW WAVENUMBER COMPONENTS OF TURBULENT BOUNDARY LAYER WALL PRESSURE FLUCTUATIONS

The principle interest in the low wavenumber components of turbulent boundary layer wall pressure fluctuations lies in their role as a source of structural excitation and subsequent reradiation of acoustic noise. This subject merits a discussion of its own, see

Leehey (1988). Here we shall simply state the main considerations. Assuming the frequencies in the audio range are of primary interest, at high flow Mach numbers it is possible to have coincidence between the convection speed of the wall pressure and the bending wave speeds of the resonant structural modes. When this occurs there is very substantial power flow from the boundary layer to the structure. This mechanism, termed hydrodynamic coincidence, is of considerable importance when dealing with aircraft cabin noise or with vibration of internal electronic components of rockets. In underwater applications, the frequencies of interest are approximately the same but the Mach number of the mean flow is two orders of magnitude less. At the same time the scantlings of a sonar dome are not terribly different from those of an aircraft hull structure. Consequently, the hydrodynamic coincidence mechanism is (ironically) unimportant in underwater applications. Since substantial sonar self-noise appears to be attributable to the turbulent boundary layer over the sonar system, we must look for another mechanism. Attention has therefore by default been directed towards the low wavenumber components of the wall pressure spectrum. They seem to be "the only game in town." As we shall see, they are exceedingly weak, approximately 35dB below the level of the convective ridge, $k_1 = \omega/U_c$, but they are capable of driving effectively both resonant and non-resonant structural modes with long bending wavelengths. In a sense, this is the converse of the resolution question discussed earlier.

The low wavenumber components have received substantial theoretical attention over the past quarter of a century. Phillips (1955) and Kraichnan (1956) considered the limit process where first the Mach number approached zero and then the wavenumber magnitude approached zero. For this limit process the wavenumber magnitude of the wall pressure spectrum approached zero as the square of the wavenumber. An equivalent statement is that the integral correlation area must vanish. Should a Taylor hypothesis based on freestream velocity be appropriate to this circumstance, the result would also say that the low frequency spectrum should vanish as the square of the frequency. Considerable experimental effort was expended early on to verify the last prediction. Indeed, in one or two cases it appeared that the frequency spectrum did tend to turn down as frequency was reduced to very low frequency. The alternative limiting process is to let the wavenumber approach zero and then let the Mach number approach zero. Compressibility remains important in this process for one passes the acoustic wavenumber

$k_{ac}=\omega/c$ on the way to zero wavenumber. In general, the limit processes are not interchangeable, see Chang and Leehey (1979) for an illustration in the closely related physical problem of predicting the radiation impedance of a resonately vibrating plate subjected to low Mach number grazing flow. Furthermore, it is the second limiting process which is physically meaningful in application. It becomes necessary to consider acoustic radiation from the boundary layer.

Powell (1960) made a prediction of the acoustic radiation to the far field away from a turbulent boundary layer over a rigid infinite plane wall. By neglecting oscillatory shear stress gradients at the wall he was able to construct a mirror image flow on the far side of the wall which permitted him to remove the wall entirely leaving two adjacent blobs of turbulence--implicitly bounded in extent. The Lighthill (1952) theory could then be applied. It states that the intensity of farfield radiation followes a M^8 law. Clearly this result does not apply to points on the image plane between the blobs, for the intensity there is not proportional to the mean square pressure, nearfield components undoubtedly dominate.

For predicting the pressure fluctuations at the wall the same formalism has been used. An important difference, however crept in. Because of the slow downstream growth of the turbulent boundary layer it has been customary to consider it as approximately homogeneous in planes parallel to the wall. Therefore to predict the pressure at a point on the wall one is led to consider excitation from a layer of turbulent fluid of finite thickness but of infinite extent over the wall. This leads one immediately to the well-known Olber paradox: Since the amplitude at the reception point is proportional to the reciprocal of the distance the source is away but the perimeter of sources increases directly as the distance, the integrated effect over the plane is to produce an unbounded pressure at the reception point, provided the sources are incoherent. It is not possible to determine quantitative levels of low wavenumber wall pressure components for wavenumbers at and below the acoustic wavenumber under these circumstances.

The last proviso is necessary, for it is quite possible to obtain bounded pressure at the reception point where there is infinite plane of coherent sources. For instance, consider the Rayleigh formula:

$$p(\vec{x},t) = \frac{\rho_0}{2\pi} \int_S \frac{v_t(\vec{y},t-r/c)}{r} d\vec{y} \; ,$$

$r = |\vec{x}-\vec{y}|$, for the pressure p at point $\vec{x}$ from a normal velocity field v over a portion S of an otherwise infinite plane rigid baffle. Move $\vec{x}$ to the plane and let S be a rigid piston of infinite extent oscillating at velocity $v_0 e^{-\iota\omega t}$. Writing $p(\vec{x},t) = p_0 \epsilon^{-\iota\omega t}$, we have

$$p_0 = \frac{-\iota\omega\rho_0 v_0}{2\pi} \int_S \frac{e^{\iota k r}}{r} d\vec{y} \; , \quad k = \omega/c \; ,$$

$$= -\iota\omega\rho_0 v_0 \int_0^\infty e^{\iota k r} dr$$

in polar coordinates. But in terms of generalized functions

$$\int_0^\infty e^{\iota k r} dr = \iota k^{-1} + \pi\delta(k) \; ,$$

Jones (1966, p. 469), hence

$$p_0 = \rho_0 c v_0$$

the expected plane wave result. Thus when distant sources, $v_0 e^{-\iota\omega t}$, are completely coherent, the pressure is bounded at the measurement point.

Bergeron (1973) showed that the wavenumber spectrum could be bounded at the acoustic wavenumber by considering a finite domain of turbulence. Ffowcs Williams (1982) extended his own (1965) results and those of Bergeron to obtain the following representations of the effect of compressibility for various ranges of low wavenumbers. The wavenumber-frequency spectrum can be written for

$$k_\alpha^2 \ll \omega^2/c^2 \ll \Delta^{-2}$$

as

$$P^*(k_\alpha,\omega) \sim \rho_0^2 U^3 \Delta^3 (U/c)^2 (\omega\Delta/U)^2 \; F_1(\Delta k_\alpha, \omega\Delta/U)$$

where Δ is the boundary layer thickness, U the free stream velocity and k_α is a two-dimensional vector in the plane of the wall. The non-dimensional spectrum F_1 is devoid of compressibility effects and is presumed non-vanishing as $k_\alpha^2 \to 0$. Then P^* is proportional to the square of the Mach number $M = U/c$ in this limit.

For $\omega^2/c^2 \ll k_\alpha^2 \ll \Delta^{-2}$,

$$P^*(k_\alpha,\omega) \sim \rho_0^2 U^3 \Delta^3 (\Delta k_\alpha)^2 F_2(\Delta k_\alpha, \frac{\omega\Delta}{U})\ .$$

This is the incompressible limit of Phillips and Kraichnan. Finally, in the neighborhood of the acoustic wavenumber ω/c,

$$k_\alpha^2 \sim \omega^2/c^2 \ll \Delta^{-2},$$

$$P^*(k_\alpha,\omega) \sim \rho_0^2 U^3 \Delta^3 \ln(\frac{R}{\Delta})(\frac{\omega\Delta}{c})^4 \delta\{(\Delta k_\alpha)^2 - (\frac{\omega\Delta}{c})^2\} F_3(\Delta k_\alpha, \frac{\omega\Delta}{c})$$

where the turbulence is limited to within a radius R and a near field of radius Δ has been excluded, both measured from the reception point. Again the non-dimensional spectra F_2 and F_3 and presumed to be innocuous.

The presence of the Dirac delta function in this last formulation has been termed by Ffowcs Williams as an expression of "the individual resonance structure of acoustically coincident elements." From another point of view, it seems that it has nothing to do with resonance, but is a evident result of the extraction of all nearfield components from the excitation. Thus we have the spatial "pure wave" equivalent of the temporal "pure tone," which must appear as a delta function in the wavenumber spectrum. It is perhaps worth mentioning that this form of the spectrum could never be measured by any line array or similar wavenumber filter as such filters measure the trace wavenumber rather than the magnitude $|k_\alpha|$ of the wavenumber and consequently will produce a result which spans all wavenumbers from zero wavenumber up to the acoustic wavenumber.

Howe (1987) has performed an independent analysis of the behavior of the spectrum in the neighborhood of the acoustic wavenumber. His results appear similar to those of Ffowcs Williams with the exception that the excluded nearfield region is of the order of an acoustic wavelength rather than of the order of a boundary layer thickness. Ffowcs Williams (1982) extended the above results to incorporate a Corcos similarity hypothesis in the analysis, applying this hypothesis to the turbulent sources rather than to the wall pressure field itself. This results in the wavenumber-frequency spectrum decaying for fixed wavenumber as ω^{-2}. As we have mentioned in the previous section, the Corcos hypothesis does not seem to be appropriate for wavenumbers of the order of the reciprocal of the boundary layer thickness. It seems that future analyses should concentrate more on this domain than simply the effects of compressibility. Howe (1987) has also considered the role of surface curvature in bounding the wavenumber spectrum in the vicinity of acoustic wavenumber accounting for the creeping transmission of grazing acoustic waves over the curved surface. As we have suggested earlier, another mechanism, that of requiring coherence of sources in the far distance would also provide a bound on the spectrum.

Attention has also been given to the heretofore neglected wall shear stress fluctuations. These seem to have a chameleon behavior and the theoretical results are a mixed bag. The shear stress fluctuations on one hand can be considered as dipole radiators with their principle axes in the plane of the wall. On the other hand, they may be considered to dissipate radiation through the mechanism of the viscous mode in the immediate vicinity of the wall. Landahl (1975) predicted them to be an important source of radiation, Howe (1979) found only the dissipation mechanism to be significant, Haj-Hariri and Akylas (1985) seem to take a position somewhat in between. They find a small contribution to low wave number wall pressure from shear stress dipoles.

There is an essential difference in the analysis of Landahl from those of Howe and HajHariri and Akylas. Landahl's analysis is constructed in such a way as to approximately determine the effect of sublayer bursting upon wall pressure. The other analyses do not specifically treat this mechanism. Although none of the analyses establish quantitative levels, the differences in approach affect the scaling laws for the wall pressure spectrum. Landahl estimates the intensity of acoustic radiation from the Lighthill stresses in the boundary layer as $I \sim \tau_w u_\tau M_\tau^5$ where M_τ is the Mach number based on friction velocity. The intensity of acoustic radiation from the wall shear stress contributions is estimated as $I \sim \tau_w u_\tau M_\tau^3$. Neither of these estimates give any indication of directivity, nor do they account for absorption of grazing acoustic waves from these sources at the wall. Neglecting these effects, we see that the contribution from distant turbulent boundary layer bursts to the mean square wall pressure must be of the order $\tau_w^2 M^4 (u_\tau/U_\infty)^4$ for the Lighthill stresses, or the quadrupole sources, and of the order $\tau_w^2 M^2 (u_\tau/U_\infty)^2$ for the fluctuating wall shear stress, or dipole sources. Since these are radiation effects, they must apply to wavenumbers at and below the acoustic wavenumber. Clearly the radiation from wall shear stress sources dominate that from Lighthill stresses at low Mach numbers. We shall refer to this scaling again in the section immediately following on experimental results for the low wavenumber wall pressure spectrum.

One especially useful theoretical result is by Ffowcs-Williams (1965) in which he points out that for sources within the boundary layer one can expect that if their wavenumbers are greater than the acoustic wavenumbers, they experience exponential decay towards the wall. This characteristic, well known in many acoustic radiation problems, is of considerable utility in dealing with problems of sonar self-noise. For example, it is not unrealistic to place what is known as an outer decoupling coating between a transducer system and the exciting wall pressure field. One can expect in the absence of other difficulties, such as discontinuities in either the inner or outer structures, that there will be a very substantial exponential decay of the exciting wall pressure fields. This together with the reduction of self-noise which is inherent in the inability of a transducer of finite size to resolve pressures from small scale turbulent eddies are perhaps the two most effective mechanisms by which sonar self-noise resulting from turbulent boundary layer excitation is presently controlled. Unfortunately neither technique is effective for excitation wavenumbers at and below the acoustic wavenumber.

Two principle techniques have been used for the measurement of the wavenumber-frequency spectrum of wall pressure. One is to mount a thin plate or membrane flush in the wall of the wind tunnel test section and to measure the response of this structure to turbulent boundary layer excitation. The response is measured at the resonant frequencies of the structure, either by a small accelerometer or by a non-contact displacement gauge. One also requires the precise knowledge of the modal pattern and the damping of that mode at resonance. Frequently the plate is elongated in the flow direction to the extent that only the zero lateral wavenumber component contributes to the response. As many as thirty longitudinal components have been measured. For use of this method, see Martin and Leehey (1977), Jameson (1970) and particularly Martin (1976). The other technique is to use a flush-mounted array of microphones, usually of 1" diameter, set as close together as possible, usually aligned in the streamwise direction. Generally the outputs are summed with alternating phase. The array typically consists of six or twelve elements. Again, in the usual measurement, only the zero lateral wavenumber contributes. For applications of this technique, see Farabee and Geib (1975) or Martini, Leehey and Moeller (1984). The structural transducer is certainly considerably cheaper than the array. It is somewhat more difficult to calibrate. The array is subject to spatial aliasing, which is only partially alleviated by using the largest possible microphone elements and setting them as close together as possible. The structure does not alias. It is equivalent in the spatial domain to a continuous analog time signal. On the other hand, the spatial side lobe structure of the microphone array can be controlled by shading the array, that is, by adjusting the sensitivities of the individual array elements. In both techniques, the resulting signal is frequency analyzed.

The major problem with any wavenumber measurement technique is spatial side lobe contamination. This takes two forms, the high wavenumber side lobes are contaminated by the wall pressure signal at the convective ridge. The side lobes below the main wavenumber side lobe are contaminated by acoustic signals. A plate transducer is particularly good at the rejection of convective ridge contamination. On the other hand, its response cannot be shaded, thus precluding the investigation of acoustic contamination problems by array shading. Finally, it is possible to insert time delays between the microphone responses of the microphone array, thus permitting array steering and the determination of localized acoustic sources in the test environment. A point to bear in mind, but one which is not of great consequency at the current state of wavenumber measurements is that the techniques are incapable of distinguishing between positive and negative wavenumbers, i.e., the measured wavenumber spectra are folded about the zero-wavenumber axis. In what follows we shall concentrate upon the longitudinal or streamwise wavenumber measurement assuming that the lateral wavenumber is set equal to zero. Further, almost all wavenumber measurements have been carried out below the convective wavenumber, as this is the domain of principle interest to structural excitation.

In all measurements to date, once the wavenumber at a given frequency is moved below the convection wavenumber, the level drops quite abruptly by as much as 35dB and then flattens off to a near constant value as far as it is possible to carry out the measurement. If one views the results at a constant wavenumber but increasing frequency there is again an abrupt fall-off from the convective ridge but a continuing dropping of level as frequency is increased further, see Figure 9. This behavior at low wavenumbers has been termed "wavenumber white." It has become customary, therefore to present data not in isocontours that are implicit in Figure 9, but collapsed in wavenumber yielding a plot of spectrum level vs. frequency. Two families of measurements presented in this way are shown in Figures 10 and 11.

The data presented in Figures 10 and 11 come from two different wind tunnels, from two different measurement techniques, both arrays and plate filters, and from widely varying test environments, ranging from closed hard-walled ducts to wall jets in a semi-anechoic chamber. One set of data is from array measurements on a buoyant body rising in water. This last set is especially significant because the environment is nearly free-field, the background noise is extremely low, and the Mach number range has been extended downward by a factor of nearly four from the wind

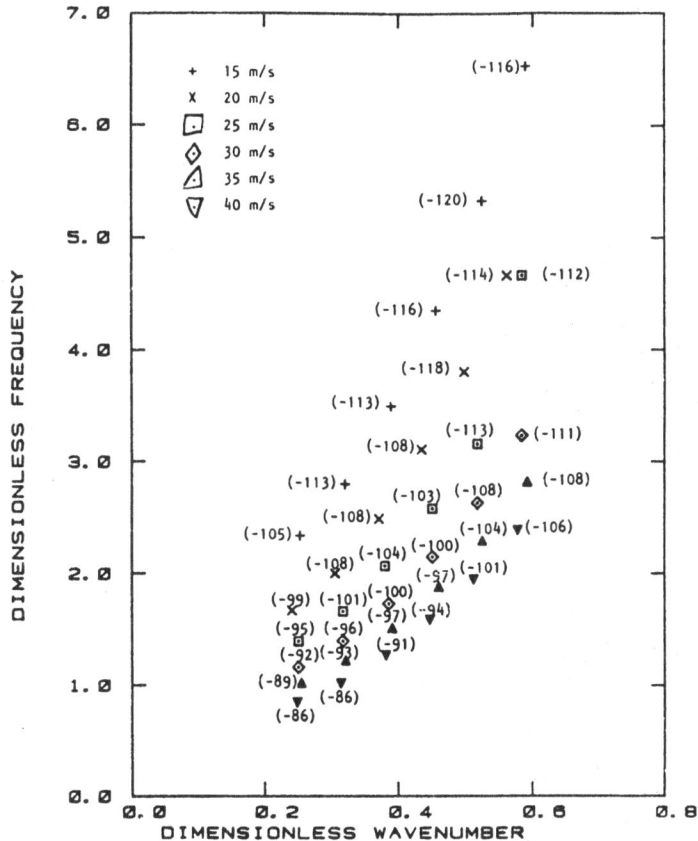

Figure 9. Low wavenumber result of Martin plate in ω-k space, from Martini, Leehey and Moeller (1984).

tunnel measurements. Considering the wide variety of test environments, it is gratifying to find that the scaling of the low wavenumber spectra on $M^2 \tau_w^2$ is valid.

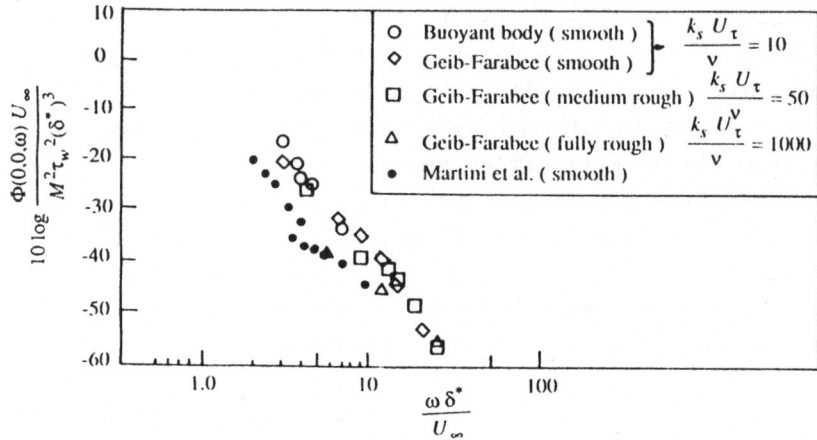

Figure 10. Wall pressure spectra at zero streamwise wavenumber. (The Mach number range extends from M=0.03 to M=0.14.)

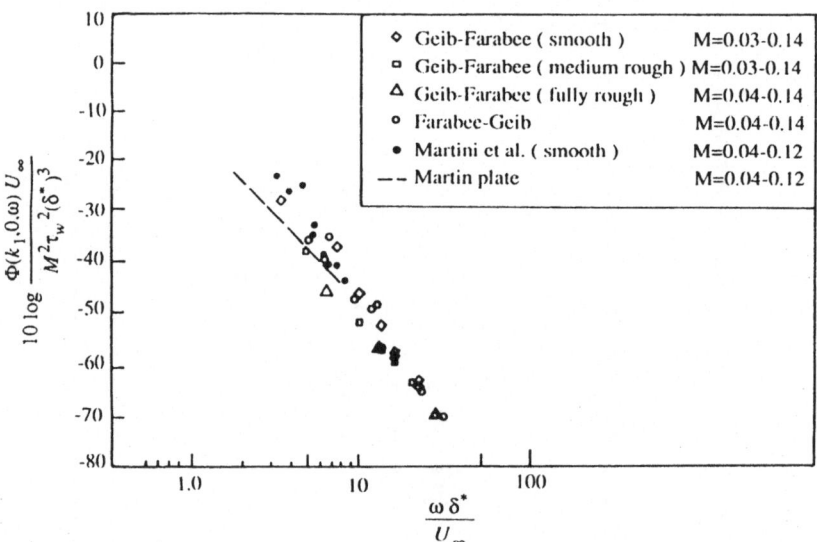

Figure 11. Wall pressure spectra for streamwise wavenumber greater than the acoustic wavenumber.

Figure 10 gives results for "summed" arrays, i.e., arrays aligned in the flow direction whose outputs are directly summed. For these arrays the outputs of the major lobes are centered on $k_1=0$ and $k_3=0$. The mean flow Mach number range extended from M=0.03 to M=0.14. It is interesting to note that the wind tunnel data of Martini et al. (1984) lie appreciably below the buoyant body data.

Figure 11 gives results for "alternating" arrays, i.e., arrays aligned in the flow direction whose transducer outputs are summed alternating in phase. Also included are results from a plate transducer, Martin (1976). Since results from a plate transducer are obtained only at resonant frequencies, a plate is similar to an "alternating" array. The major lobes for all transducers depicted in Figure 11 lay above the acoustic wavenumber.

Except for the data of Martini et al., the results for $k_1=0$ given in Figure 10 lie appreciably above those for k_1 greater than the acoustic wavenumber k_o given in Figure 11. This rise in the wavenumber spectrum below k_o may, however, be due to the greater susceptibility of the "summed" array to spatial aliasing.

It seems likely, today, that the low wavenumber measurements are, in fact, reflecting the behavior of the boundary layer itself and it behooves us to determine what is the true physical basis for these levels. Perhaps sufficient emphasis has already been given to the effects of compressibility and that we should be giving more attention to those wavenumbers that are commensurate with a very large scale eddy structures of the boundary layers and to the contributions of fluctuating wall shear stress. We have referred earlier to the uncertain state of affairs regarding the handling of wall shear stress. Burton (1974) found by measurements the relation $\overline{\tau_w^2} = 0.004\overline{p^2}$ between mean square wall shear stress and mean square wall pressure. This is a difference of 24dB. Since the wavenumber white pressure levels are some 35dB below the convective ridge, it is entirely conceivable, with some allowance for viscous mode attenuation, that the low wavenumber levels are in fact set by the grazing dipole radiation from shear stress as predicted by Landahl (1975). Clearly the best experimental scaling laws for low wavenumber spectra are identical with those developed by Landahl for the influence of oscillatory wall shear stress, i.e., scaling on $M^2\tau_w^2$. Perhaps this is an explanation as to why the scaling on M^2 extends far beyond the applicable wavenumber region suggested by the analyses of Ffowcs Williams. We note that Ffowcs Williams analyses neglected entirely the influence of fluctuating wall shear stress.

SUMMARY

We have addressed three aspects of wall pressure fluctuations created by a turbulent boundary layer. First, the resolution of high wavenumber components by very small transducers has been studied. The question of whether or not a pinhole type transducer disturbs the flow has been partially answered in the negative. Second, cross-spectral density measurements continue to show a dependence upon $\vec{r}/\delta^*$ in addition to the reduced frequency $\omega\delta^*/U_c$. These dependencies weaken the applicability of the Corcos similarity model of wall pressure especially for low wavenumber interpretations. Finally, in the matter of the direct measurement of low wavenumber components of the turbulent boundary layer wall pressure, we see that the measurement techniques and control of test environments have improved sufficiently that there is now some degree of correlation between theory and experiment. However, the essential underlying mechanism for establishing the levels of low wavenumber components remains to be determined.

We have seen that there is strong evidence to indicate the fluctuating wall shear stress may be the predominant influence in the low wavenumber wall pressure spectrum as well as in radiation from boundary layer turbulence. This places considerable importance upon quantifying experimentally the levels of wall shear stress fluctuations. In this respect the current status of research is intolerable because the ratio of rms wall shear stress to mean shear varies by a factor of approximately 10 depending on the measurement technique used.

ACKNOWLEDGEMENTS

The author acknowledges the support of the Office of Naval Research, Underwater Acoustics Program, in the preparation of this study. He also acknowledges the past support of the Mechanics Division and of the ACSAS Program of the Office of Naval Research, for those phases of work reported here that were carried out in the Acoustics and Vibration Laboratory of the Massachusetts Institute of Technology.

REFERENCES

Bergeron, R.F. (1973) "Aerodynamic sound and the low wavenumber wall pressure spectrum of nearly incompressible boundary layer turbulence," Jour. Acoust. Soc. Amer., Vol. 54, No. 1, pp. 123-133.

Blake, W.K. (1969) "Turbulent boundary layer wall pressure statistics on smooth and rough walls," Ph.D. Dissertation, Department of Ocean Engineering, M.I.T., Cambridge, Mass.

Blake, W.K. (1970) "Turbulent boundary-layer wall-pressure fluctuations on smooth and rough walls," Jour. Fluid Mech., Vol. 44, Part 4, pp. 637-660.

Blake, W.K. (1983) "Differential pressure measurement," Chapter 3 in Fluid Mechanics Measurements, (ed., R.J. Goldstein), Hemisphere Publishing Corp., Springer-Verlag, pp. 61-97.

Bull, M.K. (1963) "Properties of the fluctuating wall-pressure field of a turbulent boundary layer," NATO, AGARD Report No. 455.

Bull, M.K. (1967) "Wall pressure fluctuations associated with subsonic turbulent boundary layer flow," Jour. Fluid Mech., Vol. 28, pp. 719-754.

Bull, M.K. and A.S.W. Thomas (1976) "High frequency wall-pressure fluctuations in turbulent boundary layers," Physics of Fluids, Vol. 19, No. 4, April, pp. 597-598.

Burton, T.E. (1973) "Wall pressure fluctuations at smooth and rough surfaces under turbulent boundary layers with favorable and adverse pressure gradients," Acoustics & Vibration Lab. Report No. 70208-9, M.I.T., Cambridge, Mass.

Burton, T.E. (1974) "The connection between intermittent turbulent activity near the wall of a turbulent boundary layer with pressure fluctuations at the wall," Acoustics & Vibration Lab. Report No. 70208-10, M.I.T., Cambridge, Mass.

Chang, Y.M. and P. Leehey (1979) "Acoustic impedance of rectangular panels," Jour. of Sound and Vibration, Vol. 64, No. 2, pp. 243-256.

Corcos, G.M. (1963) "The resolution of pressure in turbulence," Jour. Acoust. Soc. Amer., Vol. 35, pp. 192-199.

Corcos, G.M. (1967) "The structure of the turbulent pressure field in boundary layer flows," Jour. Fluid Mech., Vol. 18, pp. 353-377.

Efimtsov, B.M. (1984) "Similarity criteria for the spectra of wall pressure fluctuations in a turbulent boundary layer," Soviet Physics Acoustics, Vol. 30, No. 1, Jan.-Feb.

Emmerling, R. (1973) "The instantaneous structure of the wall pressure under a turbulent boundary layer flow," Max-Planck-Inst. Stromungsforsch. No. 56-1973.

Farabee, T.M. and F.E. Geib, Jr. (1975) "Measurement of boundary layer pressure fields with an array of pressure transducers in a subsonic flow," 6th Int. Cong. on Instrumentation in Aerospace Simulation Facilities, 22-24 Sept., Ottawa, Canada; available as IEEE Publication 75CHO993-AES pp. 311-319 or DTNSRDC Report 76-0031.

Ffowcs Williams, J.E. (1965) "Surface-pressure fluctuations induced by boundary-layer flow at finite Mach number," Jour. Fluid Mech., Vol. 22, Part 3, pp. 507-519.

Ffowcs Williams, J.E. (1982) "Boundary layer pressures and the Corcos model: a development to incorporate low-wavenumber constraints," Jour. Fluid Mech., Vol. 125, pp. 9-25.

Franklin, R.E. and J.M. Wallace (1970) "Absolute measurements of static-hole error using flush transducers," Jour. Fluid Mech., Vol. 42, Part 1, pp. 33-48.

Geib, F.E. and T.M. Farabee (1985) "Measurement of boundary layer pressure fluctuations at low wavenumber on smooth and rough walls," David Taylor Naval Ship R & D Center Report No. 84-05/, Washington, D.C.

HajHariri, H. and T. Akylas (1985) "The wall-shear-stress contribution to boundary layer noise," Physics of Fluids, Vol. 28, No. 9, pp. 2727-2729.

Hakkinen, R.J. and E.J. O'Neil (1967) "On the merging of uniform shear flows at a trailing edge," Douglas Aircraft Co. Report DAC-60862.

Howe, M.S. (1979) "The role of surface shear stress fluctuations in the generation of boundary layer noise," Jour. of Sound and Vibration, Vol. 65, No. 2, p. 159.

Howe, M.S. (1987) "On the structure of the turbulent boundary layer wall pressure spectrum in the vicinity of the acoustic wavenumber," Proc. Royal Soc. London, Series A, No. 412, pp. 389-401.

Jameson, P.W. (1970) "Measurement of low wavenumber component of turbulent boundary layer wall pressure spectrum," Bolt, Beranek & Newman, Inc. Report No. 1937.

Jones, D.S. (1966) "Generalised Functions," McGraw-Hill Publishing Co., London.

Kompenhans, J. (1976) "Das akustische Verhalten uberstromter Offnungen in Abhangigkeit von der Wandgrenzschicht eine experimentelle Untersuchung, Dissertation, Gottingen.

Kompenhans, J. and D. Ronneberger (1980) "The acoustic impedance of orifices in the wall of a flow duct with a laminar or turbulent flow boundary layer," AIAA Paper 80-0990, AIAA 6th Aeroacoustics Conference, 4-6 June, Hartford, Conn.

Kraichnan, R.H. (1956) "Pressure fluctuations in turbulent flow over a flat plate," Jour. Acoust. Soc. Amer., Vol. 28, No. 3, pp. 378-

Landahl, M.T. (1975) "Wave mechanics of boundary layer turbulence and noise," Jour. Acoust. Soc. Amer., Vol. 57, No. 4, pp. 824-838.

Langeheineken, Th. and A. Dinkelacker (1978) "Wanddruckschankungen einer Ausgebildeten, Turbulenten Rohrstromung," Fortschritte der Akustik, p. 391, VDE-Berlag.

Leehey, P. (1988) "Structural excitation by a turbulent boundary layer: an overview," Jour. of Vibration, Stress and Reliability in Design, Vol. 110, April, pp. 220-225.

Lighthill, M.J. (1952) "On sound generated aerodynamically: I: General theory," Proc. Royal Soc. London, Series A., No. 1107, Vol. 211, pp. 564-587.

Martin, N.C. (1976) "Wavenumber filtering by mechanical structures," Ph.D. Dissertation, Department of Mechanical Engineering, M.I.T., Cambridge, Mass.

Martin, N.C. and P. Leehey (1977) "Low wavenumber wall pressure measurements using a rectangular membrane as a spatial filter," Jour. of Sound and Vibration, Vol. 52, No. 1, pp. 95-120.

Martini, K., P. Leehey and M. Moeller (1984) "Comparison of techniques to measure the low wavenumber spectrum of a turbulent boundary layer," Acoustics & Vibration Lab. Report No. 92828-1, M.I.T., Cambridge, Mass.

Messiter, A.F. (1970) "Boundary-layer flow near the traling edge of a flat plate," SIAM Jour. Appl. Math., Vol. 18, No. 1.

Moller, J.C. (1987) "Measurement of wall shear and wall pressure downstream of a honeycomb boundary layer manipulator, Engineer's Thesis, Department of Mechanical Engineering, M.I.T., Cambridge, Mass.

Petri, S.W. (1987) "The response of line-stiffened fluid-loaded infinite elastic plates to convecting pressure fields," Ph.D. Dissertation, Department of Ocean Engineering, M.I.T., Cambridge, Mass.

Phillips, O.M. (1955) "Surface noise from a plane turbulent boundary layer," Aero. Res. Council, London, No. 16, 963-F.M. 2099.

Powell, A. (1960) "Aerodynamic noise and the plane boundary," Jour. Acoust. Soc. Amer., Vol. 32, No. 8, pp. 982-990.

Ronneberger, D. (1972) "The acoustical impedance of holes in the wall of flow ducts," Jour. of Sound and Vibration, Vol. 24, pp. 133-150.

Schewe, G. (1983) "On the structure and resolution of wall-pressure fluctuations associated with turbulent boundary-layer flow," Jour. Fluid Mech., Vol. 134, pp. 311-328.

Sevik, M. (1985) "Topics in Hydroacoustics," Proceedings of IUTAM Symposium on Aero- and Hydroacoustics, Lyon, France, 3-6 July, Berlin: Springer-Verlag.

Willmarth, W.W. and F.W. Roos (1965) "Resolution and structure of the wall pressure field beneath a turbulent boundary layer," Jour. Fluid Mech., Vol. 22, pp. 81-94.

Willmarth, W.W. (1975) "Pressure fluctuations beneath turbulent boundary layers," in Annual Review of Fluid Mechanics, Annual Reviews, Inc., Palo Alto, Calif., Vol. 7, pp. 13-38.

The Measurement of Wall Shear Stress

by
Joseph H. Haritonidis
Department of Aeronautics and Astronautics
Massachusetts Institute of Technology
Cambridge, MA 02139

1. INTRODUCTION

Knowledge of the wall shear stress is of both fundamental and practical importance. The mean stress is indicative of the overall state of the flow over a given surface while the fluctuating stress is a "footprint" of the individual processes that transfer momentum to the wall. The measurement techniques can be divided into two main categories: direct and indirect. The indirect techniques can be further divided into momentum balance methods and correlation methods.

Correlation methods extract the necessary information to determine the shear stress from well established relations. These in turn may be divided into global and local methods. For example, the Preston tube operation relies on the law of the wall which has been determined to hold true for a great variety of flows and flow configurations. Since it usually depends on a large part of the boundary layer it would be classified as a global correlation method. It is important to mention that the outcome of a measurement with a Preston tube will depend on whether the flow does indeed conform to the law of the wall or not. In an unknown flow, there is no guarantee that the measurement will give the correct answer.

Local methods depend on the flow field in the vicinity of the device used to measure the shear stress. As in the case for global methods, local methods rely on established correlations between the flow and the shear stress. The advantage over global methods is that the flow field affecting the measurement is confined to a very small region of the flow near the wall and as such is not as sensitive to the structure of the flow away from the wall. Two examples of instruments relying on local conditions are the Stanton tube and the flush mounted hot-film or hot-wire.

Momentum balance methods involve the measurement of integral quantities to determine the wall shear stress. Two such quantities are the pressure gradient in a channel or pipe flow and the momentum thickness in a flat plate boundary layer. In the first case there is no direct dependence on the velocity profile at all while in the second there is no need for an *a priori* knowledge of the velocity profile.

Direct methods are, in principle, the ideal method for measuring wall shear stress since they sense directly the applied force to the wall by the fluid and do not depend directly on the flow field or the fluid properties. An example of an instrument of this type is the floating element sensor. Unfortunately, the advantages just mentioned are usually outweighed by a host of problems in implementing the method. Recent work, however, is addressing these problems with some success.

In addition to the above classification regarding the measurement method, shear stress measuring devices can also be classified according to whether they can measure both the mean and the fluctuating shear stress and to their sensitivity to the direction of the applied shear. These differences will be discussed in detail for each method.

The results of the different methods depend on whether the flow is compressible or not, whether there is heat transfer taking place, the presence of a favorable or unfavorable pressure gradient and whether the wall is smooth or rough among other factors. The present discussion will focus mainly on incompressible flows in gases over smooth walls though the questions raised and problems identified should be applicable to any type of flow. As there are a number of excellent review articles on wall shear stress measurement covering an exotic range of instruments and methods (Winter, 1977, Hanratty & Campbell, 1983), the emphasis here will be on how to approach wall shear stress measurement and the inherent problems in the most common techniques. The particular methods that will be discussed are those thought to be the most useful for calibration purposes and either the most reliable or holding the greatest promise for future development.

2. STRUCTURE OF TURBULENT BOUNDARY LAYERS

Since all methods depend directly or indirectly on the structure of the turbulent boundary layer, it will be useful to give a brief review based on the current understanding of the structure of these flows. When discussing laminar flow applications of the different methods, differences and similarities will be pointed out for each method separately. For most turbulent boundary layers, the law of the wall and the law of the wake (Coles, 1956) is valid for a large number of flow conditions and flow configurations. Schematically, a typical boundary layer profile appears as in Fig. 1

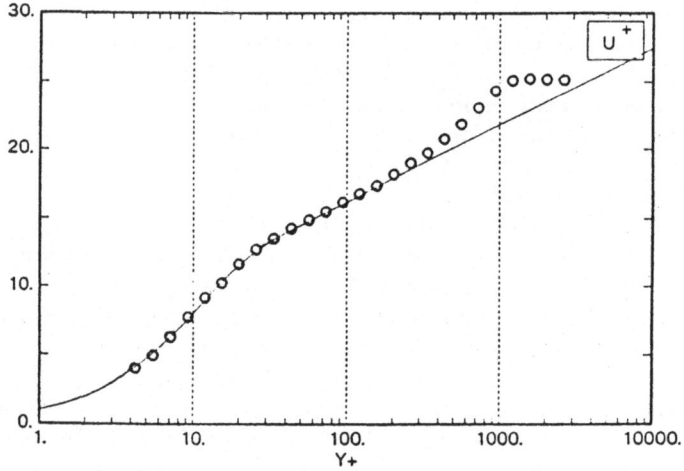

Fig. 1 Typical turbulent boundary layer profile (from Gresko, 1988).

The profile can be divided into four regions: the viscous or linear region extending from the wall to $y^+ \approx 5$, the buffer region extending from $y^+ \approx 5$ to $y^+ \approx 45$, the logarithmic region extending from $y^+ \approx 45$ to $y/\delta \approx 0.2$ and the wake region extending from $y/\delta \approx 0.2$ to δ (a + superscript implies normalization with the wall variables ν and u_τ where ν is the fluid kinematic viscosity and u_τ is the friction velocity, y is the coordinate normal to the wall, and δ is the boundary layer thickness). It is worth noting that the extent of the various regions is given in terms of wall variables, outer variables or a combination of both. This is illustrated in Fig. 2 for the case of a zero pressure gradient boundary layer.

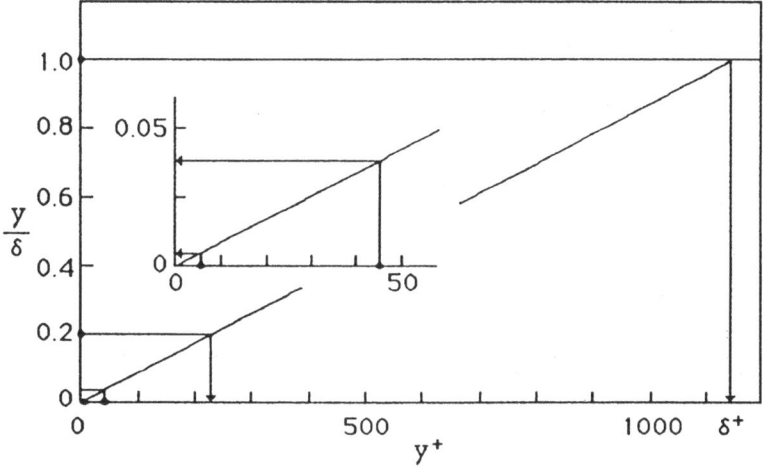

Fig. 2. Regions of a turbulent boundary layer.

The y/δ axis is mapped onto the y^+ axis through the mapping (diagonal) line and vice versa. The fixed points on the y/δ axis are the edge of the boundary layer and the end of the log region. The fixed points on the y^+ axis are the edge of the linear region and the beginning of the log region (shown in more detail in the insert). These fixed points appear to a have a very weak Reynolds number dependence. Any particular position of the mapping line is a function of the Reynolds number. As the Reynolds number increases, the mapping line tilts over so that the $y/\delta = 1.0$ point is mapped to a higher y^+ or δ^+. Thus the wake region always occupies the outer 80% of the boundary layer thickness and the buffer region ends at $y^+ \approx 45$, where the log region begins. It is evident that the extent of the log region is a function of the Reynolds number, increasing for increasing Reynolds numbers. For a low enough Reynolds number, the log region disappears. This happens when $\delta^+ \approx 225$ which corresponds, for example, to the embryonic stage of development of a turbulent spot in a laminar boundary layer. The calculations of Spalart (1986) for a turbulent boundary layer indicate that the log region disappears at $Re_\theta \approx 400$ where $\delta^+ \approx 200$, in agreement with the arguments presented above.

The outer part of the boundary layer is characterized by large scale eddies whose characteristic length and time scale are proportional to δ and δ/U_∞. These eddies reside in a region of low shear and as such have very long life-times. It is precisely for this reason that disturbances introduced into the outer part of the boundary layer take a very long time to die out. Improper tripping of a laminar flow to produce a turbulent boundary layer, using for example too large a tripping device, will result in a turbulent boundary layer, but with different properties than those of a natural boundary layer.

The log region is characterized by eddies with length scales proportional to y^+ (see Haritonidis, 1989) and time scales proportional to the wall unit of time t^*, where $t^* = v/u_\tau^2$. The eddies possess energies higher than those found in the outer layer and are the result of the interaction of the outer and the near wall flow. As the log region is in a region of higher shear it tends to adjust to the flow conditions, such as pressure gradient, much more rapidly. For our purposes, therefore, it is much more reliable as an indicator of the wall flow conditions than the outer flow.

The buffer region is very similar to the log region with the exception that the turbulent structure becomes very anisotropic (Haritonidis, 1988) and even more energetic. It is also the region from which most of the turbulent producing mechanisms originate and as such plays a very important role in the overall structure of turbulent boundary layers.

Finally, the viscous or linear region is characterized by length and time scales that scale with wall units just as the buffer and log regions do. What distinguishes the viscous region uniquely from the other ones is its ability to adjust most rapidly to changes in the wall conditions and as such is the most suitable portion of the boundary layer to infer wall conditions. In particular, it is suitable for determining the instantaneous wall stress as will be discussed in more detail in section 3.2.3. These considerations apply in the case where the wall is smooth, i.e. where the characteristic roughness height, k^+, is less than about 3. Recent work by Tani (1987), however, may require reconsideration of even this simple "established" fact. According to his analysis of Nikuradse's (1950) data, <u>drag reduction</u>, comparable in magnitude to that obtained with longitudinal riblets, is actually observed for $k^+ < 6$. If this is indeed the case, then attention will have to be paid to wall surface conditions if accurate comparisons between different wall stress measurements are desired.

The linear or viscous region is also referred to as the "laminar sublayer". We think that this is not only improper, but also wrong. While it is true that the absolute magnitude of velocity fluctuations tend to zero as the wall is approached, the local turbulence level, $u'/\overline{U}$ (a prime indicates the rms value and an overbar the average value) reaches a <u>maximum and constant value</u> in the viscous region. The measurements of Alfredsson, Johansson, Haritonidis & Eckelmann (1988) at Re_θ values of about 500 to 3,000, in boundary layers and channels, indicate that this constant value is 0.4. The numerical calculations of Kim, Moin & Moser (1987) for channel flow at $Re_\theta = 290$ show this constant to be 0.365 while those of Spalart (1986) indicate that it is about 0.37 at

$Re_\theta = 300$ and close to 0.4 at $Re_\theta = 1410$. The important point here is that $u'/\overline{U}$ asymptotes to a constant and maximum value in the viscous region, thus making it the most turbulent part of the boundary layer.

In addition to $u'/\overline{U}$, it appears that $w'/\overline{U}$, the local turbulence level of the spanwise component of velocity, also asymptotes to a maximum value of about 0.2 in the viscous region according to Alfredsson et al. (1988) and Kim et al. (1987). This becomes important when considering the magnitude and direction of the surface shear stress. The asymptotic level of the turbulence intensities, of the streamwise and spanwise velocity components, are related to the mean and fluctuating wall stress by,

$$\lim_{y \to 0} \frac{u'}{U} = \frac{\tau_x'}{\tau_x}$$

and

$$\lim_{y \to 0} \frac{w'}{U} = \frac{\tau_z'}{\tau_x}$$

It is therefore evident that any shear stress measuring device that depends on local conditions must be able to function in the presence of very large amplitude fluctuations.

Unavoidably, we shall have to consider the normal wall stress also, i.e. wall pressure fluctuations. These enter the picture on account of the fact that certain shear stress measuring devices, such as the Stanton tube, actually infer wall shear through a pressure measurement that could be affected by wall pressure fluctuations (Bradshaw & Gregory, 1959, Fulcher, 1986). Through the advent of smaller and more accurate pressure transducers as well as very quiet wind tunnels, it has been established that, at moderate Reynolds numbers (Schewe, 1986) the ratio of wall pressure fluctuations to the free stream dynamic pressure, p_w'/q, is very close to 0.01. Assuming for the moment that this ratio does not change much with Reynolds number and using the definition for the local coefficient of skin friction, $C_f = 2\overline{\tau_x}/\rho U_\infty^2$, we may write,

$$p_w' = 0.01 \frac{\overline{\tau_x}}{C_f}$$

Table I shows values of $p_w'/\overline{\tau_x}$ versus Re_θ. C_f was obtained from Re_θ using the relation (Schlichting, 1968),

$$C_f = 0.0256 Re_\theta^{-1/4}$$

TABLE I

Re_θ	C_f	$p_w'/\overline{\tau_x}$
1,000	0.00455	2.20
5,000	0.00304	3.29
10,000	0.00256	3.91
50,000	0.00171	5.84

Clearly, p_w' is larger than the shear stress, $\overline{\tau_x}$ and, as will be shown, may have an adverse effect on the measurement of the wall shear stress.

3. INDIRECT METHODS
3.1 MOMENTUM BALANCE
3.1.1 PRESSURE GRADIENT

For flows in constant area ducts, such as pipes or channels, the average wall shear stress can be obtained from the pressure gradient. The pressure drop over a given length of the duct is directly balanced by the integrated shear stress over the wetted surface of the same length. For a duct with radial symmetry, and neglecting pressure changes across boundary layers, the pressure will be the same at any azimuthal position of the duct at a given duct position. This is an important consideration since any duct (radial) asymmetry will affect the azimuthal pressure distribution. However, even if the duct possesses radial symmetry, it is important that the duct dimensions do not change along the duct length used to measure the pressure gradient.

For a duct of arbitrary cross sectional area A, the pressure drop $\overline{\Delta P}$ over a duct length L is given by,

$$\overline{\Delta P}A = \int_S \overline{\tau_x}\, ds$$

where S is the area over which the stress acts on. For a circular pipe or a channel the above equation can be written as,

$$\overline{\tau_x} = \frac{1}{2}\frac{\overline{\Delta P}}{L}\, h$$

where h is either the pipe radius or the channel height. It should be noted that this result is true regardless of whether the flow is laminar or turbulent.

Of all the methods for determining the mean wall stress, the above method is the most reliable provided that L is long enough so that an accurate measurement of the pressure drop can be made and that the flow is fully developed so that all streamwise gradients of flow quantities (except the pressure gradient of course) are zero. These requirements and the rather special types of flow that it addresses, makes the method of limited usage. For calibration purposes, however, it is ideal.

3.1.2 MOMENTUM THICKNESS GRADIENT

When considering two-dimensional laminar or turbulent boundary layers, use can be made of von Kármán's momentum theorem to relate the mean wall shear stress to the momentum thickness:

$$\frac{\overline{\tau_x}}{\rho U_0^2} = \frac{d\theta}{dx} + (H+2)\frac{\theta}{U_0}\frac{dU_0}{dx}$$

where U_0 is the velocity at the edge of the boundary layer, θ is the momentum thickness, x is the streamwise direction and H is the shape factor. It is emphasized that the above equation is valid only for two-dimensional flows and that additional terms involving turbulence quantities have been neglected. The analysis by Bidwell (1951) and Dutton (1956) indicates that when the turbulence terms become large, as in the case when a point of separation is approached, then these terms cannot be neglected. The full equation is,

$$\frac{\overline{\tau_x}}{\rho U_0^2} = \frac{d\theta}{dx} + (H+2)\frac{\theta}{U_0}\frac{dU_0}{dx} - \frac{1}{U_0^2}\int_0^\delta \frac{\partial}{\partial x}(\overline{u^2} - \overline{v^2})\,dy + \frac{1}{U_0^2}\left[\int_0^\delta\left(\int_0^y \frac{\partial^2 \overline{uv}}{\partial x^2}\,dy - \delta\frac{\partial^2 \overline{uv}}{\partial x^2}\right)dy\right]$$

where u and v are the streamwise and normal velocity fluctuations and δ is the boundary layer thickness. Dutton (1956), however, cautions that even in the case of a zero pressure gradient flow, the first integral above may contribute up to 2.5% to the skin friction coefficient.

The difficulty with this method is that, in addition to the turbulence terms, $\overline{\tau_x}$ depends on both the accurate measurement of the velocity profile, from which θ can be determined, and the determination of the streamwise gradient of θ which tends to be a rather difficult task. As a result, the overall accuracy of the method is low in addition to being a tedious and time consuming endeavor. For these reasons, its use in determining the effects of drag reduction schemes in flat plate boundary layers, for example, is suspect.

3.2 CORRELATION METHODS
3.2.1 CLAUSER PLOT

As discussed in section 2, the law of the wall is valid over a wide range of boundary conditions, such as pressure gradient or mild surface roughness for example. It is thus suitable for determining the wall shear stress once the friction velocity, u_τ, has been obtained. The Clauser plot is based on the logarithmic region of the law of the wall. Different u_τ values are used until a portion of the normalized velocity profile, $U^+ = \overline{U}/u_\tau$, falls on the line

$$U^+ = \frac{1}{\kappa}\ln y^+ + B \tag{1}$$

where κ is von Kármán's constant and B is a constant that appears to be Reynolds number dependent for certain types of flows and boundary conditions. As mentioned in the introduction, the choice of constants is as good as the correct prediction of $\overline{\tau_x}$ through u_τ. The fact that $\kappa = 0.41$ and $B = 5.0$ for the majority of flows, excluding low Reynolds numbers, makes the Clauser plot method a useful tool.

A number of points need to be made in regards to the practical application of the method:

1) Clearly, it is necessary to measure the velocity profile accurately. Hot-wires and laser-doppler anemometers (LDA) are preferable over pitot tubes

even though pitot tubes are much simpler to use. Measurements using pitot tubes to obtain the mean velocity profile are more likely than not to be in error as a pitot tube reading may be affected by large amplitude velocity fluctuations. The velocity determined from a pitot tube is obtained through the use of the Bernoulli equation, namely,

$$\overline{\Delta P} = \frac{1}{2} \rho \overline{U^2} \tag{2}$$

where $\overline{\Delta P}$ is the pressure difference between the pitot tube and the local static pressure and U is the local velocity. However, in the presence of turbulent fluctuations, the instantaneous velocity is given by $U = \overline{U} + u$, where $\overline{U}$ is the mean velocity and u is the instantaneous deviation from the mean. It is assumed in this case that the pitot tube is aligned with the mean flow direction and that transverse velocity fluctuations are small. If this were not the case, then the angular sensitivity of the pitot tube to the instantaneous flow direction would also become important (for further details, consult Chue, 1975). Inserting in eq. (2) and assuming incompressible flow, the result is,

$$\overline{\Delta P} = \frac{1}{2} \rho \overline{(\overline{U} + u)^2}$$

$$= \frac{1}{2} \rho \overline{U}^2 (1 + \frac{\overline{u^2}}{\overline{U}^2})$$

It is seen that there is now an additional contribution from the turbulence fluctuations. Neglecting higher order terms, the measured velocity, U_m, will now be,

$$\overline{U}_m \approx \overline{U} (1 + \frac{1}{2} \frac{\overline{u^2}}{\overline{U}^2})$$

The local turbulence level is about 10% at the end of the log region and becomes 40% in the viscous region (see section 2). Thus the error, in principle, varies between 0.5% and 8% very close to the wall. Given that u_τ is typically 3-4% of U_∞, the free stream velocity, it is evident that there is plenty of room for obtaining the wrong value not only for u_τ, but also for κ and B.

2) Additional errors, when using a pitot tube, can be incurred due to low Reynolds number effects connected with the tube opening (Macmillan, 1954), displacement of the effective tube height due to shear and to rectification of the pressure fluctuations, produced by the turbulent flow, by the pressure measuring system. The latter issue will be discussed in section 3.2.3 in connection with Stanton tubes.

3) A less severe source of error is related to the determination of the origin of the velocity profile. A pitot tube can easily be placed in contact with the wall while the same is not true for a hot-wire or an LDA, for example.

4) The positioning inaccuracies for hot-wires are due, among other factors, to heat transfer to the wall, probe blockage, excessive wire length or improper wire calibration at low velocities while for an LDA they are due to the difficulty of positioning the measuring volume close enough to the wall, and reflections off the wall that interfere with the signal processing.

The collective problems for all types of measuring methods below the log region, including the location of the wall, can be alleviated to a great extent through fitting the velocity profile to analytical expressions that cover the region from y = 0 to either the beginning or the end of the log region and choosing y offsets, to be added or subtracted from the nominal "y = 0" position until a best fit is obtained in the buffer and viscous regions. Van Driest (1956) derived an expression in integral form, based on a modified mixing length hypothesis, relating U^+ to y^+, namely,

$$U^+ = \int_0^{y^+} \frac{2}{1 + \sqrt{1 + 4K^2 y^{+2} \left\{ 1 - \exp(-y^+/A) \right\}^2}} \, dy^+$$

where $K = \kappa$, von Kármán's constant, and A is a constant that, together with κ, determines the offset of the logarithmic portion of the velocity profile, i.e. B in the log law.

Spalding (1961) suggested the following equation for the wall region:

$$y^+ = U^+ + \exp(-\kappa B) \left\{ \exp(\kappa U^+) - 1 - \kappa U^+ - \frac{(\kappa U^+)^2}{2!} - \frac{(\kappa U^+)^3}{3!} - \frac{(\kappa U^+)^4}{4!} \right\}$$

where κ and B are the constants in the log law. Both of the above equations give good fits, but are difficult to use. Haritonidis (1989) derived an equation, based on a modified mixing length model, for the mean velocity valid from the wall to the begining of the log region, namely,

$$U^+ = \frac{1}{\lambda} \tan^{-1} (\lambda y^+) - \frac{a}{2\lambda^2} \ln(1 + \lambda^2 y^{+2})$$

where $a = 1/h^+$, h^+ being the nondimensional channel half-height or pipe radius, and λ is a constant that is a function of a, κ and B. In the case of plane boundary layer flows, a = 0. For a = 0, κ = 0.41 and B = 5.0, λ = 0.09365 and the arctangent profile joins the log law at y^+ = 44.2.

3.2.2 PRESTON TUBE

The Preston tube appears to be the most commonly used instrument for the measurement of the average wall stress. It owes its popularity to the fact that it is very simple, it is easy to use, and does not have to be calibrated. While the first two attributes are true, the third appears to warrant some scrutiny as there are several differing calibrations available.

A schematic of a typical Preston tube is shown in Fig. 3.

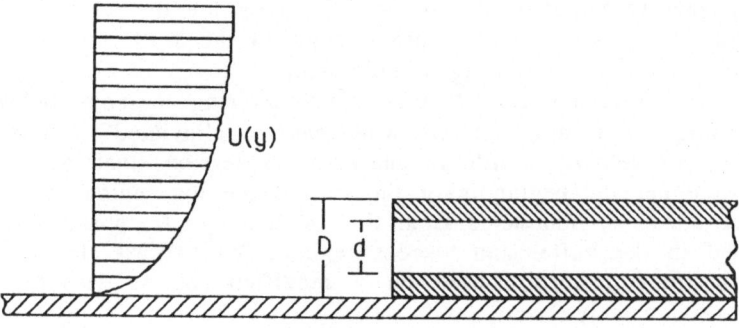

Fig. 3 Schematic of a Preston tube.

The pressure measured with a Preston tube depends on the velocity profile the tube is exposed to. If the tube is located within the log region, i.e. $D \le 0.2y/\delta$, then the pressure will be a function of the law of the wall. Through dimensional analysis, Preston (1954) showed that,

$$\frac{(P - P_r)D^2}{\rho v^2} = F\left(\frac{\overline{\tau_x} D^2}{r n^2}\right) \tag{3}$$

where P is the pressure measured with the Preston tube, P_r is a reference pressure, say the wall pressure, and D is the outer diameter of the Preston tube. The function F must be determined through calibration. His measurements in pipes and flat plate boundary layers show that the data does indeed collapse onto a unique curve, conforming to eq. (3). According to his measurements, if D^+ is greater than about 60, eq. (3) takes the form

$$\log\frac{\overline{\tau_x} D^2}{4\rho v^2} = -1.39 + \frac{7}{8}\log\frac{(P - P_r)D^2}{4\rho v^2} \tag{4}$$

In his experiments, he used tubes whose outer diameters varied by a factor of four. The ratio of the inner to outer diameter was in all cases held to 0.6. Rechenberg (1963) experimented with different ratios of inner to outer diameter and found that the results were the same provided the ratio fell in the range 0.35-0.85.

Subsequent calibrations by Rechenberg (1963), the Staff of Aerodynamics Division, NPL (1958) and Patel (1965) show that eq. (4) is an approximation to the data. Patel (1965) suggests the following empirical relations, based on measurements in pipes:

a) $0 < y^* < 1.5$ and $D^+ < 11.2$,

$$y^* = 0.037 + 0.5x^*$$

b) $1.5 < y^* < 3.5$ and $11.2 < D^+ < 110$

$$y^* = 0.8287 - 0.1381x^* + 0.1437x^{*2} - 0.0060x^{*3}$$

c) $3.5 < y^* < 5.3$ and $110 < D^+ < 1600$

$$x^* = y^* + 2.0\log(1.95y^* + 4.10)$$

where

$$x^* = \log\left(\frac{(P - P_r)D^2}{4\rho v^2}\right)$$

and

$$y^* = \log\left(\frac{\bar{\tau}_x D^2}{4\rho v^2}\right)$$

The measurements of Frei & Thomann (1980), using a cylindrical floating element in a pipe, are in excellent agreement with the above calibration over the range $2.7 < y^* < 3.5$. Given the uncertainties in determining the wall stress in a flat plate boundary layer, it is not surprising to find differences. In fact, Preston used the value of 0.42 for von Kármán's constant and 5.8 for B, both of which vary considerably from the values of 0.41 and 5.0 used today and of 0.42 and 5.45 used by Patel (1965).

So far, the discussion has been limited to Preston tubes in incompressible flows in zero pressure gradients and aligned with the flow. In high speed flows, the law of the wall does change so that the incompressible calibration is no longer valid. In addition, the actual static pressure and temperature become important parameters which must also be taken into account. For further details, the review by Winter (1977) should be consulted.

The yaw characteristics have been examined by the Staff of Aerodynamics Division, NPL (1958), Rajaratnam & Muralidhar (1968) and Prahlad (1972), among others. Their work indicates that there is no abrupt change in the measured pressure with angle, and that the error is of the order of 1% provided the Preston tube is aligned within ± 3° of the flow direction.

3.2.3 STANTON TUBE

The Stanton tube can be thought of as the limiting case of a very small Preston tube. In practice, the Stanton tube is immersed in the linear region of the velocity profile although this is not a necessary restriction. There are, however a number of fundamental differences between the two. As discussed in the introduction, the Preston tube is a global device. The pressure measured is a Bernoulli-type pressure since it results from the deceleration of fluid in front of the tube. The Stanton tube is a local device and therefore should respond much more quickly to local shear stress fluctuations. The registered pressure is both the result of the wall shear stress and the deflection of the flow due to the presence of the device itself. Which of the two dominates depends on H (see Fig. 4) and the value of the wall shear stress. Much of the discussion on the Stanton tube is applicable to similar devices such as the sublayer fence, the submerged step and others. For a more complete discussion of these devices, the reader should consult Winter (1977) and Hanratty & Campbell (1983).

A schematic of a typical arrangement of a Stanton tube is shown in Fig. 4.

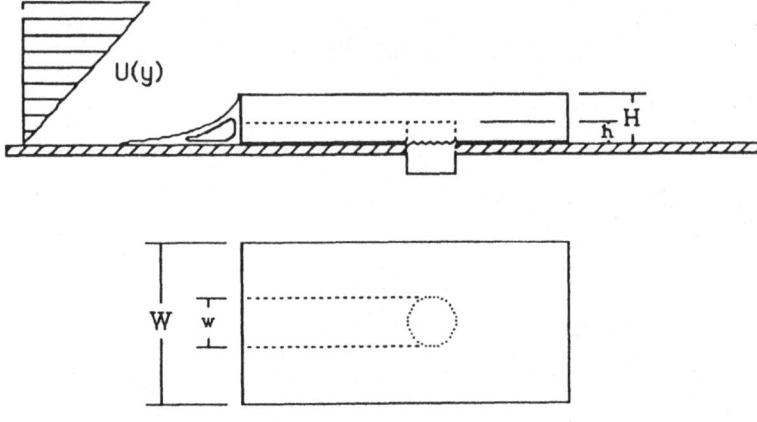

Fig. 4. Schematic of a typical Stanton tube.

Most incarnations of the device employ a razor blade or a half razor blade rather than a structure providing an upstream facing step. There does not, however, appear to be any compelling reason to use such a geometry. The key dimensions of the device are its overall height and width H and W, and the height and width of the opening, h and w. The most critical dimension is the overall height H (Fulcher, 1986). The flow separates some distance upstream from the device and the separation streamline ends up at the top of the device as shown in Fig. 4. Taylor (1938) and Trilling & Häkkinen (1955) showed that the action of the wall shear on the separation bubble is what generates a pressure on the upstream face of the device.

Trilling & Häkkinen (1955) did an analytical and experimental study of the behavior of Stanton tubes. Their analysis identifies three regions of operation of Stanton tubes depending on the value of the Reynolds number, Re_τ, based on the height H of the probe and the velocity at that height assuming that the velocity varies linearly with distance from the wall; thus,

$$Re_\tau = \frac{\overline{\tau}_x \rho H^2}{\mu^2}$$

Rearranging the above equation we obtain $Re_\tau = H^{+2}$, which is a more convenient form to work with. Depending on the value of Re_τ, they distinguish three ranges:

1) $Re_\tau << 1$. In this case Stokes flow theory would indicate that the pressure would be proportional to $\overline{\tau}_x$. The measurements of Taylor (1938) show that this is indeed the case for values of Re_τ up to about 0.1 and probably up to 0.3 if one compares his results with those of Stanton, Marshall & Bryant (1920) and Fage & Falkner (1930). Thus, for this range, H^+ should be less than about 0.5.

2) $10 < Re_\tau < 10^3$. The authors developed a theory for this range, under the assumption that the Stanton tube is embedded in the linear region, and found that $p = c\overline{\tau}_x^{5/3}$ where c is a constant that has a weak Reynolds number dependence. Specifically, it varies as $L^{1/3}$ where L is the upstream distance from the device where the flow separates from the wall. However, given that the linear region is at most 5 wall units thick implies that the upper limit of the above range should be of the order of 100 rather than 1000. This appears to be the case if we carefully examine the data presented in their Figure 3. For values of Re_τ between 10 and 100 their theory does indeed predict the correct power dependence of 5/3.

3) $Re_\tau > 10^3$. In this case, H^+ is greater than 30 and hence the pressure would be a strong function of the velocity profile outside of the linear region.

Table II summarizes the results of various investigations. The calibration equation of Bradshaw & Gregory (1959) was obtained from their data in a turbulent boundary layer.

TABLE II

Investigation	Range	Calibration Equation
Bradshaw & Gregory	$1.4 < H^+ < 63$	$y^* = -0.0786 + 0.681x^*$
East	$3 < H^+ < 100$	$y^* = -0.23 + 0.618x^* + 0.0165x^{*2}$
Pai & Whitelaw	$3 < H^+ < 14$	$y^* = 0.464 + 0.741x^*$
Taylor	$0 < H^+ < 0.3$	$\overline{\Delta P} = 1.2\overline{\tau}_x$

$$x^* = \log(\overline{\Delta P}\, H^{+2}/\,\overline{\tau}_x\,), \quad y^* = \log(H^{+2})$$

It is evident that there are substantial differences between the various calibration equations in the overlapping ranges of H^+. For example, at $H^+ = 5.0$, the first three calibration equations give the values of 5.9, 12.7 and 0.73 for the ratio of $\overline{\Delta P}/\overline{\tau}_x$. Unlike Preston tubes, Stanton tubes are characterized by many more parameters such as width of the device, overall height, height of opening etc. As shown by Fulcher (1986), the calibration constants are sensitive to changes of the various dimensions, the overall height having the greatest effect. The use of very small tubes would most certainly provide greater consistency. The problem, however, is that very small tubes result in extremely long time constants for pressure equalization thus making their use impractical.

An estimate can be made of the dependence of pressure on changes in the wall shear stress. The pressure is a function of the device geometry, the shear stress (turbulence) scales and the fluid properties. Ignoring, for the moment, the dependence on the turbulence structure, the pressure measured is a function of the device height H, and the upstream extent of the separation bubble, L, and the magnitude of the wall shear, $\overline{\tau}_x$. For $H^+ \ll 1$ and an impulsive change in wall shear, the response time, t_τ, will be of the order of the diffusion time over the distance H. Thus, in wall units,

$$t_d^+ = H^{+2}$$

In the case where $H^+ \gg 1$, inertial effects will become important. In particular, the response time will be a function of the advection time, i.e. the time formed by the velocity at the height of the device, H, and the upstream scale, $L^{1/3}$. In wall units, this time scale, t_a^+, is given by,

$$t_a^+ = \frac{L^{1/3}}{H}$$

For a spatially varying shear stress distribution due to turbulent flow, the response of the device will depend on both H^+ and L^+. Trilling & Häkkinen (1955) show that for an H^+ of the order of the viscous sublayer, the measured pressure depends on $L^{+1/3}$. They also report that available measurements indicate that L^+ is about $4\text{-}20H^+$. Thus, the relevant upstream scale is about 2-$3H^+$. If the device is to resolve the smallest spatial scale of interest, then H^+ should be less than about 3; if the device is also to resolve the smallest temporal scale of interest, then H^+ should be of order 1. It is therefore the temporal requirements that dictate the upper limit on the size of the device. The numbers estimated here are on the conservative side and perhaps slightly larger H^+'s would still provide adequate resolution.

From the discussion so far, it is evident that a Stanton tube of the appropriate design can be used to measure not only the mean but also the instantaneous shear stress. However, as will be shown, there is a relatively serious problem that must be overcome before this can be accomplished. Experiments by Bradshaw & Gregory (1959) with Stanton tubes in laminar and turbulent boundary layers showed that the measured shear stress in a tubulent boundary layer, after being calibrated in a laminar boundary layer, was about 30% higher. Conversely, for the same average shear stress, the pressure measured in the laminar flow is 20% higher than in the turbulent flow. This discrepancy was attributed to a possible influence of wall pressure fluctuations, which in fact are fairly large as pointed out in section 2. However, no definite explanation or mechanism was offered to show exactly how this influence comes about.

Experiments by Fulcher (1986) addressed this issue and the conclusion reached was that it is the nonlinear response of the entire measuring system that is responsible for the observed discrepancies. Her experiments with Stanton tubes of different geometries, similar to those shown in Fig. 4, show that the pressure and the wall shear stress may be related by the same functional form as that found by Bradshaw & Gregory (1959) or Pai & Whitelaw (1969) shown in Table II. Rearranging, we have,

$$\frac{\overline{\tau_x}\rho H^2}{\mu^2} = 10^n \left(\frac{\overline{\Delta P}\rho H^2}{m^2}\right)^m \tag{5a}$$

or

$$\overline{\tau_x} = a\overline{\Delta P}^{\,m} \tag{5b}$$

where, typically, $m \approx 3/4$. The overall height, H, was varied between 0.09mm and 0.23mm and the calibration velocity range, for both laminar and turbulent boundary layers, was between 5 and 40 m/s. It is noteworthy that Trilling & Häkkinen (1955) predict a 3/5 exponent for m in the same range of operating conditions while the measurements of Bradshaw & Gregory (1959) show very nearly a 2/3 dependence. The discrepancies may be due to differences in the probe geometries.

An analysis was performed to estimate, first, the nonlinear effects due to the functional dependence of the pressure on $\overline{\tau_x}$ and, second, the relative magnitude of wall pressure fluctuations and the pressure fluctuations induced by the fluctuating shear stress. Expressing the shear stress as a mean and a fluctuating component and making use of eq. (5b), results in an instantaneous pressure of the form,

$$\Delta P = c(\overline{\tau_x} + \tau_x)^d + p_w \tag{6}$$

where $d = 1/m$ and p_w is the wall pressure due to flow fluctuations. Averaging this equation and neglecting higher order terms yields,

$$\overline{\Delta P} = c\overline{\tau_x}^{\,d}\left(1 + \frac{1}{2}d(d-1)\,(\frac{\tau_x'}{\overline{\tau_x}})^2\right)$$

Using the value of 4/3 for d and 0.4 for $\tau_x'/\overline{\tau_x}$ gives,

$$\overline{\Delta P} = 1.036c\overline{\tau_x}^{\,d}$$

It is thus seen that the measured pressure is about 3.6% higher or that the measured shear stress would be about 3% lower than that measured in a laminar boundary layer with the same mean stress. The conclusion, therefore, is that nonlinearities alone cannot account for the observed discrepancies.

To answer the second question, we must compare the pressure fluctuations, sensed by the device, due to the flow alone and those due to the fluctuating shear stress. In section 2 it was mentioned that the wall pressure fluctuations are roughly 2-4 times the mean shear stress, at least at moderate Reynolds numbers. Solving eq. (5b) to first order for the pressure fluctuations in terms of shear stress fluctuations we obtain,

$$p'_{\tau_x} = d\overline{\Delta P}\,\frac{\tau_x'}{\overline{\tau}_x}$$

After normalizing with $\overline{\tau}_x$ and making use of eq. (5a), the result is,

$$\frac{p'_{\tau_x}}{\overline{\tau}_x} = \frac{1}{m}\,10^{-\frac{n}{m}}\,H^{+\frac{1-m}{m}}\,\frac{\tau_x'}{\overline{\tau}_x}$$

Table III shows the estimated ratio of the shear stress induced pressure fluctuations to the average wall stress for various values of H^+.

TABLE III

Investigation	H^+	$p'_{\tau_x}/\overline{\tau}_x$
Bradshaw & Gregory	3.0	1.28
"	5.0	1.63
"	10.0	2.25
Pai & Whitelaw	3.0	0.188
"	5.0	0.225
"	10.0	0.287

It is seen that in the first case the fluctuating shear stress induced pressure fluctuations are comparable in magnitude to the wall shear and thus to wall pressure fluctuations. In the second case, the shear induced pressure fluctuations are an order of magnitude smaller and thus the situation will be dominated by the wall pressure fluctuations. The difference in the pressure fluctuations measured in the two cases may be attributed to the probe geometries. The problem must therefore be related to both the wall pressure fluctuations and the pressure fluctuations induced by the fluctuating shear stress on the whole measuring system.

This idea was tested by exposing the Stanton probe to a fluctuating pressure environment produced by a speaker. While no quantitative measurements were performed, it was evident that the pressure meter, used to monitor the pressure difference $\overline{\Delta P}$, registered a significant negative <u>average</u> pressure, i.e. a negative contribution to $\overline{\Delta P}$. It seems very likely, then, that the cause of the lower pressures measured in turbulent flow is the nonlinear behaviour of the entire pressure transmitting duct from the device to the pressure meter. The above explanation should be confirmed quantitatively in addition to answering the question of why the measured pressure appears to be always lower for apparently very different measuring setups.

It is not difficult to argue that if a transducer was mounted on the upstream facing step of the Stanton tube, that the measured, average, shear stress would only be affected by the nonlinear relation between shear stress and the resultant pressure in a manner anologous to the operation of a pitot tube in a turbulent flow. Instantaneously, however, the measured pressure would be the result of the shear stress fluctuations and the wall pressure produced by the flow.

For a sufficiently small Stanton tube, completely immersed in the viscous region, one would intuitively expect it to also operate in reverse flow so that the produced pressure difference would now be negative rather than positive. This idea was tested in a flat plate boundary layer and found to be true. Thus, the Stanton tube holds promise for measuring not only the instantaneous wall stress but also its direction.

3.2.4 HEAT TRANSFER METHODS

Other methods that take advantage of the flow characteristics in the viscous region are those that depend on heat transfer from a suitable surface element. A typical device based on this principle, such as the hot-film, is shown schematically in Fig. 5.

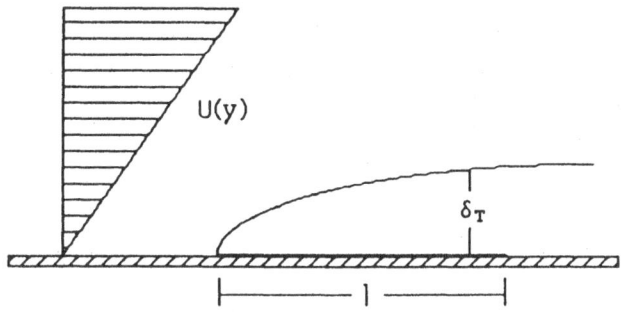

Fig. 5. Schematic of a hot-film.

The Joule heat generated electrically in the device is transfered to the fluid as well as to the susbstrate so that in general,

$$Q_J = Q_f + Q_s \qquad (7)$$

where Q_s represents the heat that is lost to the susbstrate irretrievably while Q_f represents the heat transferred to the fluid both directly from the heated surface and indirectly through the heated portion of the substrate exposed to the fluid. The fundamental assumption relating heat transfer to shear stress is that the thermal boundary layer that develops over the device lies entirely within the linear region of the velocity profile. If, in addition, it is also assumed that the velocity fluctuations grow linearly in amplitude with distance from the wall then, under certain conditions, it is possible to relate the fluctuating shear stress to the fluctuations in heat transfer. Based on all the available evidence to date, it is safe to assume that, within a substantial fraction of the viscous sublayer, the velocity field may be represented by,

$$U^+(y^+, t^+) = y^+ \left\{ 1 + f(t^+) \right\}$$

and

$$w^+(y^+, t^+) = y^+ g(t^+)$$

where f and g have zero mean. Alfredsson et al. (1988) have shown that the rms value of f is very close to 0.4 and that of g should be close to 0.2.

The relation betwen the wall shear and the heat transferred to the fluid, per unit width of the surface, is qiven by,

$$Nu = 0.807(Prl^{+2})^{1/3} \tag{8}$$

where Nu is the Nusselt number, Pr is the Prandtl number and l^+ is the streamwise extent of the heated surface normalized by wall variables (thus the shear stress dependence of Nu is "hidden" in l^+; the combination Prl^{+2} is the Péclet number, Pe, based on the streamwise length of the film and the velocity at the edge of the thermal boundary layer. We prefer to keep this term explicit so that the dependence on the Prandtl number and the probe length can be easily differentiated). The Nusselt number is defined as,

$$Nu = \frac{Q_f}{wk\Delta T}$$

where Q_f is the total amount of heat transferred from the film to the fluid, w is the width of the film (normal to the flow direction), k is the coefficient of thermal conductivity of the fluid, and ΔT is the temperature difference between the film and the fluid.

Use was made of the boundary layer approximation for the thermal boundary layer developing over the heated surface, i.e. that $\delta_T/l \ll 1$, where δ_T is the thermal boundary layer thickness at the downstream end of the surface. As will be shown later, this assumption is not always valid. This relation was derived by Ludwieg (1950) although the functional relationship between heat transfer and wall shear stress was first shown by Fage & Falkner (1931). Liepmann & Skinner (1954) arrived at the functional form of eq. (8) through dimensional arguments. In addition, they considered the effect of a pressure gradient which manifests itself in a departure from the linear velocity profile within the viscous region and the possible breakdown of the boundary layer approximation in adverse pressure gradients, i.e. when the thermal boundary layer grows to be of order l. The velocity at the downstream edge of the thermal boundary layer may be written as,

$$U(\delta_T) = \frac{\overline{\tau}_x}{\mu} \delta_T + (\frac{\partial^2 U}{\partial y^2})_w \frac{\delta_T^2}{2} + \dots \tag{9}$$

But,

$$\mu(\frac{\partial^2 U}{\partial y^2})_w = \frac{\partial \overline{\tau}_x}{\partial y} = \frac{\partial p}{\partial x} \tag{10}$$

so that eq. (9) becomes,

$$U(\delta_T) = \frac{\overline{\tau}_x \delta_T}{\mu} (1 + \frac{\delta_T}{2\overline{\tau}_x} \frac{\partial p}{\partial x} + ...) \tag{11}$$

In a favorable pressure gradient, $\overline{\tau}_x$ is large and the second term in eq. (11) is small. However, in an adverse pressure gradient $\overline{\tau}_x$ becomes small and the contribution of the pressure gradient becomes large and cannot be neglected. Clearly, as separation is approached the situation becomes even worse. If the pressure gradient term in eq. (11) dominates, then eq. (8) becomes,

$$Nu \propto (Prl^{+2})^{1/4}(\frac{1}{\overline{\tau}_x} \frac{\partial p}{\partial x})$$

A more detailed analysis on the effects of both positive and negative pressure gradients can be found in Brown (1967).

We now examine the constraints on the applicability of the boundary layer approximation to the thermal boundary layer as well as on the requirement that the thermal boundary layer lie within the linear region. The latter is important in relating the heat transfer to the wall shear while the former simply tells us whether the heat transfer rate will depend on the 1/3 power of the wall shear. From the analysis of Ling (1963) for the heat transfer from a heated strip on an adiabatic wall, the ratio of the thermal boundary layer thickness, δ_T, at the downstream end of the strip to the length of the strip, l, will be,

$$\frac{\delta_T}{l} = 3.12(Prl^{+2})^{-1/3} \tag{12}$$

The boundary layer approximation will be valid provided $\delta_T/l \ll 1$. Typical commercial films have a streamwise length of about 0.2mm so that at a free stream velocity of 10m/s, $l^+ \approx 5$, for example, in air. The ratio in this case is 1.2, hardly what we would expect. Ling (1963) estimated a correction term for the case where δ_T is of the order of l or $Pe = Prl^{+2} < 500$. The new equation for the heat transfer becomes,

$$Nu = 0.807(Prl^{+2})^{1/3} + 0.19(Prl^{+2})^{-1/6} \tag{13}$$

In the above example, $Prl^{+2} = 17.8$. Thus the relative contribution of the first and second terms are 95% and 5% respectively and consequently one would not expect to see much of a difference if the second term in eq. (13) were neglected.

However, the analysis of Ackerberg, Patel & Gupta (1978) for the same problem at low Péclet numbers, using matched asymptotic expansions, as well as their experimental results over a Péclet number range of 10^{-4} to 10^6 shows that there is a serious problem for Pe < 100. Table IV lists the experimental results of Ackerberg et al. (1978) and Ling (1963) for comparison purposes.

TABLE IV

Pe	Nu_{APG}	Nu_L	%error
10^6	80.0	80.7	0.9
10^4	17.1	17.4	1.8
10^2	3.80	3.83	0.8
10^1	2.16	1.89	-12.5
10^0	1.27	1.00	-21.3
10^{-2}	0.70	0.583	-16.7
10^{-4}	0.45	0.919	104.2

It is evident that the theory by Ling (1963) is very good down to Pe = 100, but that below that the error begins to increase rapidly. More importantly, however, the heat transfer rate does not depend any more on the 1/3 power of the wall stress; in fact, it becomes even less sensitive.

The second constraint is that δ_T should not exceed the height of the linear region, i.e. $\delta_T^+ < 5$. Solving in terms of known parameters we obtain,

$$l^+ < 4.1 Pr \tag{14}$$

It is evident once again that the streamwise extent of the device is severely limited in air while in liquids, such as water for example, the situation is not as critical. On the other hand if l^+ is too large there will a spatial resolution problem, a restriction imposed by the flow structure above the wall. In terms of importance, clearly eq. (14) is the one to pay attention to. If the thermal boundary layer extends beyond the linear region, then there is not a unique correspondence between heat transfer and wall stress. In practice, shear stress probes should be calibrated and whether they obey a simple 1/3 power law or not is immaterial.

Important aspects of heat transfer device operation are their sensitivity and frequency response. Since the most typical such device is the flush mounted hot-film, we shall limit our discussion to this device for the time being. Most hot-films are operated in the constant temperature mode and, because of their low mass, should in principle have a high frequency response. As was mentioned earlier, part of the heat generated in the film ends up directly into the fluid while the rest heats the substrate. In turn, the substrate is cooled by the flow above it. The net results of this process are:

1) The heated substrate participates in the heat transfer process to the fluid, resulting in a larger effective l^+.

2) The higher the percentage of the total heat produced that goes into the susbstrate, the lower the sensitivity of the device to changes in the wall shear.

3) The heat transfer process from the film itself has a much lower time constant than that from the substrate. As a result, the static calibration in a

laminar or turbulent flow will give the wrong results when used in a turbulent flow.

In order to address these three points it will be usefull to rewrite eq. (7) in a more explicit form for the case where the film is heated electrically. In the constant temperature mode, the temperature difference ΔT is constant and the heat input is proportional to the square of the voltage across the film. Thus, we have,

$$E^2 = A(Prl_c^{+2})^{1/3} + E_s^2 \tag{15}$$

where l_c^{+2} is proportional to the wall shear stress. The effective length, l_c, in a laminar or turbulent static calibration will be greater than the physical length because of the heat that ends up into the fluid through the substrate. In addition, the higher E_s^2 is, compared to the direct heat transfer term in eq. (15), the less E will change with changing shear stress. The magnitude of E_s^2 can be obtained from the intercept of the calibration curve with the E^2 axis i.e. at zero shear. Figure 6 shows schematically a typical calibration of a hot-film shear stress device. For simplicity, the heat transfer is shown depending on the 1/3 power of the wall shear stress. As is the usual practice, the calibration is obtained under static conditions, i.e. the device is exposed to several values of the wall stress whose average value does not change in time.

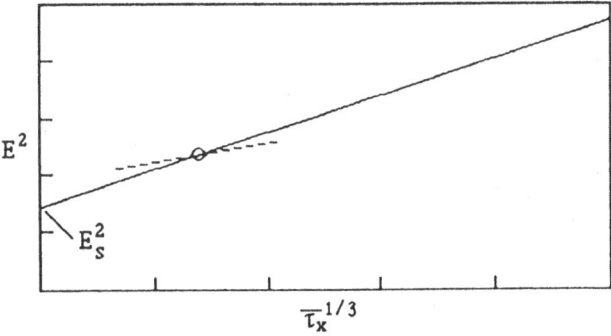

Fig. 6 Schematic of hot-film calibration.

In a fluctuating shear stress environment, the film will respond very quickly to the shear stress fluctuations while the substrate, with its much larger thermal mass, will not. The effective length of the film is now much smaller and closer to its actual value with the result that at equal instantaneous values of the wall stress to those used for the calibration, the heat transfer is less. This situation is depicted by the dotted line in Fig. 6. If we rewrite eq. (15) explicitly in terms of the effective length and the wall shear we obtain,

$$E^2 = APr^{1/3}l_c^{2/3}\bar{\tau}_x^{1/3} + E_s^2 \tag{16}$$

where the effective length is that corresponding to the calibration and hence is larger than the physical length. The sensitivity $\partial E/\partial\bar{\tau}_x$ from the static calibration is,

$$\frac{\partial E}{\partial\bar{\tau}_x} = \frac{APr^{1/3}}{6E\bar{\tau}_x^{2/3}} l_c^{2/3} \tag{17}$$

Equation (17) is usually assumed to hold in the presence of a fluctuating stress but now the effective length is smaller as discussed above. All other quantities being equal, the ratio of the fluctuating sensitivity, S_f, to the average sensitivity, S_a, will be,

$$\frac{S_f}{S_a} = (\frac{l_{cf}}{l_{ca}})^{2/3}$$

where l_{ca} is the average effective length during calibration and l_{cf} is the effective length under fluctuating stress conditions. Brown (1967), for example, reports that the effective length from a static calibration of a hot-film was about twice the physical length. If the effective length under fluctuations is taken to be equal to the physical length, then the ratio of sensitivities is only 0.63. If this state of affairs is typical, then it would go a long way in explaining the large variations of τ_x' reported in the literature (see Alfredsson et al., 1988). If the difference between the effective lengths is even larger, it becomes obvious that fluctuating shear stress measurements will be seriously in error. This will be true even if all higher moments of the voltage in eq. (17) are taken into account during calibration.

The key to getting around this problem is to either minimize the heat transfer to the substrate or to maximize the ratio of heat transferred to the fluid relative to that going into the substrate. The use of a hot-wire, in air, resting on or very close to the wall works very well as shown by Alfredsson et al. (1988). In this case it is the heat transfer to the wall that is minimized relative to a hot-film. The results of using flush-mounted, cylindrical hot-films in liquids show the right trends. The Pr number is one (water) to two (oil) orders of magnitude larger than that of air in which case the relative heat transfer, compared to air, is approximately 2 and 4.5 times larger. For example, the same flush-mounted hot-film probe used in air and oil, gives the correct answer for τ_x' in oil and is 75% too low in air. The results obtained with a cylindrical hot-film in oil are 3% lower than the numerical results of Kim et al. (1987). This difference may be attributed to the fact that the film width was equal to 25 wall units and, according to the results of Blackwelder & Haritonidis (1983) for hot-wires, some filtering of small scales may have occurred.

The response of the hot-film depends, on the low end, on the substrate material and the amount of heat lost to the substrate and, on the high end, on the size of the film and the heat transfer process relating to the film only. An estimate of the substrate role can be obtained by considering the propagation of heat waves through a semi-infinite solid slab subjected to periodic temperature fluctuations at one end (Blackwelder 1981, Carslaw & Jaeger, 1959). The wavelength, λ, of the heat wave is given by,

$$\lambda = 2(\frac{\pi\alpha_s}{f})^{1/2}$$

where α_s is the thermal diffusivity of the solid and f is the frequency of the imposed temperature fluctuation. The relevant length scale to compare to is the hot-film length. Strictly speaking, the width should also be taken into account, but we shall limit our discussion to large aspect ratio films. Taking the ratio of the two lengths and normalizing with wall variables we obtain,

$$\frac{\lambda}{l} = 2\left(\frac{\pi\tau^+}{Prl^{+2}}\right)^{1/2}(\frac{\alpha_s}{\alpha_f})^{1/2}$$

where τ^+ is the period of the fluctuation and α_f is the thermal diffusivity of the fluid. The analysis of Carslaw & Jaeger (1959) shows that the amplitude of the thermal wave will attenuate to a fraction of a percent over a distance equal to its wavelength. Thus the ratio of the wavelength, λ, to the length of the substrate is an indication of the extent to which the substrate will absorb heat from the heated surface, on the one hand, and return it to the flow, on the other. Table V shows the ratio λ/l for a number of fluids, films, wire on the wall (HWW), cylindrical hot films on the wall (HFW), streamwise lengths, and substrates for τ^+ = 1.0. The three velocity ranges correspond to experiments in a flat plate boundary layer, a channel flow with water and a channel flow with oil and are the same as those described in Alfredsson et al. (1988).

TABLE V

U_∞ m/s	Probe	Substrate	α_s m^2/s	Fluid	Pr	α_f m^2/s	l$^+$	λ/l
10	Film	Quartz	8.3×10^{-7}	Air	0.7	2.1×10^{-5}	5.3	0.16
10	HWW	Plexig.	1.7×10^{-7}	Air	0.7	2.1×10^{-5}	0.03	13.0
0.4	Film	Quartz	8.3×10^{-7}	Water	7.0	1.4×10^{-7}	5.3	0.62
0.4	HFW	Plexig.	1.7×10^{-7}	Water	7.0	1.4×10^{-7}	1.0	1.5
0.2	Film	Quartz	8.3×10^{-7}	Oil	82.0	7.3×10^{-8}	4.0	0.15
0.2	HFW	Plexig.	1.7×10^{-7}	Oil	82.0	7.3×10^{-8}	0.2	3.0

It is evident that in all the cases involving films, the ratio is less than 1.0 indicating that at high frequencies, i.e. $\tau^+ \approx 1$, the substrate will not participate in the heat transfer process and they will respond correctly to stress fluctuations. For all the other cases, i.e. the wire or cylindrical film on the wall, the indication is that there should be a problem. It should be remembered, however, that the heat transfer process from the hot-wire or the cylindrical hot-film is different from that of a plane hot-film so that a direct comparison can not be made. The fact that the results obtained with the cylindrical probes are much closer to the correct values, suggests that the heat transfer to the wall is much less that indicated in the table. This is not surprising when considering the fact that these probes are at most making contact with the wall along a "line" thus limiting the heat transfer. The period $\tau^+ \approx 1.0$ represents the highest frequency that we would expect in a wall flow. Most of the time, however, τ^+ would be one to two orders of magnitude higher in which case the ratio of λ/l would increase by about an order of magnitude. As a result, the response of the hot film will approach that of a steady flow calibration, but will deviate from the response under rapid shear stress fluctuations.

The time constant for the high frequency limit will be determined by the longer of two time constants. The first, τ_a, is the advection time determined by the time needed for a change of wall shear, initially upstream of the film, to "sweep" across the film length. The velocity at the edge of the thermal boundary layer, at the downstream end of the heated surface, is taken as the advection velocity. This time, made nondimensional with wall variables, is given by,

$$\tau_a^+ = \frac{1}{\delta_T}$$

$$= \frac{1}{3.12} (Prl^{+2})^{1/3}$$

It should be mentioned that this result, which is the inverse of eq. (12), is probably too low (see discussion following eq. 12). The second time constant, τ_d, is that associated with the adjustment of the thermal boundary layer from its old state to its new state assuming an impulsive change in wall shear, and is given by,

$$\tau_d^+ = (3.12)^2 (Prl^{+2})^{1/3}$$

This time is comparable to the time given by Bellhouse & Shultz (1968), but the numerical constant appears to be too large by a factor of three if we compare the above expression to their data, at the 3dB point. It is evident, however, that the diffusion time constant is by far the most important one and it will therefore dictate the upper frequency limit of the device.

The conclusion to be drawn from the above discussion is that traditional, flush-mounted, hot-film probes should be used with extreme caution. Regardless of the type of device used, a calibration against, at least, a known average stress is always necessary. Since not much may be known about the fluctuating stress field, it is advisable to relate the wall stress to the output voltage in polynomial form, i.e.

$$\tau_x = c_0 + c_1 E + c_2 E^2 + ... \tag{18}$$

where the stress and the voltage are <u>instantaneous</u> values. A third order polynomial will suffice for most applications. The use of eq. (16), for example, implies that we already know something about the fluctuating wall stress since we have to deal with both the average value of the wall stress <u>and</u> its higher moments. Equation (18) requires that we know only the average wall stress; the moments of the measured voltage can be easily computed.

4. DIRECT METHODS
4.1 FLOATING ELEMENT: CONVENTIONAL

Floating element techniques rely directly on the force exerted by the wall shear on an element (portion) of the wall. The wall element is mounted on a balance of which there are two basic types: the displacement balance in which the displacement of the wall element is correlated to the applied force, and the feedback balance in which the force required to maintain the wall element at its original position is equal to the applied force. Knowing the force and the area of the element allows for the calculation of the wall shear stress. A schematic of such a device is shown in Fig. 7

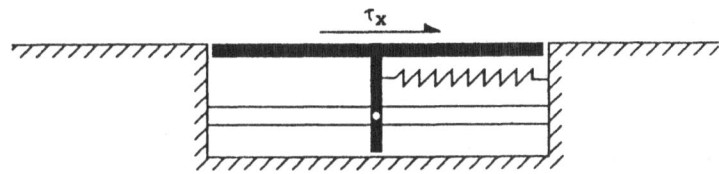

Fig. 7. Schematic of a floating element device with pivoted support.

The advantage of such a scheme is that the wall stress is measured directly without having to make any assumptions about the flow field above the device. Unfortunately, there are so many disadvantages of the traditional device that its use has been rather limited. Winter (1977) lists a number of problems relating to the use of such devices:

1) The compromise between the size of the transducer and the measurement of very small forces.
2) The effect of the necessary gaps around the element.
3) The effect of misalignment of the element with respect to the surrounding surface.
4) Forces due to pressure gradients.
5) The effects of heat transfer.
6) The use in the presence of suction or blowing.
7) The effects of gravity or acceleration if the device is to be used in a moving vehicle.
8) The effects of temperature changes.
9) The effects of leaks.
10) The effects of the whole system in the presence of large transients.

The last five problems are of a more specialized nature while the first five are generic to all such devices under normal operating conditions. To the above list we could add some additional problems or limitations of traditional devices:

11) Poor frequency responce because of their large size, thus making them unsuitable for use in measuring the fluctuating shear

12) Difficulties of handling and installation due to the very delicate nature of the device.

It is evident from the above list that there are a host of problems that affect the operation of floating element devices. It is not so much that any one of them is so detrimental, although it could be, but that collectively they make the use of existing devices cumbersome and difficult.

In terms of the mounting of the element to the supporting structure, there are two basic categories: pivoted support (Fig. 7) and parallel linkage support (Fig. 8).

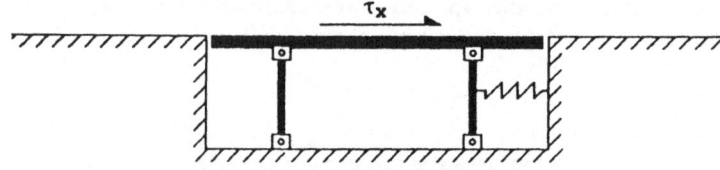

Fig. 8. Schematic of floating element device with parallel linkage support.

The pivoted support allows for the cancellation of acceleration effects through the addition of a counterweight below the pivot point. The degree to which the overall construction must be delicate can be shown through an example. At a typical free stream velocity of 10m/s in a flat plate boundary layer and an $Re_\theta = 5,000$, the wall shear is $\bar{\tau_x} = C_f q = 1.83 Pa$. If the area of the element is $1 cm^2$, then the applied force to the element is $0.000183N$ or $0.0187gr$ of force. Thus, for a resolution of 1% of this measurement, the instrument should be capable of resolving a force equal to $187 \times 10^{-6} gr$, which is a very small force indeed.

The applied force to the element is sensed through the element displacement regardless of the method of transduction. It is therefore necessary to surround the element with a small gap. The presence of the gap itself, in addition to any step (misalignment) between the element surface and the surrounding surface, will cause additional forces to act on the element. Assuming the step to be of the order of a wall unit, then we can apply the arguments presented in section 3.2.3 about the pressures generated on upstream and downstream facing steps (or recesses) in connection with Stanton tubes. The pressure exerted on the upstream or downstream facing surface of a step or recess will be of the order of the wall shear stress for $H^+ \ll 1$ and about an order of magnitude, or more, higher for $H^+ \gg 1$, where H^+ is the nondimensional step height. Thus the measured wall stress, $\overline{\tau_{xm}}$, will be,

$$\overline{\tau_{xm}} \approx \overline{\tau_x} \left(1 + c \frac{\text{step area}}{\text{element area}} \right) \qquad (19)$$

where c will be positive for a protrusion and negative for a recess and the numerical value will depend on H^+ as outlined above. The above model assumes a rectangular element whereas in practice most elements are round. For a step height, H, of the order of 0.1% of the streamwise dimension of the element, we would expect an error of the order 0.1% if H^+ is of order 1 and 1 - 10%, or even higher, for $H^+ \gg 1$.

Allen (1976, 1980) made a systematic study of the effects of steps H, gaps g, and lip sizes s, of a 12.7cm diameter floating element using both pivoted and parallel supports. These dimensions are shown in Fig. 9.

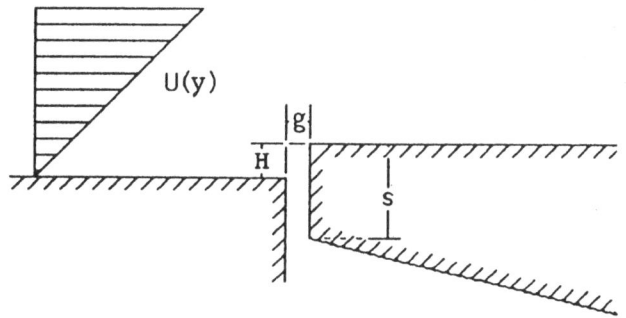

Fig. 9. Typical gap/step geometry of a floating element.

The tests were made in a supersonic flow so some caution will be necessary in extrapolating his conclusions to subsonic flows. From Allen (1976, 1977, 1980) it was estimated that one wall unit corresponds to about 10μ. His main conclusions can be summarized as follows:

1) Protrusions and recesses produce errors of the same magnitude, but of opposite sign in agreement with the discussion on Stanton tubes. However, the moments produced by these pressures on the lip surfaces and the element surface tend to cancel in the case of a recessed element and add in the case of a protruding element in agreement with the arguments given by Winter (1977). Elements supported in the pivot mode are particurlarly prone to these errors.

2) The error for a zero step size is ± 2% provided g/D is between 0.004 and 0.010 or, in wall units, between 51 and 127. For smaller values of g/D, the error can be as much as -8%. These are rather large gaps and one wonders what the effects would be in low speed flows. Smith & Walker (1959) report that at low Mach numbers (0.1 - 0.3), and for g/D = 0.005, a depression of H/D = 0.00025 did not affect the measured shear while a comparable protrusion (it is implied) caused noticeable errors.

3) Small gaps result in larger errors for the same step height. For $H^+ \approx \pm 8$, the error is less than ± 5% for g/D = 0.010 or $g^+ \approx 130$, but for g/D = 0.001 or $g^+ \approx 13$ the error is ± 50%. These results are in agreement with the arguments given in connection with eq. (19).

4) The lip size has a rather dramatic effect on the error. For $s \approx 0$, i.e. for a sharp lip, the error is at most a few percent as long as g^+ is less than about 15. As the gap gets bigger, the sensitivity to lip size diminishes, but for a given gap size the sensitivity increases as the lip size increases. The lip size appears to affect the measured force through both a pressure differential across the element and a flow through the gap. Balances of the pivot-type are more prone to this error as the differential presure around the element tends to produce a moment about the pivot point which is the same result produced by the shear on the surface of the element.

One way to get around the gap problem is to fill the gap with a liquid as proposed by Frei & Thomann (1980). This method appears to eliminate the problems just discussed, but introduces yet another complication in the whole mechanism. A different approach is to reduce the gap to dimensions that are a fraction of a wall unit thus minimizing rather than eliminating the gap problem. A first attempt at this approach is discussed in the next section.

4.2 FLOATING ELEMENT: MICROMACHINED

Schmidt, Howe, Senturia & Haritonidis (1988) have built a prototype of a micro-element using microfabricating techniques, shown in Fig. 10. The area of the element is $500\mu \times 500\mu$ and is supported by four tethers that act both as supports and restoring springs.

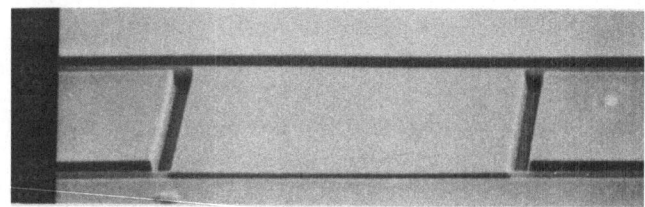

Fig. 10. Micromachined floating element. The SEM has distorted
the square element area which now appears rectangular.

The tether length is 1mm and the cross sectional dimensions are 10μ wide by 30μ high. The element, which is also 30μ thick, is suspended 3μ above the silicon substrate on which it is fabricated. The gap on either side of the tethers as well as between the element and the surrounding surface is 10μ while the element surface is flush to the surrounding surface within 1μ. The gaps in this device are about an order of magnitude smaller than those found in conventional floating elements. The element and its tethers is made of polyimide with an embedded conductor which is necessary for the operation of the differential capacitance readout scheme, shown in Fig. 11.

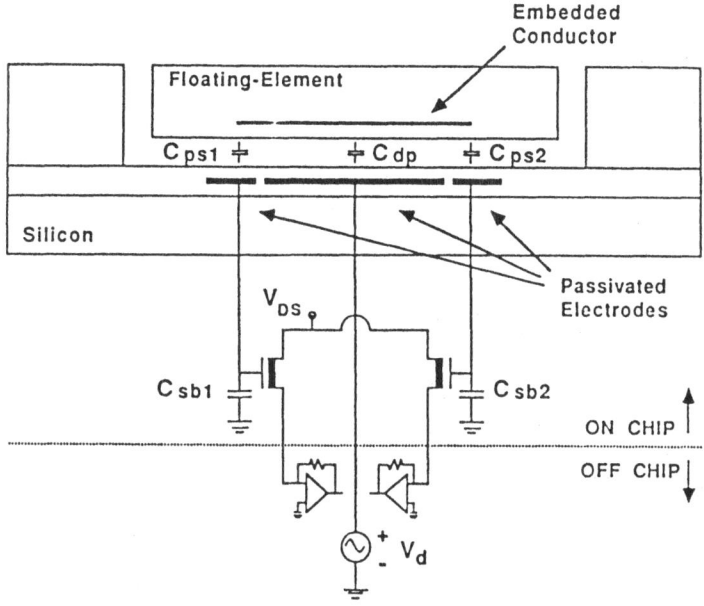

Fig. 11. Differential capacitance readout scheme.

The differential mode of operation makes the device insensitive to normal displacements of the element. Normal displacement is not expected unless wall pressure fluctuations have a scale smaller then the element area. For larger scales it is expected that the pressure will equalize above and below the element. This was verified experimentally by Schmidt (1988) who subjected the device to acoustic pressures of 84dB and frequencies of 50Hz, 100Hz, 600Hz and 3.1KHz. His results showed that the device was insensitive to these frequencies except at 3.1KHz were the plate was estimated to move less than 0.1Å.

A calculation was made of the resonance frequency of the element for the particular geometry and materials used. The resonant frequency was found to be 9.3KHz for either lateral, i.e. in the plane of the element and normal to the tethers, or normal motion. In the model for these calculations, the tethers are assumed to be wires under tension which is indeed the case. The damping of the plate in motion parallel to itself was modelled assuming Couette flow in the gap between the element and its substrate. This is justified as the diffusion time based on ν and the gap height, $t_{gap} = 3\mu$, is $0.6\mu s$. The quality factor, Q, was calculated to be 378, thus implying that the present device is capable of measuring not only the average, but also the fluctuating shear stress.

A typical calibration, in a specially designed laminar flow cell, is shown in Fig. 12.

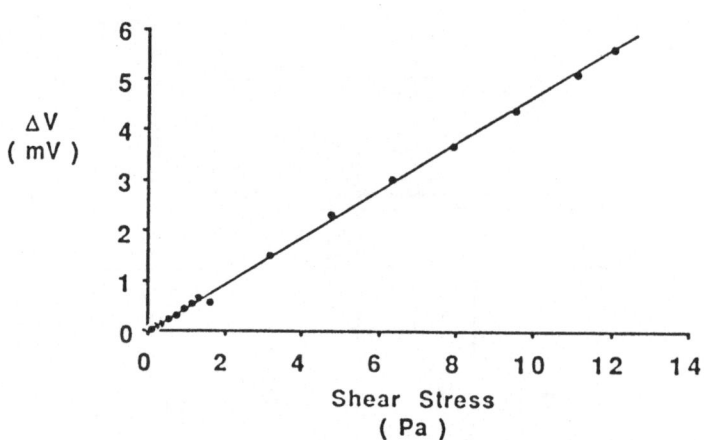

Fig. 12. Typical calibration of micromachined floating element.

No test under fluctuating stress conditions could be made because of two problems. The first was a rather high sensitivity to temperature of the specially packaged device for use in a wind tunnel turbulent boundary layer. The second was the very small signal amplitude available from the device itself.

Based on the analysis presented and the data obtained so far, it appears that there is every reason for it to perform successfully. The fact that the device can be further reduced in size by about an order of magnitude, holds promise for use in high speed flows as well as for the detailed mapping of the instantaneous spatial and temporal patterns of wall stress. Using a different geometry could provide not only the two components of wall stress, but also the normal component of vorticity at the wall.

5. CONCLUSIONS

The present survey was an attempt to discuss the problems of the most common techniques and warn of their pitfalls. Under fairly well established conditions, certain instruments, such as the Preston tube, can give reliable answers. However we would never use a Preston tube or any other instrument without first calibrating it. The need for calibration is even more accute for those instruments that are supposed to measure the fluctuating shear stress. The most satisfactory method is to monitor the velocity very near the wall, in itself a very difficult and troublesome measurement. When this is not possible, a comparison of the higher moments of the shear stress with known results is mandatory.

Given the capability of modern computational methods, it is now possible to check in detail the behaviour of hot-films on heat conducting substrates and Stanton tubes, of different geometries, in two- and three-dimensional flows. While analytical methods can provide valuable insight or trends, they can not provide the accuracy or detail needed for the evaluation of particular probe configurations or the operation of probes in a regime for which the theory becomes too complicated.

No doubt, improved or new instruments will soon appear that extend the rather poor or unreliable capabilities of existing instruments. We tend to think that micromachining techniques will play an important role in the near future.

ACKNOWLEDGMENTS

Thanks are due to Jim Wallace and Ömer Savas for reading the manuscript and offering helpful suggestions. Jean Sofronas assisted with the preparation of the manuscript.

REFERENCES

Ackerberg, R.C., Patel, R.D., & Gupta, S.K., 1978. "The Heat/Mass Transfer to a Finite Strip at Small Péclet Numbers". J. Fluid Mech., **86**, pp. 49-65.

Alfredsson, H., Johansson, A.V., Haritonidis, J.H., & Eckelmann, H., 1988. "The Fluctuating Wall-Shear Stress and the Velocity Field in the Viscous Sublayer". Phys. Fluids, **31**, pp. 1026-1033.

Allen, J.M., 1976. "Systematic Study of Error Sources in Supersonic Skin-Friction Balance Measurements". NASA TN, D-8291.

Allen, J.M., 1977. "Experimental Study of Error Sources in Skin Friction Balance Measurements". J. Fluids Eng., March 1977, pp. 197-204.

Allen, J.M., 1980. "Improved Sensing Element for Skin-Friction Balance Measurements". AIAA Journal, **18**, No. 11, pp. 1342-1345.

Bellhouse, B.J., & Schultz, D.L., 1968. "The Measurement of Fluctuating Skin Friction in Air with Heated Thin-Film Gauges". J. Fluid Mech., **32**, pp. 675-680.

Bidwell, J.M., 1951. "Application of the Von Kármán Momentum Theorem to Turbulent Boundary Layers". NACA TN 2571.

Blackwelder, R.F., 1981. "Hot-Wire and Hot-Film Anemometers", in *Methods of Experimental Physics: Fluid Dynamics,* (ed. R.J. Emrich), **18**A, pp. 259-314.

Blackwelder, R.F., & Haritonidis, J.H., 1983. "Scaling of the Bursting Frequency in Turbulent Boundary Layers". J. Fluid Mech., **132**, pp. 87-103.

Bradshaw, P. & Gregory, N., 1959. "The Determination of Local Turbulent Skin Friction from Observations in the Viscous Sub-Layer". ARC, R. & M. No. 3204.

Brown, G.L., 1967. "Theory and Application of Heated Films for Skin Friction Measurement". Proc. Heat Transfer and Fluid Mech. Inst., pp. 361-381.

Carslaw H.S., & Jaeger, J.C., 1959. *Conduction of Heat in Solids*, Oxford at the Clarendon Press, London.

Chue, S.H., 1975. "Pressure Probes for Fluid Measurement". Prog. Aero. Sci., **16**, No. 2, pp. 147-223.

Coles, D.E., 1956. "The Law of the Wake in the Turbulent Boundary Layer". J. Fluid Mech. 1, pp. 191-226.

Dutton, R.A., 1956. "The Accuracy of the Measurement of Turbulent Skin Friction by Means of Surface Pitot-Tubes and the Distribution of Skin Friction on a Flat Plate". ARC, R. & M. No. 3058.

Fage, A. & Falkner, V.M., 1930. "An Experimental Determination of the Intensity of Friction on the Surface of an Aerofoil". ARC, R. & M. 1315.

Fage, A. & Falkner, V.M., 1931. "On the Relation Between Heat Transfer and Surface Friction for Laminar Flow". ARC, R. & M. No. 1408.

Frei, D. & Thomann, H., 1980. "Direct Measurements of Skin Friction in a Turbulent boundary Layer With Strong Adverse Pressure Gradient". J. Fluid Mech., **101**, pp. 79-95.

Fulcher, K.L., 1986. "A Probe for Measuring Skin Friction in Disturbed Turbulent Boundary Layers". M.S. Thesis, Dept. of Aeronautics and Astronautics, M.I.T.

Gresko, L.S., 1988. "Characteristics of Wall Pressure and Near-Wall Velocity in a Flat Plate Turbulent Boundary Layer". M.S. Thesis, Dept. Of Aeronautics and Astronautics, M.I.T.

Hanratty, T.J. & Campbell, J.A., 1983. "Measurement of Wall Shear Stress" in *Fluid Mech. Measurements*, (ed. R.J. Goldstein). Hemisphere Publ. Corp., pp. 559-615.

Haritonidis, J.H., 1989. "A Model for Near-Wall Turbulence". To appear in Physics of Fluids.

Kim, J., Moin P., & Moser, R.D., 1987. "Turbulence Statistics in Fully-Developed Channel Flow at Low Reynolds Number". J. Fluid Mech., **177**, pp. 133-166.

Liepmann, H.W. & Skinner, G.T., 1954. "Shearing-Stress Measurements by Use of a Heated Element". NACA TN 3268.

Ling, S.C., 1963. "Heat Transfer from a Small Isothermal Spanwise Strip on an Insulated Boundary". J. Heat Transfer, **C85**, pp. 230-236.

Ludwieg, H., 1950. "Instrument for Measuring the Wall Shearing Stress of Turbulent Boundary Layers". NACA TM 1284.

Macmillan, F.A., 1954. "Viscous Effects on Flattened Pitot Tubes at Low Speeds". J. Roy. Aeronaut. Soc., **58**, pp. 837-839.

Nikuradse, J., 1950. "Laws of Flow in Rough Pipes". NACA TM 1292.

Pai, B.R., & Whitelaw, J.H., 1969. "Simplification of the Razor Blade Technique and its Application to the Measurement of Wall-Shear Stress in Wall-Jet Flows". Aero. Quarterly, **20**, pp. 355-364.

Patel, V.C., 1965. "Calibration of the Preston Tube and Limitations on its Use in Pressure Gradients". J. Fluid Mech., **23**, pp. 185-208.

Prahlad, T.J., 1972. "Yaw Characteristics of Preston Tubes". AIAA J., **10**, pp. 357-359.

Preston, J.H., 1954. "The Determination of Turbulent Skin Friction by Means of Pitot Tubes". J. Royal Aeronaut. Soc., **58**, pp. 109-121.

Rajaratnam, N. & Muralidhar, D., 1968. "Yaw Probe Used as Preston Tube". Aeronaut. J., **72**, pp. 1059-1060.

Rechenberg, I., 1963. "The measurement of Turbulent Wall Shear Stress". ZFW, **11**, No. 11, pp. 429-438.

Schewe, G., 1986. "On the Structure and Resolution of Wall-Pressure Fluctuations Associated with Turbulent Boundary-Layer Flow". J. Fluid Mech., **134**, pp. 311-328.

Schlichting, H., 1968. *Boundary Layer Theory*, McGraw-Hill, New York.

Schmidt, M.A., 1988. "Microsensors for the Measurement of Shear Forces in Turbulent Boundary Layers". Ph.D. Thesis, Dept. of Electrical Engineering & Computer Science, M.I.T.

Schmidt, M.A., Howe, R.T., Senturia S.D., & Haritonidis, J.H., 1988. "Design and Calibration of a Microfabricated Floating-Element Shear-Stress Sensor". IEEE Transactions on Electron Devices, **35**, No. 6, pp. 750-757.

Smith D.W., & Walker, J.H., 1959. "Skin-Friction Measurements in Incompressible Flow". NACA TR, R-26.

Spalart, P.R., 1986. "Direct Simulation of a Turbulent Boundary Layer up to $R_\Theta=1410$". NASA TM 89407.

Spalding, D.B., 1961. "A Single Formula for the 'Law of the Wall' ". J. App. Mech., Sept. 1961, pp. 455-458.

Staff of Aerodynamics Division, N.P.L., 1958. "On the Measurement of Local Surface Friction on a Flat Plate by Means of Preston Tubes". R. & M. No. 3185.

Stanton, T.E., Marshall, D., & Bryant, C.N., 1920. "On the Conditions at the Boundary of a Fluid in Turbulent Motion". Proc. Roy. Soc. A, **85**, pp. 365-434.

Tani, I., 1987. "Drag Reduction by Riblet Viewed as Roughness Problem". Proc. Japan Acad., **64**(B), pp. 21-24.

Taylor, G.I., 1938. "Measurements with a Half-Pitot Tube". Proc. Roy. Soc. A, **166**, pp. 476-481.

Trilling, L. & Häkkinen, R.J., 1955. "The Calibration of the Stanton Tube as a Skin Friction Meter". 50 Jahre Grenzschichtforschung, pp. 201-209.

Van Driest, E.R., 1956. "On Turbulent Flow Near a Wall". J. Aero. Sci., **23**, pp. 1007-1011 & 1036.

Winter, K.G., 1977. "An Outline of the Techniques Available for the Measurement of Skin Friction in Turbulent Boundary Layers". Prog. Aerospace Sci., **18**, pp. 1-57.

THE MEASUREMENT OF VORTICITY IN TRANSITIONAL
AND FULLY DEVELOPED TURBULENT FLOWS

John F. Foss
Michigan State University
East Lansing, MI 48824

James M. Wallace
The University of Maryland
College Park, MD 20742

ABSTRACT   The significance of vorticity and the importance of its temporally and spatially resolved measurement in turbulent flows is discussed. The measurement methods are categorized as those which use multipoint measurements of velocity to determine the required velocity gradients or circulation and those which sense these gradients directly. The characteristics and performance of these vorticity measurement methods are evaluated. Where available, measurements of some vorticity field statistics by two or more methods are compared.

LIST OF SYMBOLS

A	area
B	the magnetic field intensity
C	closed contour defining the circulation (see equation 2a)
h	hot-wire probe prong spacing
n	outward normal to surface
p	(static) pressure
t	time
u_i	velocity vector (cartesian tensor notation)
$\vec{V}$	velocity vector (vector notation)

Greek

α	hot-wire sensor angle to cross-stream plane
δ	distance around a closed contour (see equation 2a)
ϵ_{xy}	strain-rate component
ϕ	potential of the electric field
Γ	circulation (see equation 2)
γ_{xy}	hot-wire probe pitch angle in the xy plane
η	Kolmogorov dissipation length scale
ν	kinematic viscosity
Ω	vorticity ($\vec{\Omega}$ vector notation, Ω_i cartesian tensor notation; see equation 1)
$\omega_x, \omega_y, \omega_z$	cartesian vorticity components in an x,y,z reference frame
ρ	density

Subscripts

a_I the summation of the accelerative effects that permits the absolute acceleration A to be expressed as the sum of the acceleration in the local reference frame plus the Coriolis acceleration (2Ω x V), the centripetal acceleration (Ω x Ω x R) and the unsteady effect ($d\Omega/dt$ x R), (see equation 6).

S a portion of C that borders a physical surface (see equation 7)

u,v,w   cartesian velocity components in an x,y,z, reference frame

U_c convection velocity used in Taylor's hypothesis

u_r, v_θ, w cylindrical coordinate (r,θ,z) velocity components

Superscripts

     denotes normalized root-mean square values

1.0 INTRODUCTION

The use of vorticity as a primitive variable for theoretical considerations of a given flow field is well established. Lighthill (1963) and Batchelor (1967) provide extensive and quite instructive descriptions of such considerations. The observations presented in the following section are to provide new examples of these concepts which will also be used in the interpretation of the measurement techniques and some of the experimental observations.

A strong motivation for the present authors' research efforts and for their preparation of this review is predicated upon the following proposition. *The contribution of vorticity considerations to the understanding of various fluid dynamics phenomena will be strongly enhanced by the increased use of temporally and spatially resolved measurements of one or more components of the vorticity vector.*

A singularly important example of the contribution of direct vorticity measurements can be used to emphasize this general point. All turbulent shear flows that are bounded by an irrotational free stream exhibit a sharp increase in the vorticity magnitude as the observation point is advanced through the "viscous superlayer". Corrsin and Kistler (1954) have described this phenomena and provide valuable theoretical considerations regarding the relationship of the superlayer properties to the bounded turbulent motion. Numerous investigators have used a surrogate property to assess whether an instantaneous measurement is taken inside or outside the viscous superlayer; however, it is apparent that this assessment is best made by using time resolved vorticity measurements and the defining condition [$\Omega_i + \partial\Omega_i/\partial x_j$] $\neq$ 0 for fluid elements within the bounding superlayer.

The primary purpose of this paper is to provide a statement of the current state-of-the-art for such measurements and to interpret some of the contributions which have been made by them.

2.0 BASIC EQUATIONS AND DEFINITIONS

The continuously distributed three-dimensional velocity vector field can be operated upon, at a given instant in time, to provide a second three-dimensional and continuously distributed vector field

$$\vec{\Omega} = \nabla \times \vec{V} \; , \tag{1}$$

where $\vec{\Omega}$ is twice the angular velocity about the centroid of an elementary fluid particle. A companion relationship defines the circulation (Γ) as the integral values

$$\Gamma = \oint_C \vec{V} \cdot d\vec{S} = \int \vec{\Omega} \cdot \hat{n} \; dA \; . \tag{2a),(2b}$$

The NCFMF motion picture <u>Vorticity</u>[1] of A.H. Shapiro is recommended as an excellent source for visual representations of the vorticity and the circulation.

For incompressible flows of a Newtonian fluid[2], the velocity and pressure fields can be used to fully characterize the motion; the Navier-Stokes equations provide the appropriate analytical expressions for such a description. Likewise, *the vorticity and the pressure field at a physical surface* provide a complete description of a flow field (see Article II.1 of Lighthill). It is assumed that the reader is quite familiar with a flow field description that is based upon the $\vec{V}$ and p distributions; these will be used as the bases for the presentation of the analogous elements of the vorticity field.

2.1 The vorticity transport equation

The velocity-pressure description of an incompressible flow of a Newtonian fluid has the following governing equations:

Conservation of Mass $\qquad \nabla \cdot \vec{V} = 0 \tag{3}$

and

Conservation of Momentum $\qquad \dfrac{D\vec{V}}{Dt} = -\dfrac{1}{\rho} \nabla p + \nu \nabla^2 \vec{V} \; . \tag{4}$

These two equations provide a complete set for the variables $\vec{V}$ and p.

Executing the curl operation on (4) yields an expression for:

---

[1] *Available from the Encyclopedia Britanica*

[2] *Potter and Foss (1975) provide a rationale for using a description of the fluid medium: "incompressible" as a modifying adjective for "flow".*

Vorticity transport $$\frac{D\vec{\Omega}}{Dt} = \vec{\Omega} \cdot \vec{\nabla}\vec{V} + \nu \nabla^2 \vec{V} \ , \tag{5}$$

and this equation can be used to evaluate the vorticity as a function of space and time within a flow field.

By analogy with the similar equation for thermal energy, (5) provides a complete description if it is supplemented by the diffusive transport (into or out of the flow domain) of the transport quantity at the boundary. Discussions of this important effect are available from a number of sources; see, e.g., Lighthill (1963), Batchelor (1967) and Potter and Foss (1975). The flux of vorticity through a boundary surface will not be further considered in the present discussion.

The two terms on the r.h.s. of (5) provide the physical mechanisms by which the $\vec{\Omega}$ of a given fluid particle can change. Specifically, a stretching or a compression of the vorticity filament (see Section 3.3) which passes through the given particle (i.e. $\vec{\Omega} \cdot \vec{\nabla}\vec{V}$) and a net diffusion of vorticity into or out of the spatial domain of the particle (i.e., $\nu \nabla^2 \vec{\Omega}$) can cause a temporal change in the $\vec{\Omega}$ value of a given fluid element. Numerous examples of these effects are given in the following discussion.

2.2 The circulation transport equations

Forming the material time derivative of $\vec{V}$ in (2a) leads to the circulation transport equations; viz.

$$\frac{D\Gamma}{Dt} = - \oint_C \vec{a}_I \cdot d\vec{S} - \oint_C \frac{\nabla p}{\rho} \cdot d\vec{S} + \oint_C \nu \nabla^2 \vec{V} \cdot d\vec{S} \ . \tag{6}$$

Note that a_I are the accelerations that could occur if the physical problem is described in an accelerating reference frame. These effects will not be dealt with herein; the interested reader is referred to Potter and Foss (1975).

The second term on the r.h.s. of (6) is a source (or sink) of circulation if a non-barotropic condition exists in the flow, i.e. if $\rho \neq \rho(p)$. The third term is of singular importance in the vorticity/circulation descriptions of a given flow. *This term represents a source or a sink for vorticity at a physical surface* as seen by

$$\nu \nabla^2 \vec{V} \bigg]_S = \frac{1}{\rho} \ \nabla p \bigg]_S \tag{7}$$

where $\vec{J}_S$ represents the boundary surface at which $D\vec{V}/Dt = 0$. For a contour C, with one segment at the surface and the closing portions in the interior of the flow, the contour integral can be expressed as

$$\oint_C \nu\nabla^2\vec{V}\cdot d\vec{S} = \frac{1}{\rho}\oint_S \nabla p\cdot d\vec{S} + \oint_{C-S} \nu\nabla^2\vec{V}\cdot d\vec{S} \approx \frac{1}{\rho}[p(S_2) - p(S_1)] \qquad (8)$$

where the interior values of $\nabla^2 V$ are assumed to be much smaller than those at the surface and (S_1, S_2) are the two ends of the contour path that lies in the plane of the surface.

Equation (8) is for a contour that is attached to a defined circuit of mass particles. The circuit will, in general, move in space as time varies. An Eulerian form of this expression is presented by Potter and Foss (1975); viz.

$$\frac{D\Gamma}{dt} = -\int_A \nabla \times (\vec{\Omega} \times \vec{V})\cdot\hat{n}\, dA - \oint_C \vec{a}_I\cdot d\vec{s} - \oint_C \frac{dp}{\rho} + \oint_C \nu\nabla^2\vec{V}\cdot ds \ . \qquad (9)$$

It is noted that the first term on the r.h.s. of (9) may be likened to a control volume flux of the Ω component that is perpendicular to the plane of the contour C albeit the flux term is over a line and not an area. The cited reference presents a detailed interpretation of this term.

2.3 Vorticity filaments, streamlines and the solenoidal condition

At a given instant of time, there exists an infinity of lines which are everywhere tangent to the velocity vectors in a flow field, viz. the streamlines. Similarly, lines that are everywhere tangent to the vorticity vectors are termed *vorticity filaments*. It is important to recognize that both the streamlines and the vorticity filaments are, in principal, defined by an instantaneous realization of the complete flow field. Such lines can also be defined for a time (or, more generally, an ensemble) average of the instantaneous realizations. The instantaneous character of the streamlines in an unsteady flow is made quite graphic in the NCFMF motion picture Flow Visualization[1] with S.J. Kline as the film principal. As shown in this movie, the streamline through a given point in space can be dramatically altered between t and t + δt. Similarly, vorticity filaments can be dramatically reconfigured. This matter will be addressed in detail following the consideration of some basic attributes of the vorticity filaments.

Of central importance in the interpretation of vorticity filaments in a flow field is to recognize that *vorticity filaments are constrained by the solenoidal condition to appear in closed loops or to end at isolated singular points on a solid boundary.* The solenoidal condition follows from the definition of vorticity and a vector identity; specifically,

$$\nabla\cdot\vec{\Omega} = \nabla\cdot(\nabla \times \vec{V}) = 0 \ . \qquad (10)$$

The solenoidal condition can also apply to the velocity field and hence to the streamlines, if an incompressible flow is considered. Specifically, the conservation of mass equation

$$\frac{D\rho}{Dt} + \rho \vec{\nabla} \cdot \vec{V} = 0 \tag{11}$$

shows that $\nabla \cdot V = 0$ if $D\rho/Dt = 0$. Therefore, streamlines and vorticity filaments have analogous properties and these properties can be used to clarify the vorticity field for a given flow.

Figure 1 shows a cylindrical chamber with diameter D and length L and with a suspended axial flow fan at its center. Note that the fan is rotating in the horizontal midplane and inducing a flow perpendicular to this plane. The resulting flow field is easily envisioned; representative streamlines are shown. These streamlines form closed loops with the exception of the central streamline which terminates on the bounding walls.

A useful visualization of the solenoidal condition is to envision a small ring that encircles an individual filament or streamline, e.g., one of the filaments of Figure 1. The ring can be slid along the filament but it cannot slide off its end. Note that the ring can be stopped by the presence of a solid boundary. The central streamline of Figure 1 provides a good example of the latter condition.

If the fan were removed and the cylinder were rotated, viscous effects would induce a motion in the region near the walls of the container. If a steady state were allowed to form (i.e., a solid body rotation of the confined fluid) and then the cylinder were impulsively brought to rest, the same lines could be used to represent the vorticity filaments in the cylindrical container.

The concept of a streamline or a vorticity filament that "closes through infinity" is a useful one, and it is one that can be demonstrated with this example. Imagine that the upper surface is removed and that the axial flow fan is again used to induce the motion. The lower singular point is still present and the flow in the lower portion of the cylinder will be qualitatively similar to the flow of the original example. Closed streamlines are, of course, still present but these now exist above the cylinder. The central streamline, which must remain on the axis for this rotationally symmetrical flow field, is seen to extend to "infinity". The vorticity filaments, in the analog experiment, form a pattern that is similar to those of the streamlines. In both cases, if a turbulent flow were present, the rotationally symmetric condition would be valid for the time average description of the problem.

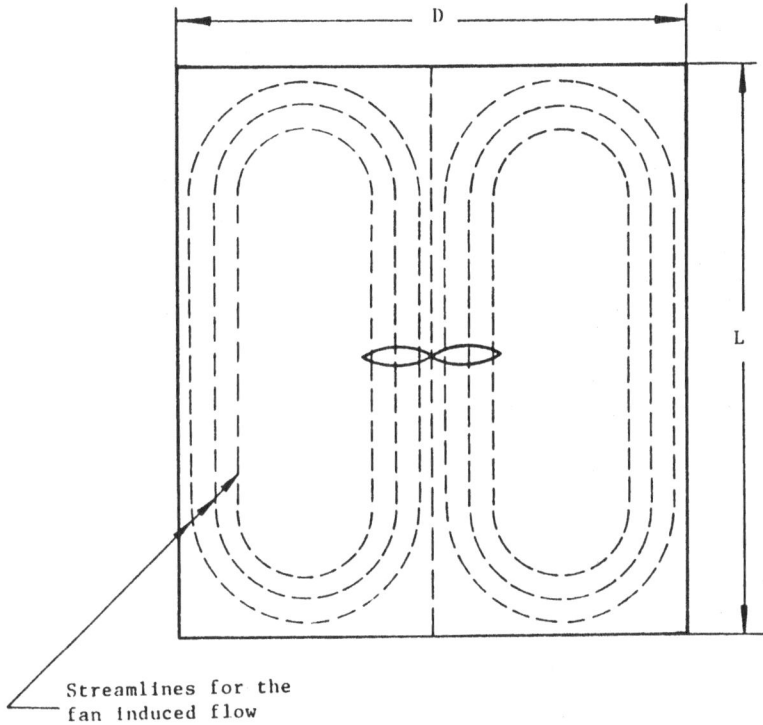

Streamlines for the
fan induced flow

or

Vorticity filaments for
the spin-down of the
forced vortex motion
(fan removed)

Fig. 1 Analogous streamlines and vorticity filaments for two demonstration flow
fields.

2.4 Vortex tubes, stream tubes and vorticies

If Figure 1 is used to represent the streamlines of the flow induced by the axial flow fan, then a stream tube may be easily defined using this figure. Consider a finite sized ring (unlike that described in Section 2.3) that is centered in the cylinder at $z = 3L/4$ and that has a cross sectional area of $\pi [0.025D]^2$. This ring defines the boundary of a stream tube whose mass flux is given as (for constant ρ)

$$\dot{m} = 2\pi\rho \int_0^{0.025D} w(r,z) \; r dr \tag{12}$$

where w is the vertical velocity component.

This mass flux value defines the stream tube boundary for all streamwise positions along the tube. Note that the cross section of the stream tube is an annulus for the down flow near the cylindrical wall of the confining vessel. The crossover point between a disc and an annulus occurs at the radial location where the bounding streamlines are closest to the top surface.

For the analog flow where the indicated lines represent vorticity filaments instead of streamlines, an exactly analogous set of definitions can be used for the identification of a vortex tube. Equation (12) becomes, for the analogous vortex tube case,

$$\Gamma = 2\pi \int_0^{0.025D} \Omega_z(r,z) \; r \; dr$$

$$= \int_0^{2\pi} v_\theta r d\theta$$

$$= [2\pi \; (0.025) \; D] \; v_\theta(0.025D, 3L/4) \quad , \tag{13}$$

and Γ = constant for all cross sections defines the vortex tube.

As clearly indicated by this example, the central streamline or vorticity filament can end at the upper or lower surface. The finite cross section stream tube or vortex tube cannot end since its conserved quantity (m or Γ) must be present at all streamwise locations.

Both tubes are defined over space at an instant of time and both tubes have a well defined relationship to the more primitive streamlines or vorticity filaments. Since the latter can be inferred from a solution to the governing equations (1), (2) or (3), the terms discussed in Sections 2.3 and 2.4 are unambiguously defined. This is typically not the case for the companion term *vortex*.

Although *vortex* is a commonly used word to describe a visually rotating mass of fluid, it is often not feasible to apply quantitative definitions to render the description precise. Görtler vorticies in a concave boundary layer, the large scale vorticies downwind of a bluff obstruction, the large scale motions of a plane shear layer, or the wing tip vortices from a lifting body are examples of important motions which do not permit simple descriptions of their boundaries, and hence of their strength, for all locations along their axes. Thus their definitions are imprecise.

One consequence of this observation is that, although a direct measurement of vorticity can be interpreted in terms of a vorticity filament and, by extension, can contribute to the inferrence of a vortex tube in a sufficiently simple environment, the vagueness that is typically associated with the identification of a "vortex" will not be removed by having access to time resolved measurements of one or more of the vorticity components.

3.0 AN EXEMPLAR FLOW FIELD

3.1 Problem description

The oblique impingement of an axisymmetric jet provides an exemplar flow field for the important concepts that are yet to be demonstrated in this paper. Foss and Kleis (1976) provide basic data on this flow field for a high Reynolds number approach jet. The qualitative attributes of this flow, that are inferred from that experimental investigation, are based upon the time averages of the velocity and pressure fields.

Figure 2 presents a schematic representation of the subject flow field. The axisymmetric approach jet is not influenced by the presence of the plate at the axial location at which the two vorticity filament loops: • - o and * - # are defined. Note that • and * are respectively selected "just" below and "just" above the stagnation streamline. It is an interesting attribute of this flow field that the stagnation streamline is positioned well below the jet centerline and that the stagnation point (x_s) is located in a region of increasing pressure; viz. $\partial p/\partial x(x_s,0,0)>0$.

3.2 Diffusive effects

Consider that the approach vorticity loops are defined at time t. If the fluid particle • is followed to a subsequent location where it is moving away from the stagnation point at time $(t_1 + \delta t)$, it is seen that the sign of its vorticity component has changed since it is now in the reverse flow boundary layer. Namely, for the material element •, $\Omega_z(t_1) < 0$ whereas $\Omega_z (t_1 + \delta t) > 0$.

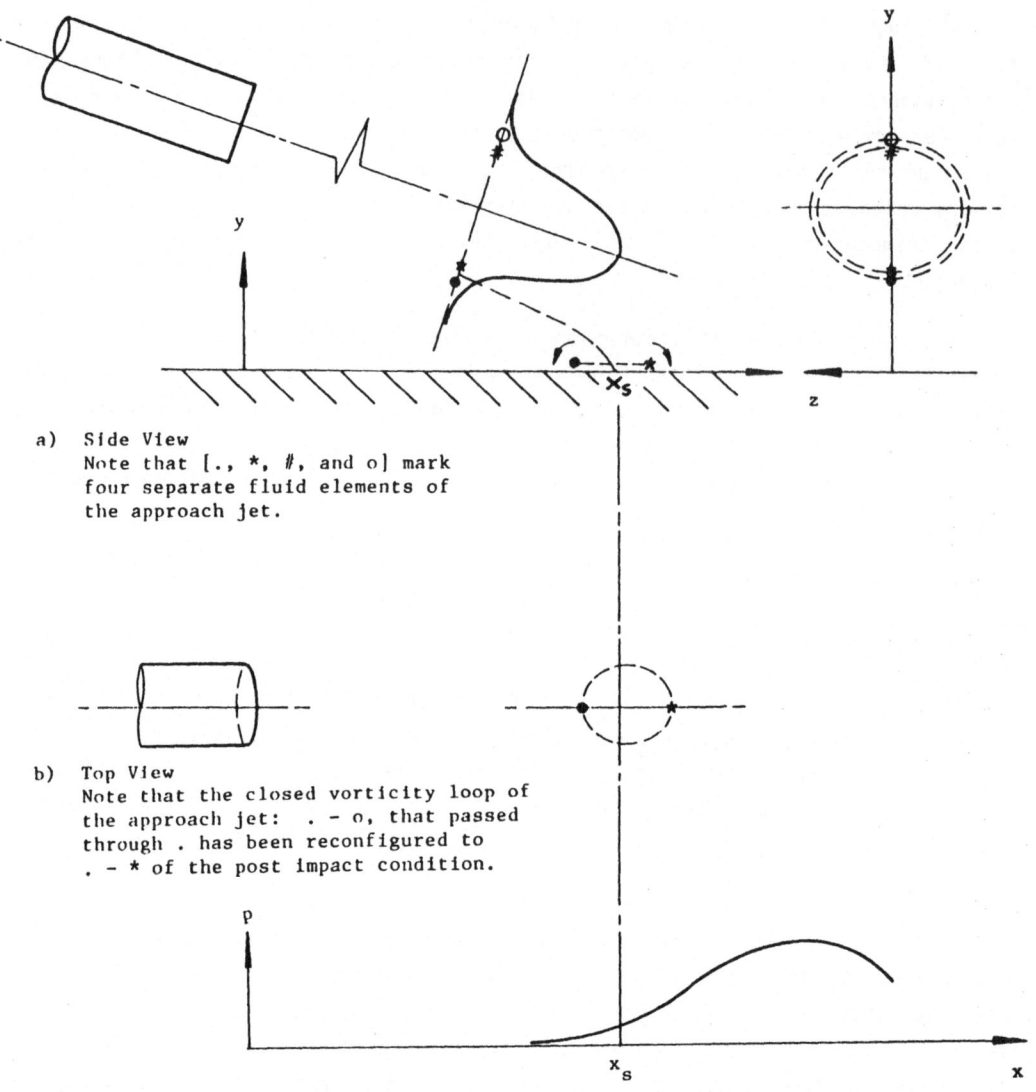

a) Side View
   Note that [., *, #, and o] mark
   four separate fluid elements of
   the approach jet.

b) Top View
   Note that the closed vorticity loop of
   the approach jet: . - o, that passed
   through . has been reconfigured to
   . - * of the post impact condition.

c) Pressure distribution p(x) at the surface of the impact plane

Fig. 2 The oblique jet impingement demonstration problem.

The surface pressure gradient near x_s is a strong source of $\Omega_z > 0$ (see Section 2.2). Once in the flow, this vorticity is diffused into the · particle as it executes the trajectory from its (t_1) position to its $(t_1 + \delta t)$ position. Note that this effect (i.e., the second term on the r.h.s. of (5)) can alter the sign of a given vorticity component. The first r.h.s. term of (5) can only alter the magnitude of the component.

Since the particles · and o now have vorticity of the same sign, the vorticity filament that passes through · at $(t_1 + \delta t)$ can no longer be connected to o. This filament can, however, be connected through * and it is arbitrarily assumed that * and δt are selected such that this is true. This dramatic reconfiguring of the filament through · emphasizes a quite general attribute of viscous flows; viz. that *at an instant of time, the vorticity filament passing through a given fluid element can be used to define a material curve coincident with that filament. However, since the $\vec{\Omega}$ value of the material elements are subject to the temporal changes suggested by (5), the vorticity filament at a subsequent time can connect a dramatically different set of contiguous material elements.* It is this dramatic reconfiguring of the vorticity filaments that is analogous to the dramatic reconfiguring of the streamlines in the unsteady flow example of the Kline movie discussed in Section 2.3.

It is instructive to consider the "material loop" that is coincident with the vorticity loop "· - o", of the approach jet in Figure 1. This contiguous set of fluid particles is quite unrelated to the vorticity loop "· - *" shown at a later time. At any instant of time, the line passing through "·" and tangent to its vorticity vector exists as a closed loop; however, this particle, and all other particles of a given loop, have time varying $\vec{\Omega}$ values given by (5). *Hence, the vorticity loops within the flow are seen to slide gracefully through the fluid medium as they connect together an ever changing set of material particles.* Only in the idealized situation of an inviscid flow is there a time invariant coincidence of a material curve and a vorticity loop. The general case is one for which the vorticity loops _cannot_ be characterized as being "frozen" to the fluid particles.

This may be the singularly most misunderstood attribute of vorticity in real flows. The clever and very instructive vortex ring experiments of Oshima and Asaka (1975) provide an eloquent statement of the "graceful slide" effect, and their motion picture is strongly recommended as a graphic representation of this phenomena. [Oshima and Asaka (1977) provide a follow-on study.] Fohl and Turner (1975) have also examined the acute intersection of two vortex rings. Their paper provides a graphic description of how the two rings are merged into a single ring (which shows the merging of the previously distinct vorticity filaments) and how this newly formed ring may evolve into two rings whose material elements represent "equal" contributions from the two original rings. Note that the phrase: "a broken vorticity ring" presents a quite different image as compared to: "a graceful sliding of the vorticity loops." Even the use of the image of "tangled spaghetti" to represent the instantaneous vorticity field of a turbulent flow can seemingly endow the flow with a "structure" that is unwarranted. It is appropriate to recall Prof. Kovasznay's admonition (1979) that one should speak of "coherent motions" and not "coherent structures" since the latter gives a misleading sense of permanence to a dynamically significant but transient attribute of the flow field.

This physical example invites the consideration of one additional analogy to describe diffusive effects on the vorticity field. Namely, the $\vec{\Omega}(t)$ history of the · particle can be likened to the temperature (T) history for the same particle in a non-isothermal flow field. If the jet was cold and obliquely impinged on a hot wall, the T(t) history of the · particle would be qualitatively influenced by the same type of diffusive effects: $\nu\nabla^2\vec{\Omega} \sim \kappa\nabla^2 T$. The property fluxes of ($\vec{\Omega}$ or T) through the surface are controlled by the gradients at the surface, albeit the vorticity flux is controlled by a gradient in the plane of the surface and the temperature (i.e., thermal energy) flux is controlled by a gradient normal to the surface.

3.3 Stretching and Reorientation Effects

Examples of the physical significance of the $[\vec{\Omega}\cdot\nabla\vec{V}]$ term can similarly be drawn from the oblique impingement flow. Consider the 4 and 8 o'clock fluid particles of the original "· - o" vorticity loop during their approach to the impact plate as shown in Figure 3. (Note that o is at 12 and · is at 6 o'clock with the clock positions defined by looking in the streamwise direction.) One can appreciate that altered values of Ω_x and Ω_z vorticity components (e.g. $\Omega_x < 0$ from the 4 o'clock and $\Omega_x > 0$ from the 8 o'clock locations) will result from the jet-plate interaction, the retardation of the lower portion of the vorticity filament loop and the consequent reorientation of the vorticity filament loop.

A second contribution of this term is the amplification (or attenuation) by stretching (or compression) of a given scalar component. The Ω_x magnitudes, noted above, will be enhanced as the material loop (which contains the 4 and 8 o'clock fluid particles) is caused to lengthen as the jet fluid spreads over the plate surface. Specifically, $\Omega_x \ \partial u/\partial x > 0$ for the 8 and $\Omega_x \ \partial u/\partial x < 0$ for the 4 o'clock segments of the loop. (Note that z components also exist for the same fluid elements). Such an effect will be present as the "post impact" vorticity loop lengthens as it moves outward from the stagnation point. It should be again noted that the $[\vec{\Omega}\cdot\nabla\vec{V}]$ term cannot change the rotational sense of Ω for a given fluid particle and that, in general, the correspondence between a vorticity loop and a material loop is only momentary.

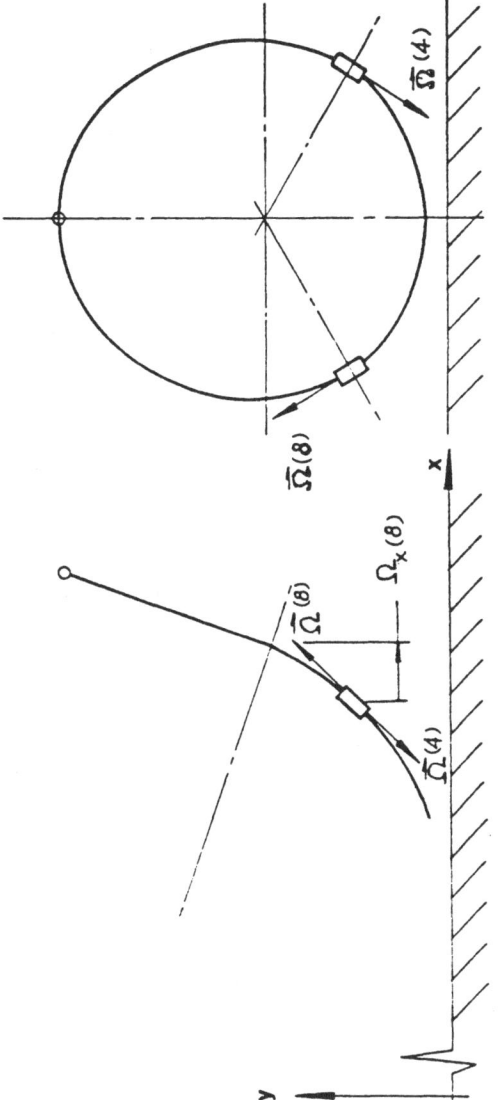

Fig. 3 Examples of vorticity reorientation effects in the oblique jet impingement flow problem.

4.0 THE TRANSPORT EQUATIONS

The preceding discussion has focussed upon the physical effects associated with an instantaneous realization of the vorticity field. It is also instructive to consider its stochastic description. Tennekes and Lumley (1972) provide an excellent discussion of such considerations. A brief summary of their discussion is provided in this section.

A Reynolds decomposition, applied to the $\Omega_i(x,y,z,t)$ field, can be directly written as

$$\Omega_i(t) = \overline{\Omega}_i + \omega_i(t) \tag{14}$$

where the index notation is used for convenience. The two non-linear terms of (5) yield "additional" terms in the time averaged equation (for a steady flow)

$$\overline{U}_j \frac{\partial \overline{\Omega}_i}{\partial x_j} = -u \frac{\overline{\partial \omega_j}}{\partial x_i} + \overline{\omega_j s_{ij}} + \overline{\Omega}_j \overline{S}_{ij} + \nu \frac{\partial^2 \overline{\Omega}_i}{\partial x_j \partial x_j} \tag{15}$$

where the mean and fluctuating strain rates are given by

$$\overline{S}_{ij} = \left[\frac{\partial \overline{U}_i}{\partial x_j} + \frac{\partial \overline{U}_j}{\partial x_i} \right] \quad \text{and} \quad s_{ij} = \left[\frac{\partial u_i}{\partial x_j} + \frac{\partial u_j}{\partial x_i} \right] \tag{16a,16b}$$

The solenoidal condition, of course, also applies to the mean vorticity field and closed vorticity loops can be imagined in such a field. (Figure 2, which represents the Foss/Kleis experiment is best considered in this manner since a high Reynolds number jet was used in that investigation and other observations suggest that the vorticity loops in such a flow are relatively small scale entities. This point is considered in more detail in a later section.)

The transport equation for the fluctuating enstrophy, $\overline{\omega_i^2}$, can also be formally stated. Its terms invite physical interpretations which have analogies to the equation for turbulence kinetic energy and the temperature fluctuation equations. The interested reader can find these interpretations in Tennekes and Lumley (1972) (and especially in the original papers by Corrsin noted therein). The fluctuating enstrophy transport equation is given by

$$\overline{U}_j \frac{\partial}{\partial x_j} \left(\frac{1}{2} \overline{\omega_i \omega_i}\right) = - \overline{u_j \omega_i} \frac{\partial \overline{\Omega}_i}{\partial x_j} - \frac{1}{2} \frac{\partial}{\partial x_j} \left(\overline{u_j \omega_i \omega_i}\right) + \overline{\omega_i \omega_j s_{ij}} + \overline{\omega_i \omega_j} \overline{S}_{ij}$$

$$+ \overline{\Omega}_j \overline{\omega_i s_{ij}} + \nu \frac{\partial^2}{\partial x_j \partial x_j} \left(\frac{1}{2} \overline{\omega_j \omega_i}\right) - \nu \overline{\frac{\partial \omega_i}{\partial x_j} \frac{\partial \omega_i}{\partial x_j}} \ . \tag{17}$$

In a recent paper Balint, Vukoslavĉević and Wallace (1988) give measurements of the terms of this equation and the corresponding equation for the mean enstrophy ($\frac{1}{2}\overline{\Omega}_i\overline{\Omega}_i$) transport for a turbulent boundary layer and discuss their interpretation.

5. MEASUREMENT STRATEGIES

Several methods which have been developed to measure various components of the vorticity vector in turbulent flows have earlier been discussed by Wallace (1986). In the remainder of this paper we will briefly review those methods and will survey methods not mentioned in that paper bringing it up-to-date.

Vorticity measurement methods can broadly be divided into two types:

• Methods which measure velocity components at closely spaced points in the flow using a variety of means as discussed below. The basic relationship between circulation and vorticity in (2b) invites a measurement strategy wherein a circulation about a small planar domain is obtained and a "local" value of the vorticity component normal to that plane is inferred. This requires obtaining spatial information on the velocity field itself V(x,y,z,t) around the planar domain. Alternatively, this velocity field data, determined at least at three spacial locations, may be finite differenced to estimate the gradients of the vorticity components. Obviously, in the limit as the circuit dimension C is reduced, the two methods are indistinguishable. It is, however, useful to distinguish experimental techniques by reference to their strategy to isolate the point function $\Omega(x,y,z,t)$.

It is also necessary to appreciate that the vorticity emphasizes the smallest scales of the motion [see e.g. the discussion by Tennekes and Lumley (1972)], and a frustrating aspect of isolating the Ω value (or one of its components) by such methods is the relative dimension in which the velocity information must be acquired. With increasing size of the circulation circuit C, the spatial average vorticity in the measurement domain represents decreasingly well the instantaneous vorticity at the center of the domain. On the other hand, a finite difference approximation of the gradients assumes that second and higher order terms are small in a Taylor expansion of the velocity field about the measurement point in the flow. This assumption becomes decreasingly valid as the spacial distance over which the gradients are estimated increases. Wyngaard (1969) has provided quite useful estimates of the theoretical (for an isotropic turbulence field) scale at which such measurements should be made and Klewicki and Falco (1988) have experimentally verified his conclusions. This important paper will be referred to in the analysis of the experimental methods.

• Methods sensing physical properties of the measurement system which vary directly or indirectly with the velocity gradients in the flow or rotation of the flow itself. Most of these methods mark the fluid in some manner and measure the subsequent changes in the marker by the velocity gradient field rather than by the velocity field itself. Markers used include diffraction gratings, volume holograms, light reflecting planar mirrors in rotating particles, and the electric potential of a conducting fluid in the presence of a magnetic field. For these methods, the spatial resolution is given by the spatial extent over which the changes in the marker, caused by the velocity gradients must be interrogated in order to recover the vorticity components. The larger the spatial extent, the poorer is the approximation of the point value of $\vec{\omega}$. An additional direct method senses the effect of fluid rotation on a zero angle of attack vane.

6. ESTIMATING VORTICITY COMPONENTS FROM VELOCITY MEASUREMENTS AT SPATIALLY SEPARATED POINTS

6.1 Streamwise Vorticity

Kovasznay (1950, 1954) was the first to suggest a means of measuring the time varying streamwise vorticity Ω_x. He proposed an arrangement of four hot-wire sensors, which act as the four legs of a Wheatstone bridge as shown in Figure 4. He proposed that the voltage between points A and C would be uniquely sensitive to the streamwise velocity and between B and D to the streamwise vorticity when this probe was operated by a constant current circuit. Kistler (1952) was the first to use this probe and to analyze it performance in detail. He found that, for a probe with all its wires geometrically and electrically identical, the galvanometric voltage across B and D in Figure 4 is given by

$$\Delta E = 2Ah(\sin\alpha)\Omega_x + A(\cos\alpha)h^2\left(\frac{\partial^2 U}{\partial y^2} - \frac{\partial^2 U}{\partial z^2}\right) , \qquad (18)$$

where the coefficient A depends on the constant operating current and on the streamwise velocity which varies in turbulent flow. The angle α and the prong separation h are shown in the figure. Obviously the probe is primarily sensitive to Ω_x when α is made as large and h as small as possible. Tangential cooling of the sensors was neglected in this analysis, which practically limits the values of α to about 60 degrees; the prong separation h is limited only by the mechanical ability to make the probe very small. Although the parasitic sensitivity to the gradients of the streamwise velocity is a second order term in (18), it will be shown below to be important. Wyngaard (1969) states, in his study of the spatial resolution characteristics of this type probe, that the Kolmogorov microscale to the active wire length ratio should not be less than 0.3. At this ratio the probe measures approximately 86% of the true mean square streamwise vorticity for an active wire length to h ratio of unity.

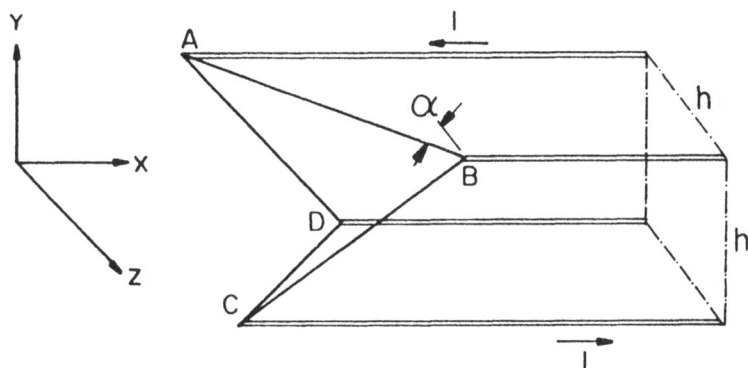

Fig. 4   Kovasznay-type streamwise vorticity probe [Vukoslavĉević and Wallace (1981)]

Of the many effects due to possible geometric and electrical asymmetries of this probe that Kistler (1952) studied, the most important parasitic sensitivity is that due to cross-stream velocity components. This sensitivity increases with increasing values of α. He, however, considered this a second order effect which could be ordinarily neglected for flows with relatively low levels of cross stream velocity fluctuations.

Kistler (1952) used this probe to measure the one-dimensional vorticity spectrum and its decay in a turbulent grid flow. Comparison with a computed spectrum obtained from a one-dimensional velocity spectrum agreed well except at low wave numbers. Uberoi and Corrsin (1951) developed a means of directly calibrating this probe by spinning it in a rotation free flow and transmitting the signals through a mercury bath commutator. Corrsin and Kistler (1954) measured the rms streamline vorticity distribution in a roughened boundary layer with the vorticity probe of Kovasznay's design; their data is compared with that of others in Section 11. Measurements in a turbulent boundary layer made by Tu and Willmarth in 1968 with a probe of this design were first published by Willmarth and Lu (1972). However, they point out that the prong spacing h, for the high speed turbulent boundary layer they studied, was about 250 viscous lengths which was obviously much too large to resolve the small scale vorticity field. In this case the probe more properly sensed the instantaneous circulation or spatially averaged vorticity rather than the point vorticity.

Kastrinakis (1976) made the first of several attempts to measure vorticity at the Max-Plank-Institut for Strömungsforschung in Göttingen. Using two Kovasznay type probes with 2.8 mm long wires and angles α of approximately 45 degrees, he measured the rms distribution of streamwise vorticity across a fully developed turbulent channel flow as well as the two point correlation coefficient for spanwise probe separation. The parasitic sensitivity of Ω_x to the streamwise velocity could be corrected by Kastrinakis (1976); the parasitic sensitivity to the cross-stream velocity components could not be corrected, however, because these components cannot be independently determined with the probe. As will be seen in Section 11, the rms values Kastrinakis (1976) measured are systematically too large when compared to later measurements. Kastrinakis, Wallace, Ghorashi and Brodkey (1977) used the same two probes to further investigate wall layer vortical structures using a pattern recognition algorithm of Wallace, Brodkey and Eckelman (1977). They assumed that the parasitic sensitivity to cross-stream velocity was averaged out to zero in the ensemble average patterns.

In a combined analytical and experimental study which confirmed what is implied in Kistler's work, Kastrinakis, Ecklemann and Willmarth (1979) showed that there is a very considerable parasitic sensitivity of values of Ω_x measured by the Kovasznay type probe to the streamwise and cross-stream velocity components. The contamination by the cross-stream velocity increases with mean velocity and is of the same order as that of the instantaneous vorticity to be measured in many turbulent flows.

Following a suggestion of Kastrinakis et al. (1979) Vukoslavĉević and Wallace (1981) built a probe with the same geometric arrangement as a Kovasznay-type but with each wire operated independently. This probe which is shown in Figure 5, has two pairs of sensors forming two orthogonal X-arrays which are heated by constant temperature circuits. Of course the disadvantage is that this design requires twice as many supporting prongs as the Kovasznay-design.

The cross-stream v and w velocity components can thus be obtained independently as long as the conventional assumption of uniform flow across the probe measuring area is made. It would thus appear that the parasitic contamination of v and w could be corrected. Vukoslavĉević and Wallace (1981) emphatically stress, however, that the usual assumption of uniform flow across the sensing area of multi-sensor hot-wire arrays is invalid in highly turbulent flows because large internal shear layers are often instantaneously present. They provide measurements of the maximum values of $\partial U/\partial y$ and $\partial U/\partial z$ in a boundary layer to support this assertion. Thus

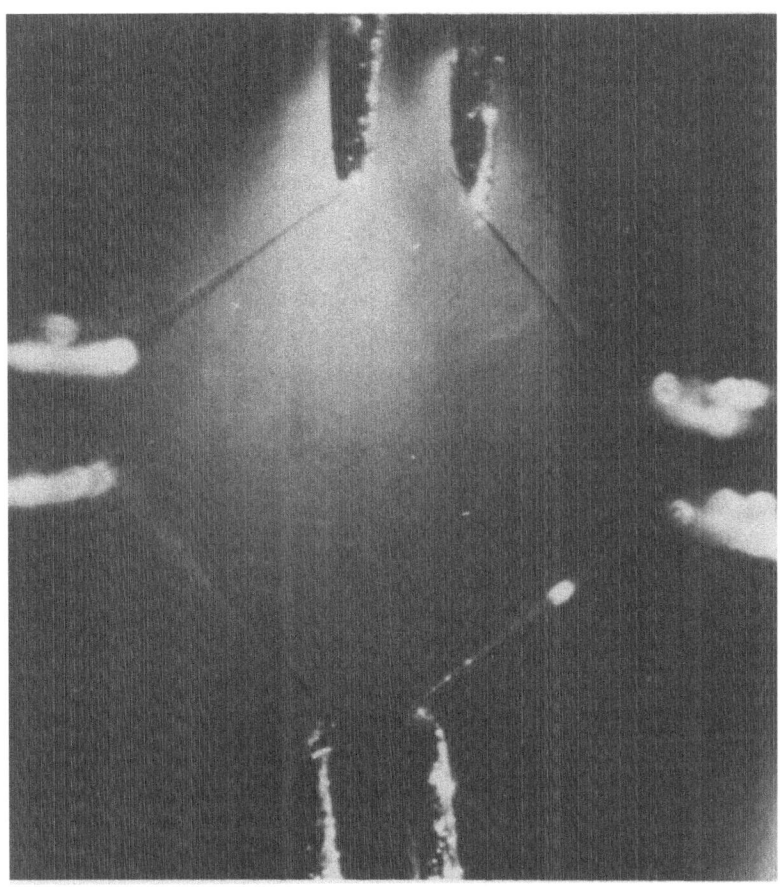

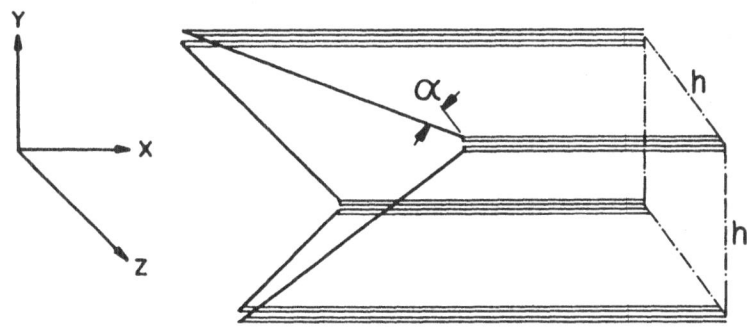

Fig. 5 Modified Kovasznay-type streamwise vorticity probe consisting of two orthogonal X-arrays [Vukoslavčević and Wallace (1981)].

instantaneously the v and w components themselves may be incorrectly determined, and the streamwise vorticity cannot be properly corrected for their parasitic contamination which, these authors show, is a first rather than a second order effect. *In essence, Vukoslavĉević and Wallace (1981) revealed the implicit and implausible assumption made in the conception of the Kovasznay-type probe, i.e. that the probe is only sensitive to the velocity gradients of the streamwise vorticity but to none of the other seven velocity gradients in the flow.* For determining $\Omega_x(t)$ instantaneously, even this modified version of the Kovasznay type probe can yield quite erroneous values. The consequences of this invalid assumption for the statistics of the streamwise vorticity component measured with a Kovasznay-type probe have yet to be determined. However, comparison with other measurements and with numerical simulations, to be discussed in Section 11, indicate that these errors, which can instantaneously be very large, may statistically be small.

Kastrinakis and Eckelmann (1983) used a modified Kovasznay-type probe to measure the basic statistics of the streamwise vorticity in a turbulent channel flow. The spacing between the sensors for this flow was 11.5 viscous lengths or about 5 times an estimation of the Kolmogorov microscale. They found that the v and w velocity components were systematically underestimated by about 20% when the probe was tested by pitching and yawing it in a uniform irrotational flow. Therefore they experimentally determined calibration factors which were used to correct the data measured in the turbulent flow. Comparison of their results with later work will be made in Section 11 below. Kastrinakis, Nychas and Eckelmann, (1983) used this same probe to examine the percentage contributions to the mean square streamwise vorticity component conditioned on the occurrence of the uv product time series in each of the four quadrants of the uv hodograph plane. They also obtained the autocorrelation of ω_x as a function of position across the channel, and, from this, they calculated the corresponding integral length scales for ω_x and ω_x^2. Nychas, Kastrinakis and Eckelmann (1985) used this data to study components of the production and dissipation terms in the fluctuating enstrophy transport equation.

In two recent papers Kasagi, Hirata and Nishino (1986) and Nishino, Kasagi and Hirata (1987) have obtained the cross-stream velocity fluctuations as well as the streamwise vorticity fluctuations using image processing of digitized hydrogen bubble photographs of the flow in the cross-stream plane. Tungsten wires of 20 μm diameter which were insulation at 2-3 mm intervals were used to generate the hydrogen bubbles. The wires were placed in staggered streamwise positions in order that the bubbles, generated at different distances from the wall, arrive at the viewing plane at the same time. This arrangement had a spanwise separation between measurement locations of $\Delta z^+ \simeq 8\text{-}12$ and a separation normal to the wall of $\Delta y^+ \simeq 7\text{-}16$. Utilizing an optical mirror far downstream, pairs of successive photographs from a 16 mm film, framed at 48 frames per second, were analyzed to determine bubble cluster displacements. From these displacements over a time interval corresponding to 1.38 viscous time units ν/u_τ^2, the v and w velocity components were determined. The data were linearly interpolated to a set of square grid points which include the wall where v and w are zero. The streamwise vorticity values were then calculated by finite difference. The spanwise velocity was overestimated in the earlier paper of Kasagi et al. (1986) when compared to the hot-film measurements of Kreplin and Eckelmann (1979). Using an automated processing method, Nishino et al. (1987) find good agreement for the velocity statistics. The streamwise vorticity statistics will be compared to data of others in Section 11.

6.2 Cross-Stream Vorticity

Foss and his co-workers (1976,1981,1987) have developed an array of four hot-wire sensors, shown in Figure 6, to measure the transverse vorticity component ω_z in turbulent flows. The response of these four sensors is used to determine the microcirculation around a nominal 1 mm² spatial domain which then yields the average vorticity within the domain. A novel scheme has been devised to obtain the magnitude and direction of the velocity component in the plane of the X-array, i.e. the wire most normal to the flow determines the magnitude and the other wire determines the direction. Foss, Ali and Haw (1987) have studied four attributes of this probe. First they look at the influence of the out-of-plane velocity component w on the pitch angle measurement $\gamma_{xy} = \tan^{-1}(v/u)$. They conclude that this contribution to the ω_z uncertainty is less important than other attributes. A more important problem is that due to the nonuniformity of the flow in the transverse or z direction. This can invalidate the assumption that the angle $\gamma_{xy}(z)$ is constant across the measuring volume in the z direction. Standard deviations of the $\delta\gamma$ population in the free shear layer indicate that relatively large errors can occur from this source, confirming the conclusion of Vukoslavĉević and Wallace (1981) mentioned earlier. Finally, the spatial dimensions of micro-circulation has been studied and shown to act as a low pass filter which is, of course, true of all multi-sensor probes.

Foss, Klewicki and Disimile (1986) have used a probe of this design to measure the velocity components u and v, the spanwise vorticity component ω_z, and the strain rate component ϵ_{xy} at the entraining boundary of a large plane shear layer. Falco (1983) has also used the probe to construct ensemble averages of the time series of spanwise vorticity in a turbulent boundary layer, and Klewicki and Falco (1986,1988) have used it to measure the statistics of the spanwise vorticity component. The latter measurements were made at R_θ = 1010-4850 in a very thick boundary layer with a 1.4 mm thick sublayer for the higher R_θ; thus the spatial resolution was quite good. These measurements will be compared with others in Section 11. The inherent problems associated with the lateral displacement between the parallel and the X-arrays of the probe in Fig. 6 stimulated the development of the probe geometry shown in Fig. 7 (the designation: "Mitchell probe", recognizes the contribution of R.A. Mitchell in threading and then attaching the final slant wire between the mounted parallel wires). This probe eliminates the $\partial\gamma/\partial z \neq 0$ error source and it permits two estimates of the pitch angle (γ) to be made. The operation of the probe and an evaluation of the possible prong interference effects will be the subject of a forthcoming publication. Results obtained with this probe have been presented by Foss and Haw (1987) and Haw, Foss and Foss (1988).

Klewicki and Falco (1988) have examined the effect of the δy spacing between the two parallel wires of the probe shown in Fig. 6. This is done by using four sensors parallel to each other and the wall, and normal to the mean flow. The inner two wires are fixed, but the outer two wires can be moved in the y-direction. The results, presented in Fig. 8, are in excellent agreement with the predictions of Wyngaard (1969) that very little attenuation in the rms gradient is observed if the spacing over which the gradient is determined is no more than 3η. Note that the spacing between the inner two wires Δy_0 of the parallel array is $1.84\nu/u^2_\tau$ or 0.94η and 1.04η for the two positions where measurements were made in this boundary layer. As the spacing of the outer two wires Δy_0 is increased, the rms value of the gradient is decreased.

Eckelmann, Nychas, Brodkey and Wallace (1977) constructed a five-sensor hot-film probe shown in Figure 9. Utilizing Taylor's hypothesis, this probe permits the measurement of both the spanwise and normal components of vorticity. Although they were able to obtain ensemble average patterns of these two vorticity components utilizing the pattern recognition algorithm of Wallace et al. (1977), this probe was not able, for $y^+ > 10$, to accurately measure the mean velocity gradient $\partial\bar{U}/\partial y$ of turbulent oil channel flow by taking U differences over the 1 mm sensor separation distance in the y direction (corresponding approximately to a Kolmogorov microscale). Although it has later been shown by Böttcher and Eckelmann (1985) that this is a very stringent test, for this reason the authors did not further examine the statistics of these vorticity components.

Hussain and Hayakawa (1987) have experimentally estimated the low wave number part of the instantaneous transverse vorticity in a turbulent plane wake 40 diameters downstream of a circular cylinder at $Re_d = 1.3 \times 10^4$. They used a rake of eight X-array hot-wire sensors with a spacing between arrays of 1 cm. From the time series of the streamwise u and transverse v velocity components they obtained estimates of the low wave number contribution to the spanwise vorticity ω_z at the midpoints between X-arrays utilizing the central difference approximation and Taylor's hypothesis to determine $\partial v/\partial x$. Of course, with this large spatial separation between measurement locations, their neglecting the nonlinearities in the velocity gradient field does not allow them to properly estimate the time series

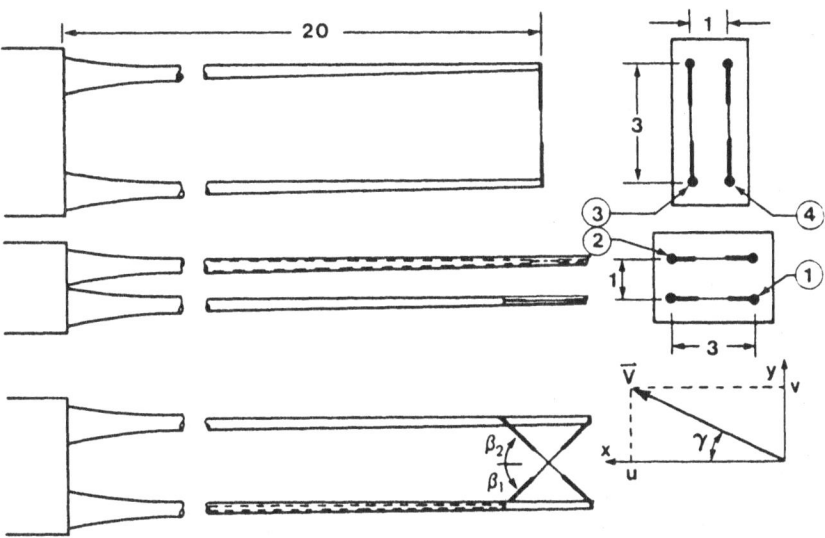

Fig. 6 Four-sensor spanwise vorticity probe [Foss (1976,1981)].

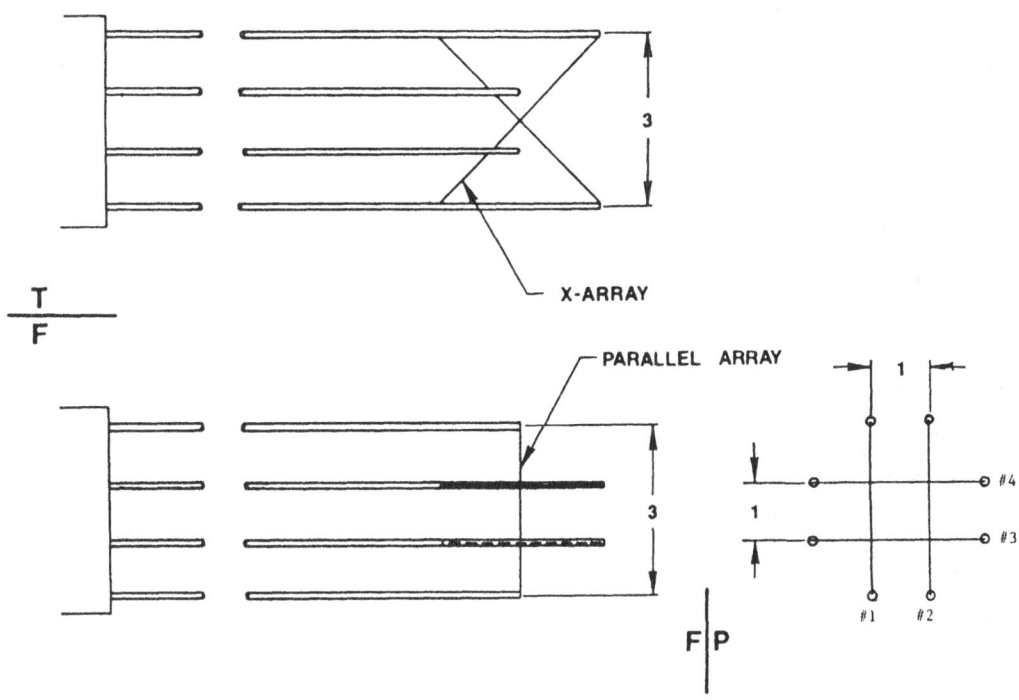

THE MITCHELL PROBE (dimensions in mm)

Fig. 7 Four-sensor (Mitchell) spanwise vorticity probe [Foss and Haw (1987) and
Haw, et al. (1988)]

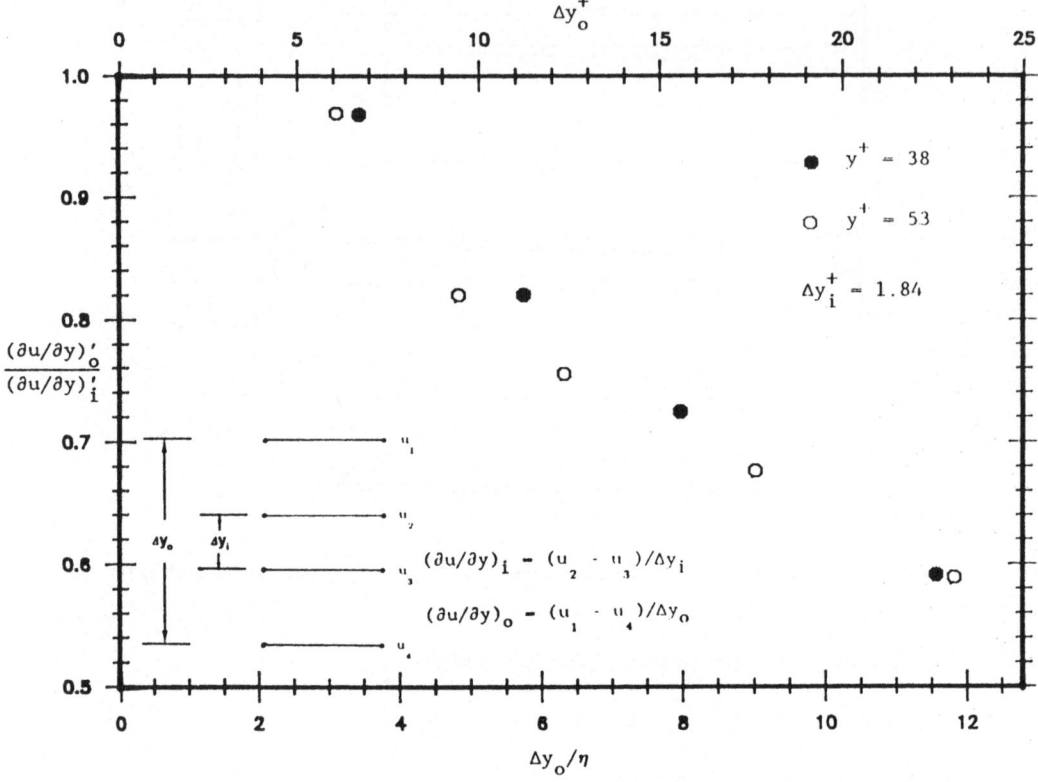

Fig. 8 An evaluation of the spacing effect on the measurement of ∂u/∂y by two separated sensors [Klewicki & Falco (1988)]

vorticity at a point in the flow. The mean velocity at the half-velocity deficit location was initially used as the convection velocity U_c in Taylor's hypothesis. For determining coherent structure properties, however, they used the measured structure convection velocity (based on vorticity correlation between two x-stations). The maps of the estimated low wave number contribution to the transverse vorticity are used to identify coherent structures and their centers.

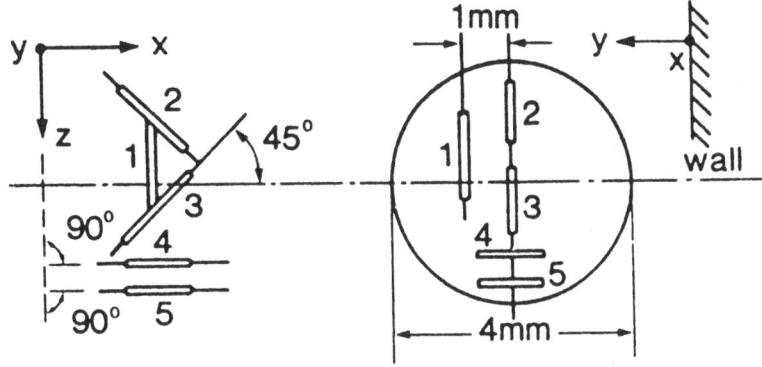

Fig. 9 Five-sensor probe to measure spanwise and normal vorticity [Eckelmann et al. (1977)]

Antonia, Browne and Shah (1988) have measured, but not simultaneously, both the normal and spanwise vorticity components ω_y and ω_z of a turbulent wake using two hot-wire X-arrays separated in the appropriate cross-stream direction and relying on Taylor's hypothesis to transform temporal derivatives into streamwise derivatives. The spacing between the X-arrays was about 1.6 mm and the wires of each array were separated by 1 mm. The Kolmogorov microscale for this flow, estimated from the average isotropic dissipation, was found to be about 0.45 mm at the centerline and 0.7 mm at $X_2/L = 2$ where L is the mean velocity defect half-width. These measurements were made in the self-preserving region of the wake at 420 diameters downstream of the cylinder. The Reynolds number based on the free stream velocity and cylinder diameter was $Re_d = 1170$. The first three statistical moments of the vorticity fluctuations have been obtained by the authors as well as probability density functions and spectra. The measured statistics show that both lateral vorticity components ω_y and ω_z behave similarly across the wake. These statistics were also estimated for the streamwise component ω_x using measurements of the two appropriate gradients. However, these gradients were not obtained simultaneously.

Kim and Fiedler (1988a) have used a six-wire probe that can be described in terms of the similar probe in Fig. 7. The vertical slant wires of the Technische Universität Berlin (TUB) probe are separated by 2w where w=1mm. Hence, these wires respond to u and v at x_0, y_0, and $z_0 \pm w$ as well as to the transverse velocity component $w(x_0, y_0, z_0 \pm w)$. The upper and lower straight wires of Fig. 7 are replaced by two X-arrays in the TUB probe. The four slant wires of these arrays, which lie in x-z planes, are separated by a distance of h=1mm between planes. These arrays respond to u, w at x_0, y_0, and z_0 as well as the transverse velocity component $v(x_0, y_0 \pm h, z_0)$.

The v(t) from the vertical wires is used to evaluate $\partial v / \partial x$ using a time resolved Taylor hypothesis and $\partial u / \partial y$ is evaluated from a 1mm² micro-circulation domain similar to that employed by Foss et al. (1987). These values permit ω_z to be evaluated as a spatial average over the domain. Measurements of the standard deviation of the transverse vorticity component in a two stream mixing layer by Kim and Fiedler (1988b) show very good agreement with the measurements of Lang (1985) described below. It is noteworthy that the relative probe dimensions were the same size in these two studies. The probe size and the shear layer momentum thickness θ(at x=100cm) dimensions were ℓ = 1.9 and 2.0mm and θ = 11.8 and 11.0mm respectively for the Lang (1985) and the Kim and Fiedler (1988b) investigations.

The streamwise vorticity ω_x can also be evaluated using this probe. The value of $\partial w / \partial y$ is obtained from the four x-z plane slant wires and the vertical slant wires provide an estimate of $\partial v / \partial z$ subject to the assumptions of a linear u(z) distribution at x_0, y_0, z_0 and a w/u value that is sufficiently small [i.e. $(w/u)^2$ is assumed to be much larger than $(w/u)^4$]. Transverse velocity corrections (i.e., w for the u, v calculation and v for u, w calculation) are included in the TUB calculations.

Lang and Dimotakis (1982) and Lang (1985) have used a laser Doppler velocimeter (LDV) to measure the u and v velocity components at spatially separated points as shown in Figure 10. By taking finite differences over the velocity measurement point separation distance h, which can be varied between 1-4 mm, they determine the spanwise vorticity ω_z utilizing a four quadrant silicon detector to collect the scattered light from the focal volumes as shown in the figure. The signal in each quadrant occurs and is sampled at random times; these signals are then converted to a continuous velocity time series using linear interpolation between the random sample times. The data was temporally filtered with a Gaussian filter profile compatible with the spatial resolution h before the spanwise vorticity was determined by finite difference. Lang (1985) has used this direct determination of $\partial v / \partial x$ to evaluate the appropriateness of the Taylor's hypothesis transformation $\partial v / \partial x = 1/U_c \, \partial v / \partial t$. Although for much of the time the two derivatives are very similar, there are occurrences of significant deviation. He also concludes that using the local mean velocity instead of the local instantaneous velocity as the convection velocity U_c in Taylor's hypothesis makes little difference. In this

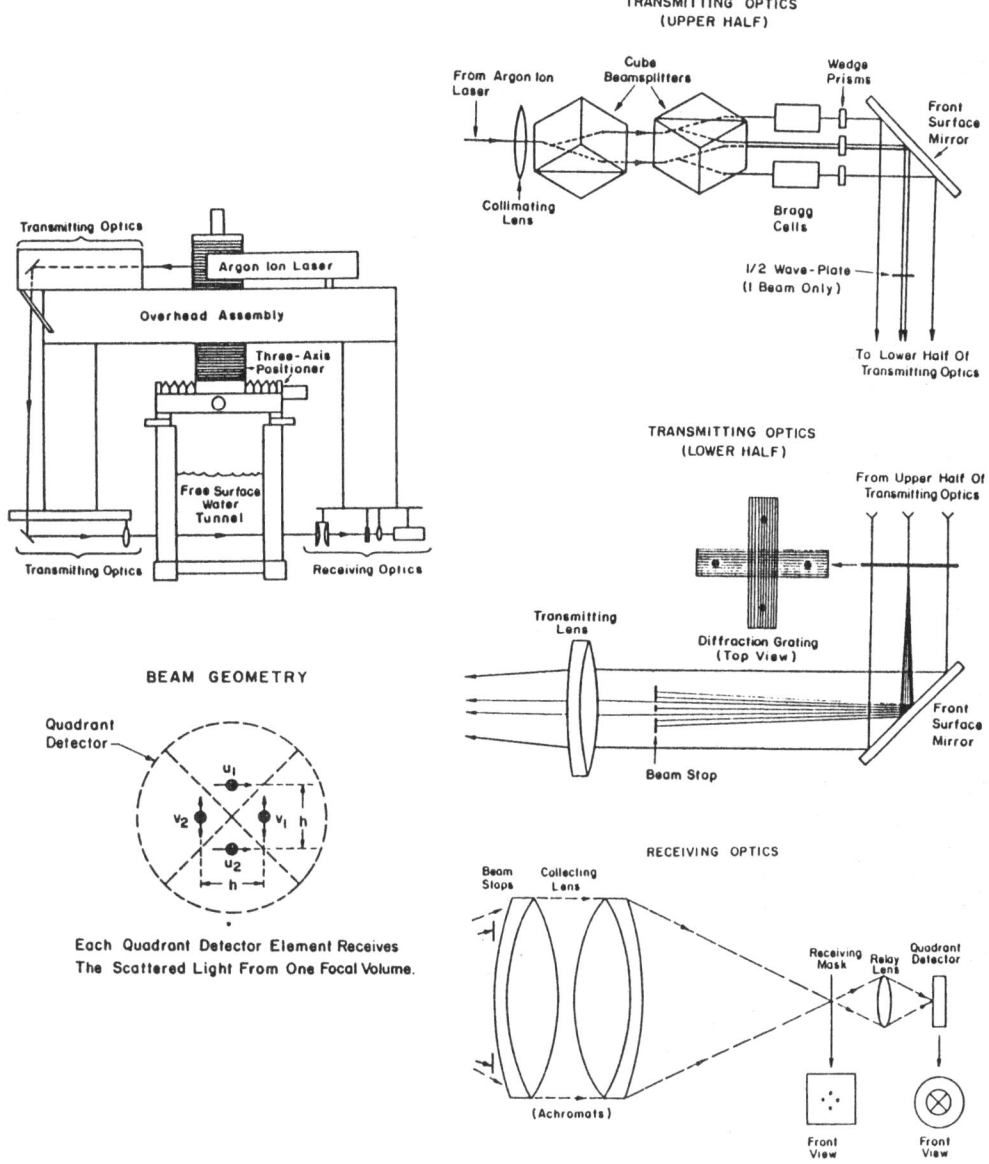

Fig. 10   LDV probe arrangement to measure spanwise vorticity [Lang (1985)]

evaluation of Taylor's hypothesis, it should be noted that the hypothesis is invoked twice: first to determine $v(t)$ at the midpoint between the two measurement locations separated in x and then to make the usual transformation to obtain $\partial v/\partial x$. As noted by Foss et al. (1987), a definitive evaluation is yet to be made.

Lang has used this multi-point LDV probe to study the two stream turbulent mixing layer. He points out that a good measure of the probe accuracy is the probability density distribution of the measured values of ω_x in the irrotational free-stream flow. His measurements of the distribution of rms ω_x across the mixing layer will be compared with recent hot-wire probe results in Section 11.

Imaichi and Ohmi (1983) have determined the two dimensional velocity field in the vortex flow behind a circular cylinder at low Reynolds number using image processing of short time exposure pathlines. Using aluminum powder particles 2-7 μm long dispersed in a water flow, the end points of the pathlines were manually located and digitized to obtain the velocity components u and v at random positions in the viewing plane. The velocity components at the nodes of a square grid were then obtained using linear interpolation. The vorticity component perpendicular to the visualized plane was determined by taking finite differences between the velocities at the node points of the grid. Figure 11 shows the pathline photographs and vorticity contours for the vortices shed behind a circular cylinder at Re_d = 100.

Sources of error considered by these authors include: 1) the ability of the particles to follow the flow [this has previously examined and questioned for the lamellar shaped particles by Maxworthy (1971) and by Heikes, Maxworthy (1982) and Savas (1985)], 2) the errors in measuring the path lengths from the two-dimensional photographic image, 3) the error in determining the pathline exposure time and 4) the numerical calculation errors including the interpolation error.

Spedding and Maxworthy (1986) and Spedding, Maxworthy and Rignot (1987) have photographed particle pathlines for flow past a rigid two-dimensional fling and for the unsteady vortex flows over a delta wing. The start and end point of each time exposed streak was digitized manually for the case of the fling and automatically in the case of the delta wing. In the latter study they used 650-850 μm polystyrene beads to mark the flow. These beads were heated to lower their sinking speed in water to 1-3 mm/s. The time exposed pathlines were composed of one bright spot, a grey central section, a second bright spot and finally a grey tail section. The lighted streamwise plane was 3 mm thick while the spanwise plane was 15 mm thick. The image processing algorithm looked for all pathlines with two bright spots and a tail. The magnitude of the velocity vector was determined by the length of the curved grey central section and its direction by the tail location.

291

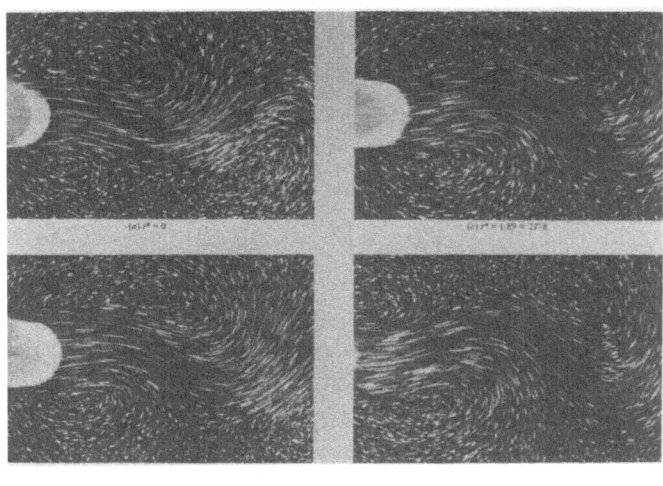

Fig. 11 1. time exposed particle traces in the wake of a circular cylinder at
Re$_d$=100
2. vorticity contours obtained from these traces [Imaichi and Ohmi (1983)]

These authors tested many different interpolation schemes to transform this randomly spaced velocity vector field to a rectangular grid. An analysis of the sources of error lead the authors to conclude that the rms error of the resulting vorticity is less than 15%.

Agüí and Jiménez (1987) have made a rigorous analysis of the errors which arise in particle tracking techniques for the measurement of velocity and vorticity fields. They found that the two dominant sources of error result from the limited resolution of the optical data acquisition system and from limited particle concentration. They have applied their analysis to a particle marked flow in the near wake of a circular cylinder. They estimate that the total visualization error is about 5%. This is composed of the resolution of the film which is related to the ability to determine the position of the end points of the particle traces, the uncertainty in the duration of the illumination pulse, and the uncertainty due to the particles not strictly following the motion of the fluid.

The sampling error depends on the particle concentration which dictates the spatial resolution of the measurements. For their wake study this spacing was 2.2 mm which (from the Nyquist sampling criteria), means that they could not recover information from wave lengths shorter than about 4.5 mm. To transform the data to a uniformly spaced grid they found that the best results are obtained using polynomial interpolations. However, they opted to use a simpler convolution interpolation scheme. Agüí and Jiménez (1987) conclude that sampling errors as large as 20-30% can occur because the length of the time exposed particle traces and the distance between them is larger than the smallest scales in the flow.

In contrast with Imaichi and Ohmi's (1983) study at low Reynolds numbers in the very near wake, the measurements of Agüí and Jiménez (1987) were made at Re_d = 2000 and further downstream. Figure 12 shows the time evolution of the spanwise vorticity field contours in this wake.

Falco and his co-workers (1987,1988) have developed an image processing technique to determine transverse components of vorticity. The method can also be used for the streamwise component if variation in the mean velocity is taken into account, and this has already been achieved by the authors. Utilizing a photochromatic chemical dissolved in kerosene, they write a grid of lines into the flowing fluid spaced at 1 mm apart with line widths of 100 μm. By photographing the distortions in images of the grid illuminated with a pulsed laser over a small time interval, the displacements of the corner of the grid squares can be determined. Figure 13 shows an excited grid which has been distorted over a long 0.4 sec interval to magnify the distortions in order to illustrate the point. The photochromatic chemical used was Kodak 1,3-trimethyl-8-nitrospiro [2-H-1

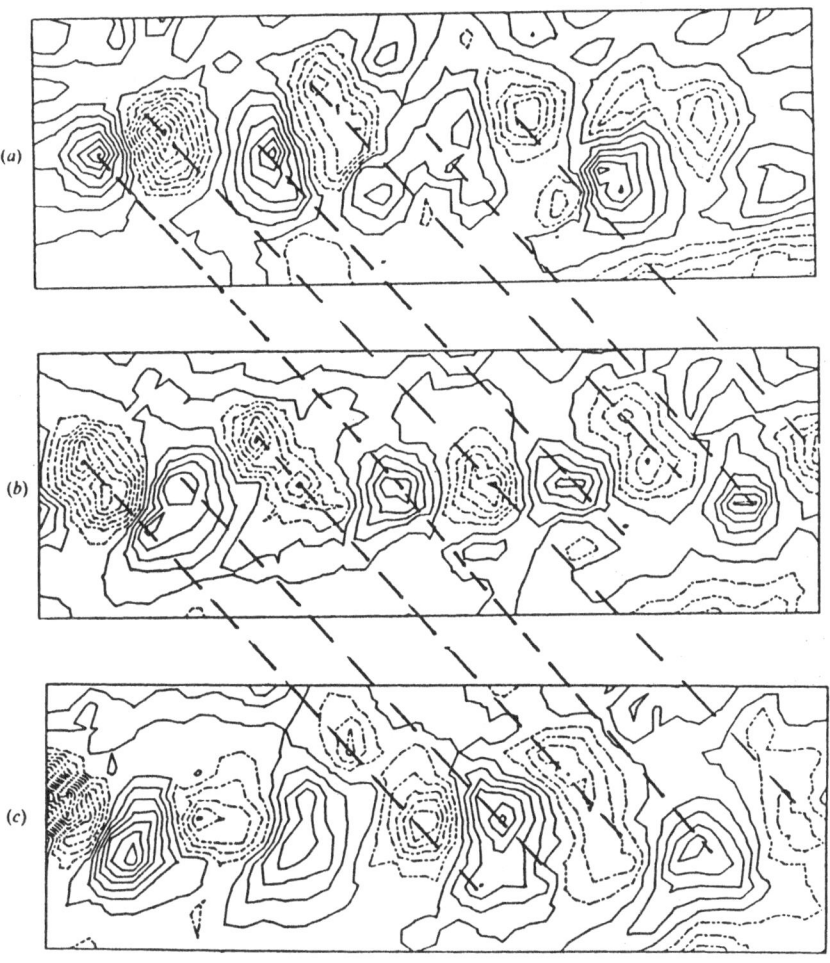

Fig. 12 Vorticity contours from particle tracking for three consecutive frames of the wake of a circular cylinder at Re_d=2000. The figure covers 5 to 23 cylinder diameters left to right. [Agüí and Jiménez (1987)]

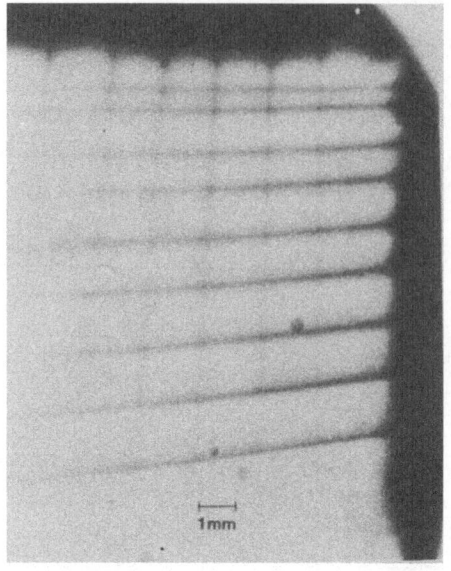

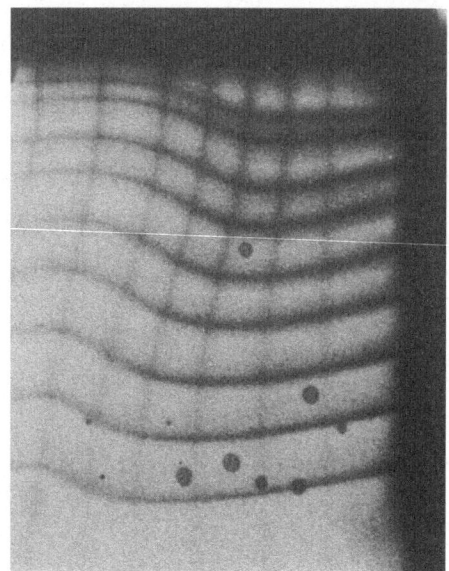

Fig. 13 Distortions of a demonstration photochromic grid over a 0.4s lapse time
[Falco & Chu (1987)]

benzyopran-2,2-indoline]. It was excited by an Excimer laser running on XeF at approximately 100 mj output. The excited chemical reflected blue light for the order of tens of seconds which was very long compared to the pulse time. An algorithm was devised to automatically find the intersection of all the lines in the grid; the displacement of these intersections determine the velocity components in the plane. The average velocity between intersections is calculated, and the integral of these average velocities around each square approximates the circulation and thus the average vorticity in the small square. The authors estimate that vorticity can be measured absolutely with this method to an accuracy of 1 sec^{-1}. Falco and Chu (1987) have tested the method by determining the mean velocity gradient in a Stokes layer with good accuracy. Falco, Chu, Hetherington and Gendrich (1988) have also measured the vortex flow of an impulsively started air foil. Although the method has not yet been tried in a turbulent flow, it appears from these tests that it will be able to accurately measure vorticity in such flows.

Adrian (1988) is developing a particle image velocimetry (PIV) system which will allow measurements of the vorticity vector perpendicular to the plane of view. Adrian and Landreth (1987) have implemented a full two-dimensional spatial correlation interrogation method for this PIV system and used it in the study of a jet impinging on a flat plate. Because of the large storage requirement (10^9-10^{10} pixels for each photograph) the photographs are interrogated over 1 mm^2 at a time and the digitized image is analyzed to obtain a velocity vector. Then the interrogation is moved to a new spot on the photograph. Obviously the method necessitates high computational speed, but it yields a robust estimate of the velocity vector field at several thousand points in the plane with an accuracy approaching that of hot-wire or LDV systems provided that the image density is high enough. The vorticity field can also be estimated from this velocity field estimate. Adrian (1988) points out that an important challenge in such methods is performing high resolution photography of very small particle images, on the order of 5 to 15 µm, over fields of view of the order of 100 mm with image resolution of about 10 µm. He believes that soon this method will be capable of application with motion pictures of two dimensional planar slices and ultimately of holographic images of fluid volumes.

6.3 The Three Components of Vorticity Measured Simultaneously

To measure all three components of the vorticity vector in turbulent flow is an exacting task which is only just now being achieved. Wassmann and Wallace (1979) designed a nine-sensor hot-wire probe to measure all three vorticity components as well as the components of the velocity and strain rate fields. The probe, shown in Figure 14, has reasonably good spatial resolution at modest Reynolds numbers and has been refined and tested by Balint, Vukoslavĉević and Wallace (1987) and

Vukoslavčević, Balint and Wallace (1988). It has three arrays with three hot-wire sensors in each array. The diameter of the total sensing area of the probe is 2.2mm with a spacing over which the velocity gradients are determined of 1.2mm. Each of the hot-wires are supported by one of the three external and the common prong within each array, and they are heated by a constant temperature anemometer bridge circuit. The common prongs are copper plated to reduce their resistance to less than 0.1 Ohms in order to prevent instabilities in the circuits and cross talk between arrays. Thermal cross talk between wires is minimized by operating the probe at a relatively low overheat ratio. Measurements have been made in a zero pressure gradient boundary layer at R_θ = 2100 and 2850 and, more recently, in a two-stream mixing layer and in a turbulent grid flow. In the boundary layer at R_θ = 2850, the distance over which velocity gradients are estimated is about 10.4 viscous lengths which is about 6.9 times the Kolmogorov microscale for this flow at y+ = 13.

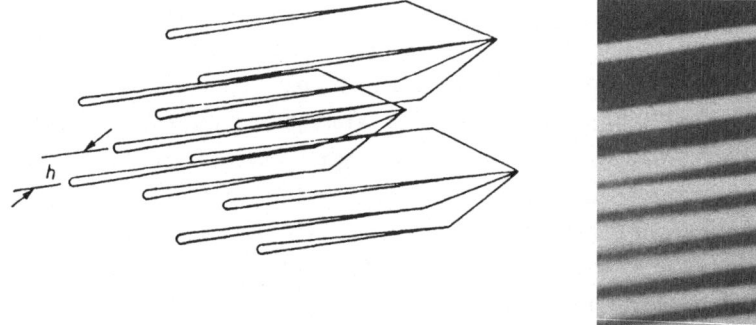
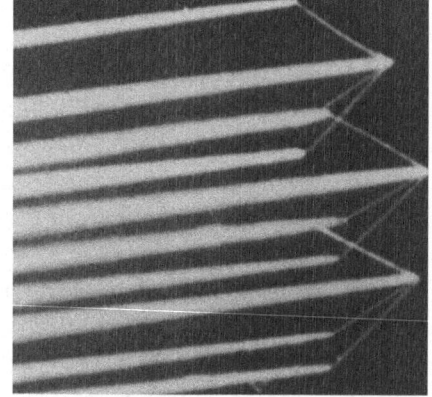

Fig. 14   Sketch and photograph of nine-sensor probe to simultaneously measure velocity and vorticity vectors [Balint, Vukoslavčević and Wallace (1987)]

Equation (19) is the algebraic form (in terms of the three velocity components at the centroid of each respective array expanded in a Taylor's series to the first order) taken by the three nonlinear expressions for the squares of the effective cooling velocities acting at the center of each of the wires of each array.

$$U_{e_{ij}}^2 = [K_{ij1}(U_0 + C_{ij1} \frac{\partial U}{\partial z} + C_{ij2} \frac{\partial U}{\partial y}) + K_{ij2}(V_0 + C_{ij3} \frac{\partial V}{\partial z} + C_{ij4} \frac{\partial V}{\partial y})]^2$$

$$+ [K_{ij3}(W_0 + C_{ij5} \frac{\partial W}{\partial z} + C_{ij6} \frac{\partial W}{\partial y})]^2 . \qquad (19)$$

The array number is indicated by the subscript i and the wire number by the subscript j. The Kijks are calibration coefficients and the Cijks are constant coefficients for a given array geometry which are positive or negative fractions of h. Solutions to this set of nine equations are the three unknown velocity components at the centroid of the probe U_0, V_0 and W_0 and the six velocity gradients in the plane transverse to the axis of the probe. The gradients in the direction of the mean flow must be obtained utilizing Taylor's hypothesis which has recently been validated by Piomelli, Balint and Wallace (1988) for bounded turbulent shear flows in the region above the buffer layer.

The statistics of the velocity field measured with this probe compared very well with previous results in the literature. The first three moments of the fluctuating vorticity components will be compared with measurements of others in Section 11. As mentioned in Section 4, the balance of terms of the mean and fluctuating enstrophy (mean square vorticity) transport equations for the turbulent boundary layer have also recently been obtained by Balint, Vukoslavčević and Wallace (1988).

Utami and Ueno (1984,1987) have utilized time exposed particle tracers in an open channel flow to determine the velocity vector field over two closely spaced planes parallel to the channel bed. By utilizing the continuity equation, the velocity component normal to the planes can be estimated between the planes, and, with this information, all three vorticity components can be estimated from the data. The method uses two cameras which alternately photograph the two planes illuminated by alternately shifting the light source up and down at 0.2 sec. intervals. The measurement planes were at $y^+ \simeq 23$ and 54 and the thickness of the viewing plane was $\Delta y^+ = 17$. Each trace is shaped to have a long segment followed by two short segments for the first camera and two short segments followed by a long segment for the second. Tracers that do not show these full sequences are ignored since they obviously leave the plane of view. The camera and lighting system are translated with the flow. With this system the authors have made interesting observations about the vortex structure of the wall region from instantaneous vorticity contours. Obviously, the resolution of this method in the normal direction only allows the low wave number contributions to the streamwise ω_x and spanwise ω_z vorticity components to be determined.

Utilizing a modification of this method Utami, Blackwelder and Ueno (1988) determined some of the basic statistics of the velocity field for a turbulent channel flow. The particle trajectories were determined using a spatial correlation of pairs of images, and the randomly spaced velocity vectors thus obtained were interpolated to a grid using a mapping algorithm based on the distance of the particles from the grid nodes. The resulting velocity statistics are in reasonable agreement with measurements using other means described in the literature. As yet statistics for the vorticity field are not available.

Although no turbulence measurements have yet been made, a very ambitious and promising method is that of Weinstein, Beeler and Lindeman (1985) called a holocinematographic velocimeter. It has recently been further described by Weinstein (1988). A high speed movie is made of single exposure holograms of particles marking a turbulent flow, and successive pairs of movie frames are analyzed to determine particle tracer movement. Weinstein, et al. (1985) estimate that one sec. of data results in 10^8 numbers showing that the data storage problem is extreme. The method is a laboratory analog to direct numerical simulations of turbulence with the advantage that it can be applied to flows with more complex geometery. Initially a flat plate boundary layer will be studied at R_θ = 1250. Marking particles are 40 μm hollow glass spheres which only slowly rise in water. Because coordinate accuracy can be determined to a far greater precision perpendicular to rather than along the focal axis, two orthogonal views will be used to obtain full accuracy in all three directions. These two views must be solved simultaneously in order to obtain the three dimensional particle coordinates. The optical set up for the in-line holography measurements are shown in Figure 15. This work and that of Adrian (1988), mentioned above, promise full three dimensional vorticity vector fields from laboratory data in the foreseeable future.

Although this paper is a survey of laboratory methods for measuring vorticity, it should be noted that an apparently quite accurate method of determining vorticity over the entire flow field for low Reynolds numbers and simple geometries is by direct numerical simulation of the Navier-Stokes equations. These simulations have given the full three dimensional vorticity vector field for fully developed channel flow [Kim, Moin and Moser (1987) with a spatial resolution of 6η, .025-2.2η and 3.5η in the x,y and z directions], for the turbulent boundary layer at R_θ up to 1410 [Spalart (1987)], for homogeneous turbulent shear flow and various irrotational straining flows [Rogers and Moin (1987)] and for the plane mixing layer [Metcalfe, Hussain, Menon and Hayakawa (1987)]. In Section 11 we will compare some of the results obtained from these simulations with those from laboratory measurements.

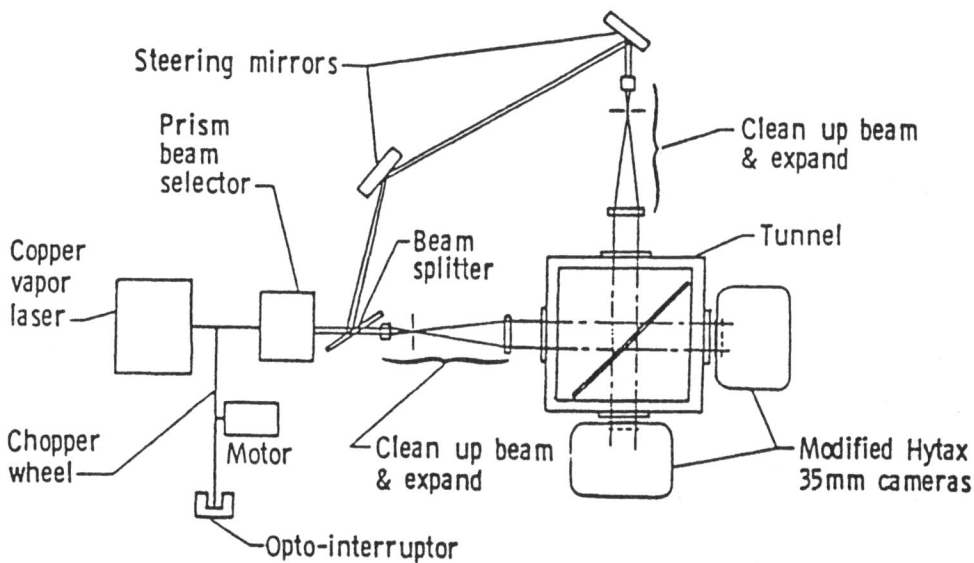

Fig. 15   High-speed holocinematographic velocimeter to obtain full field velocity
and vorticity measurements [Weinstein (1988)]

7. DIRECT ESTIMATES OF VELOCITY GRADIENTS

Gennes (1977) originally proposed the use of strophometry for the direct
measurement of the velocity gradients in turbulent flows.   Plane gratings of
parallel lines or of an array of squares may be imposed on the flow at an initial
time and the subsequent distortion of the gratings is determined by diffraction
measurements.   There are various methods of writing these gratings in the flow.
Fermigier et al. (1980,1982) produced gratings by periodically heating the fluid
with a pulsed laser which results in optical index changes.   Transmittance gratings
were produced by Cloitre and Chauveau (1983) and Charmet et al. (1984) through
changes in the photochromatic properties of the fluid.   Cloitre and Guyon (1986)
have utilized this method to study the properties of the transverse velocity
gradient $\partial u/\partial y$ in a turbulent plane Poiseuille flow.   An analysis of the motion of
diffraction spots generated by a second laser beam probe yields this gradient for a
two-dimensional grating and the transverse vorticity component ω_z for a square mesh
grating [Fermigier et al. (1984)].   The mean velocity gradient $\partial \bar{u}/\partial y$ profile in the
buffer layer and above is obtained with very good accuracy with this method.
Measurements of the transverse vorticity components have not yet been reported but
can be anticipated soon.

Agüi and Hesselink (1987,1988) and Agüi (1988) have developed a similar method
utilizing the rotation of volume holograms by turbulent convection and diffusion to
determine velocity gradients in the flow.   These holograms are written in the fluid

medium using the photochromic effect; they could also be produced using thermal absorption or other photo-induced chemical reactions. The hologram results from changes in the concentration of the species and the associated changes in the optical properties of the fluid medium. The state of the hologram is almost entirely determined by the net effect of the coupled fluid convection and diffusion. By Fourier transforming the scalar transport equation for species concentration, the deformation and translation in Fourier space can be studied as a description of the deformation and rotation in physical space. The authors conclude that photochromic holograms can be used to measure velocity gradients of the order of $3s^{-1}$ with holograms as small as 95 μm which is of the order of the Kolmogorov microscale in the channel flow and boundary layer they are investigating. This is not possible with thermal holograms because of their more rapid diffusion. To measure the vorticity component along an axis requires three writing beams to produce two orthogonal gratings which are then probed with two reading beams. Although such point vorticity measurements have not yet been made, they appear to be forthcoming soon.

8. METHODS WHICH DIRECTLY SENSE THE FLOW ROTATION

Vane vorticity meters have often been proposed as a means of measuring mean streamwise vorticity. These devices consist of a zero angle of attack propeller with very thin blades and with bearings with low friction in order to rotate with the angular rotation of the flow. McCormick, Tangler and Sherrieb (1968) have used such a meter to study the structure of aircraft trailing vortices. Holdemann and Foss (1975) have also used a vane vorticity meter in a bounded jet study. Wiegland, Ahmid and Nagib (1978) have studied the characteristics of such meters. Govinda Raju, Holla, Nigam and Verghese (1978) point out that in turbulent flow there will be a side load on the bearings which is not experienced in static calibration. For angles of attack greater than 5 degrees the friction of the device changes very significantly. Huachen and Shiying (1987) decreased the effect of bearing friction by utilizing larger vanes but, of course, the spatial resolution is decreased. The internal friction and the large moment of inertia of probes of this type usually gives very low frequency response. Hence a vane vorticity meter should normally be reserved for those applications in which only time mean $\overline{\Omega}_x$ values are desired. It is not known how this device averages an $\Omega_x(t)$ input if the second and third moments of the distribution function are relatively large.

Recently, however, Bawirzanski, Randolph and Ecklemann (1987) have designed a probe with conical bearings to minimize the internal friction. They believe the probe has an adequate frequency response for turbulent fluctuations, at least in the Göttingen oil channel flow which has a very narrow spectral range. The wake downstream of this probe, however, indicates that the presence of the probe itself significantly alters the upstream flow field. When the authors correct their measurements for this effect, they obtain rms values of streamwise vorticity in rather good agreement with those of others as discussed in Section 11.

Blackwelder (1979) has patented a probe to measure streamwise vorticity fluctuations with zero angle of attack stationary fins on a small diameter shaft. Strain gauges spirally wound around the shaft measure the torque on the fins exerted instantaneously by the flow. Strain gauges aligned along the shaft can also be used to measure the streamwise velocity. Another means of imparting the torque to the shaft, suggested by Blackwelder (1979), would be to replace the fins with a roughened sphere.

Frisch and Webb (1980) have developed an ingeneous method for sensing fluctuating vorticity directly. They use very small (25 μm), neutrally buoyant, transparent plastic particles which have planar crystal mirrors embedded within them to mark the flow. A beam of laser light is reflected by the particles as they rotate in response to the local vorticity, and this rotation rate may be detected by the optical system shown in Figure 16. The spatial resolution of this system is about 100 μm. By polymerizing methylmethacrylate, suspending this monomer in water and stirring continuously with a small amount of Mearlmaid Nacromer ZTX-B added, very small plexiglas spheres with hexagonal plate mirrors embedded within them are produced. A test fluid of dibutyl phthalate was chosen to 1) nearly match the refractive index of the spheres, 2) have no chemical reactivity with the spheres or the flow system, and 3) nearly match the density of the spheres so that they are neutrally buoyant. As shown in the figure, a pair of slits with fixed separation angles intercept the rotating beam, and photomultipliers convert this light into electrical pulses so that the time of rotation can be measured. Because of the limitations on the height of the detecting photomultiplier, the slit height only allows about 4% of the total reflections to be intercepted by the optics which is the most serious problem with this method. An additional limitation is the use of an unusual fluid which requires greater safety precautions. Measurements of the mean velocity profile in a laminar Poiseuille flow indicated good accuracy is possible. Ferguson and Webb (1983) extended this system utilizing two dimensional positioning to measure two components of vorticity.

9. A METHOD WHICH INDIRECTLY SENSES THE FLOW ROTATION

Tsinober, Kit and Tietel (1982) have developed a method, taken from magnetohydrodynamics, for measuring a component of vorticity in an electrically conducting fluid in the presence of a magnetic field. They show that

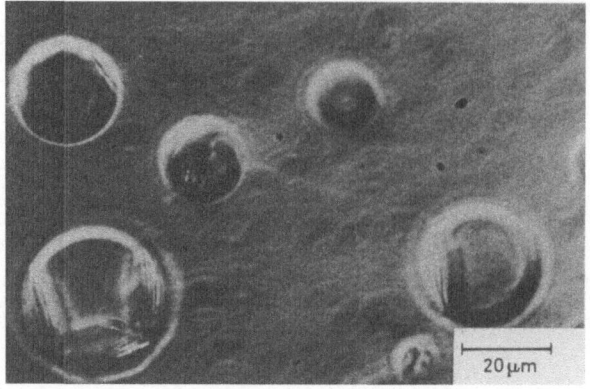

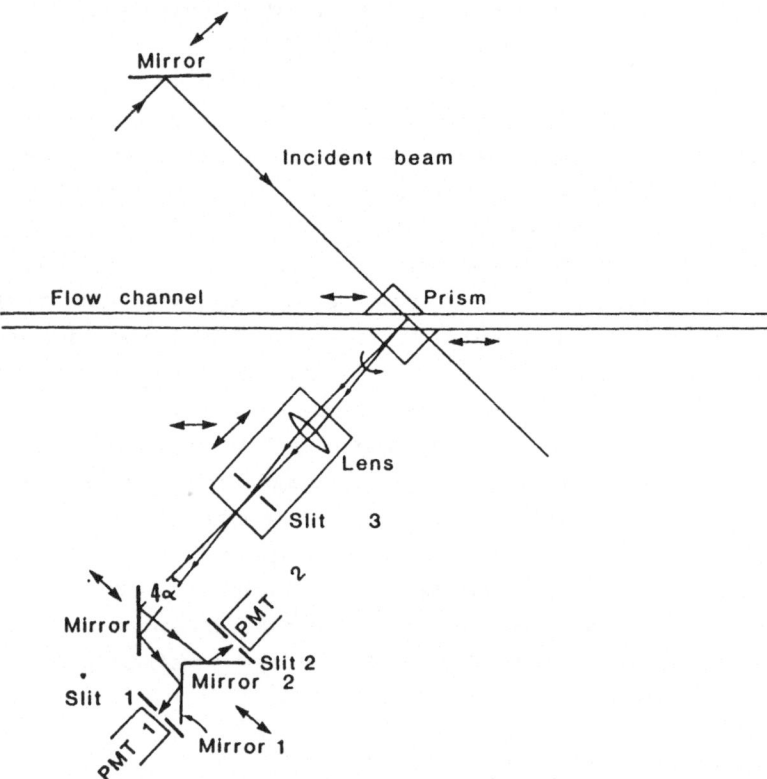

Fig. 16 Optical detection system to measure one component of particle rotation rate
[Frisch & Webb (1980)]

$$\nabla^2 \phi = \vec{B} \cdot \vec{\Omega} \tag{20}$$

where ϕ is the potential of the electric field, $\vec{B}$ is the magnetic field intensity, and $\vec{\Omega}$ is the vorticity vector. The Laplacian of the electric potential is approximated by a central difference measured with a seven electrode probe shown in Figure 17. This gives the component of vorticity parallel to the direction of the magnetic field. A very strong magnetic field and electrodes with low electro-chemical noise is needed, but the probe does not require calibration. A probe has been built with a measuring volume of 1 mm³ and measurements have been carried out in turbulent grid flows [Tsinober, Kit and Teitel (1987) and Kit et al. (1988)].

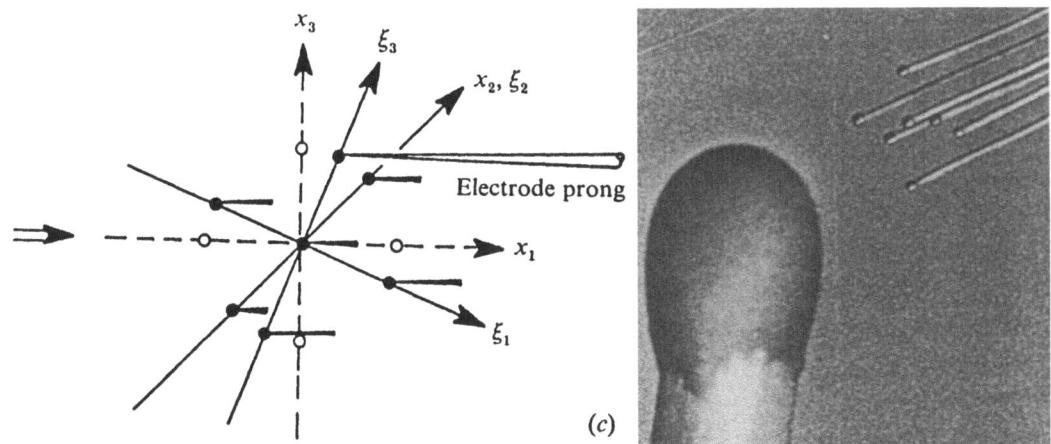

Fig. 17 Seven-sensor electrodes (compared to a match head) to measure streamwise vorticity in an electrically conducting flow [Tsinober, Kit & Teitel (1987)]

10. PHASE AVERAGED VORTICITY MEASUREMENTS

It should finally be mentioned that a number of workers have obtained phase averaged measurements of the vorticity field in a variety of transitional and turbulent flows with periodic or quasi-periodic convecting structures. These measurements are obtained by phase-locking the velocity field measurements at a spatial array of grid points with the phases of periodical or quasi-periodical motions. A detector probe and a detection method must be used to determine the phase of the flow. The vorticity is found by subsequently differentiating the phase averaged velocity field. This method has been extensively exploited by Zaman and

Hussain (1980) and Hussain and Zaman (1980,1981,1985) in a variety of turbulent free shear flows. Cantwell and Coles (1983) have obtained phase averaged spanwise vorticity contours in the turbulent near wake of a circular cylinder at Re_d = 1.4×10^5. Williams, Fazel and Hama (1984) have obtained all three phase averaged vorticity components for the transition region of a boundary layer. The latter measurements allowed the authors to quantitatively characterize vortex loops which had previously only been observed visually. Koga, Nelson and Eaton (1987) have determined phase averaged spanwise vorticity maps for unsteady separated vortices. Mathioulakis and Telionis (1987) have also used this technique in their study of a periodic separating laminar flow. A two component LDV provided the phase sampled u and v velocity components in their study.

11. COMPARISON OF VORTICITY COMPONENT MEASUREMENTS BY VARIOUS METHODS

To date there are still only a few flows in which comparable vorticity component measurements have been made. Until just recently these measurements were all in either fully developed turbulent channel flows or in boundary layers. Figures 18-20 compare distributions of the rms vorticity of the three vorticity components, ω_x, ω_y and ω_z, normalized by the viscous time scale ν/u^2_τ and as a function of nondimensional distance from the boundary surface y^+. For all three components laboratory measurements and direct numerical simulations (DNS) are compared. Several observations can be made. The channel flow DNS of Kim et al. (1987) and Kim (1988) compare very well to the boundary layer DNS of Spalart (1988) over the entire range for ω_y. This is also the case for ω_z except very near the wall where there seems to be a Reynolds number dependence as seen in the tabulation of wall ω_z values below. For ω_x, Spalart's (1988) boundary layer data at $R_\theta=660$ are consistently larger by about 35-25% over y+=1-100, than Kim et al.'s (1987) channel flow data at $R_\theta \approx 280$. Kim's (1988) channel flow data at $R_\theta \approx 660$ is closer to Spalart's data for ω_x.

From the laboratory measurements for ω_x shown in Fig. 18, Balint et al.'s (1985) boundary layer data at $R_\theta \approx 2100$ compares very well with Kim et al. (1987) and agree rather well with the outer flow measurements of Corrsin and Kistler (1954) who used the Kovasznay-type probe. For the latter data, the parasitic contamination of ω_x by the v and w fluctuations would have been less severe in this region since the intensities of these fluctuations are small there. This was not the case for the measurements of Kastrinakis (1976) who used a Kovaszvay-type probe in a channel flow. The very high values of ω_x he obtained in the buffer and log layers were undoubtedly due to this parasitic contamination. Measurements were later made by Kastrinakis and Eckelmann (1983) with the modified Kovasznay-type probe which allows correction for the v and w contamination. Their values agree quite well with the DNS data and the measurements of Balint et al. (1985). The vane vorticity meter

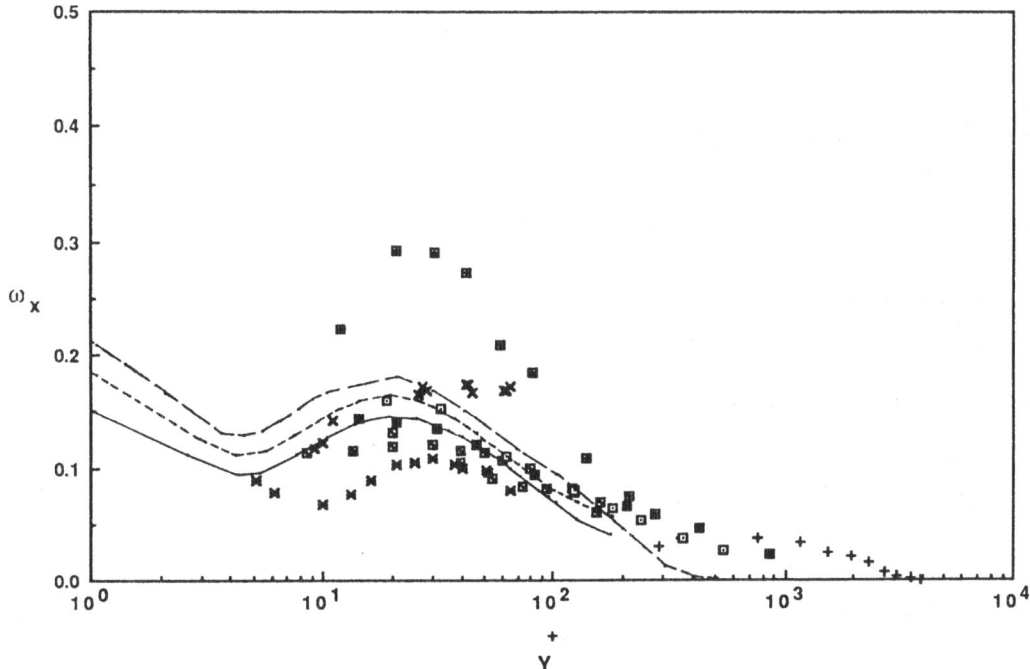

Fig.18   RMS of streamwise vorticity fluctuations [Re$_\Theta$ indicated in brackets].
Boundary layers :■ Balint et al. (1985)[2100]; + Corrsin & Kistler (1945)[7565];
— — Spalart (1987, data analysed by S. K. Robinson)[660]. Channel flows :
◪ Bawirzanski et al. (1987)[325]; x Kasagi et al. (1986)[330];⊞ Kastrinakis
(1976)[535];   ▣ Kastrinakis & Eckelmann (1983)[1085]; —— Kim et al. (1987)
[280]; ✻ Nishino et al. (1987)[300]; --- Kim (1988)[660].

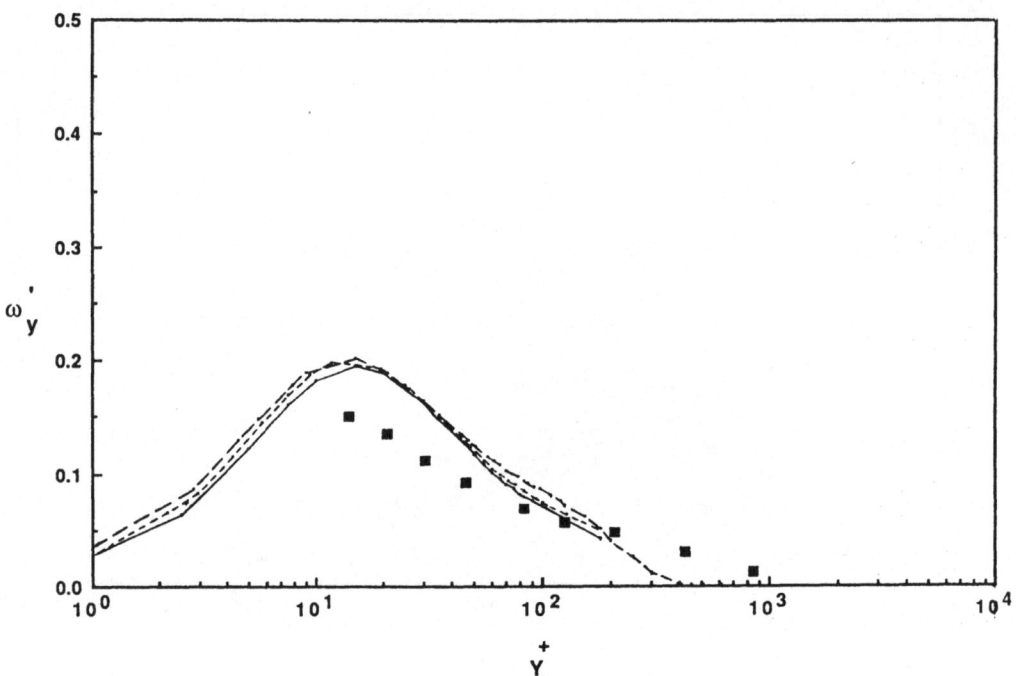

Fig.19   RMS of normal vorticity fluctuations (symbols as in fig.18).

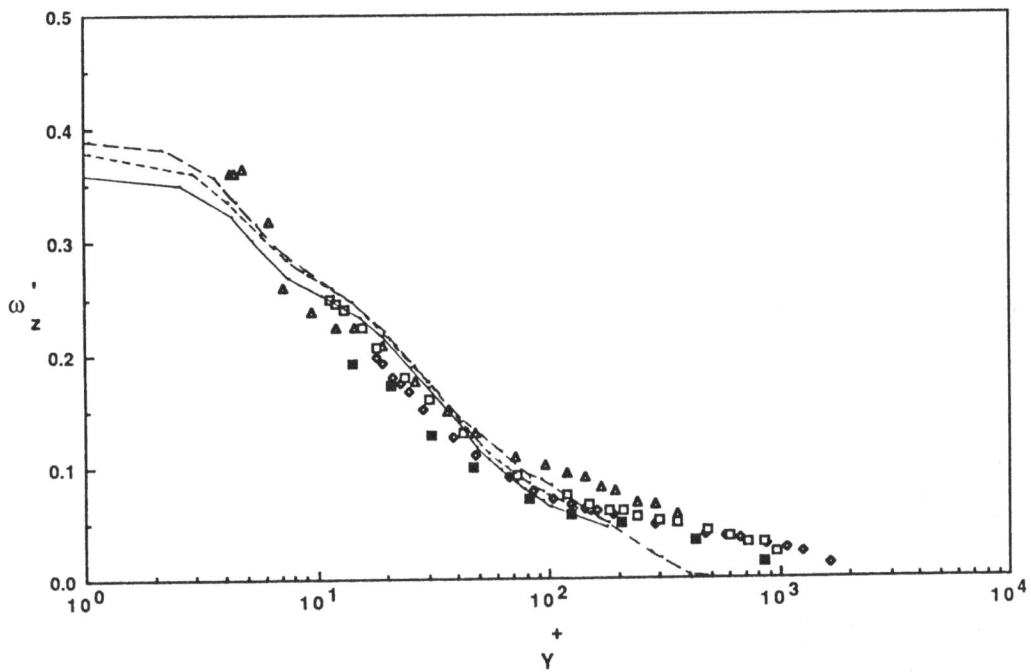

Fig.20　RMS of spanwise vorticity fluctuations. Boundary layer :
△ Klewicki and Falco (1988)[1010], □ [2870], ◊ [4850]; (other symbols
as in fig.18).

measurements of Bawirzanski et al. (1987) are a little lower in the buffer and log layers than those described above, but are surprisingly close considering the large correction applied. Nishino et al.'s (1987) data from hydrogen bubble trajectories also are low but reflects the DNS shapes at small y^+ rather well. Kasagi et al.'s (1986) hydrogen bubble trajectory data is high for $y+>20$. The latter two sets of measurements were made in the same channel flow.

Balint et al.'s (1985) ω_y measurements are somewhat smaller than all the DNS data. At y+ = 14, this difference is about 33%. The best agreement between all the laboratory measurements and all the DNS data is for ω_z as seen in Figure 20. This is not surprising because ω_z includes the largest fluctuating gradient, $\partial u/\partial y$; therefore the signal-to-noise ratio is highest for the determination of this component.

The fluctuating velocity gradients $\partial u/\partial y$ and $\partial w/\partial y$ are the limiting values of ω_x and ω_z at the wall. Measurements of the rms values ω_x and ω_z have been made at the wall using flush mounted sensors. These are given below in Table 1 where they are compared to the DNS simulated values.

	R_θ	ω_x'	ω_z'	Flow
Kim et al. (1987)	280	0.195	0.363	Channel
Kreplin & Eckelmann (1976)	330	0.065	0.250	Channel
Spalart (1987)	670	0.271	0.398	Boundary Layer
Alfredson et al. (1988)	430		0.400	Channel

Table 1. The wall values of the normalized rms streamwise and spanwise vorticity fluctuations.

In the recent paper of Alfredson et al. (1988) the authors tabulated other wall values of ω_z and concluded that the low value of Kreplin and Eckelmann (1976) is due to heat transfer to the fluid via the hot-film probe substrate. They state that the wall value of ω_z is about 0.4 based on all the available information and that the wall value of ω_x is probably about 0.2, although they have not measured this directly.

The normalized third moments (skewnesses) of the three fluctuating vorticity components are shown in figures 21-23. For boundary layers and channel flows $S(\omega_x)$ and $S(\omega_y)$ should be identically zero everywhere. Kim's (1988) DNS data satisfy this condition rather well even though the statistical sample was small which is reflected in the lack of smoothness of the curves. Balint et al.'s (1985) data also satisfy this condition rather well for $S(\omega_x)$ but somewhat less well for $S(\omega_y)$ which

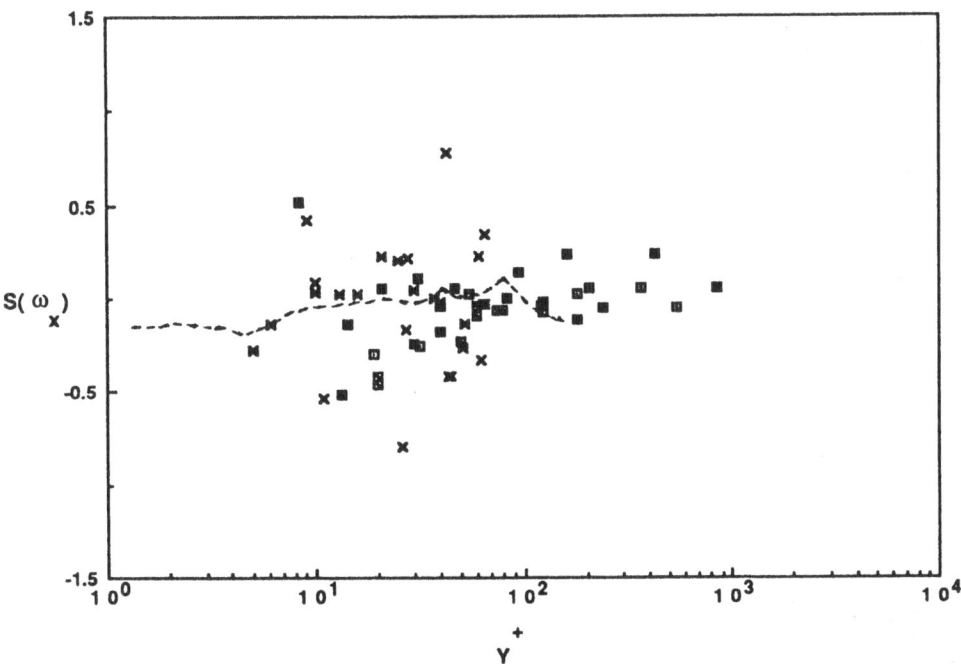

Fig.21 Skewness of streamwise vorticity fluctuations (symbols as in fig.18).

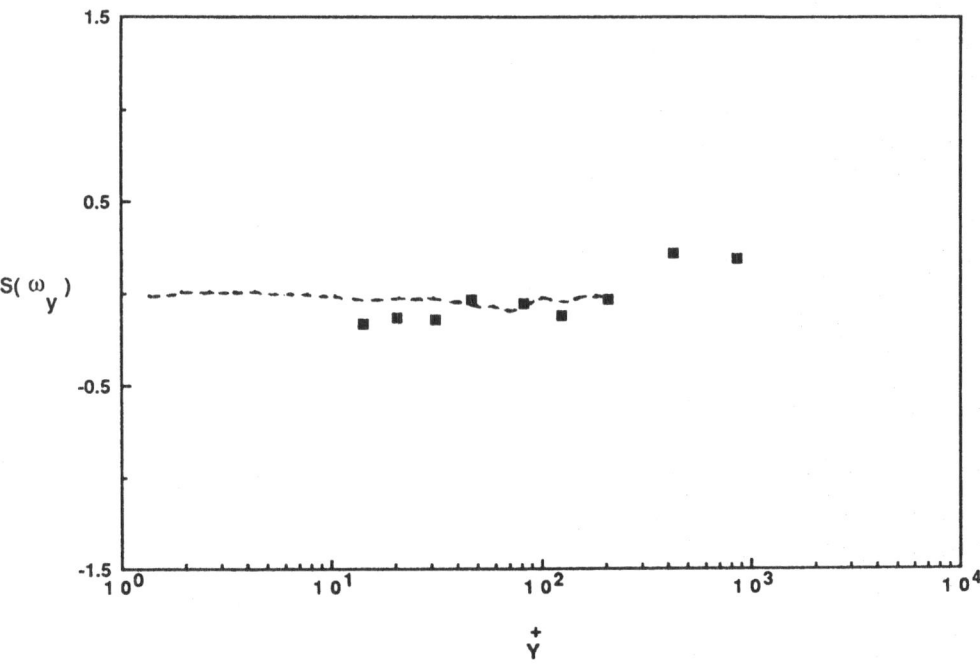

Fig.22 Skewness of normal vorticity fluctuations (symbols as in fig.18).

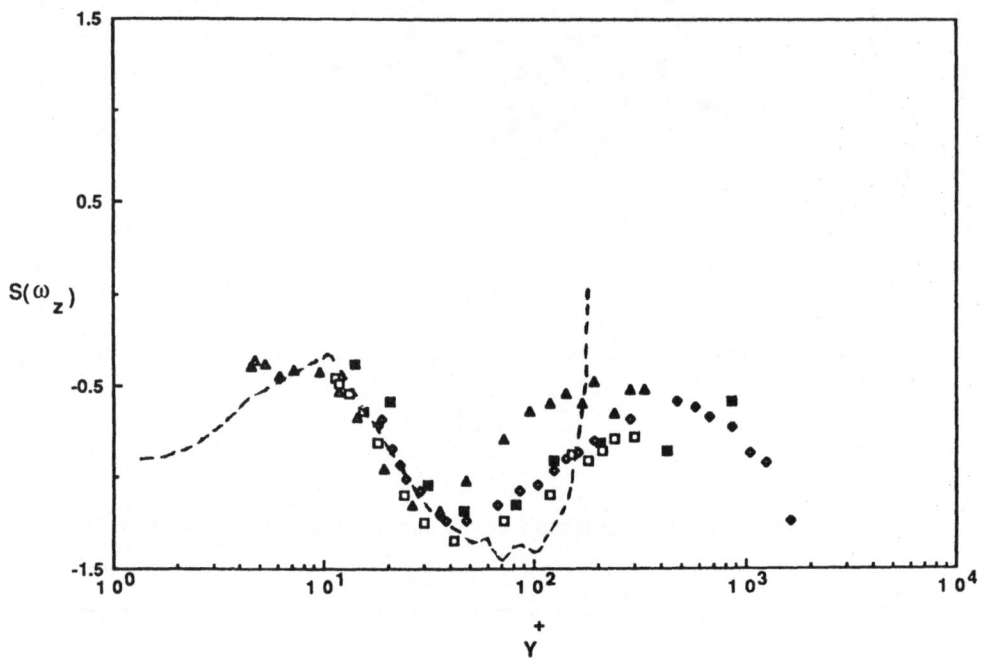

Fig.23 Skewness of spanwise vorticity fluctuations (symbols as in fig.18).

is again probably due to the poorer signal-to-noise ratio of this component. The other laboratory measurements show substantial deviation from this condition, especially near the wall. For $S(\omega_z)$ the DNS data of Kim (1988) agree well with the measurements of Balint et al. (1985) and Klewicki and Falco (1988) and show that the spanwise vorticity fluctuations are quite negatively skewed. They reach skewness values of <-1 in the lower part of the log layer. $S(w_z)$ must go to zero at the centerline of Kim's channel flow explaining the lack of agreement with the boundary layer measurements at y^+ = 180.

The normalized fourth moments (flatnesses) of the three fluctuating vorticity components are shown in figs. 24-26. The $F(\omega_x)$ data of Kasagi (1986) and Barinzanski (1987) compare well with Kim's (1988) DNS data. Surprisingly, the measurements of Balint (1985) lie significantly lower than the simulations and the measurements of Kastrinakis and Eckelmann (1983) lie much lower than those. Since the rms values from these two experiments agreed well with the DNS RMS data, this flatness difference indicates a difference in the very large amplitude values measured. Balint's (1985) $F(w_y)$ measurements agree well with Kim's (1988) DNS data for $y+\leq50$. The differences at larger $y+$ may be partly due to a Reynolds number dependence as is seen in figure 26 for $F(\omega_z)$. For the spanwise component $F(\omega_z)$, the measurements of Balint et al. (1985) and Klewicki and Falco (1988) agree well with Kim's (1988) DNS data for $y+<80$. For larger $y+$ the boundary layer measurements show a Reynolds number dependence. The channel flow simulation data, however, does not follow this dependence.

Figure 27 shows the distribution of the rms spanwise vorticity distribution ω_z in a two-stream mixing layer at a velocity ratio of 2 and a Reynolds number $(U_1+U_2)\theta/2\nu$ matched at a value of 6958 for the data of Lang (1985) and for the data of Foss and Haw (1987). The data of Balint et al. (1988) was taken at the slightly lower R_θ of 6406. The three sets of measurements essentially agree accept for the large RMS values near the centerline of the layer. In this region Foss and Haw (1987) obtain 40% higher values than Lang (1985) and 17% higher than Balint et al. (1988). When normalized by the average of the two free stream velocities, these peak rms difference are reduced. The experiments of Foss and Haw (1987) and Balint et al. (1988) were carried out in the same flow facility and under the same conditions at the University of Illinois. Temporally growing mixing layers have been simulated by Metcalf et al. (1987) and work is currently being carried out in several places to obtain a DNS of a spatially growing mixing layer. Soon, therefore, rms values as well as other statistics will be available from these simulations to compare with laboratory measurements.

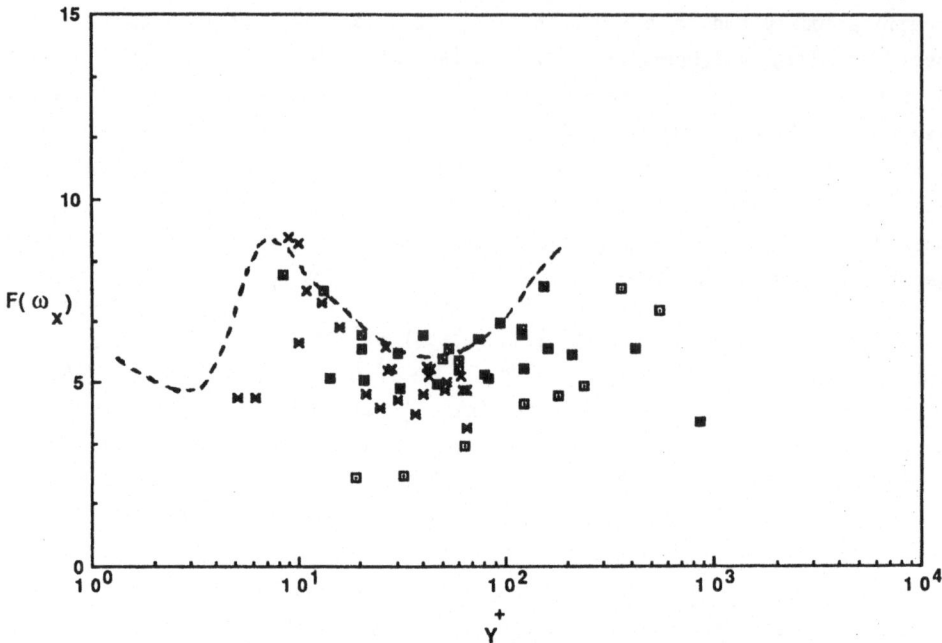

Fig.24   Flatness of streamwise vorticity fluctuations (symbols as in fig.18).

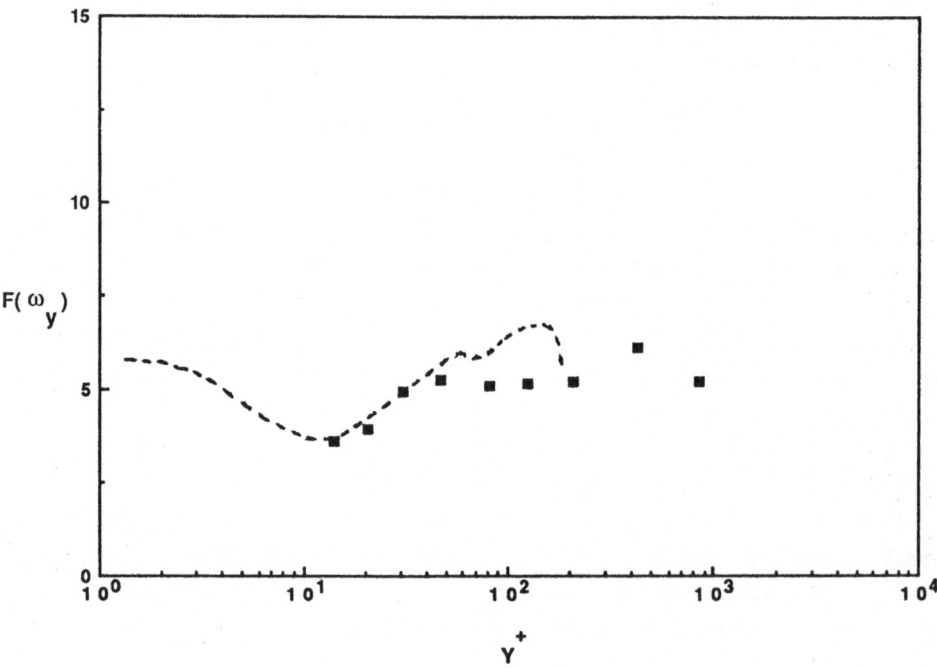

Fig.25   Flatness of normal vorticity fluctuations (symbols as in fig.18).

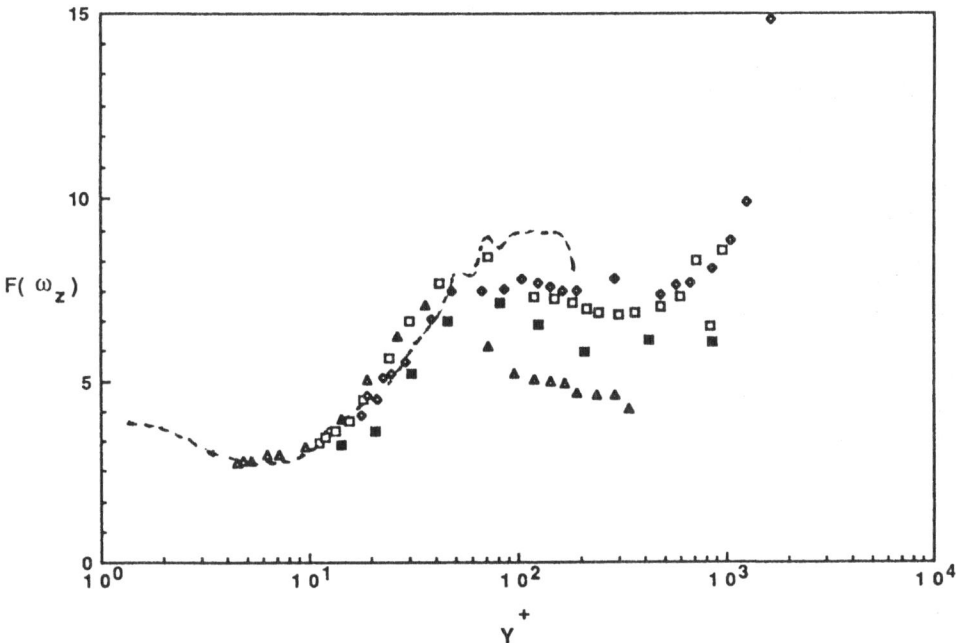

Fig.26   Flatness of spanwise vorticity fluctuations (symbols as in figs.18&20).

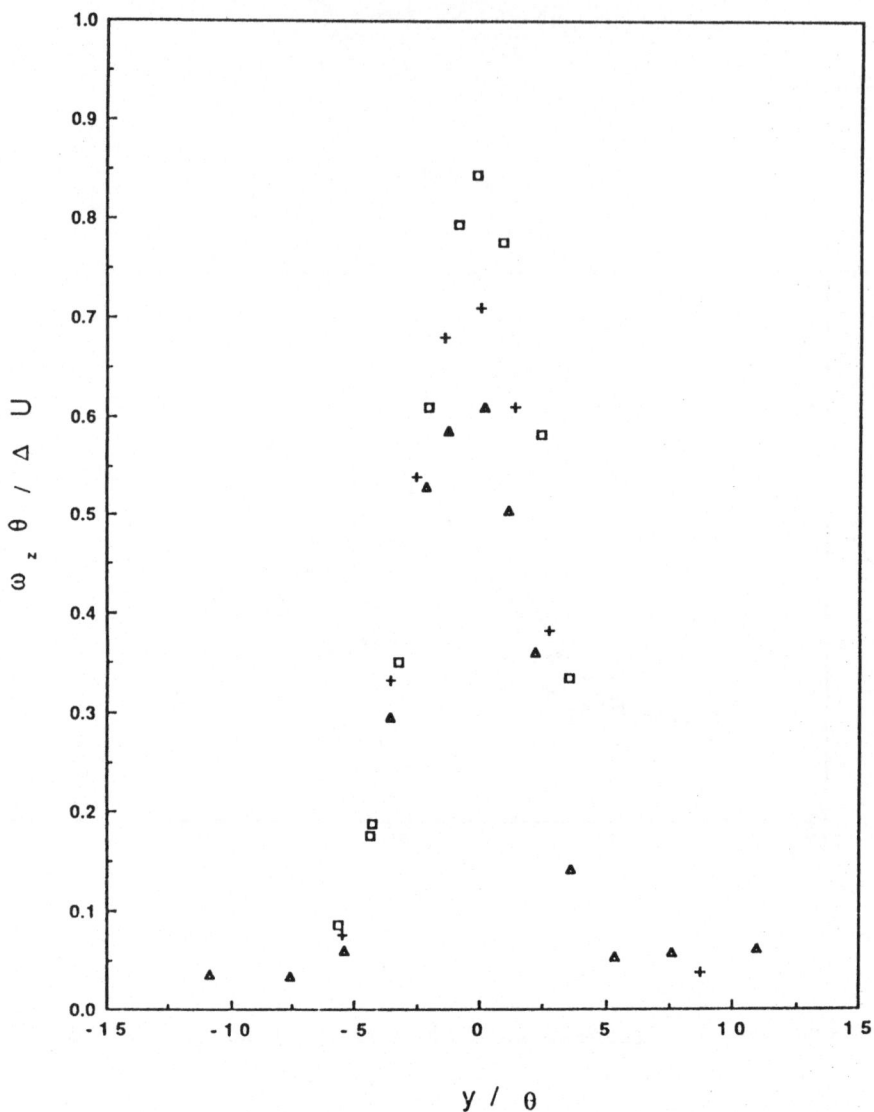

Fig.27　RMS of spanwise vorticity fluctuations [Re$_\theta$ indicated in brackets]. Two-stream mixing layer:　□ Foss and Haw (1987) [6958];　△ Lang (1985)[6958]　+ Balint et al. (1988)[6406]

12. CONCLUSIONS

There has recently been a great increase in attempts to measure one or more components of the time and space resolved vorticity vector $\Omega(x,y,z,t)$ in turbulent flows. This increase is due to improvements in methods of fabricating minature multisensor hot wire arrays, novel optical methods, and, most recently, innovative use of digital image processing of visualized turbulent flows. With the development of automatic processing of holographic images well underway, full three dimensional vorticity field data for complex turbulent flows appears within reach. Adequately resolved vorticity component measurements are always difficult; it is a testimony to the skill and perserverance of the investigators cited herein that they have recently been achieved. The authors are persuaded that these measurement methods will greatly aid us in gaining insight into the challenging dynamics of turbulence.

ACKNOWLEDGEMENTS

The authors are grateful for the support of NSF Grant No. MEA-8216946 (JFF) and DOE Grant No. DE-FG05-88ER13838 (JMW) during the preparation of this paper.

REFERENCES

Adrian, R. (1988) "Progress in particle image velocimetry for the study of fluid motion" *NSF Workshop on Imaging for Fluid Mechanics*, Ohio State Univ. Oct. 1987 (rept. ed. by Brodkey, R.S. & Guezennec, Y.G. Jan. 1988).

Adrian, R. & Landreth, C. (1987) "Interrogation of high image density particle image velocimeter photographs" *Bull. Am. Phys. Soc. 32*(10), p. 2097.

Aqüí, J.C. (1988) "Vorticity studies in flows: global three-dimensional analysis and optical point measurements" *Ph.D. Dissertation, Stanford Univ.*

Agüí, J.C. & Hesselink, L. (1987) "Holographic measurements of strain of fluids" *Bull. Am. Phys. Soc. 32(10)*, p. 2106.

Agüí, J.C. & Hesselink, L. (1988) "Holograms in motion, I: the effect of fluid motion on volume holograms" *J. Opt. Soc. Amer. 5(8), pp. 1287-1297.*

Agüí, J.C. & Jiménez, J. (1987) "On the performance of particle tracking" *J. Fluid Mech. 185*, pp. 447-468.

Alfredson, P.H., Johansson, A.V., Haritonidis, J.H. & Eckelmann, H. (1988) "The fluctuating wall-shear stress and the velocity field in the viscous sublayer" *Phys. Fluids 31(5)*, pp. 1026-1033.

Antonia, R.A., Browne, L.W.B. & Shah, D.A. (1988) "Characteristics of vorticity fluctuations in a turbulent wake" *J. Fluid Mech. 189*, pp. 349-365.

Balint, J.-L., Vukoslavčević, P., & Wallace, J.M. (1987) "A study of the vortical structure of the turbulent boundary layer" *Advances in Turbulence*, Springer-Verlag (eds. Comte-Bellot, G. & Mathieu, J.) pp. 456-464.

Balint, J.-L., Vukoslavĉević, P. & Wallace, J.M. (1988) "The Transport of Enstrophy in a Turbulent Boundary Layer" to be published in the *Proc. of the Zaric Memorial Sem. on Wall Turbulence*, Hemisphere Pub. Corp.

Balint, J.-L., Wallace, J.M. & Vukoslavĉević, P. (1988) "The statistical properties of the vorticity field of a two-stream turbulent mixing layer" to be published in the *Proc. of the Second European Turbulence Conference*, Berlin.

Batchelor, G.K. (1967) *An Introduction to Fluid Dynamics*, Cambridge University Press.

Bawirzanski, E., Randolph, M. & Eckelmann, H. (1987) "Direct measurement of streamwise vorticity fluctuations with a cylinder probe" *Proc. of 2nd Int. Sym. on Transport Phenomena in Turbulent Flows* (Tokyo), pp. 605-612.

Blackwelder, R.F. (1979) "Vorticity probe utilizing strain measurements" *U.S. Patent 4,145,921.*

Böttcher, J. & Eckelmann, H. (1985) "Measurement of the velocity gradient with hot film probes" *Exp. in Fluids*, 3, pp. 87-91.

Cantwell, B. & Coles, D. (1983) "An experimental study of entrainment and transport in the turbulent near wake of a circular cylinder" *J. Fluid Mech. 136*, pp. 321-374.

Cloitre, M, Chauveau, J. (1983) "Metal dithizonate for flash photolysis applications in hydrodynamics" *Optics Comm. 47*, pp. 42-46.

Cloitre, M. and Guyon, E. (1986) "Forced Rayleigh scattering in turbulent plane Poisseuille flows" *J. Fluid Mech. 164*, pp. 217-236.

Charmet, J.C., Fermigier, M., & Jenffer, P. (1984) "Visulization d'ecoulment et measure de gradient de vitesse par les traceurs photochrome" *Comp. Rend. Acad. Sci. 298*, pp. 103-106.

Corrsin, S. & Kistler, A.L. (1954) "The free-stream boundaries of turbulent flows" *NACA TN 3133.*

Eckelmann, H., Nychas, S.G., Brodkey, R.S, & Wallace, J.M. (1977) "Vorticity and turbulence production in pattern recognized turbulent flow structures" *Phys, Fl. 20*, pp. S225-S231.

Falco, R.E. (1983) "New results, a review and synthesis of the mechanism of turbulence production in boundary layers and its modification" *AIAA-83-0377.*

Falco, R.E. & Chu, C.C. (1987) "Measurement of two-dimensional fluid dynamic quantities using a photochromic grid tracing technique" *SPIE Photomechanics and Speckle Metrology 814*, pp. 706-710.

Falco, R.E., Chu, C.C., Hetheringtron, M.H. & Gendrich, C.P. (1988) "The circulation of an airfoil starting vortex obtained from instantaneous vorticity measurements over an area" *AIAA 88-3620.*

Ferguson, R.D, Webb, W.W. (1983) "The vorticity optical probe: a fast multicomponent model" *Proc. of 8th Biennial Symp. on Turbulence*, University of Missouri-Rolla.

Fermigier, M, Jenffer, P., Charmet, J.C., Guyon, E. (1980) "A non-perturbative anemometric method and flow visulization technique" *J. Phys. Lett. 41*, L519-L521.

Fermigier, M., Cloitre, M, Guyon, E. & Jenffer, P. (1982) "Utilisation de la diffusion Rayaleigh forcee a l'etude d'ecoulements laminaires et turbulents" *J. Mecanique Theor. Appl. 1*, pp. 123-146.

Fohl, T. & Turner, J.S. (1975) "Colliding vortex rings" *Phys. of Fluids 18(4)*, pp. 433-436.

Foss, J.F. (1976) "Accuracy and uncertainty of transverse vorticity measurements" *Bull. of the Am. Phys. Soc. 21*, pp. 1237.

Foss, J.F. (1981) "Advanced techniques for transverse vorticity measurements" *Proc. of 7th Biennial Symp. on Turbulence*, Univ. of Missouri-Rolla.

Foss, J.F., Ali, S.K. & Haw, R.C. (1987) "A critical analysis of transverse vorticity measurements in a large plane shear layer," *Advances in Turbulence*, Springer-Verlag (eds. Comte-Bellot, G. & Mathien, J.) pp. 446-455.

Foss, J.F. and Haw, R.C. (1987) "Comparitive measurement: time resolved transverse vorticity in a two-stream mixing layer" *Bull. Am. Phys. Soc.* 32(10) p. 2043.

Foss, J.F. & Kleis, S.J. (1976) "Mean-flow characteristics for the oblique imping ment of axisymmetric jet," *AIAA Jour. 14*, pp. 705-706.

Foss, J.F., Kliewicki, C.L. & Disimle (1986) "Transverse vorticity measurements using an array of four hot-wires" *NASA CR 178098*.

Frisch, M.B, Webb, W.W. (1981) "Direct measurement of vorticity by optical probe" *J. Fluid Mech. 107*, pp. 173-200.

de Gennes, P.G. (1977) "Principe de nouvelles measures sur les ecoulements par échauffement optique localise" *J. Phys. Lett. 38*, pp. L1-L3.

Govinda Raju, S.P., Holla, V.S., Nigam, A. & Verghese, M.P. (1978) "Development of a vorticity meter" *J. Aircraft 15(11)*, pp. 799-800.

Heikes, K.E. & Maxworthy, T. (1982) "Observations of inertial waves in a homogeneous rotating fluid," *J. Fluid Mech. 125*, pp. 319-345.

Haw, R.C., Foss, J.K. and Foss, J.F. (1988) "The vortical properties of the high speed region in a plane shear layer and its parent boundary layer" to be published in the *Proc. of the Second European Turbulence Conference*, Berlin.

Holdemann, J.D. & Foss, J.F. (1975) "The initiation, development, and decay of the secondary flow in a bounded jet" *J. Fluids Eng. 97(1)*, pp. 342-352.

Huachen, Pan & Shiying, Z. (1987) "Measurement of streamwise vorticity using a vane vorticity meter" *Int. Jour. of Heat and Fluid Flow 8*, pp. 72.

Hussain, A.K.M.F. & Hayakawa, M. (1987) "Education of large-scale organized structures in a turbulent plane wake" *J. Fluid Mech. 180*, pp. 193-229.

Hussain, A.K.M.F.and Zaman, K.B.M.Q. (1980) "Vortex pairing in a circular jet under controlled excitation. Part 2 - coherent structure dynamics" *J. Fluid Mech. 101*, pp. 493-544.

Hussain, A.K.M.F. & Zaman, K.B.M.Q. (1981) "The 'prefered mode' of the axisymmetric jet" *J. Fluid Mech. 110*, pp. 39-71.

Hussain, A.K.M.F. & Zaman, K.B.M.Q. (1985) "An experimental study of organized motions in the plane mixing layer" *J. Fluid Mech. 159*, pp. 85-104.

Imaichi, K. & Ohmi, K. (1983) "Numerical processing of flow-visualization pictures measurement of two-dimensional vortex flow" *J. Fluid Mech. 129*, pp. 283-311.

Kasagi, N., Hirata, M. & Nishino, K. (1986) "Streamwise pseudo-vortical structures and associated vorticity in the near-wall region of a wall-bounded turbulent shear flow" *Exp. in Fluids 4*, pp. 309-318.

Kastrinakis, E.G. (1976) "An experimental investigation of the fluctuations of the streamwise components of the velocity and vorticity vectors in a fully developed turbulent channel flow" *Dissertation, Georg-August Universität zu Göttingen.*

Kastrinakis, E.G., Wallace, J.M., Willmarth, W.W., Ghorashi, B. & Brodkey, R.S. (1977) "On the mechanism of bounded turbulent shear flows" *Lect. Notes in Phys. 75*, pp. 175-189.

Kastrinakis, E.G., Eckelmann, H, & Willmarth, W.W. (1979) "Influence of the flow velocity on a Kovasznay type vorticity probe" *Rev. Sci. Instr. 50*, pp. 759-767.

Kastrinakis, E.G. & Eckelmann, H. (1983) "Measurement of streamwise vorticity fluctuations in a turbulent channel flow" *J. Fluid Mech. 137*, pp. 165-186.

Kastrinakis, E.G., Nychas, S.G. & Eckelmann, H. (1982) "Some streamwise vorticity characteristics of coherent structures" *Structure of Complex Turbulent Shear Flow* IUTAM Symp. Springer-Verlag (eds. Dumas, R. & Fulachier, L.) pp. 31-40.

Kim, J. (1988) *Unpublished data provided to the authors.*

Kim, J., Moin, P. & Moser, R. (1987) "Turbulence statistics in fully developed channel flow at low Reynolds number" *J. Fluid Mech. 177*, pp. 133-166.

Kim, J.-H. and Fiedler H. (1988a) "Vorticity measurements in a turbulent mixing layer" *Euromech Colloquium 235 on Dynamics of Large Scale Vortical Structures.*

Kim, J.-H. and Fiedler, H. (1988b) "Vorticity Measurements in a turbulent mixing layer" to be published in *Proc. of the Second European Turbulence Conference,* Berlin.

Kistler, A.L. (1952) "The vorticity meter" *M.S. Thesis, The Johns Hopkins Univ.*

Kit, E., Tsinober, A., Teitel, M. Balint, J.-L., Wallace, J.M. & Levich, E. (1988) "Vorticity measurement in turbulent grid flows" *Fluid Dynamics Research 3*, pp. 289-294.

Klewicki, J.C., & Falco, R.E. (1986) "Spanwise vorticity distributions and correlations in a turbulent boundary layer" *Bull. Am. Phys. Soc. 31*(10), pp. 1685.

Klewicki, J.C. & Falco, R.E. (1988) "On accurately measuring statistics associated with small scale structure in turbulent boundary layers" *Report TSL-88-4*, Michigan State University.

Koga, D.J., Nelson, C.F. & Eaton, J.K. (1987) "A new program for active control of unsteady, separated flow structures" *Proc. 2nd AFOSR Workshop on Unsteady Separated Flows* (July 28-30, U.S. Air Force Acad.).

Kovasznay, L.S.G. (1950) Quart. Prog. Rep. of Aero. Dept. *Contract NORD-8036-JHB-39*, The Johns Hopkins Univ.

Kovasznay, L.S.G. (1954) "Turbulence measurements," in: *Physical Measurements in Gas Dynamics and Combustion (10)* Princeton Univ. Press (eds. Landenbuerg, R.W., Lewis, B., Pease, R.N. and Taylor, H.S.), pp. 213.

Kovasznay, L.S.G. (1979) "Panel discussion on coherent structures" *Second Turbulent Shear Flows Conference*, London, England.

Kreplin, H.-P. & Eckelmann, H. (1976) "Behavior of the three fluctuating velocity components in the wall region of a turbulent channel flow" *Phys. Fluids 22*(7), pp. 1233-1239.

Lang, D.B. (1985) "Laser doppler velocity and vorticity measurements in turbulent shear layers" *Ph.D. Dissertation, California Inst. of Tech.*

Lang, D.B. & Dimotakis, P.E. (1982) "Measuring vorticity using the laser doppler velocimeter" *Bull. of Am. Phys. Soc., 27*, pp. 1166.

Lighthill, M.J. (1963) "Introduction Boundary Layer Theory," *Laminar Boundary Layers,* Clarendon Press. (ed. Rosenhead, L.), Chap. 2.

Mathioulakis, D.S. & Telionis, D.P. (1987) "Velocity and vorticity distributions in periodic seperating laminar flow" *J. Fluid Mech., 184,* pp. 303-333.

Maxworthy, T. (1971) "A simple observational technique for investigating boundary layers, stability and turbulence" *Turbulence Measurements in Liquids,* Department of Chemical Engineering, Univ. of Missouri at Rolla, (ed. Paterson, G.K. and Zakin, J.L.).

McCormick, B.W., Tangler, J.L. & Sherrieb, H.E. (1968) "Structure of Trailing Vortices" *J. Aircraft 5(3),* pp. 260-267.

Metcalf, R., Hussain, A.K.M.F., Menon, S. & Hayakawa, M. (1987) "Coherent structures in a turbulent mixing layer: a comparison between direct numerical simulations and experiments" *Turbulent Shear Flows 5*, Springer-Verlag (Eds. Durst, F. et al.) p. 110-123.

Metcalfe, R.W., Orzog, S.A., Brachet, M.E., Menon, S. & Riley, J.J. (1987) "Secondary instability of a temporally growing mixing layer" *J. Fluid Mech. 184*, pp. 207-243.

Moin, P., & Kim, J. (1985) "The structure of the vorticity field in turbulent channel flow. Part 1. Analysis of instantaneous fields and statistical correlations" *J. Fluid Mech. 155*, pp. 441-464.

Nishino, K., Kasagi, N. & Hirata, M. (1987) "Study of streamwise vortical structures in a two-dimensional turbulent channel flow by digital image processing" *Proc. of 2nd Int. Sym. on Transport Phenomena in Turbulent Flows* (Tokyo), pp. 47-60.

Nychas, S.G., Kastrinakis, E.G. & Eckelmann, H. (1985) "On certain aspects of vorticity dynamics and turbulent energy production" *Lect. Notes in Phys. 235*, pp. 269-278.

Oshima, Y. & Asaka, S. (1975) *Natural Sci. Report, Ochanomizu Univ. 26*, p. 31.

Oshima, Y. & Asaka, S. (1977) "Interaction of two vortex rings along parallel axes" *Jour. Phys. Soc. Japan 42(2)*, pp. 708-713.

Potter, M.C. & Foss, J.F. (1975) *Fluid Mechanics*, The Ronald Press Co. (now published by Great Lakes Press Co., Okemos, MI 48864).

Piomelli, U., Balint, J.-L. & Wallace, J.M. (1988) "On the validity of Taylor's hypothesis for wall-bounded turbulent flows" to be published in, *Phys. of Fluids.*

Rogers, M.M. & Moin, P. (1987) "The structure of the vorticity field in homogeneous turbulent flows" *J. Fluid Mech. 176*, pp. 33-66.

Savas, Ö. (1985) "On flow visualization using reflective flakes" *J. Fluid Mech. 152*, pp. 235-248.

Spalart, P.R. (1988) "Direct simulation of a turbulent boundary layer up to R_θ=1410" *J. Fluid Mech. 187,* pp. 61-98.

Spedding, G.R. & Maxworthy, T. (1987) "The generation of circulation and lift in a rigid two-dimensional fling" *J. Fluid Mech. 165,* pp. 247-272.

Spedding, G.R., & Maxworthy, T. & Rignot, E. (1987) "Unsteady vortex flows over delta wings" *Proc. 2nd AFOSR Workshop on Unsteady and Separated Flows (July 28-30, U.S. Air Force Acad.).*

Tennekes, H. & Lumley, J.L. (1972) *A First Course in Turbulence,* MIT Press, Cambridge, MA.

Tsinober, A., Kit, E., & Teitel, M. (1982) "Induction velometry. Some precise relations between turbulent velocity and electrical fields" in *Metallurgical Applications of Magnetohydrodynamics* (eds. Moffat, H.K. and Proctor, M.R.E.), pp. 129-135.

Tsinober, A., Kit, E. & Teitel, M. (1987) "On the relevance of the potential difference method for turbulence measurements" *J. Fluid Mech. 175,* pp. 447-461.

Uberoi, M.S., & Corrsin, S. (1951) "Progress report on the propagation of turbulence into a non-turbulent flow" *NACA Contract NAW5504,* The Johns Hopkins Univ.

Utami, T. & Ueno, T. (1984) "Visualization and picture processing of turbulent flow" *Exp. in Fluids 2,* pp. 25-32.

Utami, T. & Ueno. T. (1987) "Experimental study on the coherent structure of turbulent open-channel flow using visualization and picture processing" *J. Fluid Mech. 174,* pp. 399-440.

Utami, T., Blackwelder, R.F. & Ueno, T. (1988) "Flow visualization with image processing of three-dimensional features of coherent structures in an open-channel flow" to be published in the *Proc. of the Zaric Memorial Sem. on Near Wall Turbulence,* Hemisphere Pub. Corp.

Vukoslavĉević, P., & Wallace, J.M. (1981) "Influence of velocity gradients on measurements of velocity and streamwise vorticity with hot-wire X-array probes" *Rev. Sci. Instrum. 52,* pp. 869-879.

Vukoslavĉević, P., Balint, J.-L. & Wallace, J.M. (1988) "A Multi-sensor hot-wire probe to measure vorticity and velocity in turbulent flows" to be published by *Jour. of Fl. Engr.* (ASME).

Wallace, J.M., Brodkey, R.S. & Eckelmann, H. (1977) "Pattern recognition in bounded turbulent shear flows" *J. Fluid Mech. 83,* pp. 673-693.

Wallace, J.M. (1986) "Methods of measuring vorticity in turbulent flows" *Exp. in Fluids 4,* pp. 61-71.

Wassmann, W.W., & Wallace, J.M. (1979) "Measurement of vorticity in turbulent shear flow" *Bull. of Am. Phys. Soc. 24,* pp. 1142.

Weinstein, L.M., Beeler, G.B. & Lindemann, A.M. (1985) "High-speed holocinematographic velocimeter for studying turbulent flow control physics" *AIAA-85-0526.*

Weinstein, L.M. (1988) "High-speed holocinematographic velocimeter for study of turbulent flow control physics" *NSF Workshop on Imaging for Fluid Mechanics,* Ohio State, Oct. 1987 (rept. ed. by Brodkey, R.S. & Guezennec, Y.G. Jan. 1988).

Wigeland, R.A., Ahmed, M., & Nagib, H.M. (1978) "Vorticity measurements using calibrated vane-vorticity indicators and cross-wires" *AIAA Journal 16*, pp. 1229-1234.

Williams, D.R., Fasel, H. & Hama, F.R. (1984) "Experimental determination of the three-dimensional vorticity field in the boundary layer transition process" *J. Fluid Mech. 149*, pp. 179-203.

Willmarth, W.W. & Lu, S.S. (1972) "Structure of the Reynolds stress near the wall" *J. Fluid Mech. 55(1)*, pp. 65-92.

Wyngaard, J.C. (1969) "Spatial resolution of the vorticity meter and other hot-wire arrays" *Jour. of Sci. Instrum. (Jour. of Phy. E.), Series 2*, pp. 983-987.

Zaman, K.B.M.Q. & Hussain, A.K.M.F. (1980) "Vortex pairing in a circular jet under controlled excitation, Part 1. General jet response" *J. Fluid Mech. 101*, pp. 449-491.

Zaman, K.B.M.Q. & Hussain, A.K.M.F. (1984) "Natural large-scale structures in the axisymmetric mixing-layer" *J. Fluid Mech. 138*, pp. 325-351.

TIME-RESOLVED HEAT TRANSFER AND SKIN FRICTION MEASUREMENTS IN UNSTEADY FLOW

T. E. Diller and D. P. Telionis
Virginia Polytechnic Institute and State University
Blacksburg, Virginia, USA

ABSTRACT

A review of heat transfer and skin friction measurement methods is presented with particular emphasis on techniques that yield details of time-resolved properties. A description of the calibration methods necessary to insure accurate measurements is included. Examples of recent unsteady heat transfer and skin friction measurements with interpretations of the meaning and importance of the results are given.

INTRODUCTION

It is usually the ultimate goal of any study of fluid flow, experimental or numerical, to generate surface distributions of quantities like pressure, skin friction, and heat transfer. Until very recently, experimental capabilities could generate information only on steady or time-averaged quantities. Theoretical and numerical methods were also limited to steady or time-averaged fields. The past few decades have seen a surge of activity in exploring the time-resolved details of unsteady flow fields. Powerful computers have made possible numerical investigation of time-dependent fields. In the experimental domain, it became quickly possible to measure surface pressures with high frequency response. The measurement of time-resolved heat transfer and skin friction proved to be more demanding tasks and this is the topic of the present paper.

Many flows that have traditionally been treated as steady actually have large components of unsteadiness. Flow unsteadiness can appear in many forms: random fluctuations of turbulence, large-scale coherent flow structures, and forced pulsations due to imposed structural or fluid forces are examples. In characterizing such flows, the details of the unsteady fluid flow are often reported, but relatively little research has been done on the details of the unsteady wall properties. The major difficulties have been due to the limitations of time-resolved measurement techniques.

Local measurements of skin friction have previously been employed to provide information on the local character of the flow. The most common example is the study of a turbulent boundary layer, in which case pressure or skin friction gages yield information on turbulent fluctuations which have correlation lengths of the order of the thickness of the boundary layer itself. In this paper such instruments are used to investigate global phenomena, like the organization of shed vorticity into large-scale vortical structures. This type of information becomes more coherent and useful if combined with simultaneous velocity measurements in the flow field under investigation.

The question here is how far within the domain of interest would a gage sense different events. It is well known that a pressure gage will respond to local fluctuations of pressure. It is also sensitive, however, to large-scale phenomena like vortex shedding, as well as acoustic waves which may propagate from distances many typical lengths away from the body. It is demonstrated here that skin friction and heat transfer gages, which have traditionally been used to disclose local features of the flow, can also be employed to provide information on larger-scale phenomena, developing further away from the body surface.

A decade ago a review of heat transfer measurement techniques for unsteady phenomena was published (Jones 1977). The capabilities for investigating the details of time-resolved heat transfer have grown considerably since that time. In this paper, the techniques that have been developed and refined in the past decade for measuring unsteady heat transfer processes are discussed. Examples of actual recordings of the fluctuating heat transfer are given for several different flows. How these detailed measurements are useful in understanding the basic transport phenomena involved is also discussed. Such measurements have opened a new window of understanding for many processes.

Hot film probes have been employed for the measurement of skin friction for more than 40 years. Discussion of the basic principles involved and reviews of earlier related contributions the reader will find in other publications (Winter 1977, Jones 1977, Hanratty and Campbell 1983). Here a short review is provided of the most recent contributions exploring the accuracy of such instruments and their limitation in providing the amplitude, frequency, and phase of unsteady motions. Examples of measurements using these instruments for unsteady flow research are provided. The discussion is limited to hot-film sensors, which are the most appropriate for high-speed response in unsteady flow.

METHODS OF HEAT TRANSFER MEASUREMENT

The methods for measuring heat transfer between a flowing fluid and a solid surface can be categorized as measurements of (1) the electric power dissipated in a heater at steady state, (2) the temperature difference across a known thermal resistance in the surface, (3) the time rate of change of thermal energy of the surface, and (4) the spatial temperature gradient in the fluid normal to the surface. The first three techniques involve measurements at the solid surface, while the last uses measurements in the fluid. Each has important advantages and limitations. Limited unsteady heat transfer measurements have been made with methods from all four categories. These will be emphasized in the following review.

Time-Averaged Heat Transfer Measurements

The usual quantities of interest are the average heat transfer from a surface (q), the local heat flux (q"), and the convective heat transfer coefficient (h). If radiation can be neglected, the local heat transfer coefficient can be easily related to the heat flux,

$$h = \frac{q''}{T_s - T_\infty}$$ (1)

where T_s represents the material surface temperature and T_∞ is the fluid temperature far from the wall. When high-speed flows are involved, T_∞ is the fluid recovery temperature. Although the experimental techniques measure either the heat transfer or heat flux, the heat transfer coefficient is usually of most engineering interest.

One of the most accurate methods for measuring convective heat transfer from a surface is by monitoring the electric power necessary to maintain an elevated surface temperature (category 1). The measurement area may be the entire surface of the model or a local region of the surface. The largest errors are due to either heat losses at the edges of the measurement region or power measurements made before thermal equilibrium has been achieved. Because the ratio of the edge area to the total surface area increases as the size of the measurement area decreases, the accuracy decreases as the size of the surface segment being measured is decreased (VandenBerghe and Diller 1988). Consequently, this technique can be quite accurate for large surface segments (Shirtliffe and Tye 1985), but is difficult to use when fine spatial resolution is required. Accordingly, few accurate measurements have been reported using local heat flux gages based on this principle of operation (Kraabel et al. 1980, Achenbach 1975, Campbell et al. 1985). An example of such a gage is shown in Fig. 1. The estimated error was less than two percent when using this gage for local convection measurements around a test cylinder (Kraabel et al. 1980).

A related technique is to measure local heat transfer coefficient values from temperature measurements of a heated thin metal film. Such a film gives a constant heat flux boundary condition over the model surface instead of the more common constant temperature surface condition. If lateral conduction can be neglected, the convective heat transfer coefficient is proportional to the known heat flux input to the surface divided by the measured temperature difference between the surface and the fluid, as shown in Eq. (1). The needed surface temperature distribution can be measured by methods such as an array of imbedded thermocouples (Dielsi and Mayle 1984, Baughn et al. 1984) or liquid crystals (Hippensteele et al. 1983, Simonich and Moffat 1984, Baughn et al. 1986). These methods, however, are limited to measuring heat flux only _from_ the surface.

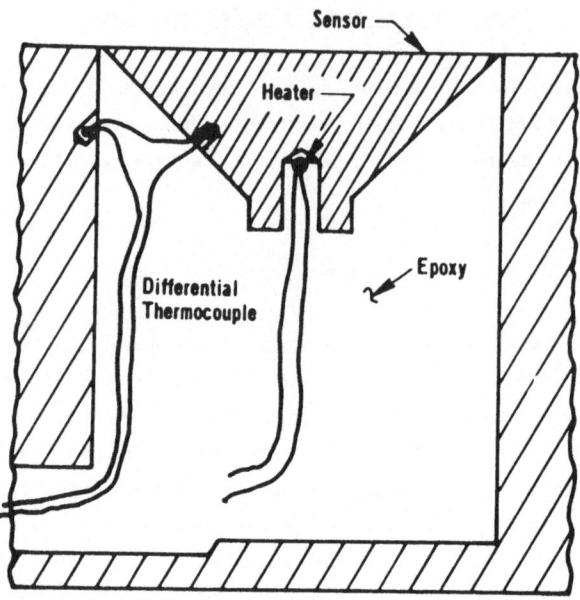

Fig. 1 Diagram of Actively Heated Sensor (Kraabel et al. 1980)

The majority of heat flux gages operate by measuring a temperature difference across a known thermal resistance (category 2). The resistance is typically an integral part of the gage, which then must be placed either on or in the model. Serious measurement problems can result from the physical and thermal disruption of the heat transfer surface caused by the presence of the gage (Diller 1985). Small physical disruptions of the surface are particularly a problem in high-speed flows. The effect of changes in the thermal boundary condition on the heat transfer coefficient have been analyzed for some specific flows by Lau and Sparrow (1980) and Moffat (1982).

The standard temperature-difference type of heat flux gages are: Gardon gages (Gardon 1953), Schmidt-Boelter gages (Kidd 1981), and standard layered gages (Bales et al. 1985). A schematic of a Gardon gage with the foil temperature distribution is shown in Fig. 2. A Schmidt-Boelter gage with a thermopile arrangement is illustrated in Fig. 3. A layered gage is illustrated in Fig. 4. In all of these gages the heat flux through the measuring element creates a temperature difference which is the measured output. The output voltage is, therefore, approximately proportional to the heat flux. Larger gages usually give a higher output voltage, but also create a larger thermal disruption of the surface. Although smaller gages are usually desirable, the accuracy of the output voltage measurement establishes a practical size limitation. One way of multiplying the output signals when thermocouples are used is to form thermopiles by placing a number of thermocouples in series across the thermal resistance. For small gages the limitation then is the physical space to place the individual thermocouple junctions.

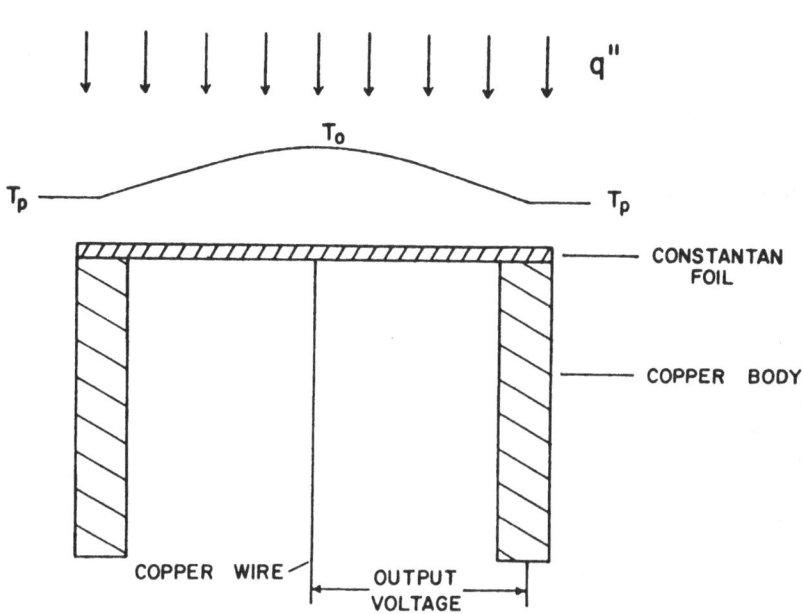

Fig. 2   A Gardon-Type Thin-Foil Heat Flux Gage

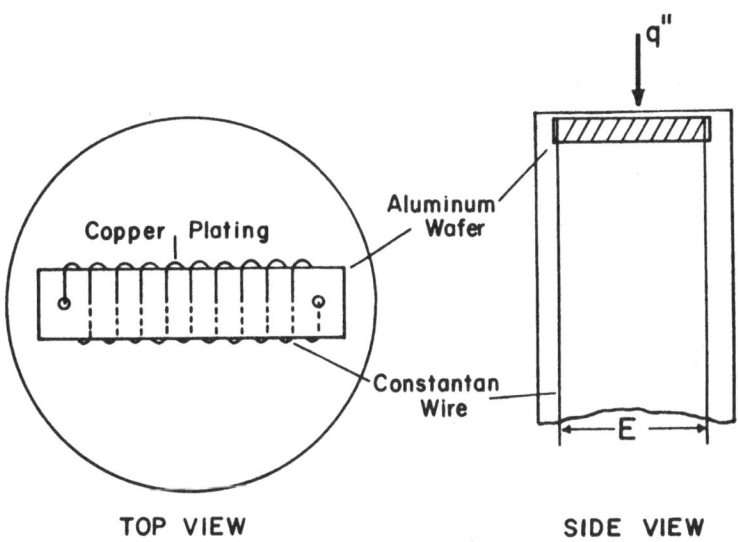

Fig. 3   Diagram of a Schmidt-Boelter Gage

The effect of the thermal disruption caused by a layered gage can be seen in Fig. 4. The surface temperature of the gage is T_1, while that of the surface of the material being measured is T_3. T_1 is closer to the temperature of the fluid than T_3, which results in a measured heat flux (q_{meas}'') lower than would occur without the gage present (q_{actual}''). Therefore, the gage acts like an additional layer of insulation placed on the material surface.

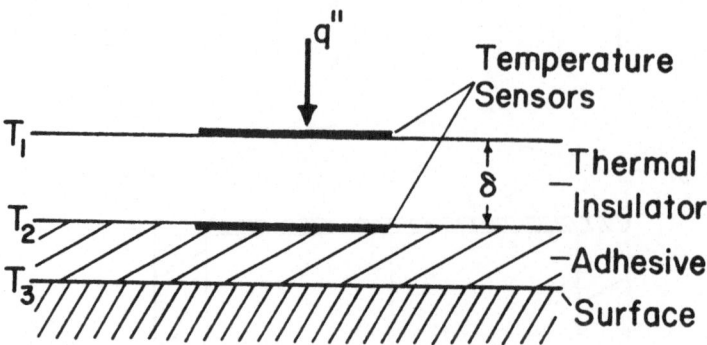

Fig. 4   Schematic Cross Section of Layered Heat Flux Gage

$$q''_{meas} = h(T_1 - T_\infty) \tag{2}$$

$$q''_{actual} = h(T_3 - T_\infty) \tag{3}$$

The measured heat flux is often used with Eq. (2) to determine the heat transfer coefficient, h.   This can result in a sizeable error.   Moreover, the change in surface temperature from T_3 to T_1 results in a change in the thermal boundary condition that the flow experiences.   This not only changes the local heat transfer coefficient (as previously mentioned), but also causes the heat transfer in the surface material to be locally nonuniform.   Therefore, to properly resolve the heat transfer under these conditions, the solution of the complete three-dimensional conjugate problem is required.   One partial experimental solution of this dilemma is to cover the entire surface with an insulating and glue layer to match that of the gage (Hollworth and Gero 1985).   This can be a burdensome process, however.   The most important criteria is still that the temperature drop across the gage and glue layer be small compared to the fluid to surface temperature difference,

$$|T_3 - T_1| \ll |T_3 - T_\infty| \tag{4}$$

for the gage measurement error to be small.   Problems with satisfying this inequality usually arise when gages are used on surfaces with heat transfer coefficients that are larger than the proper range of the gage.

Schmidt-Boelter gages have a similar sensing element as the layered gage although the gage has a different physical geometry.   The sensing element is imbedded into the top of the gage, which is then inserted through the surface material flush with the surface.   The same type of disruption of the thermal field in the material and the thermal boundary layer in the fluid results, however.

Measurement errors of Gardon gages have been investigated by Borell and Diller (1987).   As indicated in Fig. 2, the temperature of the measurement disk is inherently nonuniform for a temperature difference $T_o - T_p$ to be measured.

$$q''_{meas} = \frac{K_2}{f} (T_o - T_p) \tag{5}$$

The value of K_2/f is a constant determined by calibration. The temperature distribution for a constant heat transfer coefficient condition was calculated along with the needed corrections for the measured convective heat flux. This can be expressed in several ways, but in terms of the gage disk center temperature, T_o, the material surface temperature, T_p, and the freestream fluid temperature, T_∞, the actual heat flux from the surface is

$$q''_{actual} = q''_{meas} \left[1 + C_1 \left(\frac{T_o - T_p}{T_\infty - T_p} \right) \right] \tag{6}$$

The parameter C_1 increases as the local heat transfer coefficient and sensitivity of the gage (f) increase, but the value is normally between 0.75 and 1.0. As with the layered gage the important criterion to insure accurate measurements is that the surface temperature perturbation caused by the gage is much less than the surface to fluid temperature difference,

$$|T_o - T_p| \ll |T_\infty - T_p| \tag{7}$$

A newer variation of the layered gage (Fig. 4) that uses resistance elements across the thermal barrier to infer temperatures instead of thermocouples has also appeared (Klems and DiBartolomeo 1982, Andretta et al. 1981, Hayashi et al. 1986, Epstein et al. 1986). Its sensitivity is greater than the comparable thermocouple gage but is less than can be achieved with a thermopile. A slightly different system was developed by Liebert et al. (1985). Thermocouple junctions were sputter-coated on the inside and outside surface of the model material itself. Although this minimizes the surface disruption of the gage, the model material properties become very important and in situ calibration is difficult. For high heat fluxes, Gardon gages can be actively cooled, although this makes the analysis and calibration of the gage output more difficult (Krane and Dybbs 1987).

The third type of heat transfer measurement is to use the thermal capacitance of the gage and/or model material along with surface temperature measurements to infer the heat transfer from the time-temperature history (category 3). Although this method has been used for many years with slug calorimeters, it has become a much more useful method through the use of thin films placed directly on the material surface for measuring the time-dependent surface temperature. The sketch in Fig. 5 illustrates this type of gage. Heat transfer is inferred from the temperature measurements by using an analytical model to determine what heat transfer coefficient history is necessary to produce the observed temperature history. For the simplest case of one-dimensional heat transfer, semi-infinite

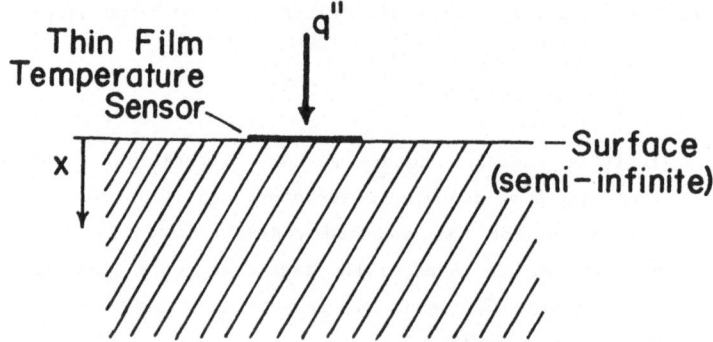

Fig. 5   Thin-Film Gage on Single Layer Semi-Infinite Substrate

substrate, and a negligible thickness of the temperature sensor, the solution for the time-dependent surface temperature is given by Arts and Camci (1985).

$$T_s - T_i = \frac{1}{\sqrt{k\rho C}} \int_0^t \frac{q''(\tau)}{\sqrt{\pi(t - \tau)}} \, d\tau \tag{8}$$

Here, T_i represents the undisturbed substrate temperature at time zero and k, ρ, and C are the thermal conductivity, density, and specific heat of the substrate. Equation (8) shows that the measured surface temperature at any time is a function of the entire heat flux history of the surface.   If the surface heat transfer coefficient is constant and the freestream temperature undergoes a step change at $t = 0$, the result becomes a standard error function-type-solution (Incropera and DeWitt 1985).

$$\frac{T_s - T_i}{T_\infty - T_i} = 1 - \left[\exp\left(\frac{h^2 t}{k\rho C}\right) \, \text{erfc}\left(\frac{h \sqrt{t}}{\sqrt{k\rho C}}\right) \right] \tag{9}$$

In practice, more complicated data processing techniques have been developed (Beck and Keltner 1987, Doorly and Oldfield 1987).   Consequently, the data analysis is much more complicated than some of the other methods discussed where the heat transfer is approximately proportional to the signal.   The other major drawback of this method is due to the uncertainty in the substrate material properties that are necessary to determine the heat transfer.

Because of the fast time response of the thin film method, it has been used extensively for measuring surface heat transfer in short duration flows.   The most popular measurement method for the surface temperature has been thin-film metal resistance layers (Gladden and Proctor 1985, Elrod et al. 1985, Kercher et al. 1983, Dunn 1986a, Consigny and Richards 1982, Nicholson et al. 1984, Tsou et al. 1983, Miller 1981,1985).   Liquid crystals have also been used with this transient temperature method (Jones and Hippensteele 1987).   Although the time response is not as fast (Ireland and Jones 1987), the temperature field over an entire surface can be simultaneously measured. Other methods of observing the surface temperature change have also been used, such as melting point coatings (Metzger and Larsen 1986).

For accurate heat transfer measurements, calibration is required for virtually every heat flux measurement system. The most often used calibration systems use either radiation or conduction. Radiation is used by the National Bureau of Standards (Kidd 1983) and by the major heat flux gage manufacturers, e.g. RdF, Medtherm, Thermogage, and Hy-Cal. Although it is a very useful technique because of the wide range of heat flux levels that can be achieved (Liebert 1985, Atkinson et al. 1985) it has several drawbacks. First, the amount of radiation being emitted from the calibration surface must be measured by a previously calibrated standard. The output from the gage currently being calibrated must then be compared to the measurement of the standard. If the standard is sufficiently accurate, the surface emissivities are well matched, and the irradiation between the two surfaces is identical, the results should be accurate for these specific test conditions of radiation heat flux to the heat flux gage. The calibration, however, may not be valid for other situations. Because there is usually negligible radiation emitted from the gage, gage temperature nonuniformities are not included in the calibration results. As demonstrated by Borell and Diller (1987), this can cause significant errors if the gage is used in convective flows where the gage temperature nonuniformities can become important. Moreover, radiation calibrations are difficult to perform for gages that are made integral to the model being tested (in situ).

An interesting variation of the radiation calibration using a pulsed laser has been reported by Epstein et al. (1986). This is a rather involved procedure, but can be summarized as follows. The model with the gage installed is placed in both a vacuum and in a liquid which is transparent to the laser light. Using an analytical model of the transient temperature rise of the two resistance elements on their layered gage for these two cases, the heat transfer to the gage is determined. This is compared to the gage output to complete the calibration. The reported accuracy of 7 percent is limited by the accuracy of the known fluid properties.

Calibrations using steady-state conduction can typically be done with very good accuracy, but they are usually limited to the lower heat flux range (Klems and DiBartolomeo 1982, Andretta et al. 1981, Leclercq and Thery 1983, Moreno et al. 1980). Therefore, they are not useful for most convective flows.

Gage calibrations performed directly in convective flows are rare. Although comparison is sometimes done with known experimental results or analyses, this requires exact duplication of the flow situation under comparison for accurate results, which is usually quite difficult. One exception is the method reported by Borell and Diller (1987), which uses a direct power measurement in an impinging air flow for comparison with the gage output. This method does not allow for in situ gage calibration, however.

The previous discussion refers to direct heat flux calibrations of gages. Such calibrations are sometimes avoided if the appropriate properties are available to the required accuracy. For the transient thin film gages (category 3) the product of k ρ C (thermal conductivity, density, and specific heat) for the substrate is needed along with a calibration of the gage resistance (or color for liquid crystals) versus temperature. The major problem is the confidence in the uniformity of the material properties of the substrate and the effect of the interface between the substrate and the film (Thompson 1981).

One of the popular substrates used is a ceramic material called MACOR™. The material properties were recently tested by Keltner et al. (1988) and compared with data supplied by the manufacturer. The values of $\sqrt{k\rho C}$ differed by up to 30 percent. An earlier comparison of this property value by Miller (1981) showed discrepancies of 14 percent. Because the measured heat transfer is proportional to this property value, it indicates one of the accuracy limits of the measurement.

Time-Resolved Heat Flux Measurements

Several methods have been used to make time-resolved heat flux measurements. Even with special techniques, however, measurements are still usually limited to low frequencies. The same categories will be used for discussing these unsteady methods as introduced at the beginning of the heat transfer measurement section.

Although the electric power measurements (category 1) usually have very long time-constants (many minutes), thin-film resistors can be used with much faster response times. Without electronic compensation, however, the substrate response still limits the total system response to a few hertz. Baughn et al. (1986) and Simonich and Moffat (1984) used liquid crystals to measure the time-dependent surface temperature along with thin-film gold heaters to supply the uniform heat flux. The time response can be enhanced analytically to some extent (Baughn et al. 1986, Kalinin and Dreitser 1970), but active control systems are really needed to significantly augment the response. This has usually been done by driving the thin-film resistance with a high quality feedback system, such as used for hot wire anemometers (Campbell et al. 1985, Suarez et al. 1983, Boulos and Pei 1974, Fitzgerald et al. 1981, Bernis et al. 1977). Because the control system tries to maintain a constant gage resistance (and consequently temperature), the voltage (gage input power) varies with the surface heat transfer. This is the same type of operation as commonly used with thin-film shear stress gages, except for the boundary condition. Shear stress gages are given a large overheat with a large adiabatic region around the active element. The fluid shear stress is then related by calibration to the heat lost from this tiny heated element. To measure the heat transfer from a surface, however, the whole surface must be maintained at a uniform thermal boundary condition because the thermal boundary layer development immediately upstream of the gage has a strong effect on the measured heat transfer. Consequently, the thin-film gage temperature must be accurately matched with the

surrounding surface temperature.   This usually necessitates using only a small
overheat (relative to the fluid temperature) and no adiabatic regions on the
surface.   Fig. 6 shows a sketch of such a thin-film gage (Campbell et al. 1985).
The small thermocouple mounted near the gage surface is used to insure that the gage
temperature is matched with the surrounding material temperature (typically within
0.2°C).

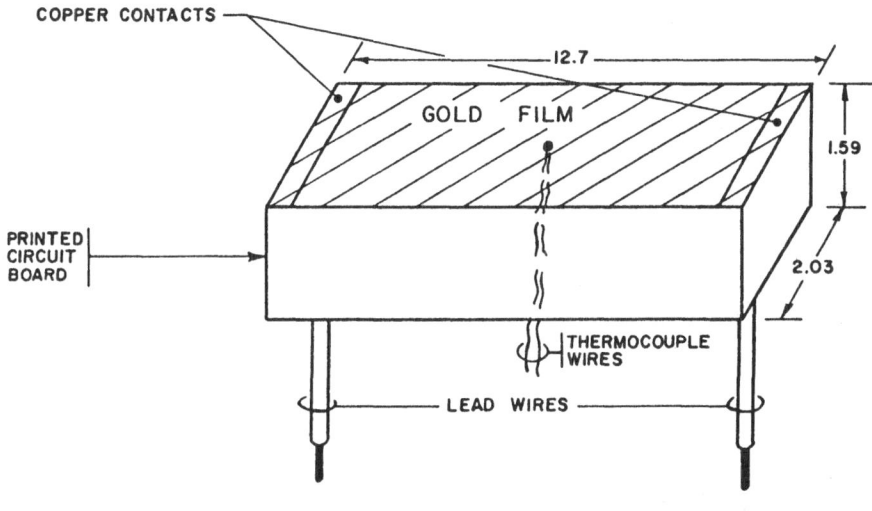

Fig. 6   Heated Thin-Film Heat Flux Gage

A hot-wire bridge circuit monitors the gage resistance and adjusts the circuit
voltage to maintain the resistance constant.   Because this operates on the error
signal, there is always some small deviation from the set point of the resistance.
For the usual temperature-resistance characteristics of thin films this offset may
be as large as 1°C, depending on the size of the gain and the settings for the
system inductance and capacitance compensation.   Although 1°C is a rather small
change in temperature, it is not negligible compared to the small overheats used for
heat flux gages.   The resulting change in heat transfer is coupled to the substrate,
and limits the frequency response of the system to about 10 to 100 hertz.   Moreover,
because the substrate acts as a thermal capacitor, it can dampen the magnitude of
the measured unsteady signal.   Experimentally this has been measured to be as large
as a factor of ten reduction in the amplitude of the unsteady signal.   If the active
resistance portion of the gage does not always cover the entire gage surface, a
nonuniform temperature results which further degrades the response of the gage
(Beasley and Figliola 1986).   Nevertheless, if used with care, these gages can still
be used for selected measurements of unsteady heat flux.

In addition to the steady calibrations as previously described, unsteady
calibrations have been performed   on these gages using a chopped light source

(Campbell et al. 1985). A schematic of the apparatus is shown in Fig. 7. This is a very useful procedure for both properly tuning the compensation of the bridge circuit and properly identifying the limitations in frequency response of the system. Quite often such direct measurements of system transient response have not been made.

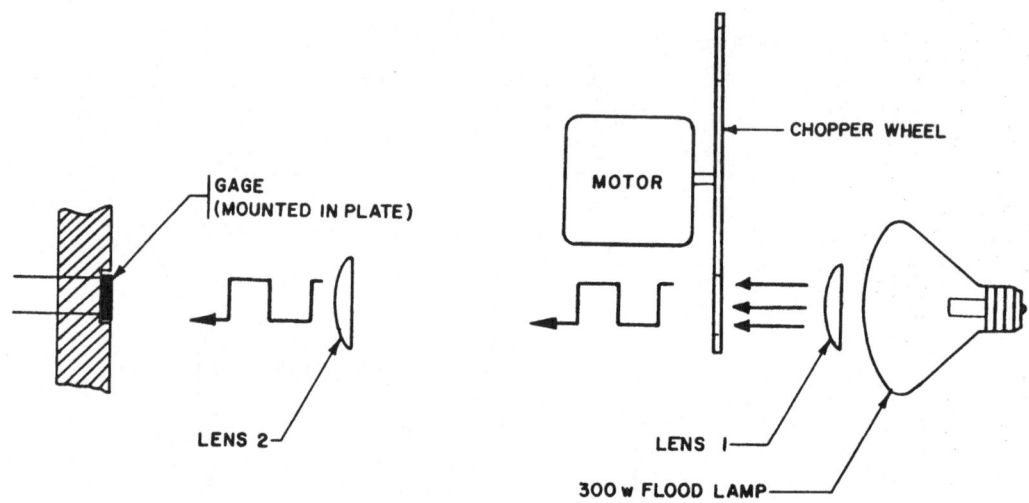

Fig. 7   The Thermal Square-Wave Apparatus for Unsteady Calibration

Gages in the second category have also been tried in unsteady flow. For Gardon gages and Schmidt-Boelter gages the best time response available is about one second (Kidd 1981, Kim et al. 1983, Keltner and Wildin 1975), which is generally too slow for measuring unsteady heat transfer phenomena. The transient response of layered gages can be analyzed with unsteady one-dimensinal heat transfer theory. The result for the time to reach 98 percent of the final temperature distribution (Hager 1965) is

$$t = 1.5 \; \delta^2/\alpha \tag{10}$$

where δ is the thickness of the gage (labelled in Fig. 4) and α is the material thermal conductivity. Using a value for α of 5×10^{-7} m^2/s, which is representative of typical insulating materials, gives a response time of 20 ms for a .075 mm (.003 in.) thick gage. The corresponding limit of the frequency response is about 50 Hertz, although with proper compensation (Malcorps 1983) the response can be enhanced up to 200 hertz (Mori et al. 1986). This is the thinnest layered gage that is standardly manufactured.

Epstein et al. (1985,1986) have handled this response problem by effectively operating the gage in two different modes. At frequencies above 1500 Hertz the temperature change of the top metal resistance layer is used to determine the heat flux from the transient response as a category 3 gage. For low frequencies (below

20 Hertz) the standard response as a category 2 gage is used. For intermediate frequencies, the heat flux is reconstructed through numerical signal processing. The laser pulse calibration (as described in the previous section) provides the material constants necessary for the two limiting solutions. Unsteady measurements at high frequency have the same difficulties as described for category 3 gages, however.

As mentioned earlier, one of the accuracy problems with using these layered gages occurs in the higher range of convective heat flux (typically heat transfer coefficients above 200 $W/m^2 °C$) where the temperature difference across the gage becomes significant compared to the total fluid-to-surface temperature difference. This causes the heat flux gage to read low because the temperature difference between the gage surface and the fluid is smaller than that between the surrounding material and the fluid. This is further complicated by the glue layer of at least 5 μm that is usually needed to attach the gage to the surface. This provides additional insulation between the gage and the surface being measured and further amplifies the error in the heat flux measurement.

One way to avoid the problem with the glue layer is to deposit the gage directly to the surface (Hayashi et al. 1986, Godefray et al. 1986). To further increase the frequency response, however, a much thinner gage is needed. According to Eq. (10), a time response of better than 5 microseconds is possible if a gage with a total thickness of 1.0 micrometer could be deposited directly on the surface. This would give a frequency response of approximately 200 kilohertz, which would be sufficient for any unsteady heat flux measurement due to fluid unsteadiness. This would also allow measurements at high heat flux because of the much smaller temperature drop across the much thinner gage. Of course, reading the gage output could also be more difficult because of the smaller output voltage due to the smaller temperature difference across the gage.

Vatell Corp. has recently produced such a thin film gage using microcircuit techniques. An increased signal is produced by a thermopile with 100 thermocouple pairs, which gives the gage sufficient output voltage for many applications. It also has the potential for high temperature operation. Although testing has not been completed, the initial results are very promising for high frequency operation (~ 100 kilohertz).

The theory for obtaining the heat transfer from the temperature signal of a thin-film temperature measurement (category 3 gage) has already been described briefly. Basically, the measured time-temperature history is used to determine what heat flux to the surface has occurred. Although there are several ways to process the signal, the heat flux is still proportional to the rate of change of the surface temperature (Jones 1977).

$$q'' = \frac{\sqrt{k\rho C}}{\sqrt{\pi}} \int_o^t \frac{\frac{dT_s}{d\tau}}{\sqrt{t - \tau}} \, d\tau \qquad (11)$$

The actual signal manipulation to obtain the heat flux can be done with either analogue circuits or by digital signal processing. Distortion is introduced into the signal by both methods, which hampers the reconstruction of the true heat flux record. It is not a simple matter, therefore, to obtain the time-resolved heat flux signal from thin-film gage measurements, as discussed by Dunn (1986b) and George et al. (1987). In spite of these problems, several investigators (Dunn et al. 1986c, O'Brien et al. 1986, Ashworth et al. 1985, Doorly et al. 1985, George 1987, Alkidas and Cole 1985) have successfully captured unsteady heat flux signals from transient temperature measurements. The applications range from gas turbine engines to space vehicles to diesel engines.

Determination of surface heat transfer from measurements of the fluid temperature gradient (category 4) are generally made for the low range of heat fluxes, particularly natural convection (Jaluria 1980, Ramachandran et al. 1985). Although the long time often needed to construct the temperature profile in the boundary layer usually limits this method to steady measurements, Mori and Tokuda (1966) successfully measured instantaneous heat transfer rates at frequencies up to 25 hertz using an optical technique.

METHODS OF SKIN FRICTION MEASUREMENT

The concept of a hot element as a tool to measure skin friction is based on the relation between local skin friction and the rate of heat transfer from a small heated element mounted flush with the surface. The long dimension of the probe is positioned normal to the flow and the leads are connected to a bridge. In hot-wire anemometry the heat transfer from a thin wire positioned in the stream is directly related to the local velocity via the well-known King's law. Somewhat surprising is the fact that if the hot element is mounted flush on the surface of the body, heat transfer is then related to the gradient of the velocity. This relationship was first studied by Fage and Falkner (1931). The electrical power, e, is related to the skin friction, τ, via the equation

$$\frac{e}{\Delta T} = A + B(\rho\tau)_w^{1/3} \tag{12}$$

where ΔT is the temperature difference, ρ is density, and w denotes evaluation on the wall. A and B are quantities which depend very mildly on ΔT. Skin friction can thus be measured simply by employing a standard hot-wire anemometer. Since one measures the power supplied to the hot film, the principle of operation of a skin friction gage is the same as that of a hot wire. The anemometer senses any change in the resistance of the film and in response supplies the appropriate current to maintain the film at constant resistance, i.e., temperature.

Good reviews of hot film gages have appeared fairly recently (Winter 1977,

Jones 1977, Hanratty and Campbell 1983). In the last review, the reader will find a good account of earlier contributions, a detailed description of the principle of operation, related mathematical formulae, and useful tips for the user. In the present short section, a brief mention is made of a few of the earlier papers pertaining mainly to unsteady flow, followed by a discussion of some contributions that appeared after the publication of the Hanratty and Campbell (1983) report. A brief note is also added in the following section on the experience of the authors, with examples of measurements over vortex shedding configurations.

Part of the energy supplied to the hot element is convected away by the flow and part of it is conducted through the substrate. It is only the first quantity which is related to the skin friction. Moreover, the conduction in the substrate makes the heated surface area of the gage larger than the thin film itself. The thermal boundary layer, which determines the local heat flux and is related to the measured skin friction, develops over the heated portion of the gage. The size of this region, however, is inversely related to the magnitude of the local skin friction. This complicates calibration, especially for the unsteady response. In addition, the thermal boundary layer must be kept small for the relation between convected heat transfer and skin friction to be valid. This requires the substrate to have a low thermal conductivity and the dimension of the film normal to the flow to be as small as possible. For example, for a turbulent boundary layer it must be smaller than the viscous sublayer.

The early versions of hot elements were relatively bulky (Fage and Falkner 1931, Ludwieg 1949, Liepmann and Skinner 1954). Great improvements were achieved by modern methods. Bellhouse and Schultz (1965,1966) used a platinum film deposited on a glass substrate while McCroskey and Durbin (1972) employed photoetching techniques. Various investigators have developed similar probes as discussed in Hanratty and Campbell (1983). Such probes have been marketed by commercial firms in the late 1970's and are now available as off-the-shelf items.

In theory, hot films appear to be versatile high-frequency-response instruments, readily available for easy use with standard anemometer equipment. However, the experience of most investigators, especially the ones who used the commercially available elements, indicates great difficulties in calibration. A serious source of error is the loss of heat in the substrate as discussed in Hanratty and Campbell (1983). For the users of the commercial units there is no room for improvement on this.

Another source of error is the mechanical position of the probe. Some commercially available probes are mounted on the end of a cylindrical rod as shown schematically in Fig. 8. If the face of the cylinder is not flush with the surface of the model, the local disturbance of the flow may introduce errors in the measurements. This is inevitable in case the flat-faced probe is mounted on a curved model. Pessoni (1974) reports that a probe protruding by 0.1 mm in an air stream may generate errors of the order of 30% to 40% of the measured value.

Similar results were obtained by Simons et al. (1979). Coney and Simmer (1979) have experimented with flexible glue to avoid the inevitable gaps generated when a flat-headed probe is mounted on a curved surface.

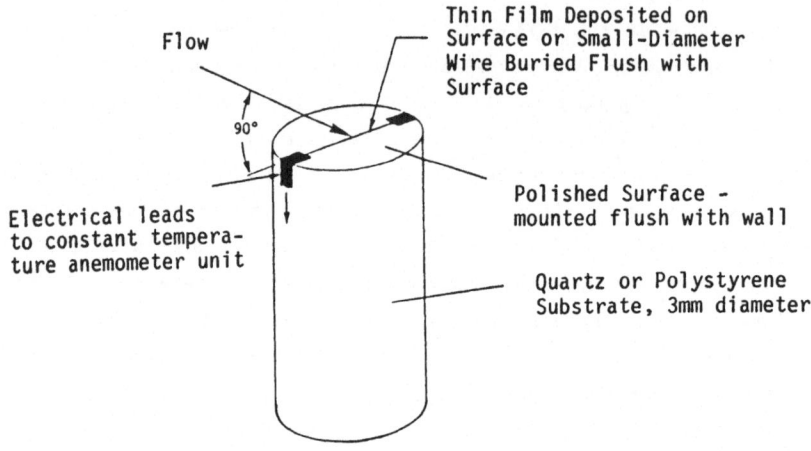

Fig. 8 Schematic Representation of a Skin Friction Probe

More recently, Lefebvre and LaPointe (1986) undertook an exhaustive study of the effects of the mounting position on commercially available skin friction sensors. They positioned a 3.175 mm diameter probe on the inside surface of a 5 cm-diameter cylindrical pipe and conducted experiments in water. Their least obstructing position was found to be the case whereby the sensor was tangent to the circle defining the cross section of the test section (Fig. 9). The spanwise ends were then recessed by 0.05 mm. This configuration was defined as the origin of positioning of the probe. Lefebvre and LaPointe (1986) then tried various recessed and protruded positions. They reported that recession of 0.0508 mm produced negligible increases in both the mean and the RMS readings of the wall shear stress. However for probes recessed by 0.0762 to 0.1778 mm, errors in $\tau_w^{-1/3}$ of up to 25% to 50% were observed, in qualitative agreement with earlier findings. Similar results were found for protruding probes.

The basic conclusion one can draw from the contributions discussed above is the following. Hot-film skin-friction instruments are very sensitive to their positioning on the surface of the model. To avoid considerable errors it was suggested by many investigators to calibrate hot films in situ. Then the instrument might be forgiving. Lefebvre and Lapointe (1986) point out that even with probes recessed at 0.0254 mm or protruding at 0.0762 mm one can have a repeatable and in fact linear calibration curve. The present authors obtained similar results with

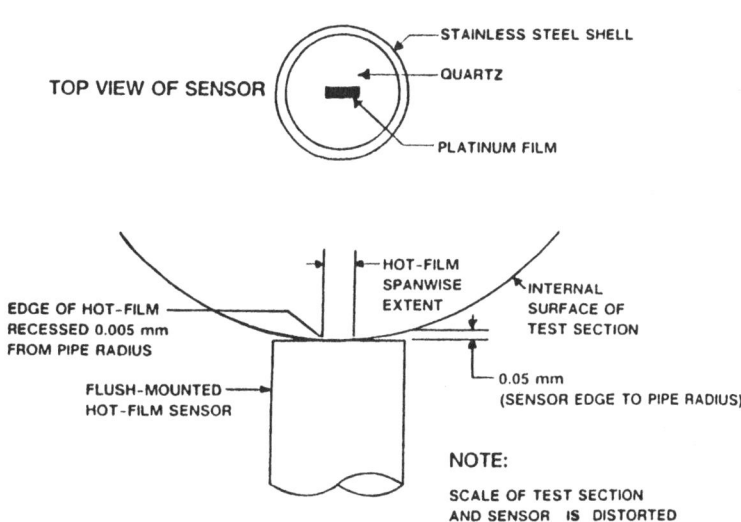

Fig. 9 Positioning of a Flat-Faced Probe on a Curved Model Surface (Lefebvre and LaPointe 1986)

sensors slightly recessed on a convex rather than on a concave body. These are experiments conducted over a circular cylinder in crossflow and will be discussed briefly in the next section.

Calibration of the magnitude of skin friction has been discussed extensively in earlier review articles (Winter 1977, Hanratty and Campbell 1983). Various methods have been used, but the most reliable experimentally is to consider the flow in a pipe, because then skin friction is directly related to pressure drop, which can be accurately measured. However, as mentioned earlier, significant errors are inevitable if the instrument is removed from the calibration rig and mounted on the model. In-situ calibration appears to be the only viable method. This requires a configuration which permits an exact analytical solution. This consideration alone practically disqualifies this instrument as a sensor of turbulent flow.

Brown and Davey (1971) devised a calibration technique involving a stationary and a rotating disk separated by a small air gap. But disturbing discrepancies were reported when different calibration curves were obtained for laminar or turbulent flow (Liepmann and Skinner 1954, Brown 1967, Pope 1972). An extensive discussion on this topic is included in Winter (1977). If both the calibrating flow and the test flow are turbulent, then the transfer of information is acceptable. Geremia (1970) showed that it is indeed possible to calibrate films in pipe flows and then obtain measurements over a flat plate.

In-situ calibration is also possible if special effects are to be tested. For example, Poll and Watson (1987) calibrated their instrument for flow over a flat plate and then changed the configuration to examine the effect of an upstream facing step on the characteristics of a turbulent boundary layer. Many investigators employed hot film gages to study laminar flows with some organization, as for example the attached flow on the front half of a circular cylinder (Barbi et al. 1986). However, again no effort was made to calibrate the instrument's response in an unsteady flow. Many investigators considered laminar and turbulent flows with a superimposed freestream pulsation and examined the response of skin friction gages (Bellhouse and Schultz 1966, Kobashi and Hayakawa 1981, Simpson et al. 1983, Ramaprian and Tu 1983, Menendez and Ramaprian 1985). Recently, Cook et al. (1986) examined carefully the phase response of hot films to periodic external disturbances. They considered periodic mean flow disturbances over a flat plate, a case for which accurate analytical and numerical solutions are available. Cook et al. found that the phase response may involve considerable errors if the overheat ratio is not greater than about 1.15 to 1.20, as shown in Fig. 10. The data presented in this figure were obtained over a flat plate at a local Reynolds number of 1.34×10^5 with a fluctuating amplitude of 13 percent. The cases listed as PF1 and PF2 represent two identical platinum film sensors and $\bar{\omega}$ is the reduced frequency. Cook et al. also found that the instruments display deviations from the theoretical values of skin friction. Measured values of phase differences with the outer flow were consistently lower than the corresponding theoretical values.

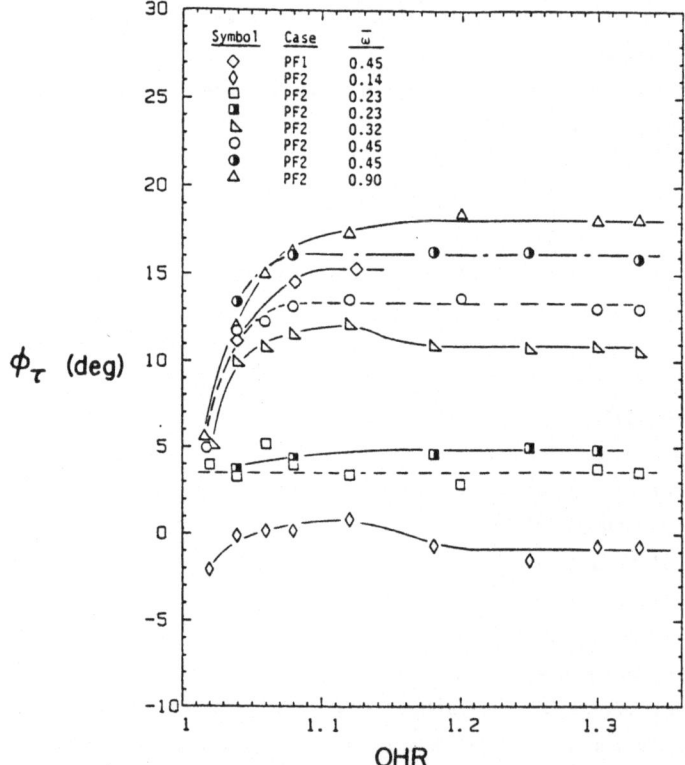

Fig. 10 The Phase of the Skin Friction Response (ϕ_τ) as a Function of the Overheat Ratio (OHR) (Cook et al. 1986)

The present group employed skin friction gauges to investigate organized shedding patterns over single and multiple cylinders. Standard off-the-shelf gages manufactured by DISA and TSI were used. These sensors were positioned on the surface of a 8.9 cm-diameter cylinder and were calibrated in situ using a numerical boundary-layer solution (Cramer et al. 1987) as well as a Navier-Stokes solution (Dhaubhadel et al. 1986) for comparison. Both theories generated identical results. For this steady flow calibration it was necessary to eliminate vortex shedding, which introduces an unwanted periodicity in the flow. This was achieved with a standard splitter plate positioned downstream of the cylinder along the direction of the flow. Typical skin friction frequency spectra at $\theta = 40°$ with and without the splitter plate are shown in Fig. 11. In this figure, it is clearly demonstrated that the splitter plate eliminates the periodic character of the flow. Moreover, simultaneous measurements with hot wire anemometers indicated that the hot film has an excellent frequency response at least for frequencies up to 40 Hz. The power supplied to the sensor for meridional positions $\theta = 0$ to 180° is shown in Fig. 12, obtained with and without a splitter plate. It is noted that like all hot-film or hot-wire sensors, this instrument is insensitive to the direction of the flow.

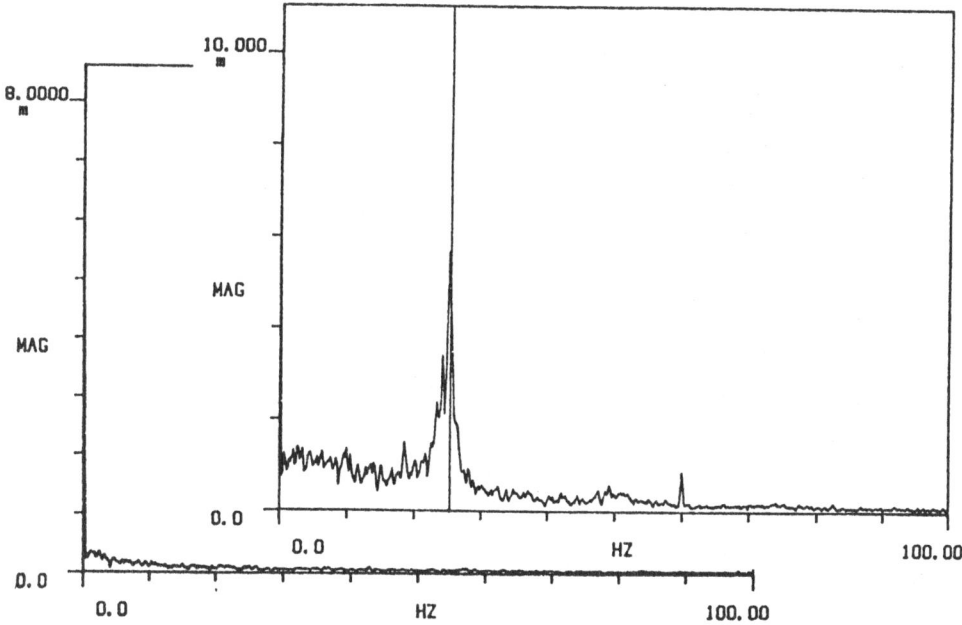

Fig. 11 Skin Friction Spectra in the Front of a Cylinder in Cross Flow With (left) and Without (right) a Splitter Plate

Hot films mounted on the model surface have been extensively used to detect transition (Bellhouse and Schultz 1968, McCroskey and Durbin 1972, Dagenhart and

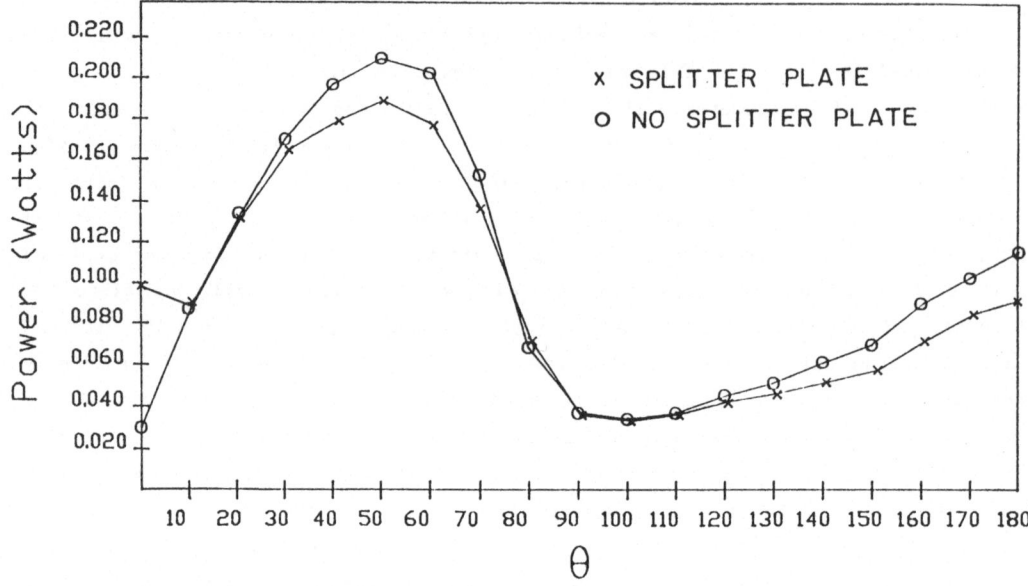

Fig. 12 Skin Friction Measurements Over a Circular Cylinder With and Without a Splitter Plate

Stack 1982, Johnson et al. 1987, Stack et al. 1988). In fact the method has been recently employed for transition measurements in flight (Bertelrud 1987, Olsson 1988). Traversing the probe slowly through transition, one can quickly detect a sudden and considerable increase of the mean magnitude.

Most recently a further development was reported by Stack et al. (1988). To study laminar separation and transition, these investigators have constructed a multi-element probe. They deposited a sequence of individual nickel films on a thin polyamide substrate, as shown in Fig. 13. Each film was 0.9 mm long and 0.15 mm

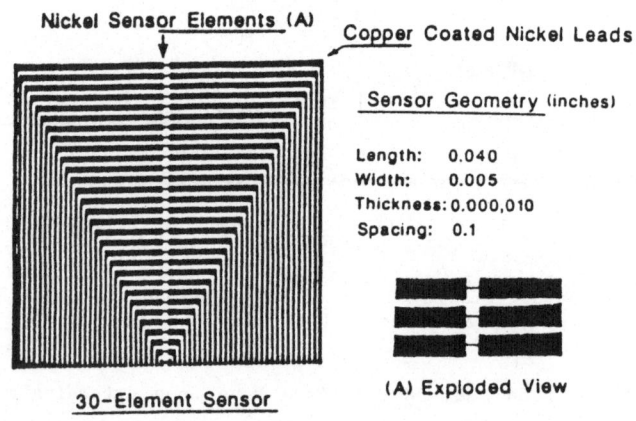

Fig. 13 The Multiple-Probe Employed by Stack et al. (1988)

wide with 50 micron copper-coated nickel leads routed to provide wire attachment downstream of the last sensor of the array. The influence of heated elements on their neighbors was examined and found negligible. Each element was operated by individual anemometers. However, only groups of sensors were operated simultaneously. The authors discovered that the information supplied by this instrument could allow detection of both laminar separation and reattachment. This is because when the signals are passed by a low-pass filter, they display a 180° phase change around the points of zero skin friction.

EXAMPLES OF UNSTEADY FLOW MEASUREMENTS

One important question of convective heat transfer is the mechanism of how unsteady fluid flow affects surface heat transfer. Although time-averaged heat transfer is usually the quantity of interest, the details of the unsteady heat transfer are most useful in understanding the basic phenomena. One example is the recent set of measurements by Doorly and Oldfield (1985) shown in Fig. 14. They used a transient thin film gage (category 3) to measure the effects of the passage of upstream rotor wakes on the heat transfer of downstream turbine blades. It has been known that the passage of wakes increases blade heat transfer, but it was thought to be due to the presence of flow unsteadiness. As illustrated in Fig. 14, this was shown not to be true. Instead, the turbulence in the wake causes a momentary large increase in heat transfer even as the velocity is lower. This appears as one of the peaks in the figure that matches the heat flux level in a

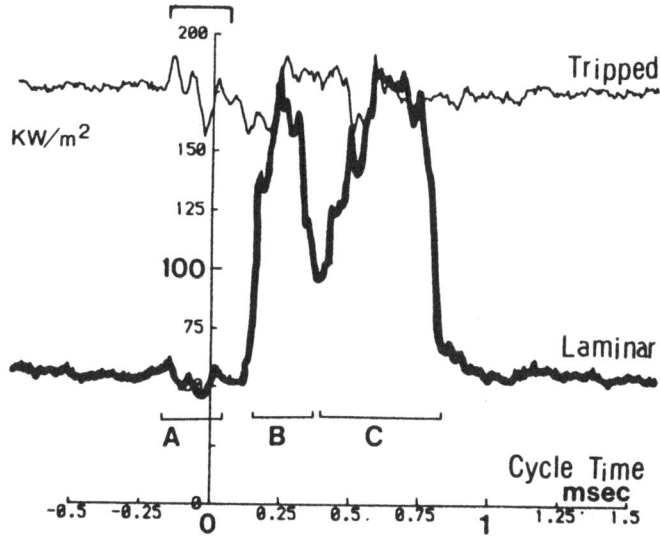

Fig. 14 Effect of Shock-Wave and Wake-Passing on Turbine Blade Heat Transfer (Doorly and Oldfield 1985)

tripped turbulent boundary layer. Averaged over the cycle, this gives the observed 10 to 15 percent increase in time-average heat transfer.

Figliola et al. (1984) used a heated thin film gage (category 1) to measure the instantaneous heat transfer to a cylinder in a fluidized bed. The time-resolved heat transfer for a bed operated in the bubbling flow regime is shown in Fig. 15. Two regions in time were found that correspond to the maximum and minimum heat flux levels measured. The minimum heat transfer was measured when a bubble was in contact with the surface, labelled h_b. When the particle emulsion was in contact with the surface, labelled h_e, the maximum heat transfer was measured. Mixing within the emulsion phase kept the heat transfer at the maximum value for the entire contact time. This is in direct conflict with the packet theories which use a penetration model that predicts the heat transfer to continually decrease in time while the emulsion phase is in contact with the surface. Consequently, the time-resolved measurements have given direct evidence that significantly alters the basic understanding of the physical phenomena in this process.

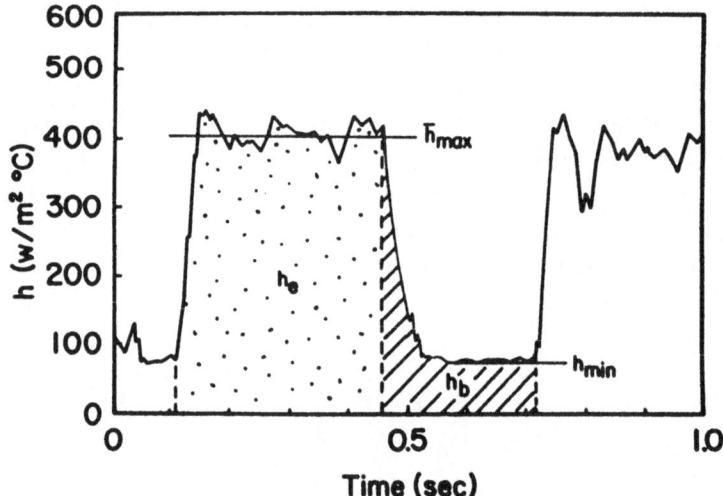

Fig. 15 Time-Resolved Heat Transfer for a Fluidized Bed Showing Emulsion and Bubble Phase Contributions (Figliola et al. 1984)

Diller (1987) also used an actively-controlled heated thin-film gage to measure the time-resolved heat transfer around a cylinder in crossflow. Although this is often considered a steady flow condition, the natural vortex shedding from the rear of the cylinder causes a periodic unsteadiness of the flow field. This is evident in the time-resolved heat transfer signal shown in Fig. 16 even near the front stagnation point ($\theta = 2°$). Although the actual signal shown is that of the bridge voltage, it is approximately proportional to the heat transfer coefficient fluctuation, with the relative amplitude indicated on the right of the figure. The shedding frequency is about 70 hertz and the amplitude of the fluctuation in heat transfer is greater than 10 percent of the mean. The amplitudes of the fluctuating heat transfer are much larger near the separation point and in the wake region on the rear of the cylinder.

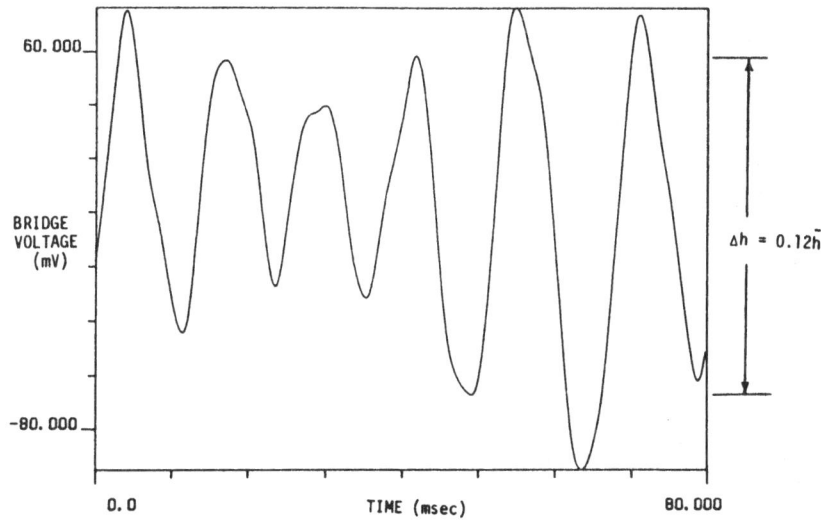

Fig. 16 Time-Resolved Heat Transfer over a Circular Cylinder at $\theta = 2°$ and Re = 65,000 (Diller 1987)

Similar measurements were also made in the wake region of a cylinder (120° from the front stagnation point) in a pulsating crossflow (Gundappa and Diller 1987). The driving frequency was set at 13 hertz to force the natural vortex shedding to lock on at one-half of the driving frequency. In addition, a hot-wire probe was placed 0.625 cm above the gage in the fluid. Although this single wire probe does not give the direction of flow, it does provide a measure of the fluid activity for comparison with the heat transfer. Both signals shown in Fig. 17 were triggered simultaneously and have been ensemble-averaged 50 times. This ability to simultaneously measure and correlate heat transfer and velocity signals is most useful. Results such as these were used to create a physical model of the steady and unsteady heat transfer in the wake of a cylinder that was able to predict the observed heat transfer increases due to pulsating flow (Gundappa and Diller 1988).

Waveforms of skin friction on the surface of a circular cylinder were obtained by Barbi et al. (1986). Their study clearly indicated that the amplitude of the fluctuation due to the natural shedding increases as separation is approached. However, this information is qualitative because in this early study there was no effort to calibrate the instruments. Skin friction and pressure measurements were also obtained along the periphery of single and multiple cylinders by VandenBerghe et al. (1988). An example of spectra of skin friction and pressure obtained at the

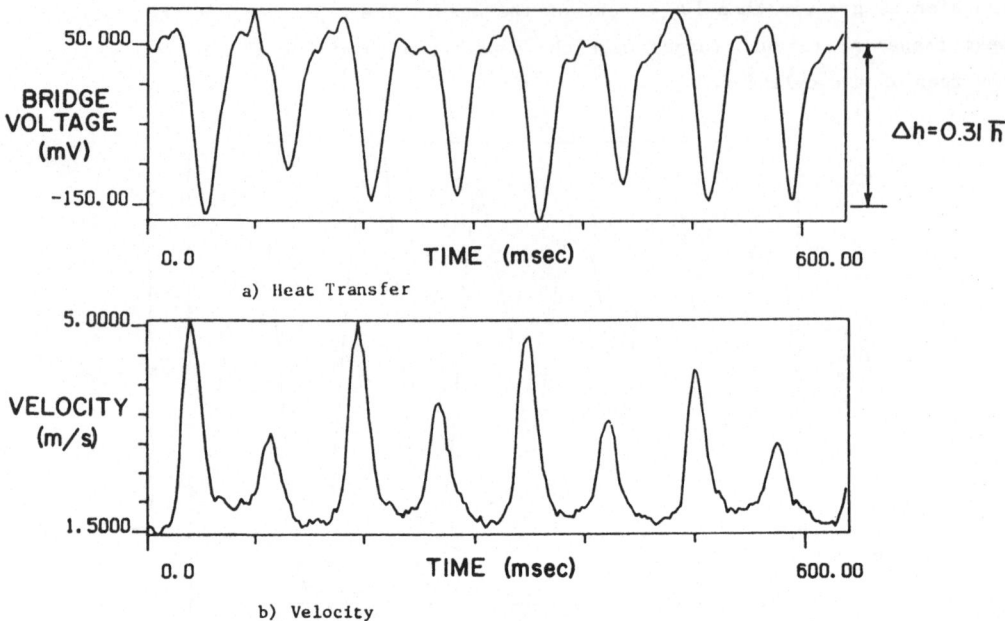

a) Heat Transfer

b) Velocity

Fig. 17   Time Records of Velocity and Heat Transfer on a Cylinder at θ = 120° in
Pulsating Crossflow, Re = 19,000 (Gundappa and Diller 1987)

same azimuthal position is shown in Fig. 18. The position is 40° downstream of the
front of the second cylinder of a triad of cylinders spaced at a distance of 1.1
diameters center to center. The unsteady flow was driven at a frequency of 18.7
hertz.

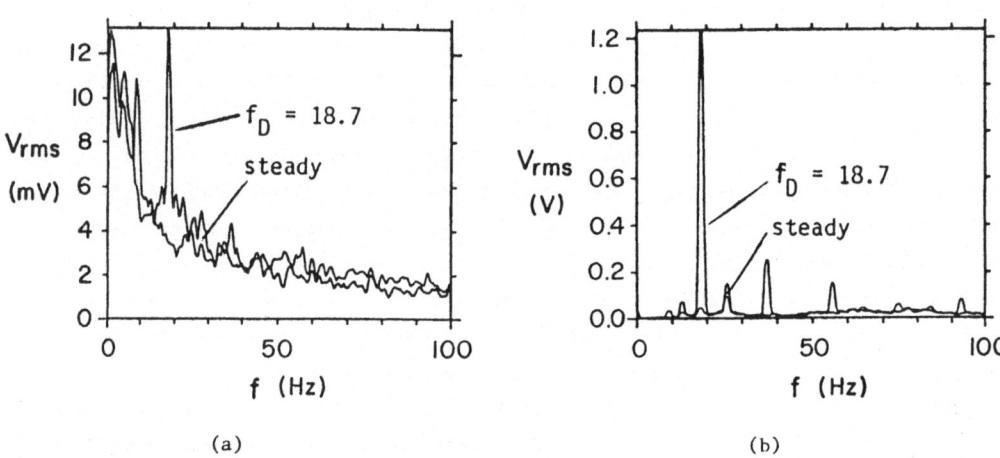

(a)

(b)

Fig. 18   Frequency Spectra of Skin Friction (a) and Pressure (b) Measured on a Triad
of Circular Cylinders, Re = 25,000

The significance of the findings presented here is that both the heat transfer and the skin friction gages have excellent frequency response, at least for the frequencies tested. This is revealed by comparing spectra of signals of hot elements obtained simultaneously with velocity or pressure as discussed above.

With further processing of signals such as those shown in Figs. 17 and 18, the amplitude and phase of the unsteady component can be identified. An example showing the change in the unsteady peak-to-peak amplitude of the heat transfer at different locations around the cylinder is shown in Fig. 19 (Gundappa and Diller 1987). The peak-to-peak amplitude of the 13 hertz driving frequency was 24 percent of the mean velocity. Although the unsteady heat transfer component was measured with the actively-controlled thin-film gage, the time-average heat transfer shown was measured with a Gardon gage. For these time-average measurements, the Gardon gage has less than one-half the error of the thin-film gage. Using two different types of gages in this fashion is one way to avoid the limitations imposed by any single type of gage.

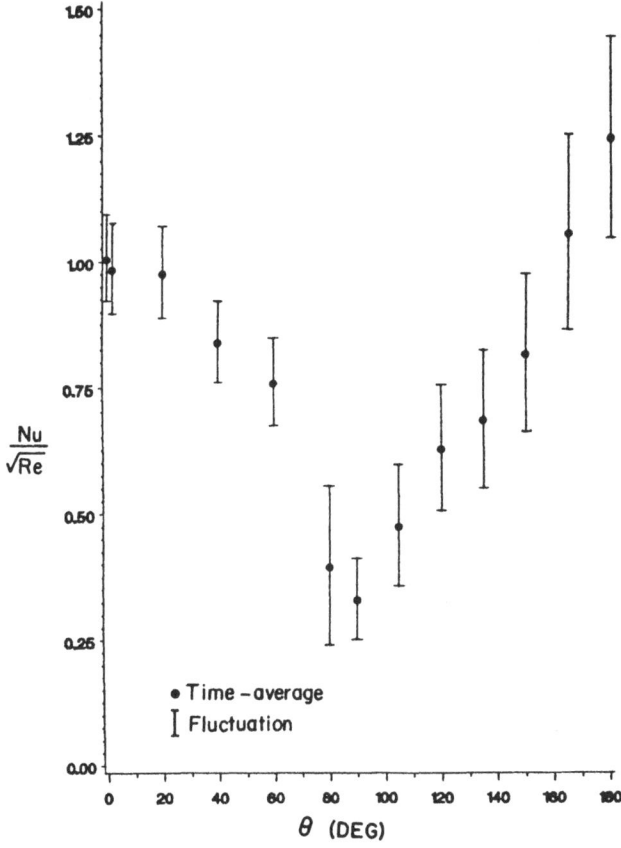

Fig. 19   Unsteady Heat Transfer Amplitude in Pulsating Flow

Unsteady mass transfer measurements have been made for the past several decades using the electrochemical technique. The details of this method are not discussed here, but the method is extensively reviewed by Mizushina (1971). Hanratty and co-workers have measured the surface mass transfer unsteadiness in turbulent flows and have demonstrated the presence of fluctuations in local mass transfer rates of greater than 30 percent of the mean (Sirkar and Hanratty 1970). Correlation of these results with measurements of the fluctuating velocity has been used along with frequency analysis of the signals to gain understanding of the turbulent mass transport process (Campbell and Hanratty 1983, Vassiliadou et al. 1986). Unfortunately, the Schmidt number of these studies has been on the order of a thousand times larger than the usual Prandtl number for heat transfer. Consequently, the mass transfer boundary layer is much thinner than the corresponding heat transfer boundary layer, and many of the results from mass transfer are not directly applicable to heat transfer.

CONCLUSIONS

A general review of heat flux measurement techniques for convective flows was presented. Potential errors in use and the importance of proper calibration were emphasized. The focus was on measurement methods that could give time-resolved heat transfer and skin friction in a useful frequency range for unsteady flows. This limits the potential techniques to a select few, which were discussed in detail. There are significant problems and limitations in using all of the current methods. New and improved instrumentation is therefore needed to make time-resolved heat flux more widely accepted.

The evidence discussed here indicates that hot-film skin friction gages can be very useful and relatively easy instruments to use. However, the data thus obtained must be viewed with caution. Calibration must be done in situ. And yet so far, no-one has attempted to check whether steady state calibration would be adequate for measuring the amplitude of skin friction fluctuations. There is convincing evidence that hot-film sensors may introduce significant deviations in the phase of the measurement, however, the frequency response appears to be excellent.

Examples were presented of the two best measurement methods available for unsteady heat transfer (catetory 1 and category 3 thin films). Results of the electrochemical method for unsteady mass transfer were also briefly discussed. These time-resolved measurements have been used to improve physical understanding of both the structure of the unsteady flow and the resulting time-average transfer. In each case discussed, the resulting knowledge was used to provide important advances in modeling the transfer process. Such encouraging results provide incentive to improve these measurement techniques and seek wider application.

REFERENCES

Achenbach, E. (1975) "Total and Local Heat Transfer from a Smooth Circular Cylinder in Cross-Flow at High Reynolds Number," Int. J. Heat Mass Transfer, Vol. 18, pp. 1387-1396.

Alkidas, A. C., and Cole, R. M. (1985) "Transient Heat Flux Measurements in a Divided-Chamber Diesel Engine," ASME J. Heat Transfer, Vol. 107, pp. 439-444.

Andretta, A., Bartoli, B., Coluzzi, B., Cuomo, V., and DeStefano, S. (1981) "A Simple Heat Flux Meter," Rev. Sci. Instrum., Vol. 52, pp. 233-234.

Arts, T., and Camci, C. (1985) "Short Duration Heat Transfer Measurements," in Measurement Techniques in Turbomachines, Vol. 2, von Karman Inst. Fluid Dynamics, Rhode Saint Genese.

Ashworth, D. A., LaGraff, J. E., Schultz, D. L., and Grindrod, K. J. (1985) "Unsteady Aerodynamic and Heat Transfer Processes in a Transonic Turbine Stage," ASME J. Engineering for Gas Turbines and Power, Vol. 107, pp. 1022-1030.

Atkinson, W. H., Cyr, M. A., and Strange, R. R. (1985) "Turbine Blade and Vane Heat Flux Sensor Development, Phase II - Final Report," NASA CR-174995.

Bales, E., Bomberg, M., and Courville, G. E., Eds. (1985) Building Applications of Heat Flux Transducers, ASTM, Vol. 885.

Barbi, C., Favier, D. P., Maresca, C. A., and Telionis, D. P. (1986) "Vortex Shedding and Lock-On of a Circular Cylinder in Oscillatory Flow," J. Fluid Mechanics, Vol. 170, pp. 527-544.

Baughn, J. W., Hoffman, M. A., Takahashi, R. K., and Launder, B. E. (1984) "Local Heat Transfer Downstream of an Abrupt Expansion in a Circular Channel with Constant Wall Heat Flux," ASME J. Heat Transfer, Vol. 106, pp. 789-796.

Baughn, J. W., Hoffman, M. A., and Makel, D. B. (1986) "Improvements in a New Technique for Measuring and Mapping Heat Transfer Coefficients," Review of Scientific Instruments, Vol. 57, pp. 650-654.

Beasley, D. E., and Figliola, R. A. (1986) "Analysis of a Local Heat Flux Probe," in Heat Transfer 1986, Vol. 2, Eds. C. L. Tien et al., Hemisphere Pub. Co., Wash., pp. 467-472.

Beck, J. V., and Keltner, N. R. (1987) "Green's Function Partitioning Procedure Applied to Foil Heat Flux Gages," ASME J. Heat Transfer, Vol. 109, pp. 274-279.

Bellhouse, B. J., and Schultz, D. L. (1965) "The Measurement of Skin Friction in Supersonic Flow by Means of Heated Thin Film Gauges," ARC R & M, p. 940.

Bellhouse, B. J., and Schultz, D. L. (1966) "Determination of Mean and Dynamic Skin Friction, Separation and Transition in Low-Speed Flow with a Thin-Film Heated Element," J. Fluid Mech., Vol. 24, pp. 379-400.

Bellhouse, B. J., and Schultz, D. L. (1968) "The Measurement of Fluctuating Skin Friction in Air Heated Thin-Film Gauges," J. Fluid Mech., Vol. 32, pp. 675-680.

Bernis, A., Vergnes, F., LeGoff, P., and Mihe, J. P. (1977) "Une Sonde a Film Chaud pour Mesure Locale et Instantanee du Coefficient de Transfert de Chaleur dan un Lit Fuidise Gaz-Solide," Powder Technology, Vol. 17, pp. 229-234.

Bertelrud, A. (1987) "Use of Hot Film Sensors and Piezoelectric Foil for Measurement of Local Skin Friction," 12th Int. Congress on Instrumentation in Aerospace Simulation Facilities.

Borell, G. J., and Diller, T. E. (1987) "A Convection Calibration Method for Local Heat Flux Gages," ASME J. Heat Transfer, Vol. 109, pp. 83-89.

Boulos, M. I., and Pei, D. C. T. (1974) "Dynamics of Heat Transfer from Cylinders in a Turbulent Air Stream," Int. J. Heat Mass Transfer, Vol. 107, pp. 767-783.

Brown, G. L. (1967) "Theory and Application of Heated Films for Skin Friction Measurement," Proceedings of 1967 Heat Transfer and Fluid Mech. Inst., Standford Univ. Press, pp. 361-381.

Brown, G. L., and Davey, R. F. (1971) "The Calibration of Hot Films for Skin Friction Measurement," Rev. Sci. Instrum., Vol. 42, pp. 1729-31.

Campbell, D. S., Gundappa, M., and Diller, T. E. (1985) "Design and Calibration of a Local Heat-Flux Measurement System for Unsteady Flows, submitted to ASME J. Heat Transfer, also in Fundamentals of Forced and Mixed Convection, Eds. F. A. Kulacki and R. D. Boyd, ASME, pp. 73-80.

Campbell, J. A., and Hanratty, T. J. (1983) "Mechanisms of Turbulent Mass Transfer at a Solid Boundary," AIChE Journal, Vol. 29, pp. 221-229.

Coney, J. E., and Simmers, D. A. (1979) "The Determination of Shear Stress in Fully Developed Laminar Axial Flow and Taylor Vortex Flow, Using a Flush-Mounted Hot Film Probe," DISA Information Bulletin, No. 24, May, pp. 9-14.

Consigny, H., and Richards, B. E. (1982) "Short Duration Measurements of Heat-Transfer Rate to a Gas Turbine Rotor Blade," ASME J. Engng. for Power, Vol. 104, pp. 542-551.

Cook, W. J., Giddings, T. A., and Murphy, J. (1986) "Response of Hot Element Wall Shear Stress Gages in Laminar Oscillating Flow," AIAA Paper No. 86-1100.

Cramer, M. S., Kim, B. K., VandenBrink, D. J., and Telionis, D. P. (1987) "Unsteady Heat Convection Over Circular Cylinders," AIChE Journal, Vol. 33, pp. 19-25.

Dagenhart, J. R., and Stack, J. P. (1982) "Boundary-Layer Transition Detection Using Flush-Mounted Hot-Film Gages and Semiconductor Dynamic Pressure Transducers," AIAA Paper No. 82-0593.

Dhaubhadel, M., Reddy, J. M., and Telionis D. P. (1986) "Penalty-Finite-Element Analysis Coupled Fluid Flow and Heat Transfer Over In-Line Bundles of Cylinders in Cross Flow," Nonlinear Mechanics, Vol. 21, pp. 361-373.

Dielsi, G. J., and Mayle, R. E. (1984) "A Composite Constant Heat Flux Test Surface," in Heat and Mass Transfer in Rotating Machinery, eds. D. Metzger and N. H. Afgan, Hemisphere Pub. Co., Washington, pp. 259-268.

Diller, T. E. (1985) "Heat and Mass Transfer Measurements in Unsteady Flow," in Forum on Unsteady Flows in Biological Systems, Eds. M. H. Friedman and D. C. Wiggert, ASME, pp. 71-74.

Diller, T. E. (1987) "Heat Transfer in Unsteady Flow," in Proceedings of the International Conf. on Fluid Mechanics, Ed. S. Yuan, Peking Univ. Press, Beijing, pp. 1217-1222.

Doorly, D. J., Oldfield, M. L. G., and Scrivener, C. T. J. (1985) "Wake-Passing in a Turbine Rotor Cascade," AGARD CP-390.

Doorly, D. J., and Oldfield, M. L. G. (1985) "Simulation of the Effects of Shock Wave Passing on a Turbine Rotor Blade," ASME J. Engineering for Gas Turbines and Power, Vol. 107, pp. 998-1006.

Doorly, J. E., and Oldfield, M. L. G. (1987) "The Theory of Advanced Multi-Layer Thin Film Heat Transfer Gauges," Int. J. Heat Mass Transfer, Vol. 30, pp. 1159-1168.

Dunn, M. G. (1986a) "Heat-Flux Measurements for the Rotor of a Full-Stage Turbine: Part 1 -- Time-Averaged Results," ASME J. Turbomachinery, Vol. 108, pp. 90-97.

Dunn, M. G. (1986b) "Time-Resolved Heat-Flux Measurements for the Rotor Blade of a TFE-731-2 HP Turbine," in Convective Heat Transfer and Film Cooling in Turbomachinery, von Karman Inst. for Fluid Dynamics, Rhode Saint Genese.

Dunn, M. G., George, W. K., Rae, W. J., Woodward, S. H., Moller, J. C., and Seymour, P. J. (1986c) "Heat-Flux Measurements for the Rotor of a Full-Stage Turbine: Part II -- Description of Analysis Technique and Typical Time-Resolved Measurements," ASME J. Turbomachinery, Vol. 108, pp. 98-102.

Elrod, W. C., Gochenaur, J. E., Hitchcock, J. E., and Rivir, R. B. (1985) "Investigation of Transient Technique for Turbine Vane Heat Transfer Using a Shock Tube," ASME Paper 85-IGT-17.

Epstein, A. H., Guenette, G. R., and Yuzhang, C. (1985) "High Frequency Response Heat Flux Gauge for Metal Blading," AGARD CP 390.

Epstein, A. H., Guennette, G. R., Norton, R. J. G., and Cao, Y. (1986) "High-Frequency Response Heat-Flux Gauge," Review of Scientific Instruments, Vol. 57, pp. 639-649.

Fage, A., and Falkner, V. M. (1931) "On the Relation Between Heat Transfer and Surface Friction and Laminar Flow," Aero. Res. Counc., Lond., R&M, No. 1408.

Figliola, R. S., Beasley, D. E., and Subramaniam, C. (1984) "Instantaneous Heat Transfer Between an Immersed Horizontal Tube and a Gas Fluidized Bed," ASME Paper No. 84-HT-111.

Fitzgerald, T. J., Catipovic, N. M., and Javanovic, G. N. (1981) "Instrumented Cylinder for Studying Heat Transfer to Immersed Tubes in Fluidized Beds," Ind. Eng. Chem., Fundam., Vol. 20, pp. 82-88.

Gardon, R. (1953) "An Instrument for the Direct Measurement of Intense Thermal Radiation," Review of Scientific Instruments, Vol. 24, pp. 366-370.

George, A. H. (1987) "A Transducer for the Measurement of Instantaneous Local Heat Flux to Surfaces Immersed in High-Temperature Fluidized Beds," Int. J. Heat Mass Transfer, Vol. 30, pp. 763-769.

George, W. K., Rae, W. J., Seymour, P. J., and Sonnenmeier, J. R. (1987) "An Evaluation of Analog and Numerical Techniques for Unsteady Heat Transfer Measurement with Thin Film Gauges in Transient Facilities," in Proc. of the 1987 ASME/JSME Thermal Engineering Joint Conf., Eds. P. J. Marto and I. Tanasawa, ASME, NY.

Geremia, J. O. (1970) "An Experimental Investigation of Turbulence Effects at the Solid Boundary Using Flush-Mounted Hot Film Sensors," Ph.D. Thesis, George Washington University.

Gladden, H. J., and Proctor, M. P. (1985) "Transient Technique for Measuring Heat Transfer Coefficients on Stator Airfoils in a Jet Engine Environment," AIAA Paper No. 85-1471, July.

Godefray, J. C., Francois, D., Gageant, C., Miniere, F., and Portat, M. (1986) "Thin Film and High Temperature Thermal Sensors Deposited by RF Cathodic Sputtering," ONERA TP No. 1986-28.

Gundappa, M., and Diller, T. E. (1987) "Unsteady Heat Transfer Measurements Around a Cylinder in Pulsating Crossflow," in 1987 ASME/JSME Thermal Engineering Joint Conference, Eds. P. J. Marto and I. Tanasawa, ASME, NY, pp. 629-634.

Gundappa, M., and Diller, T. E. (1988) "An Analytical Model for the Unsteady Heat Transfer in the Wake Region of a Cylinder in Crossflow," accepted for 1988 Winter Annual Meeting of ASME, Chicago, Nov. 27-Dec. 2, 1988.

Hager, N. E., Jr. (1965) "Thin Foil Heat Meter," Rev. Sci. Instr., Vol. 36, pp. 1564-1570.

Hanratty, T. J., and Campbell, J. A. (1983) "Measurement of Wall Shear Stress," in Fluid Mechanics Measurements, Hemisphere Publishing Co., Washington, DC.

Hayashi, M., Sakurai, A., and Aso, S. (1986) "Measurement of Heat-Transfer Coefficients in Shock Wave-Turbulent Boundary Layer Interaction Regions with a Multi-Layered Thin Film Heat Transfer Gauge," NASA-TM-77958.

Hippensteele, S. A., Russell, L. M., and Stepka, F. S. (1983) "Evaluation of a Method for Heat Transfer Measurements and Thermal Visualization Using a Composite of a Heater Element and Liquid Crystals," ASME J. Heat Transfer, Vol. 105, pp. 184-189.

Hollworth, B. R., and Gero, L. R. (1985) "Entrainment Effects on Impingement Heat Transfer: Part II - Local Heat Transfer Measurements," ASME J. Heat Transfer, Vol. 107, pp. 910-915.

Incropera, F. P., and DeWitt, D. P. (1985) Fundamentals of Heat and Mass Transfer, Second Ed., John Wiley & Sons, New York, pp. 202-206.

Ireland, P. T., and Jones, T. V. (1987) "The Response Time of a Surface Thermometer Employing Encapsulated Thermochromic Liquid Crystals," J. Phys. E: Sci. Instrum., Vol. 20, pp. 1195-1199.

Jaluria, Y. (1980) Natural Convection Heat and Mass Transfer, Pergamon Press, Oxford, pp. 307-313.

Johnson, C. B., Caraway, D. L., Stainback, P. C., and Fancher, M. F. (1987) "Transition Detection Study Using a Cryogenic Hot-Film System in the Langley 0.3-meter Transonic Cryogenic Tunnel," AIAA Paper No. 87-0049.

Jones, T. V. (1977) "Heat Transfer, Skin Friction, Total Temperature, and Concentration Measurements," in Measurement of Unsteady Fluid Dynamic Phenomena, Hemisphere Pub. Corp., Washington, DC, pp. 63-102.

Jones, T. V., and Hippensteele, S. A. (1987) "High-Resolution Heat-Transfer-Coefficient Maps Applicable to Compound-Curve Surfaces Using Liquid Crystals in a Transient Wind Tunnel," in Developments in Experimental Techniques in Heat Transfer and Combustion, Eds. R. O. Warrington et al., ASME, pp. 1-9.

Kalinin, E. K., and Dreitser, G. A. (1970) "Unsteady Convective Heat Transfer and Hydrodynamics in Channels," Advances in Heat Transfer, Vol. 6, pp. 367-502.

Keltner, N. R., and Wildin, M. W. (1975) "Transient Response of Circular Foil Heat Flux Gages to Radiative Fluxes," Rev. Sci. Instrum., Vol. 46, pp. 1161-1166.

Keltner, N. R., Bainbridge, B. L., and Beck, J. V. (1988) "Rectangular Heat Source on a Semi-Infinite Solid - An Analysis for a Thin Film Heat Flux Gage Calibration," ASME J. Heat Transfer, Vol. 110, pp. 42-48.

Kercher, D. M., Sheer, R. E., Jr., and So, R. M. C. (1983) "Short Duration Heat Transfer Studies at High Free-Stream Temperatures," ASME J. Eng. for Power, Vol. 105, pp. 156-166.

Kidd, C. T. (1981) "A Durable Intermediate Temperature, Direct Reading Heat Flux Transducer for Measurements in Continuous Wind Tunnels," AEDC-TR-81-19, November.

Kidd, C. T. (1983) "Determination of Experimental Heat-Flux Calibrations," Arnold Engineering Center, AEDC-TR-83-13, August.

Kim, B. K., Borell, G. J., Diller, T. E., Cramer, M. S., and Telionis, D. P. (1983) "Pulsating Flow and Heat Transfer Over a Circular Cylinder," in Proceedings of the Symposium on Nonlinear Problems in Energy Engineering, DOE CONF-830413, pp. 96-101.

Klems, J. H., and DiBartolomeo, D. (1982) "Large-Area, High Sensitivity Heat-Flow Sensor," Rev. Sci. Instrum., Vol. 53, pp. 1609-1612.

Kobashi, Y., and Hayakawa, M. (1981) "Structure of Turbulent Boundary Layer on an Oscillating Flat Plate," in Unsteady Turbulent Shear Flow, IUTAM Symposium, Springer.

Kraabel, J. S.; Baughn, J. W., and McKillop, A. A. (1980) "An Instrument for the Measurement of Heat Flux from a Surface with Uniform Temperature," ASME J. Heat Transfer, Vol. 102, pp. 576-578.

Krane, M., and Dybbs, A. (1987) "Numerical Study of the Effects of Boundary Conditions on the Measurement and Calibration of Gardon Type Heat Flux Sensors," in Structural Integrity and Durability of Reusable Space Propulsion Systems, (NASA), pp. 51-57, N87-22775.

Lau, S. C., and Sparrow, E. M. (1980) "Average Heat Transfer Coefficients for Forced Convection on a Flat Plate with an Adiabatic Starting Length," ASME J. Heat Transfer, Vol. 102, pp. 364-366.

Leclercq, D., and P. Thery (1983) "Apparatus for Simultaneous Temperature and Heat-Flow Measurements under Transient Conditions," Review of Scientific Instruments, Vol. 54, pp. 374-380.

Lefebvre, P. J., and LaPointe, K. M. (1986) "The Effect of Mounting Position on Hot-Film Wall Shear Stress Sensors," AIAA Paper No. 86-1101.

Liebert, C. H. (1985) "Heat Flux Sensor Calibration," in Structural Integrity and Durability of Reusable Space Propulsion Systems, NASA Conf. Pub. 2381.

Liebert, C. H., Holanda, R., Hippensteele, S. A., and Andracchio, C. A., (1985) "High-Temperature Thermocouple and Heat Flux Gauge Using a Unique Thin Film - Hardware Hot Junction," ASME J. of Engineering for Gas Turbines and Power, Vol. 107, pp. 938-944, also NASA TM 86898.

Liepmann, H. W., and Skinner, G. T. (1954) "Shear-Stress Measurements by Use of a Heated Element," NACA TN 3269.

Ludwieg, H. (1949) "Ein Gerat zur Messung der Wandschubspannung Turbulenter Reibungschichten," Ing. Arch. 17, pp. 207-218. "Instrument for Measuring the Wall Shear Stress of Turbulent Boundary Layers," NACA TM 1284 (1950).

Malcorps, H. (1983) "Method to Increase the Bandwidth of Heat Fluxometers," Review of Scientific Instruments, Vol. 54, pp. 381-384.

McCroskey, W. J., and Durbin, E. J. (1972) "Flow Angle and Shear Stress Measurements Using Heated Films and Wires," J. Basic Eng., Vol. 94, pp. 46-52.

Menendez, A. N., and Ramaprian, B. R. (1985) "The Use of Flush Mounted Hot-Film Gages to Measure Skin Friction in Unsteady Boundary Layers," J. Fluid Mech., Vol. 161, pp. 139-159.

Metzger, D. E., and Larson, D. E. (1986) "Use of Melting Point Surface Coatins for Local Convection Heat Transfer Measurements in Rectangular Channel Flows with 90° Turns," ASME J. Heat Transfer, Vol. 108, pp. 48-54.

Miller, C. G. (1981) "Comparison of Thin-Film Resistance Heat-Transfer Gages with Thin-Skin Transient Calorimeter Gages in Conventional Hypersonic Wind Tunnels," NASA TM 83197.

Miller, C. G. (1985) "Refinement of an 'Alternate' Method for Measuring Heating Rates in Hypersonic Wind Tunnels," AIAA J., Vol. 23, pp. 810-812.

Mizushina, T. (1971) "The Electrochemical Method in Transport Phenomena," in Advances in Heat Transfer, Vol. 7, pp. 87-161.

Moffat, R. J. (1982) "Turbulent Boundary Layer Heat Transfer," in Film Cooling and Turbine Blade Heat Transfer, Vol. 1, von Karman Inst. for Fluid Dynamics.

Moreno, J., Jimenez, J., Cordoba, H., Rojas, E., and Zamora, M. (1980) "New Experimental Apparatus for the Study of the Benard-Rayleigh Problem," Rev. Sci. Instrum., Vol. 51, pp. 82-85.

Mori, Y., and Tokuda, S. (1966) "The Effect of Oscillation on Instantaneous Local Heat Transfer in Forced Convection from a Cylinder," Proc. Third Int. Heat Transfer Conf., AIChE, NY, pp. 49-56.

Mori, Y., Uchida, Y., and Sakai, K. (1986) "A Study of the Time and Spatial Structure of Heat Transfer Performances Near the Reattaching Point of Separated Flows," in Heat Transfer 1986, Vol. 3, Eds. C. L. Tien et al., Hemisphere Pub. Co., Wash., pp. 1083-1088.

Nicholson, J. H., Forest, A. E., Oldfield, M. L. G., and Schultz, D. L., (1984) "Heat Transfer Optimized Turbine Rotor Blades - An Experimental Study Using Transient Techniques," ASME J. Eng. Gas Turbines Power, Vol. 106, pp. 173-182.

O'Brien, J. E., Simoneau, R. J., LaGraff, J. E., and Morehouse, K. A., (1986) "Unsteady Heat Transfer and Direct Comparison to Steady-State Measurements in a Rotor-Wake Experiment," in Heat Transfer 1986, Vol. 3, Eds. C. L. Tien, et al., Hemisphere Pub. Co., Wash., pp. 1243-1248.

Olsson, P. J. (1988) "Transition Measurements in Flight," AIAA Paper No. 88-2107.

Pessoni, D. H. (1974) "An Experimental Investigation into the Effects of Wall Heat Flux on the Turbulence Structure of Developing Boundary Layers at Moderately High Reynolds Numbers," Ph.D. Thesis in Mechanical and Industrial Engineering, Univ. of Illinois, Urbana, Appendix B.

Poll, D. I. A., and Watson, R. D. (1987) "On the Relaxation of a Turbulent Boundary Layer After an Encounter with a Forward Facing Step," AGARD Conference Proceeding No. 365, Paper No. 18.

Pope, R. J. (1972) "Skin Friction Measurements in Laminar and Turbulent Flows Using Heated Thin-Film Gauges," AIAA Journal, Vol. 10, pp. 729-730.

Ramachandran, N., Armaly, B. F., and Chen, T. S. (1985) "Measurements and Predictions of Laminar Mixed Convection Flows Adjacent to a Vertical Surface," ASME J. Heat Transfer, Vol. 107, pp. 636-641.

Ramaprian, B. R., and Tu, S. W. (1983) "Calibration of a Heat Flux Gage for Skin Friction Measurement," ASME J. Fluids Engng., Vol. 105, pp. 455-457.

Shirtliffe, C. J., and Tye, R. P., Eds. (1985) Guarded Hot Plate and Heat Flow Meter Methodology, ASTM, Vol. 879.

Simonich, J. C., and Moffat, R. J. (1984) "Liquid Crystal Visualization of Surface Heat Transfer on a Concavely Curved Turbulent Boundary Layer," ASME J. Eng. Gas Turbines Power, Vol. 106, pp. 619-627.

Simons, D. B., Li, R. M., and Schall, F. D. (1979) "Spatial and Temporal Distribution of Boundary Layer Shear Stress in Open Channel Flows," NSF Final Report, Civil Engineering Dept., Colorado State Univ., p. 57.

Simpson, R. L., Shivaprasad, B. G., and Chew, Y.-T. (1983) "The Structure of a Separating Turbulent Boundary Layer, Part 4: Effects of Periodic Free-Stream Unsteadiness," J. Fluid Mech., Vol. 127, pp. 219-261.

Sirkar, K. K., and Hanratty, T. J. (1970), "Relation of Turbulent Mass Transfer at High Schmidt Numbers to the Velocity Field," J. Fluid Mech., Vol. 44, pp. 589-603.

Stack, J. P., Yeaton, R. B., and Dagenhart, J. R. (1987) "Predicted and Hot-Film Measured Tollmien-Schlichting Wave Characteristics," Symposium of Natural Laminar Flow Control Research, NASA Langley Research Center, Hampton, VA.

Stack, J. P., Mangalam, S. M., and Kalburgi, V. (1988) "The Phase Reversal Phenomenon at Flow Separation and Reattachment," AIAA Paper No. 88-0408.

Suarez, E., Figliola, R. S., and Pitts, D. R. (1983) "Instantaneous Azimuthal Heat Transfer Coefficients from a Horizontal Cylinder to a Mixed Particle Size Air-Fluidized Bed," ASME Paper No. 83-HT-93.

Thompson, W. P. (1981) "Heat Transfer Gages," Ch. 7 in Fluid Dynamics Methods of Experimental Physics, Vol. 18B, Ed. R. J. Emrich, Academic Press, N.Y., pp. 663-685.

Tsou, Fu-Tang, Chen, S. J., and Ko, S.-Y. (1983) "Measurements of Heat Transfer Rates Using a Transient Technique," ASME Paper No. 83-HT-87.

VandenBerghe, T., and Diller, T. E. (1988) "Analysis and Design of Experimental Systems for Heat Transfer Measurement from Constant Temperature Surfaces," submitted to Experimental Thermal and Fluid Sciences.

Vassiliadou, E., McConaghy, G. A., and Hanratty, T. J. (1986) "Effect of Drag-Reducing Polymers on Mass Transfer Fluctuations," AIChE Journal, Vol. 32, pp. 381-388.

Winter, K. G. (1977) "An Outline of the Techniques Available for the Measurement of Skin Friction," Prog. Aerospace Sci., Vol. 18, pp. 1-57.

SCANNING LASER ANEMOMETRY AND OTHER
MEASUREMENT TECHNIQUES FOR
SEPARATED FLOWS

Roger L. Simpson
Department of Aerospace and Ocean Engineering
Virginia Polytechnic Institute and State University
Blacksburg, VA 24061

1. Introduction

The title of this chapter reflects the fact that the behavior of separated flows is sufficiently different from unseparated flows to warrant special attention to experimental techniques. Simpson (1985) has reviewed the physical phenomena of separated flows with backflow. One purpose here is to review some recent advances in the methods for measuring velocity fields in the presence of the separation process. Another purpose is to discuss in some detail rapidly scanning laser anemometry, which permits the measurement of entire velocity profiles in a short period of time.

If we have two-dimensional mean flow separation either in a laminar or turbulent flow, it is well known that there is a region of backflow in the separation zone. For a steady laminar flow separation bubble, which may occur in ideal cases, interpretation of constant temperature hot-wire anemometer signals may be easy and backflow velocities can be accurately measured, providing probe disturbance effects are small. Sandborn (1972), Comte-Bellot (1976), and Perry (1982) summarize the general knowledge of the physical behavior of hot-wire and hot-film transducers. However, for most practical cases that involve turbulence, this technique and associated data interpretation procedures are not so simple to use. A hot-wire anemometer is not directionally-sensitive to the cooling velocity so one cannot deduce the flow direction from one sensor. Since relatively high turbulence intensities are encountered in separated flows and calibration results are free convection dependent near zero velocity, the hot-wire sensor fixed relative to the test surface is not an acceptable velocity measuring transducer for separated turbulent flows. Furthermore, it is likely that such a probe disturbs the flow to some degree. Even so, several recent modified hot-wire techniques have been developed and are discussed below.

The pitot-static tube is another classical velocity and static pressure measuring device that is commonly used for unseparated flows but should not be used in flows with flow reversal. It not only disturbs a separated flowfield but is sensitive to many factors that prevent reliable interpretation of the data: pressure measuring system response, yaw and pitch, turbulence, low Reynolds number effects, and flow reversal.

Pulsed-wire anemometry permits the magnitude and direction of the flow to be determined. A pulse of current through a heater wire warms the gas flowing by. Resistance thermometers located a distance up and downstream of the heater detect the passage of this heated fluid. The speed and direction of this fluid is determined through the use of an electronic circuit. A directionally-sensitive corona discharge anemometer can determine the flow direction and magnitude since the paths of ionized particles are changed by gas flow. The use of these devices is discussed below.

Since the introduction of the laser in the early 1960's, laser anemometry (LDA) or, equivalently, laser velocimetry (LDV) has been developed as a nonintrusive quantitative method for flow measurements. Here we will discuss application of this modern technique to separated flows, particularly those with flow reversal. Finally, rapidly-scanning laser anemometry will be discussed.

2. Hot-Wire and Hot-Film Techniques

As mentioned in the Introduction a hot-wire or hot-film velocity probe is insensitive to the flow direction and thus a single sensor fixed relative to the test surface is almost useless when the instantaneous flow direction is changing, as in a turbulent separated flow. For locations with backflow, rectified signal voltage probability diagrams are produced. As pointed out by Simpson (1976), there is no unique way to relate the output of a hot-wire anemometer to the time flow behavior obtained by a directionally-sensitive laser anemometer when substantial backflow is present.

A rectified hot-wire anemometer signal will produce high mean velocities U and low mean square fluctuations $\overline{u^2}$, as shown in Figure 1. When backflow occurs less than 10% of the time, mean velocities can be deduced from the signal. Reliable mean square fluctuations can be obtained with backflow less than 5% of the time (Simpson, 1976). If one knows the shape of the actual velocity probability distribution, then corrections can be made to hot-wire signals as long as there is a very small amount of flow reversal (Simpson, 1976; Dengel and Vagt, 1982). Figure 1 shows that when negative velocity samples of a pulsed-wire anemometer are omitted, the results agree with the erroneous hot-wire results. The correction procedures of Bradbury (1976) and Tutu and Chevray (1975) also resolve the large difference between hot-wire and pulsed-wire results when local turbulence intensities $\sqrt{\overline{u^2}}/U$ are below about 0.5.

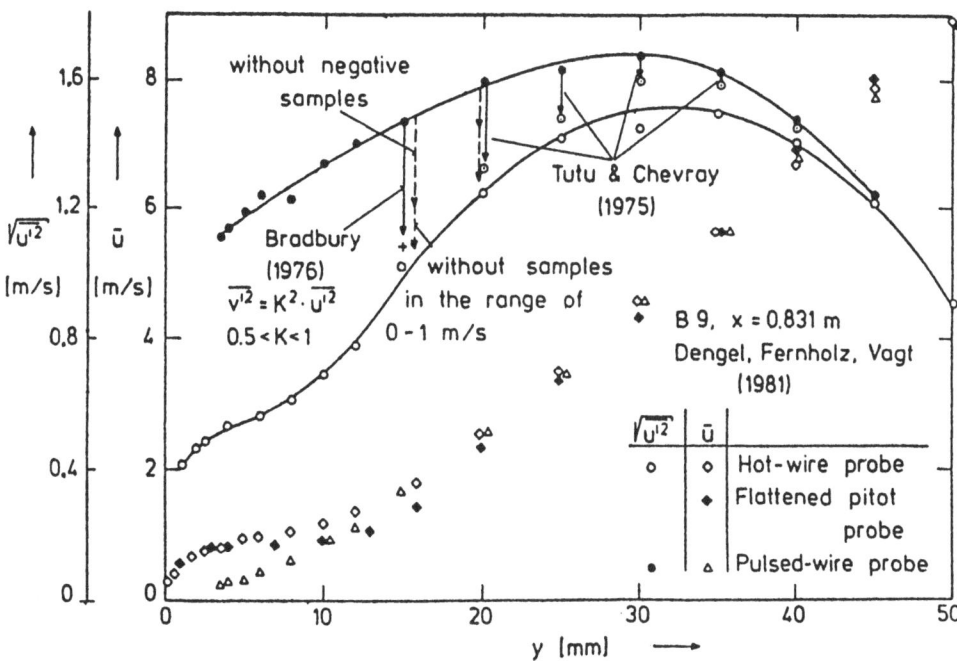

Figure 1   A comparison between pulsed-wire and hot-wire measurements in a strong adverse pressure gradient boundary layer.   Effect of hot-wire corrections.   Dengel and Vagt (1982).

Several attempts to alleviate this directional insensitivity difficulty have been made. Coles, Cantwell, and Wadcock have used a "flying hot-wire" (Coles and Wadcock, 1979), in which a hot-wire is swung in a circle through the flow at a known velocity and position as a function of time. Basically this introduces a sufficiently high bias velocity to the hot-wire so that the flow with respect to the wire is in an approximately known direction. Subtracting the bias velocity from the signal velocity determines the unknown fluid velocity contribution. Standard commercially-available X-array probes were used to determine the U and V mean velocity components and the respective u and v instantaneous velocity fluctuations.

This technique has several special features. One good feature is that data are obtained along a line rather than at a point. A negative feature is that the wake of the whirling arm is a substantial moving disturbance in the flow. Another good feature is that the hot-wire probes are inherently self-calibrating in pitch. The hot-wire signals must be transmitted through slip rings from the whirling arm. One must make many passes through the separated flow with the "flying hot wire" in order to have enough data samples at each physical location so time-averaged turbulence quantities and mean velocities can be obtained. Thus the reproducibility of the swung apparatus position is important and a sophisticated data acquisition, retrieval, and processing computer system must be used. A linear version of the flying hot wire has been used by Perry and Watmuff (1979).

Recently, Thompson and Whitelaw (1984) presented a flying-hot-wire anemometer design in which the hot-wire probe is moved in a known non-circular path by a crank and slider mechanism. No slip rings are required. The path and speed of the hot-wire can be adjusted for the region containing flow reversal by the mechanism.

The relatively small split-film hot-film probe (TSI Model 1287) has two equal and constant temperature sensors on one quartz substrate, each covering 170° of the periphery. Because the local convective heat transfer coefficient around a circular cylinder varies with angle, the heat transfer to the fluid from each sensor is dependent on the incident flow direction. Thus, in principle, both the magnitude and direction of the flow perpendicular to the substrate axis can be determined, providing one knows whether the flow comes from downstream or upstream. Several investigations have attempted to use this probe with varying degrees of ease and success for unseparated flows (Spencer and Jones, 1971; Blinco and Sandborn, 1973; Fuller, 1974; Browand and Weidman, 1976; Simpson and Shackleton, 1977; Ho, 1982). Young (1976) indicates that hot-film probes should be used with caution, since measured fluctuation quantities that are 30% too low can be obtained.

The most critical factor in using the split-film probe is obtaining a thermal balance between the two sensors. If the two sensors are not closely at the same temperature, thermal interference of one sensor with the second will occur. Individual constant temperature anemometer sets are used to power each sensor and the balancing resistance in the opposing bridge leg of each anemometer must be adjustable within 0.01 ohms for a 15 ohms operating resistance. Anemometer electronic drift must be minimized and circuit resistance values must be stable.

To use this split-film sensor in a separated flow with changing flow direction, which apparently has not yet been done, would require keeping track of the instantaneous flow direction since there are two flow directions that would produce the same signal. For a flow direction detector one could locate downstream a single hot-wire sensor perpendicular to the streamwise direction and substrate axis to sense the thermal wake of the split film. When the flow is in a downstream direction, the hot-wire would cut the thermal wakes of the split-film sensors and sense a higher fluid temperature; for flow in the reversed direction the unheated fluid temperature would be sensed. Thus for the case when the split is parallel to the test wall one could determine uniquely the instantaneous incident flow direction in the U-V plane.

One can determine whether there is forward flow or backflow near the wall with a split-film. If the split between the films is perpendicular to the test wall, then the difference between the two output voltage signals would indicate whether the flow is moving downstream or upstream, even though there would still be the ambiguity of whether the flow was moving away from or toward the wall. A three sensor split-film (DISA Model 55R92) with each film having 120° of the perimeter can be used to eliminate the directional ambiguity of the instantaneous velocity in the U-V plane (Jorgensen, 1982).

The future of the split-film for separated flow is questionable. If a styrene substrate can be used then the thermal conductivity of the substrate would be 1/7th that of the current split-film probes. Thermal interference of the two or three sensors could be minimized with less sensitivity to small operating temperature differences. If a thermal-wake-detecting hot-wire must be used to determine flow direction, then one should first consider pulsed-wire anemometry, described in section 3 below, and avoid the quantitative uncertainties associated with the split-film probe.

Ligrani et al. (1983) reviewed work on multiple hot-wire velocity and direction probes. In some designs, one constant-temperature wire detects the velocity magnitude and acts as a heater of the flow over this wire. One or more wires operated as resistance thermometers are placed downstream and/or upstream to sense the thermal wake of the hot-wire and, therefore, determine the flow direction. In others, two or more constant-temperature wires are operated at different overheat ratios; the ratio of the apparent measured velocities from these wires can indicate when the flow moves in the nominal downstream direction. For example, when the flow moves downstream, the thermal wake of the upstream wire passes over the lower overheat downstream wire, using it to indicate a lower apparent measured velocity. The ratio of the apparent velocities from the upstream and downstream wires is greater with downstream flow than with upstream flow.

Müller and Schon (1987) recently described a hot-wire technique designed for three-component measurements of instantaneous velocity vectors in high-intensity turbulent flows. By means of a thermal wake detector added to a triple-sensor probe, velocity vectors lying within a specified angular acceptance domain were identified and rectification was eliminated. Conditional averages of mean velocities and products of the fluctuations up to third order were determined. The final results were computed as the sum of weighted averages from several measurements covering the complete angular range of velocity fluctuations. Qualification measurements in low- and moderate-intensity flow compared well with results of standard hot-wire anemometry. Measurements in a complex flow with angular fluctuations up to ±55° were reported.

3. Pulsed-Wire Anemometry

The pulsed-wire anemometer (Bradbury and Castro, 1971) is capable of determining the flow velocity and direction within certain limitations. Their probe, shown schematically in Figure 2, has three fine wires. The central wire is pulsed with a short duration voltage pulse which heats the fluid passing over it at that time. This heated fluid is convected away with the local instantaneous velocity of the flow. The other two wires on the probe are operated as resistance thermometers. They are used to measure the time for the heated fluid tracer to travel from the pulsed wire to one of the other wires. The component of velocity that is perpendicular to all three wires (U cos θ in Figure 2) is measured, in principle being the distance S between each wire divided by the pulse travel time. In practice one must calibrate this probe because of the diffusion of thermal energy in the flow. The flow direction is determined by which wire detects the thermal pulse. This probe has a "blind spot" for instantaneous flow directions where the heated wake does not pass over a detector wire.

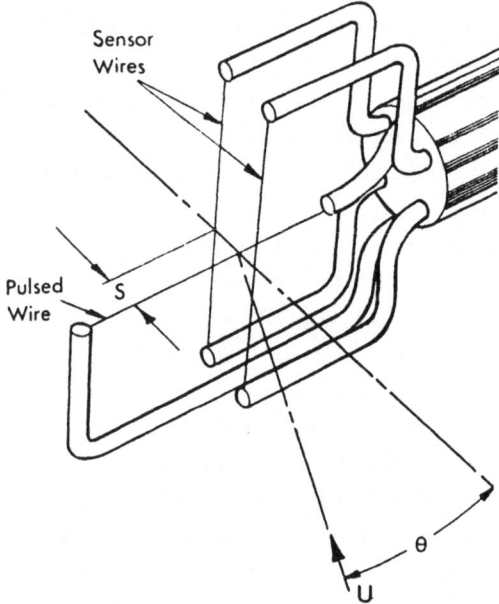

Figure 2 Schematic of the pulsed-wire anemometer. Pela Inst. Ltd., U.K.

There are several advantages to this instrument in addition to being directionally sensitive: there is no upper restriction on measurable turbulence level, it is usable in variable density flows, the calibration is dependent only on the probe geometry and thermal diffusion effects and wire fouling is generally not a significant problem. Castro (1971) measured u', v', and w' rms turbulence fluctuations as well as the mean velocity for a recirculating separated flow. To make such measurements, mean square fluctuation data from three probe orientations must be made and these quantities deduced from algebraic equations. It is not possible to measure u, v, and w fluctuations simultaneously.

Disadvantages to this commercially-available probe include the limited usable velocity range (0.25 - 15 m/sec), limited mean velocity accuracy (1 to 5%), non-continuous signal, and large size (1 cm long wires). Measurements of low turbulence intensities are relatively uncertain; turbulence intensities less than 2% cannot be measured. However, Castro and Cheun (1982) conclude that, in flows of such high intensity that hot-wires would be useless, measurements of all the Reynolds stresses can be made with an accuracy probably better than 30% (for v') or even 15% (u' and $\overline{uv}$). In the medium-intensity range (10-30%, say) it has been shown that pulsed-wire measurements can be as accurate as hot-wire measurements, provided the yaw response extends to large enough angles. However, in this case the errors in $\overline{uv}$ and v' are rather greater than at higher intensities and depend critically on the extent of the yaw response. Eaton and Johnston (1980) also used the Bradbury and Castro probe with little calibration drift and about the same level of measurement uncertainty as reported above.

Skinner et al. (1982) developed a low-speed probe design (0.12 - 2 m/sec). Only one detector wire was used, so backflows cannot be detected. Tomback (1973) discussed application of this instrument to inhomogeneous flow. J. Kielbasa (1975) reported a system in which the heated wire is periodically pulsed and the frequency of thermal pulses received is directly proportional to the velocity perpendicular to the wires. The usable velocity range is between 10 cm/sec and 20 m/sec and the claimed accuracy is 1.5%. Only two wires are used and an electronic relay and a flip-flop circuit are used to decide the flow direction.

In more recent work, Castro and Haque (1987) showed that the viscous wake of the upstream sensor wire has a measurable effect on the yaw response of the standard probe shown in Figure 2. A probe in which the sensor wires lay in a plane about 30° to the pulsed wire axis apparently reduces this viscous wake effect and produces a cosine-law yaw response up to ±80°. Similar results were obtained by Jaroch (1985) who also found that such a probe has an equally satisfactory pitch response. Jaroch also tested three different types of pulsed-wire probes with respect to their yaw - and pitch - response behavior and showed that they can be designed to have large acceptance cone angles (about ±70°) and small deviations from the ideal cosine law. Jaroch (1985) also reported for two highly turbulent shear flows that the differences between hot-wire and pulsed-wire measurements of U and u' are of the order estimated by Tutu and Chevray (1975), where applicable.

The measurements of the quantities V, v' and $\overline{uv}$ with pulsed-wire probes and with single hot-wire probes show large scatter, which in the case of V measurements, is so high that the data cannot be used. Part of this scatter can be explained by the fact that three measurements have to be made to calculate V, v' and $\overline{uv}$, and equations have to be used which are very sensitive to small errors in the measured data. In the case of pulsed-wire probes, however, another factor must be taken into account. This is the quality of approximation of the calibration curve and its storage in the microcomputer. If only data measured with the sensor with the largest and most symmetric acceptance cone, the smallest and most symmetric critical region, and the best approximation quality of the calibration curve are taken into account, the scatter in the measurements of v' and $\overline{uv}$ is very small. The pulsed-wire data differ by no more than 7% from the hot-wire data including plausible error estimates, although the exact nature of the errors in pulsed-wire measurements of v' and $\overline{uv}$ is not known.

Because of the physical size of the typical probe, it cannot be used much closer than 3-4 mm from a solid boundary. Devenport (1985) used a probe shown in Figure 3

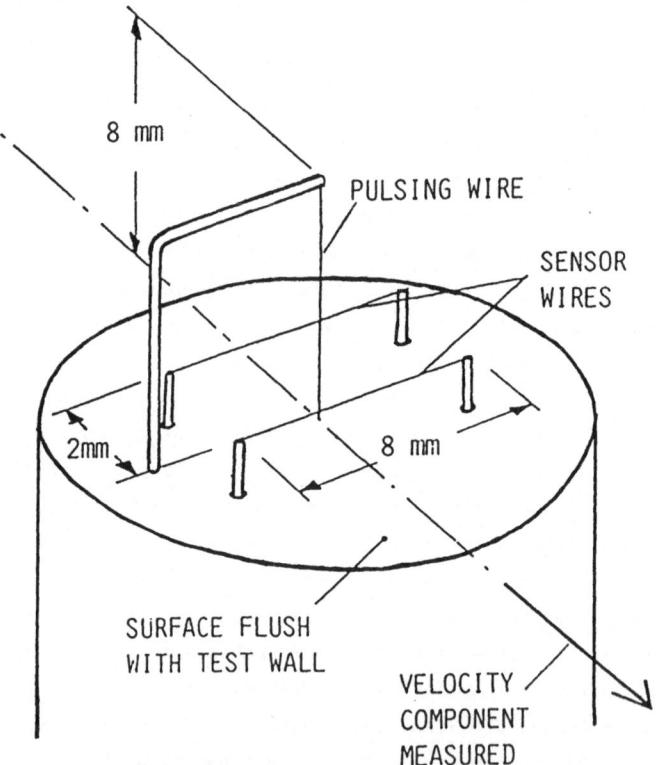

Figure 3 The near-wall pulsed-wire probe. Devenport (1985)

with the fixed pulsed wire vertical and the traversable sensor wires parallel to the solid surface. The pulsing wire and sensor wires were 8 mm long and made from 9 μm and 5 μm diameter tungsten wire, respectively. Half millimeter diameter phosphor-bronze wire was used for the prongs. Accurate measurements of mean velocity down to a y^+ of less than 2 were made without the necessity for corrections.

Preliminary measurements with the near-wall pulsed-wire probe showed that small construction differences between the sensor wires caused large errors in measurements made closer than 1 mm from the wall. Since two sensor wires could not be made sufficiently identical only one was used. The better wire was selected and two traverses were made at each profile location, one with the selected wire facing downstream and one with it facing upstream. Each pair of traverses was then combined to give profiles of the axial components of time-mean velocity and turbulence intensity.

Like standard probes the near-wall pulsed-wire probe was limited to the measurement of instantaneous velocities of less than about 20 m/s. The finite length of the pulsing wire (8 mm) prevented measurements further than about 5.5 mm from the wall. In an equilibrium turbulent boundary layer the near-wall pulsed-wire probe was found to measure mean velocity to an accuracy of 3% or better, and turbulence intensity to an accuracy of 7% or better, when compared with a single hot-wire anemometer.

Castro and Dianat (1987) also developed a near wall probe to allow measurements between $1 < y^+ < 100$. Figure 4 shows their probe with the pulsed wire and sensor wires parallel and perpendicular to the wall, respectively. The plane of the sensor wires was inclined 30° to the pulsed wire to maximize the accuracy of the cosine yaw response, as shown by Jaroch and Castro and Haque and discussed above. Unlike Devenport's work with a normal to the wall pulsed wire, thermal diffusion effects caused significant errors for $y^+ < 7$, which appear to be different for turbulent or laminar boundary layers. To estimate the diffusional effects theoretically, Castro and Dianat assumed that the heat tracer follows a path determined by vertical diffusion, which requires more time to reach the sensor wire than without diffusion. Using this model corrections to laminar and turbulent boundary layer data produced very good agreement with accepted near wall velocity profiles, although different correction constants are required for laminar and turbulent flows.

Westphal et al. (1981) and Eaton et al. (1982) also used a modified thermal tuft probe that had a 12.5 μm diameter nickel central heater wire. A probe (Figure 5)

Figure 4   Through wall pulsed-wire probe of Castro and Dianat (1987).

with traversable wires was developed and used to make turbulent flow velocity profile measurements within 0.2 mm of the wall.   Mean skin friction values inferred from these measurements were within 2% of independently determined skin friction values

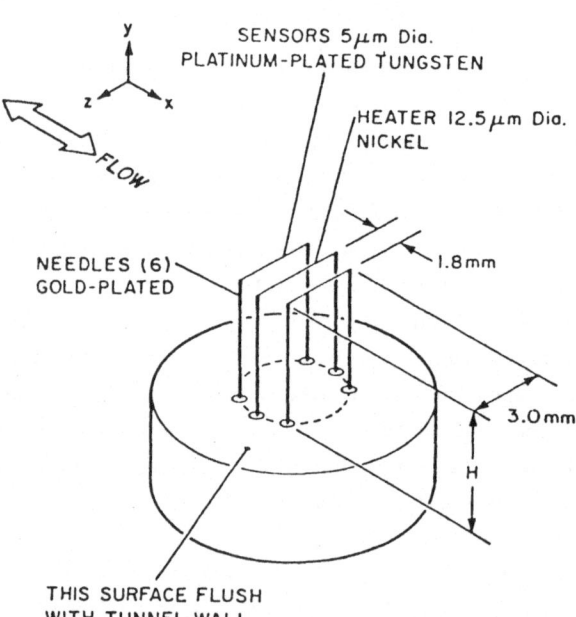

for forward flow.   Ginder and Bradbury (1973) and Ruderich and Fernholz (1983) used a pulsed-wire probe to measure the skin friction.   Dianat and Castro (1983) also report a similar skin friction probe that measures velocities down to 0.08 mm above the surface.   Reliable skin friction probability diagrams can be obtained with these probes.   Even though time-resolved skin friction values can be measured using this probe, measurable spectral frequencies are less than half of the reciprocal of the time of flight of a thermal pulse (Nyquist criterion).

Figure 5   Sketch of pulsed-wall probe.   Eaton, Westphal, Johnston (1982)

In more recent work on the measurement of skin friction, Castro, Dianat, and Bradbury (1987) showed that the time of flight of a heat tracer is critically dependent on the form of the upstream velocity profile. The difference between flight times obtained in cases of different velocity profiles having the same surface velocity gradient is greater than that anticipated solely on the basis of the difference in convection velocity at the height of the heat source. Calibrations made in a laminar channel flow produce too high measured turbulent skin friction values. There is no simple way of correcting a laminar calibration for use in turbulent flows. Errors are functions of the Peclet number, based on U_τ and the wire spacing S, and the geometry, i.e., S/H, and S/(channel height) (Figure 5). Calibration in a turbulent channel flow is the common practice by the Surrey (Castro), Stanford (Westphal, Eaton), and Berlin (Fernholz, Jaroch) groups. Devenport (1985) calibrated his standard PELA Instruments probe in an equilibrium turbulent boundary layer against a Preston tube.

Both Castro et al. (1987) and Devenport (1985) had sensor and pulsed wire lengths of about 2 mm fixed parallel to one another 0.5 mm apart and about 0.05 mm from the surface. A 9 μm diameter tungsten pulsed-wire and 2.5 μm diameter sensor wires were used. These probes obey a cosine law response. Castro et al. (1987) showed that the measured shear stress intensity increases with decreasing sensor wire length ℓ_s since longer lengths do not respond to the spanwise sublayer structure as well, i.e., $\ell_s U_\tau / \nu < 30$ is required for good estimates.

4.   Corona-Discharge Anemometer

A directionally-sensitive corona discharge anemometer has recently been developed by Durbin et al. (1987). The idea of ionizing air and determining its velocity from the current produced by ion convection is at least 60 years old, as discussed by Durbin et al. Compared to a hot-wire probe, the corona anemometer is quite rugged.

The principle and design of the anemometer probe are illustrated by Figure 6. A needle is raised to sufficient electric potential to ionize gas in a neighborhood of its tip: in air voltages of order 2-3 kV are required for this particular device. The collector needles are held at essentially ground potential. The electric field of the needles drives ions across the air gap between high and low voltage. In the absence of gas flow and if the probe is completely left-right symmetric, equal currents will flow to the two collectors: the currents are of order 1 μA. When gas flows through the gap between the high and low voltage tips, ions will be convected with the flow and more will be collected by the downstream collector tip than by the other. The difference between the currents collected by the tips provides a measure of gas velocity. Because the current difference can take either sign (or be zero) the probe can follow changes in flow direction.

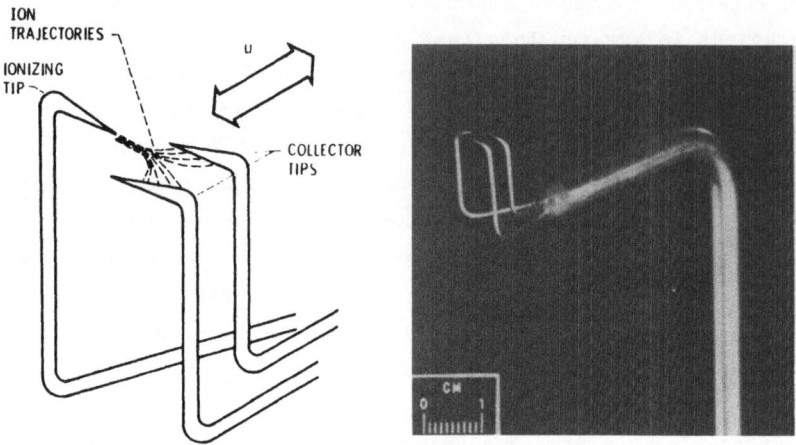

Figure 6   Sketch illustrating corona discharge probe and photo of probe.   Durbin et al. (1987)

If the tips are arranged symmetrically, then the probe will be sensitive only to the single component of velocity directed through the gap between the high and low voltage tips, in the plane of the tips. In the design in Figure 6 the tips lie at the corners of an equilateral triangle whose side is 3mm. The component of velocity to which the probe is sensitive is indicated in Figure 6. A linear relationship exists between $(i_1 - i_2)/(i_1 + i_2)$ and this velocity component. The yaw response approaches the ideal cosine law as the probe asymmetry is reduced.

There are several disadvantages of this anemometer. A large (3mm) minimum gap and associated measurement volume are required to permit stable operation without sparking and to keep achievable tolerances of construction. The velocity components not being measured must be small compared to μE ($= 75$m/s for the probe shown in Figure 6), where μ is the ion mobility and E is the electric field.   At frequencies above 1kHz the response of this probe begins to fall off, agreeing with the estimate of $|U|$/(gap length) for the upper limit to good frequency response.

Durbin et al. suggest several other probe designs. Probes with larger gaps could be designed for large-scale flows. An increase in collector needle spacing, with a fixed gap between the high and low voltage sides, could extend the measurable velocity. A probe with four collector tips on the corners of a square, with the emitter facing the center of the square could measure simultaneously two velocity components. A surface-mounted probe could measure the state of a separated flow.

5. Laser Anemometry

Laser anemometry is one experimental velocity measurement technique that can be used for separated flow studies to determine precisely the flow magnitude and direction without calibration. In essence the principle of this experimental method is the scattering of closely monochromatic laser light by moving particles in the flow. Since the Doppler-shifted frequency of the scattered light is velocity dependent, fluid velocities can be deduced if the scattering particles follow the flow. Naturally there is no disturbance to the flow except additional particle seeding, if any, so there is an obvious advantage to using this method in separated flows. Mazumder et al. (pp. 234-269, Thompson and Stevenson, 1974) note that one can easily supply small enough particles that follow the flow to most flowfields of interest. Several general references on laser anemometry with many other cited references include: Stevenson and Thompson (1972), Thompson and Stevenson (1974, 1979), Durst, Melling, and Whitelaw (1981), Durão et al. (1982, 1984, 1986) and the publications of the equipment manufacturers DANTEC and TSI. Here we will not review the many contributions made to laser anemometry in the last quarter century since its introduction, but will discuss the practical application of this technique to turbulent separated flows with flow reversal.

If one wishes to map a moderately-sized flowfield, the LDA focusing and receiving optics should be on one side of the flow for stable alignment. Otherwise, the repetitive realignment of the optics at each new spatial focusing position becomes tedious and very time consuming. Unfortunately, for Mie scattering of light (particle size > wavelength of light) the intensity of light backscattered toward the focusing optics is about 1/100th of the intensity of forward scattered light (Van de Hulst, 1964). Thus, having decided on a backscattering arrangement one must use large receiving lens and a high-powered laser in order to obtain an adequate signal.

If one simply collects the backscattered light from one focused incident beam and beats it with a reference frequency beam of unscattered light, he will find that his signal frequency is dependent on the angle of backscattering. In other words there is receiver aperture broadening of the signal frequency and the solid angle of the receiving optics must be decreased to eliminate this ambiguity. Since this step will reduce the received signal power to an unacceptably low level, one must abandon this approach and use a dual-scattering fringe system (Durst et al., 1981).

In the dual-scattering fringe system two equally intense beams of wavelength λ are focused at an angle θ to one another with each beam waist in the focal volume of interest. Fringes of spacing $\lambda_f = \lambda/2\sin(\theta/2)$ are formed because of interference of the light wavefronts. Now if a particle crosses these fringes with a component of velocity u perpendicular to the plane of a fringe, light is scattered in all directions with its intensity varying with frequency U/λ_f. Aside from the fact that the signal-to-noise ratio (SNR) is much greater in this arrangement (Mazumder, 1970), one can use a very large receiving lens with no additional aperture broadening. Only one component of velocity can be measured with a single pair of beams, so two other beams of different colors or wavelengths must be used to form another set of fringes which is perpendicular to the velocity component V.

If only stationary fringes from two beams with the same frequency are used, a given speed particle moving upstream or downstream produces the same signal. Thus, if we wish to determine the flow direction with LDA, we must shift the frequency of one beam of a pair so that fringes in the focal volume are moving with time as the waves in the two beams beat together. As long as the frequency difference between the two beams is more than twice the frequency change produced by the scattering particles U/λ_f, signals from particles moving in all directions will be obtained (Durst et al., 1981). There are several ways of changing the frequency of an incident laser beam, but the most straightforward method uses an acoustic-optical Bragg cell (Buchhave, 1975). Basically, a Bragg cell uses an ultrasonic driver of frequency f_0 acting on a material to diffract incident light of frequency f passing through the material. The most intense diffracted beam has a frequency $f + f_0$ (Williard, 1949). Not only is the laser output beam efficiently split by a Bragg cell, but equal beam intensities, path lengths and like polarization can be easily achieved. In the two-component backscattering fringe-type anemometer used by this writer (Simpson and Chew, 1979), a dual Bragg cell shifts the horizontal first-order diffracted beam by 25 MHz and the vertical beam by 15 MHz. Signals of $(U-V)$ as well as U and V are available.

To measure W one can use a single incident beam with a third wavelength λ along the lens axis which measures only the velocity component parallel to that beam (Munoz et al.; 1974). Basically, the backscattered light is received back through the focusing lens and is beat with a reference beam to produce the doppler signal. Kreid and Gram (1976) appear to have increased the SNR ratio for this arrangement and use a Bragg cell to make the arrangement directionally sensitive. However, they require an etalon to increase the coherence length of the laser and a good optical table for precise alignment of the received signal and the reference beam. Shiloh and Simpson (1982) used a separate fringe-type anemometer to measure W whose beams entered the tunnel 90° to the U and V measuring beams. This arrangement has all of the advantages of a fringe system over a reference beam system, but with the disadvantage of a separate access window. Meyers (1985) also concluded that a fringe-type optical arrangement should be used instead of a reference beam system. His review of various fringe systems led to the conclusion that all surveyed arrangements utilized a component angled about the sample volume to obtain a measure of a portion of the W-component velocity. The orthogonal configuration, such as used by Shiloh and Simpson, is the limiting case and yields the most accurate measurements.

Signal processing and interpretation of LDA signals is still in a state of refinement (Thompson and Stevenson, 1979; Buchhave et al., 1979; Durão et al., 1982, 1984, 1986). Without going into great detail, the several types of available signal processing techniques will be outlined. The frequency tracker has been used when so much seeding from natural impurities is present that a nearly continuous signal can be obtained. Unfortunately when the signal is not continuous, as for nearly all gas flows, the tracker can lock onto noise or become unlocked altogether. The validity of the output signal becomes dependent on signal dropout levels and the special circuits that are used to keep the tracker locked onto the frequency of one signal burst to the next (Buchhave et al., 1979). In high turbulence intensity flows such as in separated flows, the tracker can more easily lose the signal due to dropout (p. 184, Thompson and Stevenson, 1974). With a relatively noisy high-powered argon-ion laser, it is very difficult for a tracker to remain locked on the signal even for a continuous signal.

The frequency counter basically filters, amplifies, validates, and counts each burst of LDA signal (Durst et al., 1981). Unfortunately the SNR ratio must also be large (45 dB) to obtain unambiguous results since the electronics must pick a single frequency from each signal burst. If one can see the signal distinctly on an oscilloscope then there is some hope of using a frequency counter for processing.

The photon-correlation technique (Durst et al., 1981) uses the auto-correlation of the received signal to determine the frequency. A number of workers are using it because virtually no added seeding particles are required. Naturally occurring sub-micron-sized particles are sufficient for light scattering, although relatively long record and correlation times are required for good results (Cummins and Pike, 1977; Mayo and Smart, 1980). No instantaneous velocities can be measured; only mean velocities and rms fluctuations for low turbulence intensity flows can be reliably obtained. Thus, this technique is not very satisfactory for reversing flows. Brown (1986) used polarization differences in the two incident beams to create a phase difference between two signals for the same velocity component received by different detectors. A positive or negative phase difference determines a downstream or upstream flow direction without any frequency shifting while employing a photon-correlator.

Fast spectrum analysis (Simpson and Barr, 1975) uses a swept-filter spectrum analyzer and pulse-shaping and sample-and-hold circuits. Sampling frequencies up to 8kHz can be obtained, making possible measurement of turbulence frequencies up to 4kHz. Very noisy signals (15 dB SNR) with a high level of signal dropout and high frequency signals normally associated with Bragg cell frequency shifted systems can be processed. In practice the broadband noise must be only low enough that the signal frequency can be discriminated. The faster the sweep rate the greater SNR required. For example a 25 dB SNR continuous signal can be discriminated at a 1800 Hz sweep rate.

This latter technique was used by Simpson et al. (1977, 1981) in their separated turbulent boundary layer measurements. This technique was chosen for economy, simplicity, and ease of validating data. Simpson (1976) reported on the interpretation of LDA signals near separation and compared these results with hot-film anemometer results. Quantities such as mean velocities, turbulence intensities, Reynolds shearing stresses, third and fourth moments of turbulent fluctuations (skewness and flatness), and the fraction of time that the flow is in one direction can be obtained. Figure 7 is a typical histogram of particle velocities in the separated flow region. For frequencies greater than 25 MHz, the horizontal flow was downstream while lower frequencies corresponded to backflow upstream.

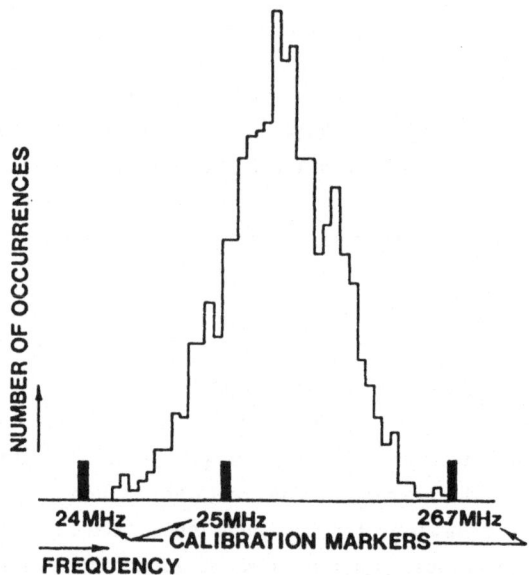

Figure 7 Typical histogram of laser anemometer data samples in a separating flow. Zero velocity at the 25 MHz Bragg cell shifted frequency; positive velocities at higher frequencies. Simpson (1976)

After more than a decade of using these signal processing techniques, a new generation of more robust signal processing methods are under development and use because of advances in computer technology. High speed digitizers (> 200 Msamples/sec, 8 to 12 bits) with hardwired memories (> 16 Mbytes) presently permit the acquisition of poor SNR signals for processing in the frequency domain. The prognosis for further hardware improvements is quite good. Digitizers with 1 gigasamples/sec and 12 bit resolution are expected to have gigabyte memories in a few years.

Standard fast-Fourier-transform (FFT) techniques can be applied to these data to process SNR signals of -5 to -10 dB, as shown in Figure 8 (b, c), as well as high SNR counter-quality signals (Figure 8 (a)). Discrete-Fourier-transform (DFT) algorithms

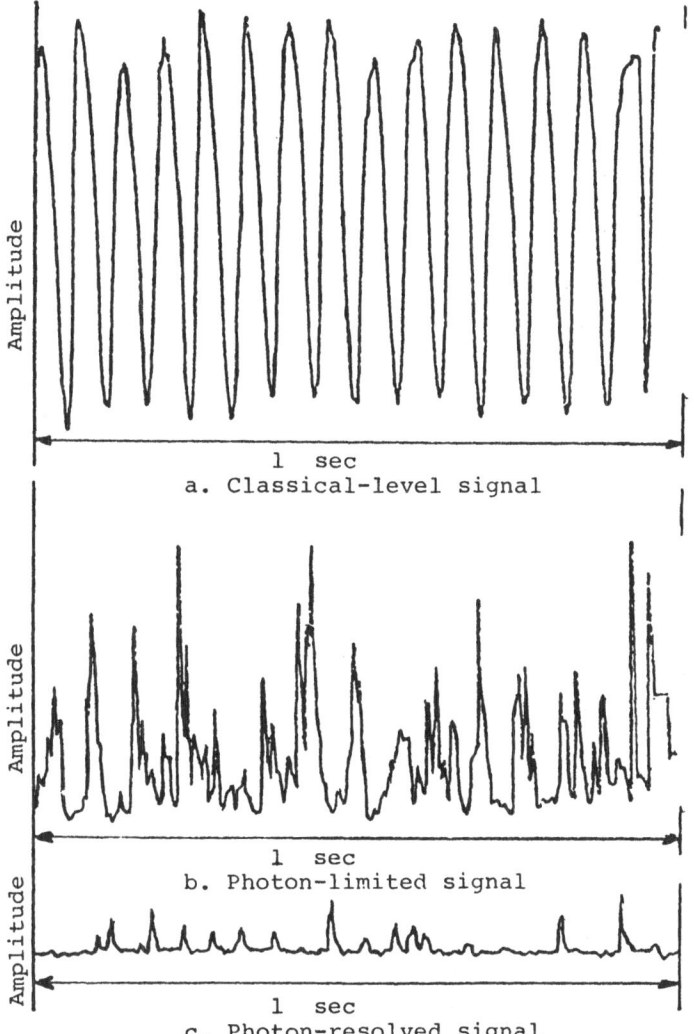

a. Classical-level signal

b. Photon-limited signal

c. Photon-resolved signal

Figure 8   Signal types used in laboratory evaluation of the Discrete Fourier Transform Processor (Kalb and Crosswy, 1983):   a) Counter-quality signal; b) and c) low·SNR signals requiring frequency-domain processing.

have been applied to reduce software processing time in determining signal frequencies (Kalb and Crosswy, 1983).   Basically, the spectral energy content of only a few frequencies is determined around the narrow-band signal frequency domain.

Frequency interpolation schemes seem to determine closely the signal frequency when compared to some computationally intensive FFT techniques. The DANTEC Burst Spectrum Analyzer is a hardwired frequency domain processor that performs a DFT on a set of samples (Lading, 1987). Signals with SNR of -3 dB have been processed with a data rate of valid demodulated signals of several kHz.

Efforts to improve the counter type of processor have also been made. The TSI "IFA" uses zero-crossings of bandpass filtered signals to validate counter-measured frequencies and produce higher data rates for lower SNR cases (Jenson et al., 1988). The bandpass filter frequency range with the highest validated data rate is automatically selected. Apparently a SNR of 10 dB can be processed.

The seeding of separating flows determines the signal quality and data rate to a large extent. For liquid flows, an average concentration of particles with a narrow distribution of diameters can be maintained to provide at most one particle in the measurement volume at any given time. By filtering the liquid, sub-micron particles that reduce the SNR can be eliminated, thus permitting a high enough SNR for counter or tracker signal processing.

For gaseous flows, the particle seeding situation is worse. Sub-micron-sized particles scatter background light and reduce the SNR. When using a frequency counter processor one must eliminate all sources of noise, so efforts have been made to eliminate sub-micron particles. Bachalo et al. (1977), Seegmiller et al. (1978), and Driver et al. (1987) used 0.5 µm polylatex solid spherical particles that were carried by an alcohol solvent into the test plenum where the alcohol evaporated. Driver et al. obtained about 50 signals/sec in the backflow and 2000 signals/sec in the outer region. Johnson (1981) and Délery (1983) report that good seeding particles (d < 1 µm) can be produced by the natural condensation of oil vapor in transonic tunnels.

When a poly-disperse aerosol is used, many sub-micron particles are generated, leading to an unacceptably low SNR for using a frequency counter processor. In the Simpson et al. (1977, 1981) experiments, the sub-micron particles actually produced a white spectrum background noise that permitted more sensitive signal discrimination in fast spectrum analysis processing. About 400 signals/sec were obtained throughout the detached flow, except nearest the wall where a lower rate was achieved. Crabb et al. (1981) report 20 dB SNR signals produced by the condensation of kerosene vapors into particles less than 5 µm in diameter. A recent workshop on seeding problems (Hunter and Nichols, 1985) dealt with the design and user experience of a number of different seeders in a number of different flow facilities.

6. Rapidly-Scanning Laser Doppler Anemometry

As pointed out above the laser Doppler anemometer (LDA) is widely used today since it can measure the direction and magnitude of the instantaneous flow velocity accurately and nonintrusively. Commercial LDA optical systems are available with specially designed slow-moving traversing mechanisms to obtain point-by-point measurements in large wind tunnels. Although a rapidly scanning laser Doppler anemometer has various useful applications, the work by this author and his colleagues was motivated by the desire to learn more about the flow structure of a separated turbulent flow from almost instantaneous velocity profiles. The objective of their work has been the development and use for wind tunnel experiments of a forward-scattering scanning LDA system that scans along a straight line perpendicular to the mainstream velocity component and is capable of scanning over distances of 40 cm or longer at high scan frequencies. Before describing rapidly-scanning optical systems, some slow-scan and discrete multi-point systems will be mentioned, which are basically slowly traversed point-by-point measurement systems.

a. Slow-Scan and Multi-Point LDA Systems

There have been a few scanning LDA systems designed and used in the past. The proper operation of these systems was either limited to low frequencies or short scan ranges. Bendick (1971) described an on-axis scanning LDA system that used a translational oscillating mirror. Measurements were made across a 6 mm I.D. circular tube with a scan speed of the measurement volume of approximately 0.4 m/sec. The proper operation of this design at higher scan speeds and larger scan ranges is limited because of the inertia of the moving optics.

A two-color dual-beam backscatter scanning LDA system was reported by Grant and Orloff (1973). The scan was accomplished by translating a focusing lens in the direction of the optical axis. Scan ranges from 60 cm to 200 cm with variable scan speeds of the measurement volume from less than 1 cm/sec up to 1.5 m/sec were reported. Faster scan rates are also limited for this translational system because of inertial considerations. For more information concerning the application of this design see Orloff and Biggers (1974) and Orloff et al. (1975).

A backscatter scanning system was reported by Rhodes (1976) with a focal length of 3 m, a scan range of 30 cm, and a scan frequency of 30 Hz. Velocities at 16 discrete positions along the optical axis were measured using a large rotating wheel containing 16 ports, 15 of which held glass plates of different thickness that produced different effective focal lengths. The residence time at each position was 2 msec. For more information see Gartrell and Jordan (1977) and Meyers (1979).

All of the above scanning LDA systems scan along the optical axis which bisects the angle between incident laser beams. The design of Durst et al. (1981) uses a relatively large rotationally-oscillating mirror in front of a conventional LDA backscatter optical system to scan the measurement volume perpendicular to the incident laser beams along an arc. The image of the measurement volume for all the positions falls on one physical point on the photomultiplier tube. The position transducer and PM-tube signals were simultaneously digitized by a two-channel transient recorder which was triggered only when a Doppler burst was present. The mean and rms velocity profiles taken by the scanning system agreed with the pointwise measurements for low scan frequencies. The inertia of the scanner mirror limited scan frequencies to about 15 Hz. Cheung and Street (1986) also used a relatively large rotationally-oscillating mirror in front of a conventional 2 velocity component backscatter LDA (Figure 9) to make scanning measurements along an arc at 1 Hz, although measurements at higher scan rates were possible.

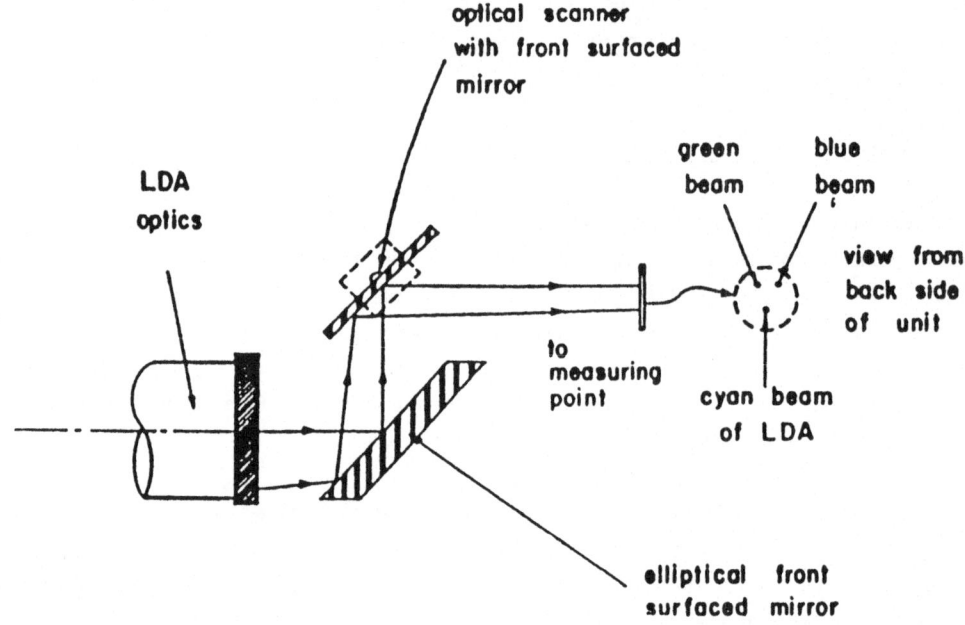

Figure 9 Optical wave-follower set up. Cheung and Street (1986)

A multiple-measurement-point optical system has been reported by Nakatani et al. (1978) which is capable of measuring instantaneous velocity profiles. In contrast to the above-mentioned designs, there is no moving or scanning device. Spherical lenses are used to increase the beam diameter of the two horizontal incident laser beams. After passing through a converging cylindrical lens, these intersecting beams form a vertical measurement volume along a straight line. The scattered light from this vertical measurement volume is received by a series of optical fibers connected to individual photodetectors, each collecting data simultaneously from a different position of the measurement volume. Nakatani <u>et al.</u> (1985 a, b, c; 1986) also have built multi-point LDA systems that have employed Bragg cells and diffraction gratings, or combination mirrors, or multi-branched fiber optics to produce a number of separate measurement volumes. Multiple detectors with multiplexed signals or a mechanical scanning switch were employed. These designs would be expensive to construct for a large scan range with good spatial resolution and sufficient laser power, so that these designs are relatively impractical for most applications.

b. <u>One-Velocity-Component Rapidly-Scanning Systems</u>

A main requirement of the optical design described by Chehroudi and Simpson (1983, 1984) was to have rapid predictable measurement volume movement over a distance of 40 cm for measuring the component of the instantaneous velocity that is perpendicular to the scan direction. In this case the two incident laser beams form fringes parallel to the scan direction so that the measured flow velocity is independent from the velocity of the scanning measurement volume. To avoid the severe scanning inertial limitations of previous investigations, a relatively small but rapidly moving beam deflection device was required for both beams. Only one moving part was used, thus minimizing the number of moving components, avoiding operational and synchronization problems, and keeping the design simple and inexpensive.

Figure 10 is a schematic diagram of the transmitting optics. An Argon-Ion laser with 1.4 watts at 514.5 nm. was used. The spherical lens combination L1 and L2 was

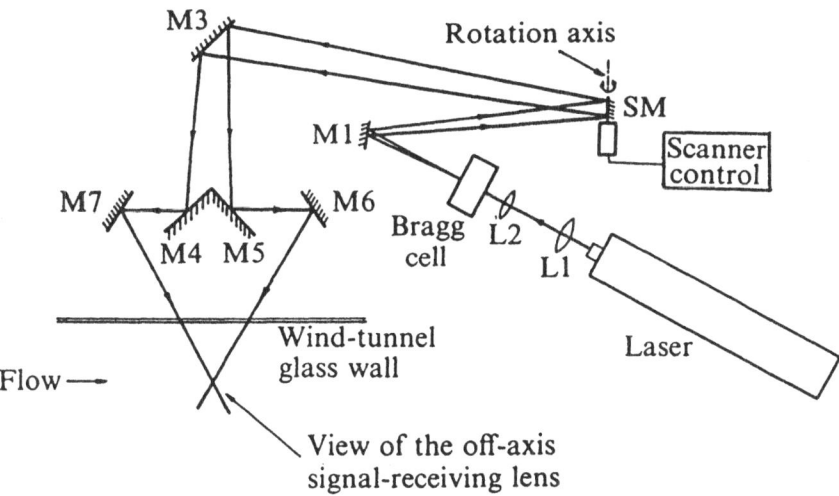

Figure 10 Schematic diagram of incident beam optics. M1-M7, fixed mirrors; L1-L2; lens; SM, scanning mirror. Chehroudi and Simpson (1984)

employed to decrease the diameter of the parallel-ray beams to 2 mm at the measurement volume. To make the system directionally sensitive a Bragg cell with f_B = 15.2 MHz was used to split and frequency shift the incoming laser beams.

A small oscillating galvanometer scanner with open-loop control was used. The reflected beams from the scanner mirror SM move perpendicular to the plane of Figure 10. Consequently the scanning action of the probe volume is in that same direction. The scanner control unit has a voltage output proportional to the angular position of the scanner mirror, which in this design is also proportional to the position of the measurement volume. In the experiments reported by Chehroudi and Simpson (1983, 1984, 1986) the scanner was operated with its internal sawtooth waveform for angular position vs. time. With this waveform there are no inertial limitations for scan frequencies up to 150 Hz. For a sinusoidal waveform, the maximum scan frequency for an adjustable frequency scanner is about 300 Hz. The performance of this scanner is described further in section 6.e below dealing with signal processing. Resonant oscillating galvanometer scanners with sinusoidal waveforms at fixed frequencies of up to 4 KHz are commercially available at reasonable cost.

Five relatively long front surface mirrors were used for mirrors M3 to M7 to permit large scan distances. The mirror flatness and the method of attachment of these mirrors was found to be a critical aspect of this design so that the two beams cross and produce a well-defined measurement volume at all scanning volume positions (Chehroudi, 1983). For a simple flatness test, a horizontal laser beam was reflected from various points on each mirror to a graph paper target a long distance away as the mirror mount was precisely traversed in a vertical direction. It was found that hanging the mirrors from their top edges on the mounts minimized flatness changes, especially changes due to room temperature variations. For inexpensive mirrors the minimum limit for the intersecting beams diameters was about 2 mm so that the two beams cross and form a well-defined measurement volume for any position within the scan range. In order to reduce this minimum diameter to about 3/4 mm, coated front surface mirrors on thick substrates were obtained that are flat to within a fraction of a wavelength per cm of length. It is worth mentioning that the intersection angle between the beams can be varied easily by mirrors M6 and M7 to select the best fringe spacing for the type of seeding material used.

As shown in Figure 11, the receiving optics consist of two cylindrical lenses CL1 and CL2, a long rectangular mirror M8 to reflect the light, and a photomultiplier tube with an adjustable width slit. The receiving optics which are located off-axis of the incident beams are covered by a light-absorbing enclosed frame to avoid stray scattered light from outside the measurement volume.

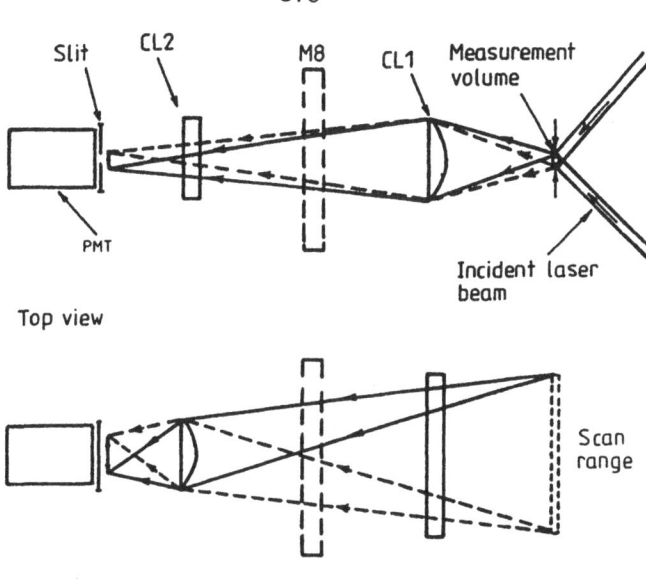

Figure 11 Schematic of ray paths through the receiving optics. M8 is ray-turning mirror. Chehroudi and Simpson (1984)

The first cylindrical lens CL2 focuses the image of the streamwise dimension of the probe volume onto the slit. As also shown in Figure 11, a 2:1 enlargement of the image is achieved, making the system performance more insensitive to the slit width adjustment. The second cylindrical lens CL2 is used to collect the scattered light from the measurement volume at all scanning positions and focus it onto the slit. For a 12 cm movement of the measurement volume, the image moves 1 cm along the slit. A photomultiplier tube with a 46 mm diameter end window was used. The position of each component of the optics is adjustable.

There are specific advantages of this design, such as: the ability to vary the fringe spacing from 1 μm to 4 μm; inexpensive optical components; straight line scans rather than along an arc; system insensitivity to slit width and vibration; and the capability to scan up to 150 Hz (Chehroudi, 1983).

Owen (1984) developed a single velocity component scanning LDA and made measurements in both air and water flows. A six-sided rotating mirror polygon was used although the scan rate was limited to 125 Hz due to restrictions on the data acquisition rates and the number of points required for one profile.

An oscillating mirror scanner is superior to the much more expensive rotating mirror polygon scanner even though the latter can achieve adjustable scan frequencies of up to (500 N) Hz, where N is the number of sides or facets of the polygon and may be as large as 48. This apparent advantage is more than offset by disadvantages. The rotating polygon has many different mirror facets with minute individual imperfections that lead to differences in reflected beam positions. The reflected beams are in unwanted directions 90% of the time. The oscillating mirror scanner reflects the laser beams from a single mirror surface that is close to the axis of rotation and can produce no dead time of the laser beams outside the measurement region of interest. Hino et al. (1986) support this conclusion.

Econonou (1986) and his advisor Williams have extended the design of Chehroudi and Simpson to permit scans in the plane perpendicular to the mainstream flow velocity. As shown in Figure 12, the light beam, after leaving the laser passes through two bi-convex lenses TL1 and TL2, in order to collimate and reduce its diameter at the location of the measuring volume. Next it goes through the Bragg cell where it is split into two equal intensity beams, one of which is frequency shifted by 40 MHz. The beams coming out of the Bragg cell diverge at a 0.2 degree angle. To increase their separation distance, the two beams are first reflected by two prisms P1 and P2 at right angles to the optical axis, and are then reflected again by the two mirrors M1 and M2 in order to approach the first scanner mirror at a sufficiently large angle.

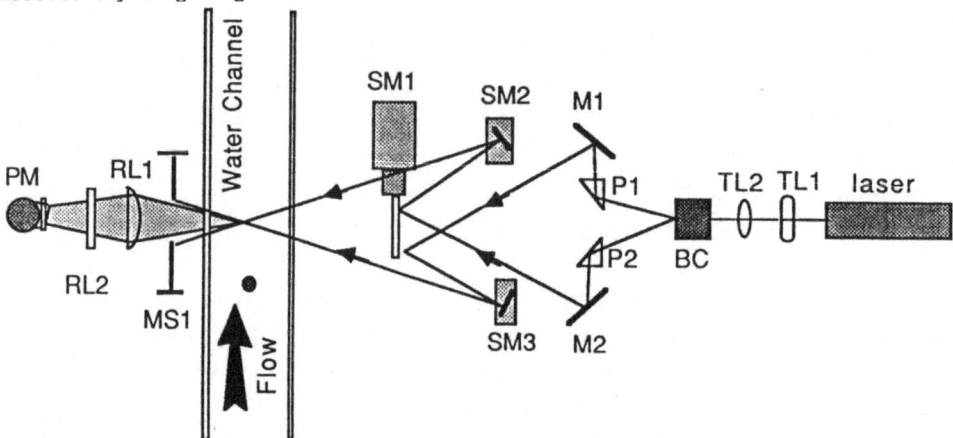

Figure 12 Optics set up of Economou (1986). Notation: TL - transmitting lens; BC - Bragg cell; P - prism; M - mirror; SM - scanning mirror; TS - test section; RL - receiving lens; PM - photomultiplier tube.

Next the beams are reflected from the oscillating mirror scanner SM1 and subsequently are individually reflected from the oscillating mirror scanners SM2 and SM3 to form the actual measurement volume at a new point of intersection in the flow field. The scanners are like the one used by Chehroudi and Simpson with the same capability. Scanner mirror SM1 moves the beams so that the measurement volume moves at right angles with respect to the flow direction, enabling the system to make measurements along a line in the y-direction which is out of the page in Figure 12. The two scanning mirrors SM2 and SM3 change the angle of the intersecting beams.

The forward-scatter receiving optics are very similar to those used by Chehroudi and Simpson. A mask MS1 is used to block the incident beams and two cylindrical lens RL1 and RL2 are used to focus the measurement volume image on a slit in front of the PM tube. A short-coming of their design is that signals (and noise) are received from all along the long probe volume that is produced by the crossing incident beams. This disadvantage must be tolerated if the receiving optics are to receive signals from anywhere in the measurement plane without using a moving optical component. Chehroudi and Simpson used off-axis receiving optics (Figure 10) that better defined the length of the measurement volume and improved spanwise spatial resolution of measurements, but they were only scanning along a single line.

An improvement in spanwise spatial resolution would result if a scanner mirror SM4 synchronized with SM2 and SM3 were used to reflect off-axis scattered light into the receiving optics. Normally SM2 and SM3 are scanned at a lower rate than SM1, so SM4 (Figure 13) could use a large mirror to collect more scattered light. Large f/number receiving optics need to be used to keep the image of the measurement volume in focus over the spanwise distance.

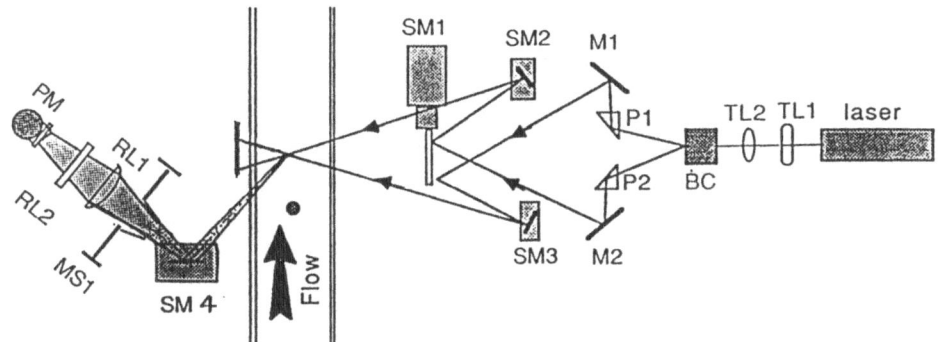

Figure 13 Receiving optics with scanner SM4 for scanning across test section.

c. A Three-Velocity-Component Rapidly Scanning Optical System

Based on the one-velocity-component design of Chehroudi and Simpson, a three-velocity-component transmitting optics system has been developed by the author and his students. Figure 14 shows a top-view schematic of the transmitting optics of this three-velocity-component system. A 5 W argon ion laser with 514.5 nm wavelength is used. The laser beam passes through a lens combination L_1 and L_2 to decrease its diameter at the measurement volume, and is split and frequency-shifted by a dual transducer Bragg cell to have a directionally-sensitive system. The four outcoming beams from the Bragg cell are directed towards a rotationally-oscillating scanner mirror SM by a small mirror M1. Two beams are in a vertical plane shown as the middle beam that leaves the Bragg cell in Figure 14. To equalize the path lengths of these two beams and the other two to the measurement volume, additional mirrors between M1 and SM are used for the central two beams. The beams are then reflected to the very flat last-stage mirrors M4 - M7 by a long fixed mirror M3. As the scanner mirror SM oscillates the two beams in the vertical plane cross along an arc whereas the other two beams cross along a straight line perpendicular to the plane of Figure 14.

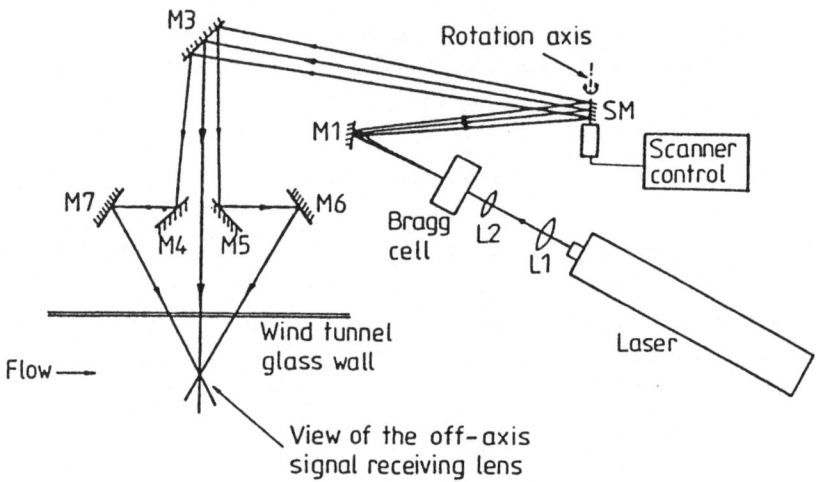

Figure 14 Schematic diagram of incident optics. Three-velocity component rapidly-scanning LDV.

This does not present a problem because the incident beams have slowly convergent rays and the beams in the vertical plane are at a small angle and have a long crossing volume. As long as a portion of this volume is coincident with the beam crossing volume from the other two beams, the off-axis receiving optics of Chehroudi and Simpson described above can be used to define the measurement volume which moves perpendicular to the plane of Figure 14.

Figure 15 shows the four beams which cross in the measurement volume to form six moving fringe patterns, which are employed to make three-velocity-component measurements. This figure shows the fixed angle α between beams A and D and beams A and B, the fixed angle β between beams A and C, and the instantaneous angle γ that the plane of beams A, B, and D makes with the U, W plane. In terms of these angles the following equations are the principal relations for determining the velocity components.

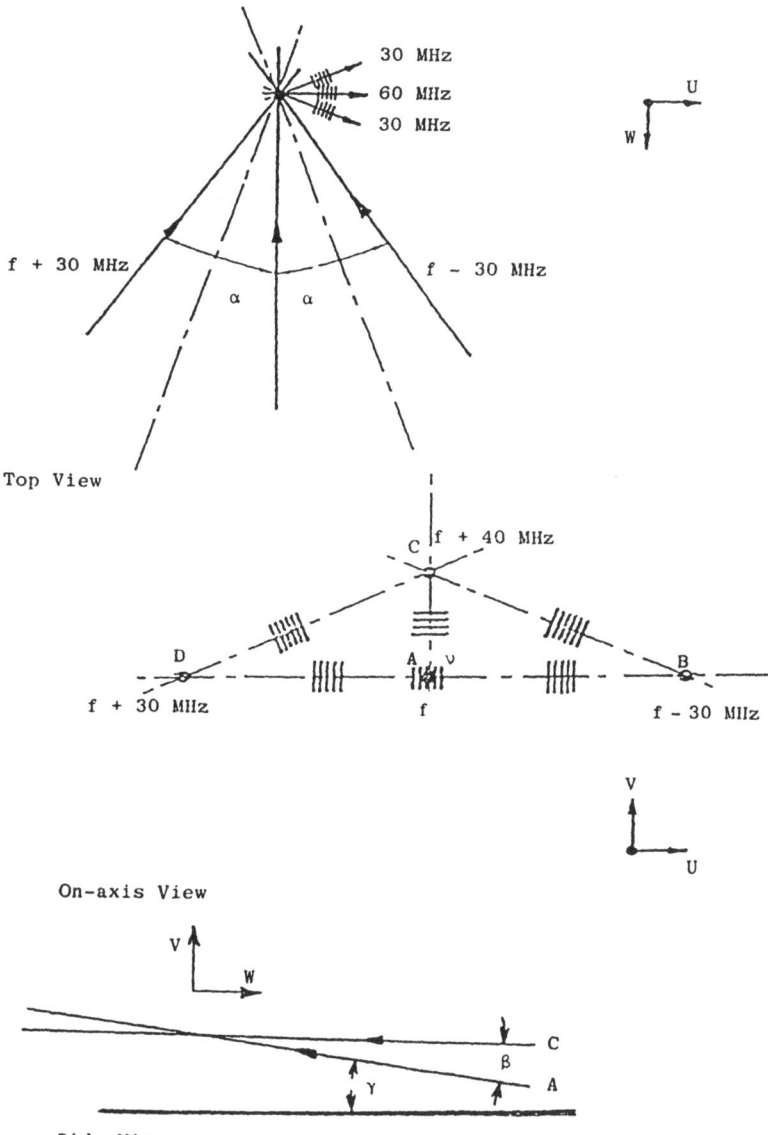

Figure 15 Beam arrangement, three-velocity component rapidly-scanning LDV.

$$\lambda_{BD}(f - 30 \text{ MHz}) = - U \tag{1}$$

$$\lambda_{AD}(f - 15 \text{ MHz}) = - U \cos (\alpha/2) - W \cos (\gamma) \sin (\alpha/2) \\ + V \sin (\alpha/2) \sin (\gamma) \tag{2}$$

$$\lambda_{AB}(f - 15 \text{ MHz}) = - U \cos (\alpha/2) + W \cos (\gamma) \sin (\alpha/2) \\ - V \sin (\alpha/2) \sin (\gamma) \tag{3}$$

$$\lambda_{AC}(f - 21.5 \text{ MHz}) = V \cos (\gamma - \beta/2) - V_s + W \sin (\gamma - \beta/2) \tag{4}$$

where

$$\lambda_{BD} = \lambda/2 \sin \alpha$$
$$\lambda_{AD} = \lambda_{AB} = \lambda/(2 \sin (\alpha/2))$$
$$\lambda_{AC} = \lambda/(2 \sin (\beta/2))$$

The signal frequency f around 30 MHz is used to determine U directly from equ. (1) as in a one-velocity-component system and is independent of γ and β. The frequency difference between the two signals around 15 MHz is related to the V and W velocity components as given by the difference between equs. (2) and (3). γ is small and can be accurately determined by the scanner position. Because of the scanning velocity V_s of the measurement volume, an additional frequency shift of V_s/λ_{AC} is produced as reflected in equ. (4). Using the experimentally determined values of V_s and γ, V and W can be determined from equs. (2-4). Other moving fringe patterns occur near 6.5 MHz and 36.5 MHz, are related to U, V, and W, and can be used to check the values determined by equ. (1-4).

All velocity component signals are contained in the one photomultiplier tube signal. By careful selection of the fringe spacings and the direction of scan, the ambiguity of having signals in overlapping frequency domains can be avoided. For example, if λ_{BD} = 3.1 µm, then unambiguous U-velocity-component frequencies between 25 MHz to 42 MHz with a 30 MHz frequency shift can be obtained, i.e., U-component-velocities between -17 mps and +37 mps. Above this velocity, signals between beams BC and BD would overlap. V-velocity-component frequencies between 20 MHz and 25 MHz can be maintained by scanning in the positive V direction. Scanning in the opposite direction will lower the signal frequency range. The signals containing the W-velocity-component are less than 15 MHz and should not overlap with the near 6.5 MHz signals produced by beams C and D.

In a recent development, we use an electro-optic laser beam deflector (Pockel cell) just before the Bragg cell (not shown in Figure 14) which does not introduce additional frequency shifting. This permits us to choose between two parallel scan lines that are separated ΔX (about 6 mm) in the free-stream flow direction. By synchronizing this deflector with the up-and-down scanning of the measurement volume, we are able to scan up (or down) at one X_1 location while scanning down (or up) during the retrace at the other X_2 location. This permits us to obtain information on streamwise changes in the flow structure. Two separate slits and two photodetectors are used for collecting signals from the two X locations.

d. Some Theoretical Considerations

i. Number of fringes crossed by a particle in a scanning LDA

It is important that a scanning LDA be operated such that each scattering particle crosses a sufficient number of fringes to generate a valid signal for a signal processor. A simple equation can be derived to represent the number of fringes crossed by a particle as it passes through the measurement volumes. This analysis is based on the assumptions that the probe volume is two dimensional and circular in shape, the particle motion is dominated by its streamwise velocity component U, and the particle is smaller than the fringe spacing.

The number of fringes that a particle crosses is equal to the residence time for the particle in the measurement volume divided by the time for the particle to travel one fringe spacing, which is given by the following equation (Chehroudi and Simpson, 1984).

$$\frac{N_u \lambda_{BD}}{d} = \frac{-\left| \frac{U}{V_s} - \frac{U_f}{V_s} \right| \cos(\theta + \phi)}{\left[(\frac{U}{V_s})^2 + (1 - \frac{V}{V_s})^2 \right]^{1/2}} \tag{5}$$

Here N_u is the number of fringes crossed, d is the effective diameter of the probe volume from which signals are obtained, λ_{BD} is the fringe spacing, V is the normal velocity, and $U_f = f_B \lambda_{BD}$ and V_s are the velocity with which fringes appear to move and the upswing scan velocity, respectively. As shown in Figure 16, θ is the angle between the particle relative velocity and that of main flow directions, while ϕ is the angular position of a particle entering the measurement volume. In order to capture signals from particles moving in all directions, $U_f \geq 2U$ (Durst et al., 1981), which restricts the useful velocity range for a scanning LDA.

For any particle entering from the position angle ϕ, the non-dimensional factor of $\frac{N_u \lambda_{BD}}{d}$ attains its maximum value when $\cos(\theta + \phi)$ is equal to -1, which physically means that the particle passes through the center of the measurement volume. Figure 16 shows the maximum of $\frac{N_u \lambda_{BD}}{d}$ versus U/V_s for three different U_f/V_s ratios. As shown in Figure 16 the maximum number of fringes crossed by a particle decreases as the scan velocity increases.

The maximum number of cycles per signal burst used by the counter type processor is 32. This simplified model was used by Chehroudi and Simpson (1984) to insure that there was a sufficient number of fringes crossed for various scan speeds and velocities in their boundary layer. For example, with a maximum streamwise velocity of 17 m/sec, a fringe spacing of 3.1 μm, a fringe velocity of 47 m/sec and a small effective streamwise measurement volume dimension of 3/4 mm viewed by the slit, a

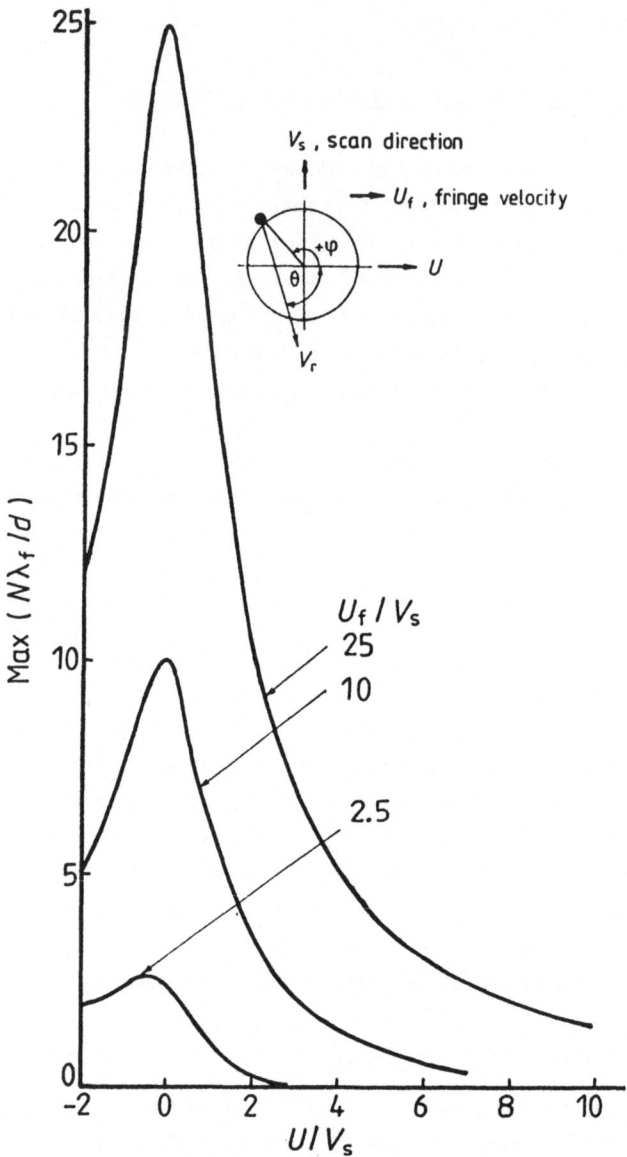

Figure 16   Number of fringes crossed by a particle as a function of U_f/V, and U/V, for $V = 0$; from equation (5). Chehroudi and Simpson (1984)

particle will cross no more than 32 fringes with an upper limit scan speed of approximately 200 m/sec. To use a higher scan velocity and still obtain a sufficient number of fringe crossings, a higher Bragg cell frequency should be used.

For a multi-velocity-component scanning system Antoine and Simpson (1986) examined the number of horizontal fringes cut by a particle when no Bragg cell frequency shift was used between beams A and C in Figure 15. When a Bragg cell frequency shift is used between beams A and C, the number of fringes N_v cut by a particle is given by the following equation that was derived by the same reasoning used to obtain equation (5).

$$\frac{N_v \lambda_{AC}}{d} = \frac{-\left| \frac{V}{V_s} - 1 + \frac{V_f}{V_s} \right| \cos(\theta + \phi)}{\left| (\frac{U}{V_s})^2 + (1 - \frac{V}{V_s})^2 \right|^{1/2}} \qquad (6)$$

Here N_v is the number of fringes crossed, $V_f = \lambda_{AC} f_{BV}$, and f_{BV} is the Bragg shift frequency for beam C. To capture signals from particles moving in all directions $V_f \geq 2(V_s - V)$.

ii. <u>Velocity Bias Considerations</u>

According to the velocity biasing analysis of McLaughlin and Tiederman (1973), the data rate in a pointwise or stationary measurement volume LDA system is proportional to the magnitude of the instantaneous velocity vector. In such a case, more high velocity signals are obtained and an average of these signals would produce mean and rms velocity values that are biased or in error a small amount. Similarly, the data rate in a scanning LDA system in which the measurement volume itself moves is proportional to the magnitude of the instantaneous <u>relative</u> velocity of the fluid with respect to the measurement volume (V_r in Figure 16).

Based on the biasing analysis of Buchhave, George and Lumley (1979), the percentage biasing error decreases as the parameter Q and the turbulence intensity $\sigma/\bar{U}$ decrease (see their Fig. 3-5). The Q parameter is the ratio of the minimum number of zero crossings required for a valid measurement to the maximum number of fringe crossings available ($0 \leq Q \leq 1$). Q has a relatively weak effect on the percentage bias error, especially for $\sigma/\bar{U} < 0.2$. With the same σ or rms velocity fluctuations, the biasing effect decreases using a scanning LDA system since the <u>relative</u> velocity of the fluid with respect to the moving measurement volume V_r should be used in the turbulence intensity instead of the mean flow velocity and since $V_r > U$. Consequently, the biasing effect decreases as the scan velocity increases. From another viewpoint, since signals present at a given spatial location are sampled at equal time intervals due to scanning, such a method of signal sampling of an ergodic flow leads to no bias of statistical averages of turbulence phenonena (Simpson and Chew, 1979; Adams <u>et al.</u>, 1984).

For example, the required number of zero crossings for a burst counter measurement was 32 and the effective total number of fringes available was about 250 for the work of Chehroudi and Simpson. The Q factor mentioned above was about 0.1 and had little effect on the bias error. Should a high scanning velocity cause Q

to be unacceptably large, the Bragg cell frequency should be increased to produce a larger maximum number of fringe crossings available.

e. Data Acquisition Methods and Some Experimental Results

For a scanning laser anemometer system one needs to know the exact position of the measurement volume at the time when a given validated velocity signal is detected. Several approaches for obtaining this information simultaneously are possible. The most straightforward approach is to use two A/D converters of the analog velocity and scanner mirror position signals or use digital signals directly and record simultaneously sampled data. Chehroudi and Simpson (1984) found that the switching time between acquiring data from two separate channels was significant for their computer. During this switching time no valid velocity data can be recorded. This problem is more severe with more rapid scan velocities. In an alternate approach Chehroudi and Simpson (1984) multiplexed the scanner orientation signal which is proportional to the measurement volume position and the velocity signal to eliminate the time lost when switching from one signal to the other.

Figure 17 is a block diagram of the data acquisition equipment used by Chehroudi and Simpson while Figure 18 shows several signals used in the multiplexed data acquisition and processing logic. This data processing method relies heavily on the linearity of the upsweep ramp of the scanner internal sawtooth waveform (Figure 18a)(Chehroudi, 1983; Brosens, 1976). To trace each cycle of scanning, a pulse marker is used with a voltage level exceeding the A/D saturation voltage. As shown in Figure 18c, this pulse marker is a stretched SYNC pulse from the scanner drive unit. It is added to the analog output of the LDA burst counter processor and the result is sampled by a fast A/D converter 1 with controllable sampling period, T, from a pulse generator (see Figures 17 and 18). The sampling period T is adjusted such that at least one and at most two A/D samples are obtained during the stretched SYNC pulse. Since the height of this pulse is always greater than the saturation voltage level of the A/D converter, one has an eventmarker for each scan cycle in the data record.

A statistical analysis of the internal sawtooth waveform (Figure 18a) of the scanner drive unit indicated that the position voltage in the upsweep ramp section of the waveform is repeatable at any given phase in each cycle and does not vary more than ±2% of the peak-to-peak scanner voltage. In addition this analysis determined the slope and intercept of the upsweep ramp section of the least-squares fit to the ensemble-averaged waveform (Figure 18e) that is used with the position calibration line (Figure 18d).

389

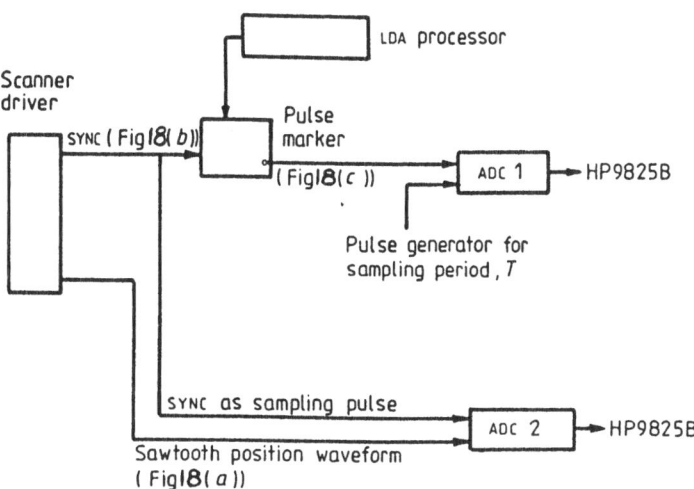

Figure 17 Block diagram of data-acquisition equipment. Pulse marker circuit stretches scanner SYNC pulse and adds analogue output from counter. Chehroudi and Simpson (1984)

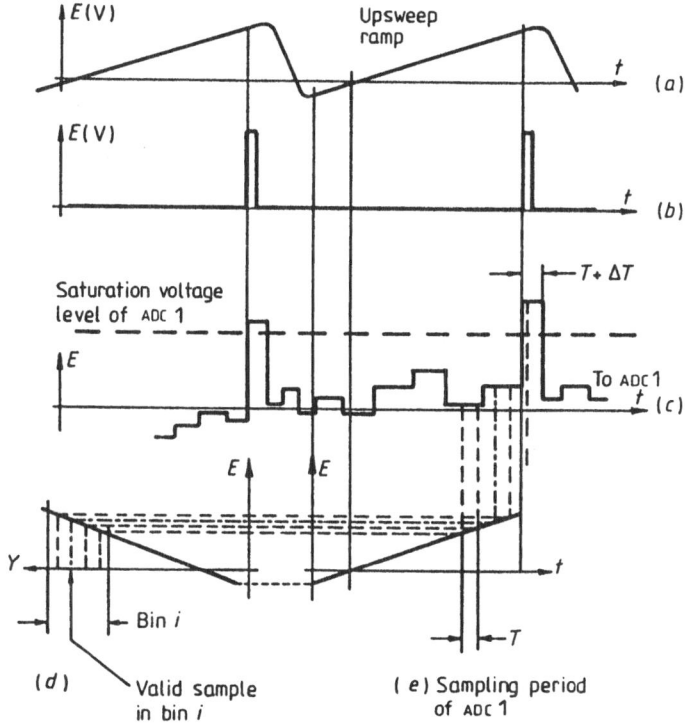

Figure 18 Data-acquisition and signal-processing signals: (a) scanner internal sawtooth waveforms; (b) SYNC pulse from scanner; (c) sum of analogue output from LDA processor and the stretched SYNC pulse; (d) scanner voltage plotted against measurement volume position calibration line; (e) sampled data points from (c) with corresponding position voltages on the linear least-square fit waveform.

This calibration line was obtained by placing an opaque mask along the scanning measurement volume path with small slits at a number of known y locations. Thus, the resulting PM-tube signal contained pulses that could be used to trigger sampling of the scanner voltage to produce the E vs. y calibration shown in Figure 18d.

Although the shape, frequency, and E vs. y calibration of the sawtooth waveform are quite stable, the voltage level of the waveform (Figure 18a) may drift slightly up or down. This means that the intercept of the voltage versus time line in Figure 18a can change. The value of this intercept at the time when each data record is taken is important for accurately determining the position. For this reason a second A/D converter, which is externally triggered by the SYNC pulse from the scanner driver, is used to monitor intermittently this voltage drift.

The total scan distance range is subdivided into a number of position bins. Signals generated at measurement volume positions within each bin are grouped together for statistical averaging purposes. The uncertainty in measurement volume position ΔY contributes an apparent turbulence fluctuation proportional to (Δy) $(\partial U/\partial y)$, where U is the instantaneous velocity. The computer data acquisition program logic determines new counter validated samples when the signal voltage changes for two consecutive computer samples (see Figure 18c). Therefore, the maximum position uncertainty for each signal is determined by the sampling period and is equal to the distance traveled by the measurement volume during that time. Since the vertical dimension of the measurement volume in the Chehroudi and Simpson case was about 2 mm, the sampling period T for each scan speed was chosen to be about 100 μsec such that the distance traveled in that period was less than 2 mm. With such a measurement volume size there is little point in making the measurement volume position uncertainty much smaller than the particle position uncertainty within the measurement volume. Such measurement volume position uncertainties can be almost eliminated by using a high sampling rate A/D converter, say 1 Msample/sec, although the computer memory is filled much faster.

Since an LDA burst counter signal processor can only produce valid results for high SNR signals, it is of decreasing value in a rapidly-scanning system because the SNR and the number of cycles in a burst decrease with scanning rate. The data rate limitations of a scanning LDA is not entirely due to the lack of proper seeding. As the measurement volume scans the flow the processor receives variable signal-to-noise ratio signals due to the probe volume movement and variations in seeding density as noted by Orloff et al. (1975). Thus, unlike pointwise measurements where the processor can be adjusted for that signal alone, in scanning measurements the processor must be adjusted to accommodate signals from all across the scanned flow. Using a sinusoidal scanner waveform instead of a sawtooth would permit the measurement volume to spend more time at the extremes of the scan range and improve the data rates there. Since this scanner can produce a repeatable sinusoidal waveform up to 300 Hz, a much higher scan rate is possible.

Currently, the author and his colleagues are using a 200 Msample/sec transient recorder with a 16 Mbyte memory to acquire multiplexed burst signals from the photodetector and scanner mirror orientation SYNC pulses. Poor SNR signals can be processed in the frequency domain to produce a higher data rate and obtain almost instantaneous velocity profiles with 30 to 60 points per scan. At a scan rate of 150/sec one can measure entire velocity profiles with statistically low uncertainty mean and rms values during less than 1 minute of data acquisition. Even with a low data rate of 500 sec^{-1}, space-time correlations all across a separating turbulent boundary layer were obtained (Chehroudi and Simpson, 1985). Several almost instantaneous velocity profiles were obtained, demonstrating the potential of rapid-scanning LDA as improved signal processing produces higher data rates.

7. Concluding Remarks

Laser anemometry and rapidly-scanning LDA, in particular, are powerful velocity measuring techniques that should be developed and exploited further to provide needed information on the structure of turbulent separated flows. Continued improvements in the acquisition and processing of low SNR signals are needed. While traditional hot-wire anemometry has a less useful future for separated flows, pulsed-wire anemometry appears to be useful, especially for low velocity flows or at least low velocity backflows.

ACKNOWLEDGMENTS

The author would like to thank the Office of Naval Research and the Air Force Office of Scientific Research for support of his rapidly-scanning LDA research. Dr. W.J. Devenport and Mr. K.A. Shinpaugh of VPI&SU provided constructive comments on this manuscript.

REFERENCES

Adams, E.W., Eaton, J.K. and Johnson, J.P. (1986), "An Examination of Velocity Bias in a Highly Turbulent Separated and Reattaching Flow," pp. 21-37, Laser Anemometry in Fluid Mechanics - II, Selected Papers of Durão et al. (1984).

Antoine, M. (1985) "A Rapidly Scanning Three-Velocity-Component Laser Doppler Anemometer," M.S. Thesis, Aerospace and Ocean Engineering Department, VPI&SU.

Antoine, M. and Simpson, R.L. (1986), "A Rapidly Scanning Three-Velocity-Component Laser Doppler Anemometer," J. Physics E: Sci. Inst., 19, pp. 853-858.

Bachalo, W.D., Modarress, D. and Johnson, D.A. (1977), "Experiments on Transonic and Supersonic Turbulent Boundary Layer Separation," AIAA-77-47, 15th Aerospace Sciences Meeting.

Benedick, P.J. (1971), "A Laser Doppler Velocimeter to Measure Instantaneous Velocity Profiles," Proc. Flow Symposium, May 10-14, Pittsburgh.

Blinco, P.H. and Sandborn, V.A. (1973), "Use of the Split-Film Sensor to Measure Turbulence in Water Near a Wall," pp. 403-413, Proc. of Third Symposium on Turbulence in Liquids, G.K. Patterson and J.L. Zakin, editors, Dept. Chem. Engrg., Univ. Miss.-Rolla.

Bradbury, L.J.S. (1976), "Measurements with a Pulsed-Wire and a Hot-Wire Anemometer in the Highly Turbulent Wake of a Normal Flat Plate," J. Fluid Mech., 77, pp. 473-497.

Bradbury, L.J.S. and Castro, I.P. (1971), "A Pulsed-Wire Technique for Velocity Measurement in Highly Turbulent Flow," J. Fluid Mech., 49, pp. 657-691.

Brosens, P. (1976) "Scanning Accuracy of the Moving-Iron Galvanometer Scanner," Optical Engineering, 15, No. 2, pp. 95-98.

Browand, F.K. and Weidman, P.D. (1976), "Large Scales in the Developing Mixing Layer," J. Fluid Mech., 76, pp. 127-144.

Brown, R.G.W. (1986), "Combined Transform Doppler Anemometry," pp. 39-50, Laser Anemometry in Fluid Mechanics - II, Selected Papers of Durão et al. (1984).

Buchhave, P. (1975), "Laser Doppler Velocimeter with Variable Frequency Shift," Optics and Laser Tech., pp. 11-16, February 1975.

Buchhave, P., George, W.K. and Lumley, J.L. (1979), "The Measurement of Turbulence with the Laser-Doppler Anemometer," Annual Review of Fluid Mechanics, 11, pp. 443-503.

Castro, I.P. (1971), "Wake Characteristics of Two-Dimensional Perforated Plates Normal to an Airstream," J. Fluid Mech., 46, pp. 599-609.

Castro, I.P. and Cheun, B.S. (1982), "The Measurement of Reynolds Stresses with a Pulsed-Wire Anemometer," J. Fluid Mech., 118, pp. 41-58.

Castro, I.P. and Dianat, M. (1987), "A Pulsed Hot-Wire Probe for Near Wall Measurements," 6th Symposium on Turbulent Shear Flows, Toulouse, France, September 7-9.

Castro, I.P., Dianat, M. and Bradbury, L.J.S. (1987), "The Measurement of Fluctuating Skin Friction with a Pulsed Wire Wall Probe," Turbulent Shear Flows V, p. 278, ed. F. Durst, B. Launder, F. Schmidt, J. Whitelaw, Springer-Verlag.

Castro, I.P. and Haque, A. (1987), "The Structure of a Turbulent Shear Layer Bounding a Separation Region," J. Fluid Mech., 179, pp. 439-468.

Chehroudi, B. and Simpson, R.L. (1985), "Space-Time Results for a Separating Turbulent Boundary Layer Using a Rapidly-Scanning Laser Anemometer," J. Fluid Mech., 160, pp. 77-92.

Chehroudi, B. and Simpson, R.L. (1984), "A Rapidly Scanning Laser Doppler Anemometer," J. Physics E: Sci. Inst., 17, pp. 131-136.

Chehroudi, B. and Simpson, R.L. (1983), "Scanning Laser Doppler Anemometer and Its Application in Turbulent Separated Flow," Report WT-7, Dept. Civil/Mechanical Engrg., Southern Methodist Univ., Dallas, TX 75275; DTIS Report.

Cheung, T.K. and Street, R.L. (1986), "A Wave-Following Laser Doppler Anemometer," Laser Anemometry in Fluid Mechanics - II, Selected Papers from the 2nd Inter. Symp. on Appl. Laser Anemometry to Fluid Mech., Adrian et al., editors, pp. 123-140, LADOAN - Instituto Superior Tecnico, Lisbon.

Coles, D. and Wadcock, A.J. (1979), "A Flying Hotwire Study of Two-Dimensional Mean Flow Past an NACA 4412 Airfoil at Maximum Lift," AIAA Journal, 17, pp. 321-329.

Comte-Bellot, G. (1976), "Hot-Wire Anemometry," Annual Review of Fluid Mechanics, 8, pp. 209-231, M. Van Dyke, et al., editors, Annual Reviews Inc., Palo Alto, California.

Crabb, D., Durao, D.F.G. and Whitelaw, J.H. (1981), "Velocity Characteristics in the Vicinity of a Two-Dimensional Rib," Third Symposium on Turbulent Shear Flows, pp. 16.5-16.10.

Cummins, H.Z. and Pike, E.R., editors (1977), Photon Correlation Spectroscopy and Velocimetry, Plenum Press, New York (NATO Advanced Study Institute, July 26-August 6, 1976).

Délery, J.M. (1983), "Experimental Investigation of Turbulence Properties in Transonic Shock/Boundary Layer Interactions," AIAA Journal, 21, pp. 180-185.

Dengel, P. and Vagt, J.D. (1982), "A Comparison Between Hot-Wire and Pulsed-Wire Measurements in Turbulent Flows," Institutsbericht 01/82, Technische Universität Berlin; Fourth Symposium on Turbulent Shear Flows, pp. 15.12-15.16 (1983).

Devenport, W.J. (1985), "Separation Bubbles at High Reynolds Number: Measurement and Computation," Ph.D. dissertation, Cambridge Univ.

Dianat, M. and Castro, I.P. (1983), "Fluctuating Surface Shear Stresses on Bluff Bodies," J. Wind Eng. Ind. Aerodyn., 17, p. 133.

Driver, D.M., Seegmiller, H.L. and Marvin, J. (1987), "Time-Dependent Behavior of a Reattaching Shear Layer," AIAA Journal, 25, pp. 914-919.

Durão, D., Adrian, R., Asanuma, T., Durst, F. and Whitelaw, J.H., editors (1986), Third International Symposium on Applications of Laser-Doppler Anemometry to Fluid Mechanics, LADOAN - Instituto Superior Tecnico, Lisbon, July 7-9.

Durão, D., Adrian, R., Asanuma, T., Durst, F., Whitelaw, J.H., editors (1988) Fourth International Symposium on Applications of Laser-Doppler Anemometry to Fluid Mechanics, LADOAN-Institute Superior Tecnico, Lisbon, July 11-14.

Durão, D., Adrian, R., Durst, F., Mishina, H. and Whitelaw, J.H., editors (1982), International Symposium on Applications of Laser-Doppler Anemometry to Fluid Mechanics, LADOAN - Instituto Superior Tecnico, Lisbon, July 5-7.

Durão, D., Adrian, R., Durst, F., Mishina, H. and Whitelaw, J.H., editors (1984), Second International Symposium on Applications of Laser-Doppler Anemometry to Fluid Mechanics, LADOAN - Instituto Superior Tecnico, Lisbon, July 2-5.

Durbin, P.A., McKenzie, D.J. and Durbin, E.J. (1987), "An Anemometer for Highly Turbulent or Recirculating Flows," Exp. Fluids, 5, pp. 184-188.

Durst, F., Lehmann, B. and Tropea, C. (1981), "Laser Doppler System for Rapid Scanning of Flow Fields," Rev. Sci. Instrument., 52, pp. 1676-1681.

Durst, F., Melling, A. and Whitelaw, J.H. (1981), Principles and Practice of Laser-Doppler Anemometry, 2nd Ed., Academic Press, New York.

Eaton, J.K. and Johnston, J.P. (1980), "Turbulent Flow Reattachment: An Experimental Study of the Flow and Structure Behind a Backward-Facing Step," Report MD-39, Thermosciences Div., Dept. of Mechanical Engineering, Stanford University.

Eaton, J.K., Westphal, R.V. and Johnston, J.P. (1982), "Two New Instruments for Flow Direction and Skin-Friction Measurements in Separated Flows," ISA Trans., 21, pp. 69-78.

Econonou, M. (1986), "Design and Performance of a Scanning Laser Doppler Velocimeter," M.S. Thesis, Mech. Engr., Ill. Inst. Tech., Chicago.

Fuller, W.R. (1974), "Calibration of a Split-Film Sensor," M.S. Thesis, School of Engrg., Univ. Southern California.

Gartrell, L.R. and Jordan, F.J. (1977), "Demonstration of a Rapid-Scan Two-Dimensional Laser Velocimetry in the Langley Vortex Research Facility for Research in Aerial Applications," NASA TM-74081.

Ginder, R.B. and Bradbury, L.J.S. (1973), "Preliminary Investigation of a Pulsed-Gauge Technique for Skin Friction Measurements in Highly Turbulent Flows," ARC 34448.

Grant, G.R. and Orloff, K.L. (1973), "Two-Color Dual-Beam Backscatter Laser Doppler Velocimeter," J. of Applied Optics, 12, pp. 2913-2916.

Hino, M., Nadaoka, K., Kobayashi, T., Sato, Y., Muramoto, T. and Hironaga, K. (1986), "A Beam-Scan Type Laser Doppler Velocimeter for Simultaneous and Continuous Measurement of Velocity Profiles," Third Inter. Symposium on Applications of Laser Anemometry to Fluid Mechanics, Lisbon, July 7-9, Durão et al., editors, Paper 8.5.

Ho, C.M. (1982), "Response of a Split-film Probe Under Electrical Perturbations," Rev. Sci. Inst., 53, pp. 1240-1245.

Hunter, W.W. and Nichols, C.E. (1985), "Wind Tunnel Seeding Systems for Laser Velocimeters," NASA CP-2393.

Jaroch, M. (1985), "Development and Testing of Pulsed-Wire Probes for Measuring Fluctuating Quantities in Highly Turbulent Flows," Expts. in Fluids, 3, pp. 315-322.

Jenson, L., Menon, R.K. and Fingerson, L.M. (1988), "An Automatic Signal Processor for LDV Systems," Fourth International Symposium on Applications of Laser Anemometry to Fluid Mechanics, Lisbon, July 11-14, Durão et al., editors, paper 2.20.

Johnson, D.A. (1981), "Laser Velocimetry Applied to Transonic Flow Past Airfoils," 1st Symposium Num. and Phys. Aspects of Aero. Flows, Cal. State - Long Beach, Springer-Verlag.

Jorgensen, F.E. (1982), "Characteristics and Calibration of a Triple-Split Probe for Reversing Flows," DISA Information, No. 27, pp. 15-22.

Kalb, H.T. and Crosswy, F.L. (1983), "Discrete Fourier Transform Signal Processor for Laser-Doppler Anemometry," AEDC-TR-83-46, Arnold Engineering Development Center, AAFS, Tennessee.

Kielbasa, J. and Rysz, R. (1975), "Eine Oszillationsmethode zur Messung der Strömungsgeschwindigkeit von Gasen," contribution to the 50th Anniversary Volume of Max-Planck-Institut für Strömungforschung, Göttingen, West Germany; also MPI Report 112 (1973).

Kreid, D.K. and Grams, G.W. (1976), "Confocal LDV Utilizing a Decoupling Beam Splitter Combiner," Applied Optics, 15, pp. 14-16.

Lading, L. (1987), "Spectrum Analysis of LDA Signals," Dantec Information, 5, pp. 2-8.

Ligrani, P.M., Gyles, B.R., Mathioudakis, K., and Breugelmans, F.A.E. (1983), "A Sensor for Flow Measurements Near the Surface of a Compressor Blade," J. Physics E., Sci. Inst., 16, pp. 431-437.

Mayo, W.T. and Smart, A.E., editors (1980), Photon Correlation Techniques in Fluid Mechanics, Proceedings from the 4th International Conference, August 25-27, 1980, Stanford Univ. Dept. of Aeronautics and Astronautics.

Mazumder, M.K. and Wankum, D.L. (1970), "SNR and Spectral Broadening in Turbulence Structure Measurements Using a CW Laser," Applied Optics, 9, pp. 633-637.

McLaughlin, D.K. and Tiederman, W.G. (1973), "Biasing Correction for Individual Realization of Laser Anemometer Measurements in Turbulent Flows," Physics of Fluids, 16, pp. 2082-2088.

Meyers, J.F. (1985), "The Elusive Third Component," Int. Symp. Laser Anemometry, A. Dybbs and P. Pfund, ed., ASME Winter Annual Meeting, Nov. 17-22, FED-33, pp. 247-254.

Meyers, J.F. (1979), "Application of Laser Velocimetry to Large Scale and Specialized Aerodynamic Tests," TSI Quarterly, 5, Issue 4, pp. 5-12.

Müller, U.R. and Schon, Th. (1987), "Developments Towards a Three-Component Hot-Wire Technique for High-Intensity and Separated Turbulent Flows," Paper 6-4, 6th Symposium on Turbulent Shear Flows, Sept. 7-9, Toulouse, France.

Munoz, R.J., Mocker, H.W. and Koehler, L.E. (1974), "True Airspeed Measured by Airborne Laser Doppler Velocimeter," NASA Tech. Brief 73-10506 (ARC-10763).

Nakatani, N., Maegawa, A., Izumi, T. and Yamada, T. (1986), "Advancing Multipoint Optical Fiber LDVs - Vorticity Measurement and Some New Optical Systems," Third Inter. Symposium on Applications of Laser Anemometry to Fluid Mechanics, Lisbon, July 7-9, Durao et al. editors, Paper 8.4.

Nakatani, N., Nishikawa, T. and Yamada, T. (1980), "LDV Optical System with Multifrequency Shifting for Simultaneous Measurement of Flow Velocities at Several Points," J. Physics E: Sci. Instrum., 13, pp. 172-173.

Nakatani, N., Tokita, M., Maegawa, M. and Yamada, T. (1985a), "Multipoint LDVs Using Phase Diffraction Gratings and Optical Fiber Probes - Measurement of Coherent Structures in Premixed Flame," Proc. of the 2nd Inter. ASME Laser Anemometry Symp., Miami (ASME) FED-33, p. 111.

Nakatani, N., Tokita, M., Maegawa, M. and Yawada, T. (1985b), "Simultaneous Measurement of Flow Velocities Variations at Several Points with the Multipoint LDV," Proc. of the Inter. Conference on Laser Anemometry Advances and Applications, Manchester (BHRA), p. 401.

Nakatani, N., Tokita, M., Izumi, T. and Yamada, T. (1985c), "LDV Using Polarization-Preserving Optical Fibers for Simultaneous Measurement of Multidimensional Velocity Components," Rev. Sci. Instrum., 56, p. 2025.

Nakatani, N., Yorisue, R. and Yamada, T. (1978), "Simultaneous Measurement of Flow Velocities in Multipoint by the Laser Doppler Velocimeter," Proc. of the Dynamic Flow Conf., Marseille-Baltimore, pp. 583-590.

Orloff, K.L. and Biggers, J.C. (1974), "Laser Velocimeter Measurement of Developing and Periodic Flows," Proc. 2nd Int. Workshop on LDV, 2, pp. 143-168, Purdue University.

Orloff, K.L., Corsiglia, V.R., Biggers, J.C., and Ekstedt, T.W. (1975), "The Accuracy of Flow Measurements by Laser Doppler Methods," Proc. LDA Symp., Copenhagen, pp. 624-643.

Owen, F.K. (1984), "A Scanning Laser Velocimeter for Turbulence Research," NASA Contractor Report 172493.

Perry, A.E. and Watmuff, J.H. (1979), "Phase-Averaged Large-Scale Structures in Three-Dimensional Wakes," Report FM-12, Dept. of Mechanical Engineering, Univ. of Melbourne, Australia.

Perry, A.E. (1982) Hot-Wire Anemometry, Oxford, Clavendan Press.

Rhodes, D.B. (1976), "Optical Scanning System for Laser Velocimeter," SPIE, 84, pp. 78-84.

Ruderich, R. and Fernholz, H.H. (1983), "An Experimental Investigation of the Turbulent Shear Flow Downstream of a Normal Flat Plate with a Long Splitter Plate Modification of a Model," Fourth Symposium on Turbulent Shear Flows, pp. 19.7-19.12, Karlsruhe, F.R.G., Sept. 12-15.

Sandborn, V.A. (1972), Resistance Temperature Transducers, Metrology Press, Fort Collins, Colorado.

Seegmiller, H.L., Marvin, J.G. and Levy, L.L., Jr. (1978), "Steady and Unsteady Transonic Flows," AIAA Journal, 16, pp. 1262-1270.

Shiloh, K. and Simpson, R.L. (1982), "A Laser Anemometer for Crossflow Velocities," J. Physics E., Sci. Insti., 15, pp. 428-431.

Shiloh, K., Shivaprasad, B.G. and Simpson, R.L. (1981), "The Structure of a Separating Turbulent Boundary Layer: III, Transverse Velocity Measurements," J. Fluid Mech., 113, pp. 75-90.

Simpson, R.L. (1985), "Two-Dimensional Turbulent Separated Flow," AGARDograph AG-287-Vol. 1.

Simpson, R.L. (1976), "Interpreting Laser and Hot-Film Anemometer Signals in a Separating Boundary Layer," AIAA J., 14, pp. 124-126.

Simpson, R.L. and Barr, P.W. (1975), "Laser Doppler Velocimeter Signal Processing Using Sampling Spectrum Analysis," Rev. Sci. Inst., 46, 7, pp. 835-837.

Simpson, R.L. and Chew, Y.-T. (1979), "Measurements in Highly Turbulent Flows: Steady and Unsteady Separated Turbulent Boundary Layers," pp. 179-196, Laser Velocimetry and Particle Sizing, Hemisphere, Thompson and Stevenson, ed.

Simpson, R.L., Chew, Y.-T. and Shivaprasad, B.G. (1981a), "The Structure of a Separating Turbulent Boundary Layer: I, Mean Flow and Reynolds Stresses," J. Fluid Mech., 113, pp. 23-51.

Simpson, R.L., Chew, Y.-T. and Shivaprasad, B.G. (1981b), "The Structure of a Separating Turbulent Boundary Layer: II, Higher Order Turbulence Results," J. Fluid Mech., 113, pp. 53-74.

Simpson, R.L. and Shackleton, C.R. (1977), "Laminariscent Turbulent Boundary Layers: Experiments on Nozzle Flows," Project SQUID Report SMU-2-PU; NTIS-AD-A037441/3GI.

Simpson, R.L., Strickland, J.H. and Barr, P.W. (1977), "Features of a Separating Turbulent Boundary Layer in the Vicinity of Separation," J. of Fluid Mech., 79, pp. 553-594.

Skinner, G.T., Dunn, M.G. and Hiemenz, R.J. (1982), "A Low-Speed Heat-Pulse Anemometer," Rev. Sci. Inst., 53, pp. 342-348.

Spencer, B.W. and Jones, B.G. (1971), "Turbulence Measurements with the Split-Film Anemometer Probe," pp. 7-15, Proc. of Second Symposium on Turbulence in Liquids, G.K. Patterson and J.L. Zakin, editors, Dept. Chem. Engrg., Univ. Miss.-Rolla.

Stevenson, W.H. and Thompson, H.D., editors (1972), The Use of the Laser Doppler Velocimeter for Flow Measurements, Proceedings of a Workshop Co-Sponsored by Project SQUID and the U.S. Army Missile Command, March 9-10, School of Mechanical Engineering, Purdue University.

Thompson, B.E. and Whitelaw, J.H. (1984), "Flying Hot-Wire Anemometry," Experiments in Fluids, 2, p. 47f.

Thompson, H.D. and Stevenson, W.H. (1979), Laser Velocimetry and Particle Sizing, Hemisphere, New York.

Thompson, H.D. and Stevenson, W.H., editors (1974), Proceedings of the Second International Workshop on Laser Velocimetry, March 27-29, Two Volumes, Engrg. Exp. Station Bulletin 144, Purdue University.

Tombach, J.H. (1973), "An Evaluation of the Heat Pulse Anemometer for Velocity Measurement in Inhomogeneous Turbulent Flow," Rev. Sci. Inst., 44, pp. 141-148.

Tutu, N.K. and Chevray, R. (1975), "Cross-Wire Anemometry in High Intensity Turbulence," J. Fluid Mech., 71, pp. 785-100.

Van de Hulst, H.C. (1964), Light Scattering by Small Particles, Wiley, New York.

Westphal, R.V., Johnston, J.P. and Eaton, J.K. (1984), "Experimental Study of Flow Reattachment in a Single-Sided Sudden Expansion," NASA CR 3765.

Westphal, R.V., Eaton, J.K. and Johnston, J.P. (1981), "A New Probe for Measurement of Velocity and Wall Shear Stress on Unsteady Reversing Flow," J. Fluids Engrg., 103, pp. 478-482.

Williard, G.W. (1949), "Criteria for Normal and Abnormal Ultrasonic Light Diffraction Effects," J. Acoustical Soc. Am., 21, pp. 101-108.

Young, M. (1976), "Calibration of Hot-Wires and Hot-Films for Velocity Fluctuations," Report TMC-3, Thermosciences Div., Dept. of Mechanical Engineering, Stanford University, Stanford, California.

MEASUREMENT TECHNIQUES IN LABORATORY ROTATING FLOWS

Patrick D. Weidman
Department of Mechanical Engineering
University of Colorado
Boulder, CO 80309

1. INTRODUCTION

The purpose of this chapter is to survey classical and modern measurement techniques used in rotating flow experiments. Since the measurement of rotating flows is now a broad and rapidly developing art, it is clear that only a summary of the essential features of each measurement system can be given. Here we will be primarily concerned with flows driven by rotating boundaries, rotational flows produced inside stationary boundaries, motion driven by thermal heating of a fluid in a basic state of rotation and other flows for which rotational forces are of primary importance. Free vortex flows in the absence of rotation, wind tunnel flow over rotating bodies, and rotating liquid helium flows which necessitate specialized measurement techniques are excluded. Equations and diagrams are incorporated only when they add substantially to the understanding of the operation of a particular measurement system. An attempt is made to include historical perspective and development. Also, I felt it was important to provide an extensive bibliography to exhibit the range of applications of each measurement technique and thereby provide guidance for the experimentalist interested in rotating flows.

Since the early investigations by Thompson (1855) on fluid resistance incurred by rotating disks and by Plateau (1863) on the behavior of rotating liquid drops, the field of rotating fluid mechanics has attracted the attention of engineers and scientists alike. Progress in the development of experimental detection systems has originated from three separate communities: (i) engineers concerned with rotating fluid machinery such as fluid transmission drives, rotating electrical motors, and rotating disks for magnetic or optical storage of binary data; (ii) geophysical fluid dynamicists interested in laboratory simulations of convective motion in rotating flows, rotating stratified flow over topographical features, or the fluid motion created by a solid body translating through a rotating fluid; and (iii) experimentalists concerned with a fundamental understanding of fluid stability.

The true father of the science of rotating fluid dynamics is G. I. Taylor. His experimental and theoretical papers in the early 1900's brought a fundamental understanding to the often curious behavior of rotating flows. For example, Taylor

(1917) showed that a neutrally buoyant cylinder towed horizontally through a fluid rotating about its vertical axis traveled in a straight line while a sphere "violently deviated" from the straight line path along which it was towed. In another investigation Taylor (1922) demonstrated experimentally that a solid sphere initially rotating at the angular speed of the surrounding fluid in solid body rotation would cease its rotation the moment it was towed along the axis of fluid rotation. In yet another experiment, Taylor (1923a) produced dye lines above a cylinder traversing a fluid normal to its vertical axis of solid body rotation to show the existence of a stagnant fluid column well above the cylinder. These stagnant regions produced above and below horizontally moving obstacles in a rapidly rotating fluid are now referred to as Taylor columns. Taylor's (1923b) seminal work on the stability of fluid motion between co- and counter-rotating concentric cylinders is often cited as the first successful analysis of the stability of a viscous fluid. The toroidal vortices that appear at the onset of instability now bear his name. Experimenters are encouraged to peruse these early papers for they will certainly be inspired by the ingenious laboratory mechanisms devised by Taylor to demonstrate fundamental behavior in rotating fluids.

Throughout this article certain canonical experimental flows will be referred to: (i) One is the convective flow generated in a rotating fluid annulus with inner and outer cylindrical walls maintained at different temperatures. This fundamental geophysical flow apparatus simulates global atmospheric motion driven by equator-to-pole temperature differences, and circles concentric with the annulus are analogous to latitudinal circles on the globe. This experimental facility is called the *laterally heated rotating annulus*. (ii) The flow inside a rotating cylinder with an inclined bottom surface produced by a differentially rotating upper lid is used to simulate ocean gyres driven by wind shear stress. This geophysical apparatus has come to be known as the *sliced-cylinder experiment*. (iii) Another rotating flow that continues to provide an important testing ground for stability theory is the flow produced in the annular gap between concentric, differentially rotating cylinders. Additional complications include superposed axial flow through the annulus or differential heating between the inner and outer cylinders. The stable flow in this *Taylor-Couette facility* is referred to as *circular Couette flow* and the unstable motions produced therein are collectively referred to as *Taylor vortex flow*.

Large rotating tables designed for investigation of geophysical flows have been constructed at different research institutions, notably in Germany, France, England and the United States. It appears that the first major facility was constructed in 1926 at the Kaiser-Wilhelm Institute for Flow Research in Gottingen under the direction of Ludweig Prandtl. This facility was created with the idea that the experimenter should ride with the rotating laboratory to conduct experiments. Adolf Busemann was in charge of the construction and provides this historical note (Busemann, 1971): "This project proceeded rather well until the day arrived when the walls were installed all around. The afternoon before that day it was a great pleasure to try out the rotating platform at all speeds provided by the driving motor. As soon as the walls were closed and the internal lights turned on, the situation was quite different. It was nauseating in the true meaning of the word. The chamber was not the expected tool to study meteorology in an easy chair." Prandtl, however, had foreseen that the observer inside the *rotierendes Zimmer* might become uncomfortable at high rotation speeds and had prepared for the overhead installation of a large rotoscope (see §3.3) for outside viewing. The largest by far is the 14 m rotating table in active use at the Institut de Mechanique in Grenoble.

The presentation given here follows the following outline. A review of the deleterious effects of intrusive probes in §2 sets the scene for a discussion of measurement techniques in rotating fluid flows. Flow visualization described in §3 are of primary importance for understanding rotating flow behavior. Pressure, temperature and velocity measurement techniques are outlined in §4, §5, and §6, respectively. A discussion of shear stress probes in §7 is followed by a presentation of heat and mass transfer measurement techniques applied to rotating flows in §8 and §9, respectively. Torque and novel force measurement systems are covered in §10. Density gradients play a crucial role in geophysical flows and their formation and measurement in rotating systems are discussed in §11. Interfacial height detectors are surveyed in §12 and some specialized stability flow detectors are discussed in §13. A final warning about some rather subtle probe interference problems is given in §14 and the review is ended with brief concluding remarks in §15.

2. PROBE INTERFERENCE

Many rotating flow fields are characterized by nearly-closed streamlines and the disturbing effects induced by inserting measurement probes in such a flow has long been recognized although perhaps not fully appreciated. Indeed, the desire to probe a rotating flow in a nonintrusive manner has guided the development of new measurement systems and has led to some recent innovative techniques.

G. I. Taylor (1935) was perhaps the first to recognize and deal with the problem of probe interference in rotating flow. He designed an experiment to measure the azimuthal velocity distribution of turbulent flow in the gap between an inner rotating cylinder and a stationary outer cylinder. Small Pitot tubes were constructed, mounted through the outer cylinder wall and traversed radially across the gap. Taylor noted that special difficulties arise in making these measurements "because the wake behind any measuring apparatus, such as a Pitot tube which is inserted into the annular space between the cylinders is carried round with the fluid and forms a ring-shaped region where the velocity is different from that which would exist in the absence of the Pitot tube." In fact, he obtained different pressure distributions with different diameter probes. The interesting part of this study was Taylor's perception that the velocity deficit in the wake of the probe was proportional to the drag on the probe. Using known cylinder drag coefficients, he was able to derive a formula to extrapolate the pressure measurements to that of a probe of zero diameter. The velocity profiles derived from the corrected measurements were in excellent agreement with theoretical predictions.

Much use has been made of dyed fluid to visualize the motion of the bulk fluid into which it is injected. In this sense the dye stream may be thought of as a probe marking the streakline fluid motion away from the dye source. In rotating source-sink simulations of ocean basin flow it is natural to introduce dye at the source to visualize the basin currents. Faller (1960) reported the formation of unexpected Eastern boundary currents in a simulation of the intrusion of Mediterranean water into the Atlantic Ocean. Repeat experiments conducted by Faller and Porter (1975) revealed that the Eastern boundary currents observed 15 years earlier were due to a mismatch between the temperature of the injected dyed water and the interior bulk fluid. This example of "probe interference" exhibits the strong influence that weak buoyancy forces can have on slow motions in a rotating fluid system.

A probe disturbance which has an important bearing on the measurement of vortex-breakdown phenomena was reported by Hummel (1965). He inserted a 3 mm diameter half-sphere yaw probe in the vortex formed by leading edge separation from a delta wing at angle of attack in a wind tunnel. The vortex was observed to "burst" and in planes normal to the delta wing downstream of the burst (vortex breakdown) flow direction and static and total pressures were measured. Hummel commented that such measurements posed difficulties because the inserted probe caused a pressure increase in the vortex. The pressure increase in turn caused the burst to move upstream a distance that depended on the size and orientation of the pressure probe. He noted that the probe influence was especially strong if the burst happened to occur close to the delta wing's trailing edge.

A well-documented report on hot-wire probe interference in a large circular Couette facility with air as the test fluid is given by Coles and Van Atta (1966). The outer 38-inch diameter cylinder rotated while the inner 32-inch diameter cylinder was held stationary. A small hot-wire protruding from the inner cylinder fixed at mid-radius detected the wake of a moveable hot-wire probe mounted on the opposite side of the cylinder. One percent variations in the azimuthal velocity were recorded by the fixed probe as the companion probe traversed the radial gap between cylinders 180° upstream.

Probe interference in temperature measurements were first reported by Kaiser (1969) in a rotating laterally heated annulus. The temperature probes were made by mounting thermocouple junctions at the end of small diameter hypodermic tubes. He observed that as one probe traversed through the fluid, the temperature of the other fixed probe changed by a significant amount, on the order of 15%. Kaiser spent considerable effort mapping the temperature distortion field induced by the the probes and used this data to correct measured radial temperature profiles. In a follow-on study Fultz and Kaiser (1971) found that the velocity field, determined by time-of-flight measurements of neutrally buoyant ink dots between two fixed probes, was also significantly affected by the presence of the probes. They noted that velocity and temperature measurements in the rotating laterally heated annulus experiment "had been in progress for nearly twenty years before any quantitative indication of probe problems were reported in the literature."

Thin wires are often strung through a fluid flow as a cathode for electrolytic visualization techniques or to support thermistor beads. Analyses by Schraub, et al (1965) and Davis and Fox (1967) have shown that the velocity defect behind the wire interferes with hydrogen bubble timelines for velocity measurement. Wortmann (1953) has likewise analyzed this problem with application to the tellurium dye method. These represent basic problems associated with the introduction of single wires in both rotating and nonrotating flows; specific problems associated with the use of a large number of wires supporting thermistor beads in confined rotating fluid experiments are reported in §5.3.

A very careful study of probe self-interference has been reported by Cerasoli (1975). In rotating source-sink annulus experiments he noted that hot-wire probe supports can give rise to low wavenumber, azimuthally propagating disturbances which may be falsely interpreted as Ekman layer instabilities. After acquiring extensive data on wave frequencies and magnitudes, wavenumbers and phase relationships, it was concluded that wave motions interpreted as Type II Ekman instabilities and inertial eigenmodes by Tatro and Mollo-Christensen (1967) and others were actually probe-induced instabilities. Further discussion of hot-wire probe support interference is given in §6.3.

Hide, et al (1977) scanned arrays of 32 and 64 thermocouples spaced at equal intervals along a circular ring located at mid-radius and mid-height in a laterally heated rotating annulus to obtain wavenumber spectra. In order to minimize flow interference, the array consisted of a single constantan wire ring with individual copper wire junctions soldered around its perimeter. The tiny copper wires ran vertically through the upper half of the fluid volume to an electronic scanner circuit on the rotating table. In another configuration the thermocouple ring was again located at mid-level, but adjacent to the surface of the outer cylinder, and the copper wires ran radially out small holes drilled through the cylinder wall. Stable azimuthal velocity profiles measured by following neutrally buoyant polystyrene spheres illuminated with a horizontal light sheet at the mid-plane level were made with and without the thermocouple arrays. A 10% change in the stable velocity field induced by the first thermocouple array was observed and only a 3% change was induced by the second. However, the amplitudes and drift rates of instability waves measured by the two arrays differed as much as 20%. The important conclusion here is that unstable flows can be more profoundly disturbed by probe insertion than stable flows.

These studies clearly document the adverse effects incurred by placing fixed objects in a rotating flow to extract flow field information. Probe insertion fundamentally alters the velocity, pressure, and temperature in both stable and unstable flow. More subtle probe interference effects will be discussed in §14.

3. FLOW VISUALIZATION

Isothermal flows are characterized primarily by streamlines and velocity profiles. These quantitative features can be determined with certain flow visualization methods if one has proper understanding of pathlines, streaklines, and streamlines. A *pathline* is the trajectory of an isolated particle moving through the fluid. A *streakline* is the locus of points connecting all particles released from a fixed point in space. A *streamline* is the curve everywhere tangent to the velocity vector. For example, the curve traced by a neutrally buoyant particle released from a point and advected with the flow marks a pathline. Ink continuously released from a probe marks a streakline. For steady flow the three lines are coincident and hence a streamline may be determined from either a single particle path or a dye streak. In unsteady flow, except in special cases, the three lines are markedly different from each other. Furthermore, pathlines, streaklines, and streamlines observed in a moving frame of reference generally bear no resemblance to those observed in a stationary frame. Another curve characterizing fluid motion is the so-called *timeline*. The simultaneous release of particles from a curve in a flow will trace out a timeline at a later instant in time. A timeline becomes useful when the release curve is normal to the flow. For example, if the release curve is positioned in a line normal to a steady one-dimensional flow, the timeline ideally provides a direct replication of the local velocity profile. In this section we divide the discussion into liquid flows, air flows, optical viewing techniques, refractive index sensitive visualization systems and naturally visible flows.

3.1 Liquid Flows

The use of dyed fluid particles has long been an important method of visualization for liquid flows. Numerous methods are presently available for producing a dyed line, a dyed surface, or a dyed volume in the fluid. These include dissolution from dye crystals, direct liquid dye injection for producing streaklines, electrolytic techniques for producing timelines, electrochemiluminescent visualization and various radiation-induced flow visualization techniques. Internal and free surface particle suspensions offer another type of visualization which permit viewing of large sections of the liquid flow field with proper lighting. These methods are discussed in turn below.

3.1.1 Inks and Dyes

Dye Crystals

A simple and often effective means by which liquid motion may be observed is to drop solid dye crystals into the rotating flow. Dissolution of dye crystals is particularly useful in geophysical flow experiments for which the relative speed of fluid motion is low. In this case small crystals will fall vertically through the rotating fluid leaving a colored trace in its path. Furthermore, the undissolved crystals will remain on the bottom of the tank and thereby provide visualization of the local boundary layer flow and associated instabilities. By far the most popular crystal is potassium permanganate which is orange-red in color.

Prandtl (1937) sprinkled small crystals of potassium permanganate in a tank positioned in the center of his rotating room. When the rotation rate was slightly changed, spiraling inflow or outflow was observed by the dye streaks on the bottom of the tank. The deflection of the streaks was measured to be 45° in laminar flow in agreement with Ekman boundary layer theory, but less than 45° when the boundary layer was turbulent. Faller (1963) produced steady Ekman boundary layer flow in a large cylindrical rotating tank by withdrawing water from the center and introducing it tangentially at the outer rim. Potassium permanganate crystals sprinkled in the fluid at the outer rim of the tank sank to the bottom and formed a sheet of dye which, when convected into the unstable flow region at smaller radii, permitted visualization of nearly stationary Type I Ekman instability waves. Faller and Kaylor (1966) noted that the rapidly moving Type II Ekman instability waves could not be visualized by this method because the dye did not sufficiently fill the boundary layer to mark the unstable flow. A critical discussion of the use of dye crystals for Ekman boundary layer visualization studies is given in that paper. Whitehead (1980) used dye columns formed by sinking potassium permanganate pellets to identify the region of dynamic withdrawal in his experiments on selective withdrawal from a stratified rotating fluid.

Liquid Dye Injection

A common method to visualize liquid flow patterns is to inject a foreign dyed liquid from a probe immersed in the flow or from a small orifice on the surface of a body in contact with the flow. Dye lines advected with the fluid are truly *streaklines* and must be interpreted as such. The injected dye stream can be balanced in either density or viscosity with the bulk fluid, but not both. Because buoyancy affects the motion of the dye line, density is the most important fluid property to match, especially in low-speed laboratory flows. Density matching is also important in rapidly rotating flows for which high centripetal accelerations would centrifuge differentially heavy fluid parcels outward or differentially light ones inward. Density matching to water flow is often achieved by mixing the prepared dye with water-alcohol mixtures. However, these mixtures are generally not a true solution and therefore the different components can be acted upon in different ways by inertial or centrifugal forces. Furthermore, in certain liquid mixtures supporting the dye, double-diffusive instabilities may be encountered when the dye stream comes in contact with the bulk flow.

The rapid dispersion of a dye line into a turbulent flow restricts its use to laminar flow visualization. Even in laminar flow one must be concerned with the diffusion of the dye into the bulk fluid. If the diffusion is rapid the flow can be visualized only a short distance downstream of the injection point. Werle (1960) has shown that dye filaments may be greatly stabilized by mixing the dye with milk, presumably because the fattiness of the milk retards diffusion of the dye solution into the surrounding stream. Werle (1973) has presented several beautiful photographs of vortex flows using multiple, separately-colored dye streaks stabilized by milk. A dye stream injected in a confined rotating flow increasingly contaminates the fluid and therefore limits the duration of the experiment. Since the bulk fluid is usually water, this only represents an inconvenience rather than an economic concern. Another consideration is visibility of the dye streak relative to the background against which the flow is being observed. Thus the choice of the dye color depends on the particular experimental facility and available lighting. Examples of dyes used in rotating water flows include fluorescein, eosine, rhodamine, negrosine, potassium permanganate, crystal violet, carmine red, malachite green, biological stains and food coloring.

It is advantageous to have a dye injection system for which the coloring can be erased either because the test fluid is expensive to replace or because it is inconvenient to replace the fluid in a complicated experimental setup. In silicone oil Wimmer (1976) used Ceresblau GN and noted that clear silicone oil may be recovered by the addition of a animal charcoal and subsequent filtering. Recently, Sommeria, et al (1989) have reported a simple dye system in which a solution of ferric ammonium sulfate is injected into the water bulk fluid which contains ammonium thiocyanate. When these chemicals react a dark red dye is formed, but it can be erased by adding ascorbic acid. We have tried this visualization system in our laboratory and found that it works equally well in glycerol solutions.

Some applications of dye visualization in rotating flows will now be given. Taylor (1923) injected eosin and fluorescein dye solutions through six small orifices around the circumference of an inner rotating cylinder before it was set into motion in his classical study of the onset of instability in circular Couette flow. After the inner cylinder began to rotate, the dye was observed to uniformly coat the cylinder wall at subcritical speeds. Taylor hypothesized that once the vortex instability set in, the dye would be swept around the outer margin of the vortex in the axial plane because it would tend to follow the $\psi = 0$ steamline from where it originated. This was indeed the case and he obtained photographs of the instability using an eosine dye mixed in alcohol. Taylor further noted that the dye injections did not always behave consistently. In discussing the dye emanating from an orifice on the inner cylinder, he wrote: "It frequently happened in fact that the colored liquid formed two columns, one going up and the other down, when strong eosin solution mixed in alcohol was used." Taylor offered no explanation for this behavior but it appears he had encountered the effects of what is now known as double-diffusive instability.

Baker (1968) used fluorescein dye to measure the slow axial displacement of fluid during a spin-down experiment. An eye dropper full of dye is squirted into the central axis of a cylindrical vessel of water while it is spinning up to solid body rotation. The dye fell to the bottom and spread out in the Ekman layer. After the fluid achieved solid body rotation the angular speed of the cylinder was suddenly reduced by about 10%. In a few seconds the thin layer of ink moved off the floor and rose slowly upward due to the weak divergent flow in the Ekman layer. Consistent with linear spin-up theory, the entire layer of dye remained exactly horizontal as it propagated upward through the liquid. This simple experiment shows how dye visualization may be used to demonstrate complicated aspects of rotating fluid flow.

Fultz (1949) injected a solution of carmine red in ammonia to visualize convective fluid motion in a rotating hemispherical liquid shell induced by thermal heating at the south pole. Stommel, et al (1958) made extensive use of dye injection to visualize western boundary currents and other stationary flow patterns in laboratory experiments designed to simulate planetary flow in bounded basins. Injecting negrosine ink into a circular Couette flow with differential heating between the inner and outer cylinders permitted Karlsson and Snyder (1965) to observe a stationary cat's eye instability. Maxworthy (1968) injected dye near the north pole of a hollow sphere moving along the axis of a rotating fluid to obtain a beautiful photograph of the flow in the Ekman boundary layer and its ejection at the south pole in the form of a helix. Boyer (1971) injected a solution of rhodamine in a rotating water channel flow to measure the lateral deflection of streamlines over long shallow ridges. Maxworthy, et al (1985) injected dye along the axis of intense isolated vortices in a rotating fluid to observe and measure characteristic features of kink-waves, helical waves, and axisymmetric solitary waves traveling along the vortex cores.

Multi-color dye injection is often useful in visualizing complicated flow patterns. Sarpkaya (1971) injected two separately colored dye streams into a confined swirling laminar flow to elucidate the differences between stationary and travelling vortex breakdown phenomena. The dye streams issued through the end plane of the cylindrical vortex tube, one on center and the other off center. These remarkable color photographs are well worth examining. Escudier, et al (1980) also employed two-color dye visualization in a study of vortex breakdown in turbulent flow. Their photographs exhibit the Reynolds number evolution of complicated vortex structure that could not be readily discerned using multiple dye streams of identical color. Whitehead (1985) used multi-colored dye injection to visualize the transport and mixing of internal and free-surface fluid layers in a stratified flow simulation of the Mediterranean outflow.

Dye may also be used to label vortex rings, intrusions and other isolated fluid structures. For example, Taylor (1917), in some early studies on rotating fluids, filled a vortex generator chamber with a solution of fluorescein dye. An impulsive displacement of the fluid from the chamber generated a vortex ring clearly marked by the dye so that its curved trajectory through the ambient rotating fluid could be photographed. Flierl, et al (1983) used dye injection to label and follow the trajectory of modons generated by a turbulent jet issuing at depth in a uniformly rotating fluid. Joseph, et al (1984a) mixed dye in highly viscous immiscible fluids to observe roller instabilities induced by a horizontal shaft rotating at the level of the quiescent liquid-liquid interface. Whitehead and Chapman (1986) photographed a dyed gravity current intruding into clear water along a sloping bottom in a rotating tank. Measurements of the propagation speed of the gravity current were made by timed displacements of the visible dye front. Griffiths and Hopfinger (1987) injected different colored dyes into the center of two like-signed vortices separately generated in a rotating fluid in a study of the coalescence of geostrophic vortices. Not only did the dyed fluids serve to mark the subsequent shape and position of the vortices, but the fluid in the single merged vortex was color labeled to the vortex from which it had originated. Mori, et al (1987) observed the dynamics of coherent baroclinic eddies generated on a sloping bottom in a rotating fluid by following the motion of a dyed saline solution released from a cylinder into fresh water.

Critical discussions on the use of dye injection for particular measurements have been reported by a number of authors. Discussions of problems associated with dye visualization of Taylor vortex cells are given by Taylor (1923) and Snyder and Karlsson (1964). Schwarz, et al (1964) criticize the use of ink injection as a detector for the onset of instability. Boyer and Guala (1972) present an interesting visualization for which part of the dye line is representative of streamline motion and the remaining part is not. This serves to caution the experimentalist in the proper interpretation of dye-line visualization which mimics streamlines only in steady flow.

Liquid Drops

In some experiments small, neutrally buoyant liquid drops can provide visualization and quantitative measurement. Weidman and Johnson (1982) describe a method of generating a carpet of liquid drops of uniform density in a salt-stratified system using a kerosene-Freon mixture. Droplet size control and prevention of density deterioration through the rapid evaporation of the more volatile Freon are attained by pressurizing the mixture in a closed reservoir and ejecting it through a rake of twelve small hypodermic tubes towed slowly along the tank just below the liquid surface. The drops sink to a uniform depth predetermined by the density of the mixture. The only problem with this form of visualization is that the initial 0.5 mm diameter drops coalesce to form larger drops upon accidental contact during the settling process or when they get jostled around by a passing wave. The layer of neutrally buoyant drops follow their disturbed pycnoclines and mark passing waves. Maxworthy (1983) used this form of visualization to observe internal Kelvin waves in a rotating channel containing salt-stratified fluid. He determined the position, amplitude, and phase of mode-two waves from timed photographs following distorted pycnoclines marked by layers of colored drops strategically placed above and below the density interface.

In a reproduction of Plateau's (1863) original experiments, Tagg, et al (1980) examined the behavior of neutrally buoyant rotating silicone oil drops attached to a rotating shaft immersed in a water-methanol bath. Visualization of the drop shape was enhanced with the addition of soluble dye, but the rotation rate of the drops could not be discerned by this uniform coloring. Estimates of the radial distribution of azimuthal velocity inside the drops were obtained by injecting tiny neutrally buoyant, nearly opaque liquid drops of water-methanol into the primary silicone oil drop and following their motion with moving picture photography. In a spacelab experiment Wang, et al (1986) followed the motion of immiscible tracer drops embedded in an acoustically rotated primary drop to determine the drop rotation rate using cine photography.

3.1.2 Particle Tracers

Seeding liquids with tiny reflectant particles readily provides the experimentalist with a qualitative feeling for the nature of a laboratory flow. With little effort one can often discern whether the flow is quiescent or moving, steady or unsteady, periodic or modulated, laminar or turbulent. In confined rotating geometries one can usually make an accurate measurement of the transition between flow states and observe the morphology of those states. Internal suspensions and particle suspensions on a liquid interface are considered separately in the following sections.

Internal Suspensions

Light reflected from tiny particles suspended in liquid flows have been used for visualization and quantitative measurement since the earliest recorded fluid dynamic experiments. Specific materials used in the past include aluminum flakes, magnesium powder, burnt lime, mica flakes, iriodin powder, sawdust, guanine platlets, and essence of pearl developed for the manufacture of cosmetics.

Although aluminum flakes have enjoyed a relatively long era of popularity, there has been a recent trend to use guanine platlets (sold under the trade name Kalliroscope) and pearlessence. The advantage of these modern flakes is their low settling rates in common test fluids such as water, glycerol and silicone oil, each having a density in the neighborhood of 1 g/cm^3. Aluminum flakes have a density of 2.47 g/cm^3 compared with 1.62 g/cm^3 for guanine platlets and about 3.0 g/cm^3 for natural essence of pearl (Mearl Corporation, private communication). It is clear from these figures that the reduced settling speeds of the latter two visualization particles are not due to a match between particle and test fluid densities. In fact, for a variety of practical reasons, most experimentalists do not try to match the density of the working fluid with that of the visualization particles. Guanine platelets and pearlessence stay in suspension longer because of their reduced size. For example,

Matisse and Gorman (1984) report the size for guanine platelets to be 6 X 30 X 0.07 μm.

Proper interpretation of light patterns reflected from suspended flake-like particles has always proved difficult. Until recently there has been little quantitative treatment on the orientation of flake suspensions in fluid flow. A rule of thumb often quoted is that the particles align themselves with the local shear. This rule probably stemmed from the work of Jeffery (1922) who analyzed the motion of an isolated neutrally buoyant ellipsoidal particle in uniform shear flow. As the ellipsoid approaches a flat elliptical disk, Jeffery's results do show that these idealized tracers rotate about a diametrical axis in line with the axis of shear. Goldsmith and Mason (1962) used microphotography to track the angular rotation and the axial spin of single disks typically 600 μm in diameter with thickness/diameter ratios in the range 0.05 to 0.25. In a constant shear flow they found that the predictions of Jeffery (1922) hold correct if an equivalent axis ratio is used in lieu of the measured axis ratio. They also showed that not only do thin disks tend to align themselves with the axis of shear, but they also spend a major fraction of their rotation period with their flat faces nearly aligned with the stream surface. In a given experiment with circular disks of equivalent axis ratio near 0.1 subjected to a 0.6 sec^{-1} shear, the disk spent 70% of a rotation period within $\pm 10°$ of being parallel to the stream surface and rotated 160° during the remaining 30%. This was also substantiated by Coles (1965) who viewed the motion of thin aluminum flakes with axis ratios on the order of 0.01 in very slow laminar flow through a microscope. He observed that the flakes maintained constant orientation with their flat surfaces nearly parallel to the streaming flow except for an occasional rapid rotation by 180°.

In a flow visualization of Taylor cells using a suspension of aluminum flakes Euteneuer and Reimann (1971) noticed that the outflow boundaries between counter-rotating vortices were substantially brighter than the inflow boundaries when viewed head-on with oblique lighting. Apparently unaware of the earlier results by Jeffery (1922), they formulated a simple model which assumed that the thin aluminum flakes, taken to be rectangular plates of finite thickness, exactly followed stream surfaces edge-on. The Taylor cells were modeled as perfect tori and the authors calculated the spiraling trajectory of the particles over the toroidal surface. Using Lambert's law for reflection, the relative intensity that would be observed for a uniform suspension of spiraling, non-interfering flakes subject to uniform lighting was calculated. Although there is much to criticize about the simplicity of the model, it indeed predicted the relatively brighter intensity at the outflow boundaries that was experimentally observed.

In order to assess the response of suspended aluminum flakes to an approaching cylindrical shear front, Weidman (1976) made quantitative estimates of the time it would take a particle to rotate 45°. When viewed through the circular glass endplates of the rotating cylinder spun up from rest, the flake suspension in water took on a matted appearance in the undisturbed region ahead of the wave front and a granular appearance in the sheared flow behind the front. The calculation was carried out using an exact solution by Greenspan for the shear velocity generated by impulsive spin-up modeled by a one-dimensional Wedemeyer equation. Using Jeffery's (1922) results for an idealized flat circular disk particle, the angle of particle rotation was calculated by a time integration of the shear field at a fixed point in space. The results showed that the particles respond to very weak shear and this fact was corroborated by laser Doppler velocity measurements of the time-dependent flow.

Savas (1985) derived some important statistical properties of flake suspensions using a probability-density function for the orientation of the flakes defined by its spherical polar angles (θ, ϕ). For a given initial distribution of the the probability density function $f(\theta_0, \phi_0)$ its evolution $f(\theta, \phi)$ was evaluated using the results of Jeffery (1922). In the limiting case of zero flake thickness, Savas found that a particle will move edge-on through the fluid with its axis at a given inclination to the shear, not necessarily along the direction of principal strain. A convincing test of the analysis was made by applying the probability formulation to a given time-dependent flow field illuminated and observed at fixed angles. The flow was spun-up from rest in a closed cylinder and the evolution of the probability density function was evaluated using measured results of the local strain rate. The radial positions of two visible peaks observed in the experiment were in excellent agreement with the positions predicted by the time-dependent probability density function. An important conclusion of this study is that suspended reflective flakes are not well-suited for visualization of low-amplitude waves in high shear flows. Having read Savas's (1985) paper, Andereck, et al (1986) deduced the following rule of thumb for interpreting photographs of Taylor vortex flows visualized with particle flake suspensions: a dark area indicates flow along the line of sight of the observer and a light area indicates flow perpendicular to the line of sight. In view of the above-cited observations by Euteneuer and Reimann (1971) of markedly different light intensities at the inflow and outflow boundaries of the Taylor cells, this rule should be exercised with care.

As pointed out by Coles (1965) the density of aluminum pigment is generally higher than that of the host fluid and centrifuging effects at high rotation rates limit the usefulness of the visualization technique. Euteneuer (1969) and Weidman (1976) used this to some advantage when investigating nonstationary Gortler-Taylor instability during spin-down from a state of uniform rotation in a cylindrical container. After spin-up to a high rotation rate, the aluminum particles uniformly coated the cylindrical wall. Stroboscopic illumination of the clear interior fluid through glass end plates provides easy visualization of the changing wavelengths of the instability during spin-down as the cellular motion erodes the thin aluminum coating. Use of this visualization method to determine the onset of Gortler instability, however, inevitably gives a delayed value of the onset time because the instability must grow to finite amplitude in order to scour the pigment from the walls.

Probably the earliest recorded use of particles to visualize rotating fluid flow for scientific purposes can be attributed to Wilcke (1780). As a model of cyclonic motion in the atmosphere, he formed a spiral vortex in water, contained in a cylindrical vessel, by mechanically stirring the upper liquid layer with an eccentrically mounted thick steel wire. The addition of burnt lime particles mixed in the fluid provided the first visualization of the phenomena now known as vortex breakdown. Shultz-Grunow (1956) was the first to employ aluminum flake suspensions (AFS) for visualization of Taylor vortices. Ludwieg (1964) used a suspension of iriodin powder to successfully test his theory for the stability of flow in the gap between a stationary cylinder and an inner spiraling shaft. Rossby (1969) produced photographs of some remarkable instability flow patterns in rotating Rayleigh-Benard convection. Versteergen and Jankowski (1969) measured neutral stability curves for the flow between eccentrically rotating cylinders by visualization studies with AFS. Also using AFS, Euteneuer (1970) observed that Gortler-Taylor instabilities during spin-down in a cylinder were followed by the onset of Tollmien-Schlichting waves. Vogel (1971) used aluminum particles to photograph the cross-section of a vortex breakdown bubble. Koschmieder (1972) made extensive use of AFS in water to study the time-dependent fluid motion produced by sudden cooling or heating of the cylindrical walls in a rotating annulus. Burkhalter and Koschmieder (1973) and Koschmieder (1979) measured the evolution of Taylor vortex wavelengths at supercritical Taylor numbers well into the turbulent regime using aluminum flakes variously suspended in water, machine oil, or glycerol mixtures. Weidman (1976) measured the stability boundary for Type II Ekman spirals at negative Rossby numbers with AFS visualization of the unsteady boundary layer flow at the endwall of a cylinder during spin-up from rest. In a similar manner Wimmer (1978) measured the neutral stability curve for Type I Ekman spirals at positive Rossby numbers. Two new forms of stable secondary modes in low aspect ratio Taylor-Couette flow were discovered by Benjamin (1978) using a pearlessence suspension in glycerol. Takeuchi and Jankowski (1981) employed AFS to measure the axial wavelength and angle of inclination of spiral vortex cells at the onset of instability in spiral Poiseuille flow. Wimmer (1981) determined the neutral stability boundary for flow in the gap between co-rotating and counter-rotating concentric spheres with an aluminum flake suspension. Buhler (1983) measurered the stability boundary for axial throughflow between concentric spheres with the inner sphere stationary and the outer one rotating; excellent photographs of this flow made visible with AFS may be found in Buhler (1986). Wimmer (1983) observed nonuniform wavelength toroidal cells for flow in the gap between truncated rotating cones with AFS visualization.

Hopfinger, et al (1982) produced streak photographs of the organized vortical motions generated from grid turbulence in rotating flow with the aid of trace amounts of wood particles measuring approximately 300 μm in diameter. Heikes and Maxworthy (1982) illuminated a suspension of aluminum particles in a vertical plane to detect inertial waves generated by uniform horizontal flow over two-dimensional ridges and axisymmetric obstacles in a homogeneous rotating fluid. Their full-field visualization permitted measurement of the group-path angle for the fundamental wavenumber excited by the cylindrical ridge and also the inclination of Taylor columns above the axisymmetric obstacles. This study by Heikes and Maxworthy (1982) is particularly instructive because it demonstrates the abundance of quantitative information that can be gained from proper use of particle flow visualization.

Benjamin and Mullin (1982) and Mullin (1982) illuminated a suspension of pearlessence with a light sheet to observe the cross-section of toroidal vortices in Taylor-Couette flow. Accurate wavelength measurements of certain anomalous modes were made in this steady flow with the aid of a cathetometer. Lorenzen (1985) continued these experiments using a suspension of iriodin pearlessence. Savas (1983) discovered an instability of unsteady Bodewadt flow in the form of concentric circular waves. With an aluminum particle suspension, he observed the wave structure in both planform and cross-section and made wavelength measurements at the onset of instability. Dickinson and Long (1983) observed a turbulent front produced by an oscillating grid in a rotating fluid with AFS and measured its propagation speed with cine photography. Andereck, et al (1983) reported five new Taylor vortex instability flows observed with a water suspension of guanine platlets. Dominguez-Lerma, et al (1986) studied the problem of wave-number selection in a Taylor-Couette facility with a spatially ramped outer cylinder using guanine platelets suspended in glycerol. Hathaway and Fowlis (1986) presented some interesting pictures of flow regimes in a shallow rotating cylindrical annulus driven by a vertical temperature gradient using Kalliroscope. They reported that the best contrast for visualizing the instabilities was achieved with back lighting through the fluid medium, with side lighting through the outer cylinder walls, or with glancing forward illumination, depending on the particular flow regime being observed. This high sensitivity to illumination for the different flow fields is in accord with the analysis by Savas (1985).

Particle suspensions in liquids used in other creative ways will be described in forthcoming sections. These include streakline photography (§6.1), laser speckle and particle image displacement velocimetry (§6.5), and fluid stability detection through the measurement of the intensity of laser beams transmitted through or reflected from a liquid suspension of particles (§12.1 and §12.2).

Interfacial Suspensions

Ahlborn (1902) was perhaps the first to report using aluminum powder and lycopodium powder sprinkled on the the free surface of a water flow for the scientific purpose of flow visualization. Prandtl (1904) used mica flakes and aluminum powder in some of his early experiments to visualize boundary layer flow around solid bodies. A number of flow visualization photographs using aluminum flakes suspended on a liquid surface are given in Tietjens (1934). Certain difficulties with this visualization technique have long been understood. For studies of the flow around solid bodies the model is generally towed through still water to minimize free-stream turbulence. The water must be meticulously clean in order to avoid capillary effects from distorting the free surface and disrupting a two-dimensional flow simulation. Tietjens (1934) suggested the following test for cleanliness of the water surface. Sprinkle some aluminum powder on the free surface and gently blow vertically down on it to clear a circular area of the surface. If the aluminum particles remain where they are the surface is clean, but if the circle closes on itself the surface is contaminated. Even with clean water the curvature of the liquid at a solid body due to non-normal fluid contact will steer free surface particle suspensions away from the body. A 90° contact angle can be maintained by coating the walls of the model with paraffin. Another problem is that tiny flakes suspended by surface tension tend to agglomerate on the free surface. Particle agglomeration can be substantially reduced by adding a surfactant to the liquid. For example, Ibbetson (1967) added a few drops of 'Photo-flo' to water in his experiments on Rossby wave propagation in a rotating annulus to inhibit free surface particle coalescence.

Hide (1958) made extensive use of aluminum powder for free-surface flow visualization in some laterally heated rotating annulus experiments. From the free-surface patterns he recorded stability boundaries, mode numbers, wave amplitudes and drift velocities of azimuthal instability waves. Faller and Mooney (1971), in a study of Ekman layer motion over a regular array of hills, used polyethylene rods 12 mm in diameter and 30 mm long for surface tracers. Each rod was weighted with a tack at one end to keep the axis vertical and nearly totally submerged. The authors photographically recorded three successive positions of the rods at equal time intervals to estimate the circumferentially-averaged surface stress. Stern, et al (1982) dropped small paper floats on the surface of a radially collapsing free-surface density current in a rotating tank. Analysis of their motion helped delineate regions of reverse flow in the main current from those in eddies shed from the density current. Flierl, et al (1983) also used paper pellets to visualize the epicycloidal motion induced at the free surface by asymmetric modons produced at depth in a rotating fluid. Velocity vector fields were obtained using time lapsed photography. Boyer, et al (1984a) used surface floats downstream of a vertical circular cylinder piercing the free surface in a rotating channel flow to determine the trajectories of the centers of shed vortices and estimate the vorticity distributions. Strong asymmetries in the vorticity fields between cyclones and anticyclones were determined by this simple observational technique. Narimousa and Maxworthy (1985) placed neutrally buoyant particles at the interface between two liquid layers to record flow patterns using streak photography in an experiment on shear-driven coastal upwelling. The particle luminescence method of Gharib, et al (1985) to be described in §3.1.4 has also been used for flow visualization and velocity measurement on a liquid free surface.

Two novel free surface tracers have been reported in rotating flow experiments. Faller (1963) floated small light bulbs 2.5 cm long shining vertically upward to record with time lapse photography the relative angular velocity generated by source-sink flow in a rotating tank. It was reported that the movement of the light bulbs were unaffected by the deleterious effects of surface tension and the wind stress set up at the surface of the rotating liquid because the tracers were well embedded in the upper fluid layer. Sambuco and Whitehead (1976) tethered four wooden logs to float in place on the crest of water flowing over a weir in a rotating hydraulics experiment. The local streamline deflection of the fluid flowing across the weir was determined by the angle of inclination of the floating logs.

3.1.3 Electrolysis Techniques

Tellurium Dye Method

An electrolytic technique for flow visualization using tellurium metal was developed by Wortmann (1953) for use in water flow. The method has been reviewed by Merzkirch (1987) from which the following description is taken. An electrolytic cell is made by running a d-c current through a tellurium cathode immersed in water to a distant metal anode. The cathode may be pure tellurium wire or a platinum wire electroplated with tellurium. With sufficient voltage applied between the electrodes, a complicated electrochemical reaction takes place in which tellurium ions are separated from the cathode and brought into a state of colloidal suspension with oxygen from the water. The result is a black dye that marks the fluid. A small amount of sodium chloride improves the water conductivity and a few drops of hydrogen dioxide provide additional oxygen molecules to enchance the formation of the colloidal suspension. A pulsed tellurium filament stretched normal to a stream provides a dark timeline which may be used for velocity measurement in steady flow if the colloidal state of the dye is stable. The suspension stabilizes when the water pH is elevated to a value above 7 by the addition of small amounts of potassium hydroxide. Since tellurium is actually given off in suspension, the dye is not a true solution and the tellurium wire has a limited lifetime. Wortmann (1953) demonstrated the technique by pulsing current through a wire strung normal to the flow over a flat plate. The dye trace formed a timeline indicative of the laminar velocity boundary layer profile. Eichorn (1961) has developed small tellurium probes for streakline visualization in low-speed natural convection flows.

The only reported use of the tellurium dye method in rotating flow appears to be that of Benjamin and Barnard (1964). They investigated the flow created by pulling the plug on fluid contained in a horizontal pipe rotating about its central axis. For high rotation rates the ensuing motion is one of an axisymmetric air cavity propagating down the pipe to replace the annular fluid flow discharging from the pipe. The water flow ahead of the cavity was measured by stringing fine tellurium plated platinum wires diametrically through the liquid at three upstream locations in the pipe. Contact with a metal bearing served as the anode and the three cathode wires were connected to a 200 V supply in the laboratory through individual slip-rings. A one-second pulse released tellurium dye whose motion was recorded with a movie camera. Centerline displacements of the fluid ahead of the air cavity were measured and compared with the displacement of the tip of the intruding cavity.

Hydrogen Bubbles

A popular method for visualizing water flow is the electrolytic technique for producing tiny hydrogen bubbles first used by Geller (1955). In its simplest form, a fine platinum wire oriented normal to the flow upstream of the region to be visualized serves as the cathode in a d-c circuit. The anode may be any convenient conducting part of the apparatus or a metal electrode inserted in the flow. With the proper level of current passing through the wire, small hydrogen bubbles are generated at the wire surface and swept downstream by the oncoming flow. Like

Wortman, Geller (1955) demonstrated the hydrogen bubble visualization technique in a laminar boundary layer flow.

Schraub, et al (1965) studied the technique in detail and reported various methods of utilizing hydrogen bubbles for quantitative velocity measurements in time-dependent water flows. The goal is to insulate the cathode at uniform sections along its length and produce short pulses of bubbles separated in time, thereby combining streakline and timeline information. They reported a simple pulsing circuit tunable for various operating conditions. More sophisticated control circuits have been reported by Clutter, et al (1961). Sodium sulfate is added to enchance the conductivity of soft water, but generally no added electrolyte is needed for hard water. They found best results for wires in the range 0.0005-0.002 inches in diameter, platinum being the obvious choice for strength and corrosion resistance.

A valuable contribution by Schraub, et al (1965) is their careful analysis of errors in velocity measurement due to bubble rise velocity, velocity defect behind the wire, measurement of particle displacement from photographic film, etc. Using a three-dimensional horizontal flow containing large-scale fluctuations as a worst case example, they estimated a 4-5% uncertainty in velocity measurement for speeds greater than 20 cm/sec, but reported that a much better accuracy could be expected for simpler flows. A lower bound on the usefulness of the method due to bubble buoyancy effects was placed at a *horizontal* speed of 2 cm/sec. Low speed *vertical* velocities, on the other hand, may be accurately measured if the bubble buoyancy is known (Clayton and Massey, 1967).

Use of hydrogen bubbles in rotating flows has been somewhat limited. In laboratory simulations of tornado and hurricane type flows, Maxworthy (1967, 1972) pulsed horizontal and vertical hydrogen bubble wires to obtain measurements of azimuthal and radial velocity profiles, respectively. Maxworthy (1970) also used hydrogen bubbles to visualize the complicated flow field induced by the motion of a sphere along the axis of a rotating fluid. Accurate measurement of radial distributions of both axial and azimuthal velocities relative to the rotating frame were obtained using closely spaced multiple-pulsed timelines emitted from a single wire mounted upstream of the moving sphere. Hydrogen bubbles pulsed simultaneously from three wires mounted across the flow, at three forward locations, gave a clear picture of the upstream wake and delineated a conical forward stagnation zone attached to the sphere. Downstream visualization with hydrogen bubbles exhibited the complicated structure of the rearward wake and revealed similarities with the flow of a stratified fluid over an obstacle.

The Baker Method

A pH indicator technique introduced by Baker (1966) permits both flow visualization and velocity measurement in low speed aqueous flow. The method is based on the fact that a current passed between two electrodes placed in an indicator solution locally changes the pH by a proton transfer reaction near the anode. If the solution is titrated to its end point on the acid side, the reaction will cause a color change in the fluid as it locally crosses to the basic side of the end point. Since the densities of the two forms of solution are exactly equal, the dyed fluid provides an excellent flow marker. The persistence of the colored liquid depends on the rate at which the autoprotolysis equilibrium is restored, which in turn depends on the dispersive and diffusive characteristics of the indicator. If the persistence is relatively short, the technique is limited to low fluid velocities, but has the advantage that it may be run continuously in a closed rotating flow system. Baker (1966) used thymolsulphonephthalein (thymol blue) because it is one of the few indicators readily soluble in water and has high acid-base color contrast. The yellow acidic solution changes to dark blue in the basic state and this contrast can be enhanced by illuminating the fluid with a yellow sodium light. Recently, Focke and Knibbe (1986) have reported successful use of o-cresolphthalein. This indicator has high color contrast since it is colorless in the acidic state and red-purple in the basic state. Its color retention is longer than thymol blue so that it may be used in higher speed flows. However, its low solubility in water requires the use of another test fluid. Focke and Knibbe used a mixture of equal volumes of water and ethanol.

The anode in these titration visualization methods is generally a single wire or a network of orthogonal wires stretched across the flow. Platinum is desirable because of its strength and resistance to electrolytic corrosion. The current between the electrodes must be low enough to prevent the generation of hydrogen bubbles at the anode. Velocities measured from timelines using thymol blue are limited to about 5 cm/sec according to Baker (1966). The lower limit, set by diffusion, is about 0.01 cm/sec.

Thymol blue (synonymous for the Baker method) has been used in numerous rotating flow experiments, especially those related to geophysical fluid dynamics. Baker (1967) found his method ideal for measurement of the low speed velocity profiles in the $O(E^{1/3})$ and $O(E^{1/4})$ Stewartson layers. Baker and Robinson (1969) used Thymol blue to obtain excellent velocity measurements and visualization of oceanic gyres in a rotating spherical annulus experiment designed to simulate the β-effect. Hide, et al (1968) used timelines to measure the tilt angle of Taylor columns produced by low Rossby number flow past spheres and cylinders. Malkus (1968) followed the motion of a radial thymol blue dye line produced by liquid motion in a rotating precessing sphere modeling the Earth's fluid core. Traces of the observed dye movement in a meridional plane demonstrated solid body interior motion bounded by complicated reverse flow shear layers near the outer spherical boundary. Beardsley (1969) used the Baker method for measurement of intensified western boundary currents and reported that velocities less than 0.1 cm/sec could be detected. Maxworthy (1968) observed the complicated meridional flow field around a sphere slowly translating along the axis of a rotating fluid using thymol blue. In a similar experiment Pritchard (1969) reported the detection of fluid particle displacements less than 0.01 inch using the Baker method. Hart (1971) stretched three vertical wires across a rotating channel flow to visualize predicted cross-channel roll instabilities using thymol blue dye streaks. Hseuh and Leguckis (1973) used thymol blue timelines to measure velocity profiles in a rotating water channel developed to simulate western boundary currents. Firing and Beardsley (1976) measured the angular rotation of an isolated eddy produced by the vertical retraction of a cylinder mounted flush with the lower boundary of a rotating fluid. This simple experiment exhibits the usefulness of the Baker method for both flow visualization and quantitative measurement. Finally, we mention the work of Koschmieder and Lewis (1986) who have presented some striking thymol blue photographs of Hadley circulations generated by nonuniform heating in a rotating tank.

Electrolytic Precipitation

This electrolytic precipitation visualization technique was introduced by Taneda and Honji (1971). The method is based on the observation that if a small potential difference is applied between a solder cathode and a metal (brass) anode in an electrically conducting liquid, a white colloidal cloud is released from the solder. In water the colloidal suspension is slightly negatively buoyant. Dye lines produced from solder wires in a streaming flow appear as dense white streaklines which are readily photographed against a dark background. For proper application of this technique the electrical conductivity of the working fluid must be carefully controlled. Tap water may be used, although better control of the visualization is obtained by using distilled water to which small amounts of common salt and sodium carbonate are added (Boyer and Davies, 1982). Well defined dye lines are produced by optimizing both the applied voltage and the salt concentration. Too much salt results in polarized bubbles and too little produces no tracer. Similarly, the voltage to the electrodes needs to be adjusted for effective dye production and yet avoid bubble formation. Degradation of the surface of the solder with time through the action of electrolysis mandates periodic replacement of the cathodes.

Presumably this visualization technique could be operated in a pulsed mode as with hydrogen bubble visualization, but to date it has only been used to produce continuous streaklines. Boyer and Davies (1982), in their study of rotating flow past a cylinder on a β-plane, produced bright streaklines from a series of vertical solder wires located upstream of the obstacle. A solder cathode mounted flush around the circumference of the cylinder gave good visualization of the separation bubble and unsteady dynamics at the rear of the cylinder which prevail at higher Rossby number. In the regime for which the separation bubble is steady, quantitative measurements of flow separation angle and bubble length was made using electrolytic precipitation visualization. In the unsteady regime the dye lines follow the shed vortices and can be used to measure the shedding frequency. Boyer, et al (1984b) have also used electrolytic precipitation to observe f-plane rotating flow of a homogeneous liquid past disks and cylindrical depressions. The latter two applications utilized flow visualization in a horizontal plane. Electrolytic precipitation has also been proven successful for the visualization of lee waves in a vertical plane produced by stratified flow over long ridges (Boyer and Biolley, 1986) and over isolated three-dimensional topography (Boyer, et al, 1987a).

Electrochemiluminescence

Electrochemiluminescence is the generation of light by electrochemical means. Harvey (1940) noted that when a voltage is applied across two electrodes immersed in the flow field of a chemiluminescent solution, a blue glow appears at the surface of the anode. Howland, et al (1962) first recognized its use for flow visualization and showed that the glow is generated within a few wavelengths of the anode surface, a distance much smaller than the boundary layer thickness for flow over the anode body. This led Howland, et al (1966) to realize that electrochemiluminescence is well-suited for visualizing flow phenomena that occur at the wall, particularly flow separation. The chemiluminescent solution studied in detail consisted primarily of luminol dissolved in distilled water. Since luminol only exhibits chemiluminescence above a pH of 8, potassium hydroxide is added to adjust the solution pH level. Hydrogen peroxide is the active electrolyte whose concentration strongly affects the light intensity and potassium chloride is the supporting electrolyte. Satisfactory glow can also be obtained using methanol, acetone and glycerol mixtures. The d-c circuit consists of a low resistance voltage supply, an anode in the shape of the model about which the flow is to be studied, a cathode and an additional saturated potassium chloride-calomel electrode. This latter electrode is placed near the model anode in a location where it does not disturb the flow, and is used to measure the potential between the anode and the solution. Of the many anode metals tested, only platinum gave a satisfactory luminescent glow. For typical flow velocities in the range 2-30 cm/sec it was found that no glow appeared at potentials less that about 0.2 V and the glow left the surface of the anode at voltages above 1.5 V. In the intermediate voltage range the glow stayed attached to the anode surface and the potential was set to obtain maximum glow contrast.

Howland, et al (1966) produced a number anode shapes for flow visualization purposes in a rotating flow of the chemiluminescent solution. The boundary of flow separation from cones, cylinders and curved plates immersed in the rotating fluid appeared as dark blue lines of concentrated glow intensity. At voltages above 1.5 V the wake of the test models were illuminated by the glow field swept downstream of the anode, but the authors gave no practical application for this effect since the glow lines could not be identified as either streamlines, streaklines or pathlines. An alternative use of electrochemiluminescence for mass transfer measurement is reported in §8.

3.1.4 Radiation Induced Visualization

Laser Induced Fluorescence

Laser induced fluorescence (LIF) as a tool for liquid flow visualization was first reported by Dewey (1976). Fluorescence is a process by which the molecules of a substance absorb photons from an external energy source and then reradiate some of the absorbed energy. Small concentrations of sodium fluorescein or rhodamine 6G dissolved in water and excited with radiation from an Argon ion laser (514.5 nm) produce a remarkably strong fluorescence because of the high quantum-yield of the dyes. Also, these dyes fluoresce with a very short response time which makes them suitable for visualization of complicated time-dependent fluid motion. Dewey (1982) has shown that the luminescence intensity is linearly proportional to the dye concentration and the local laser power. LIF requires a match between the radiation wavelength of a laser with the peak of the fluorescence response curve of the dye. The only problem is to find matched wavelength dyes which are soluble in common liquids. For He-Ne lasers (632.8 nm) we have had some success with Oxazine 720 dissolved in methanol.

For visualization purposes the dye is mixed in dilute solution with the working fluid and injected into the flow stream from orifices along flow boundaries or from injection tubes placed in the flow field. The dye stream is generally excited by a sheet of laser light to identify flow structure in the plane of illumination. In two-stream liquid phase flows, such as a free shear layer formed downstream of a splitter plate or a jet penetrating into a quiescent fluid, it is useful to let one of the streams be the dilute dye solution. Then laser sheet illumination provides a clear picture of the mixing process at liquid-liquid boundaries (Dimotakis, et al, 1983).

In rotating flows Escudier (1982, 1983) and Bornstein and Escudier (1984) have made considerable use of LIF using fluorescein dye excited by Argon ion laser light sheets. Escudier (1982), in particular, has presented some remarkable photographs of the structure of five different types of vortex breakdown observed in swirling flow confined within cylindrical tubes. Bornstein and Escudier (1984) used LIF to pinpoint the location of wandering vortex-breakdown bubbles at the time of a laser Doppler velocity scan. In the flow produced in a closed cylindrical container by rotation of one endplate, Escudier (1984) observed three vortex bubbles aligned in succession along the cylinder axis. The intricate detail of the flow structure within the vortex bubbles made visible by LIF testify to its usefulness as a flow visualization tool.

Particle Luminescence

Nakatani, et al (1971) proposed a technique for measuring velocity in a liquid flow seeded with luminescent particles such as zinc sulfide or cadmium sulfide available with characteristic diameters on the order of 15 μm. Excitation of these particles with visible or ultraviolet radiation gives rise to a very short electronic transition (fluorescence) followed by a slow exponentially decaying luminescence due to lattice vibration (phosphorescence). It is well-known that the wavelength of emission is longer than that of excitation. Hence there is no diffusion of the excited light and marked particle boundaries remain sharp over the lifetime of emission. Nakatani, et al (1971) discussed two ways by which this method may be used for velocity measurement of flow in a rectangular channel moving in the z-direction. In the first method, a thin plane normal to the flow at $z = 0$ is continuously illuminated and the intensity of particle emission at downstream position $z = d$ is measured to determine the time of travel to that position by use of the emission decay rate formula with known time constant. This method has not yet been developed. In the second method, a sheet of light is pulsed over a very short time and the marked particles form a timeline that can be used to deduce velocity profiles in steady flow. This technique has been developed by Nakatani, et al (1975) to measure average velocity distributions across a diverging channel. The authors also report the lifetimes of various luminescent materials which range from 40 ms for a zinc sulfide/cadmium sulfide mixture to 1500 ms for pure calcium sulfide.

Gharib, et al (1985) have demonstrated an alternate method of utilizing particle luminescence for flow visualization and velocity measurement. They coated 200 μm polystyrene spheres with submicron size zinc sulfate crystals which have an absorption wavelength of 400 nm and an emission wavelength of 530 nm. These particles were sprinkled over the surface of a rotating liquid and hit with two successive excitation light pulses during the lifetime of photoemission. What one observes on a photograph is a bright spot due to the intense short-lived fluorescence ($10^{(-5)}$ sec) followed by a decaying afterglow which persists for about 100 ms for these zinc sulfate crystals. The velocity is then determined by measurement of the distance between spots and the direction of fluid motion is unambiguously determined by the direction of intensity decay during luminescence. Gharib, et al (1985) demonstrated their particle luminescence technique in a rotating flow by measuring time-dependent free-surface velocity profiles for water in a cylindrical container set into motion by a paddle rotating near the bottom boundary. Particles sprinkled on the water surface were hit with two 20 μs flashes separated in time by 250 ms. Photographs of these double pulse records were taken every 3 sec during fluid spin-up. The dot pairs in a given photograph were analyzed by a computer trace-following-algorithm (see §6.1) and radial profiles of the azimuthal velocity in this axisymmetric flow were generated. Although this technique is presently under development, it shows great promise for full-field two-dimensional flow visualization and velocity measurement.

Photochromism

Photochromism, or coloration by light, is the reversible change of a single chemical species between two states having different absorption spectra. The organic dyes typically turn magenta, blue, purple, or green upon exposure to ultraviolet (UV) radiation and then clear again when the radiation is removed. Fowlis (1979) has performed experiments on the thermal fade rate of 2 Benzo Pyrane dissolved in dilute concentrations in silicone oil exposed to 50 mw UV laser power for 3 seconds. He found that the relaxation time is a strong function of the oil temperature in that it disappears much more quickly at higher temperature. Typical fade rates ranged from 1-10 min at 45 °C and 10-200 min at 30 °C, depending on the dye concentration. The visualization technique was first reported by Popovich and Hummel (1967) for use in a study of the viscous sublayer in turbulent pipe flow. Further studies by Frantisak, et al (1969) and Iribarne, et al (1972) showed how photochromic dye may be used to make quantitative measurements in laminar and turbulent flow. In these applications a very low concentration of the photochromic compound is dissolved in the test fluid and illuminated by a UV laser beam normal to a transparent wall. For high transmittance fluids the laser beam will excite a uniformly bright timeline across the flow section and velocity profiles can be measured if the pathlines of the individual particles are known.

Larsen and Roesner (1982) showed how pathlines could be measured using highly saturated solutions of photochromic dye in the parent liquid. In this case light is absorbed along only a very short distance from the point of illumination which makes it possible to create small colored-fluid spots by introducing the UV light into the fluid through a glass-fiber probe. For example, if the fiber probe is moved at constant speed along a line in the fluid, colored fluid volumes can be generated at equal space intervals by pulsing the UV light through the fiber probe at equal time intervals. The investigators demonstrated the use of this method by measuring pathlines of convective fluid motion in a differentially-heated rectangular box.

Photochromic dye visualization is particularly attractive for experiments in confined rotating fluids because the external laser excitation is virtually disturbance-free and contamination is not a problem --- the dye solution may be used repeatedly after each thermal relaxation has occurred. Bar-Yoseph, et al (1986) employed photochromic dye visualization in a study of fluid motion in the gap between eccentrically rotating spheres. A UV flash from inside the hollow inner sphere illuminated photochromic molecules adjacent to the wall of that sphere in the fluid gap. The colored particles in the boundary layer were convected around the gap to visibly mark asymmetric vortex patterns above and below the equator. In a recent publication Roesner (1988) has presented beautiful photographs of two-cell vortex breakdown patterns using photochromic dye visualization.

3.2 Air Flows

In the following sections we describe methods for visualizing rotating flows with air as the test fluid. Surface flow patterns may be visualized with China clay and interior fluid motion is observed by injecting smoke or helium bubbles into the fluid stream. A new technique incorporating soap films provides distinctive visualization of two-dimensional air flow instability patterns in confined geometries.

3.2.1 China Clay

The China clay method was developed through collaborative efforts of the National Physics Laboratory and the Chemical Research Laboratory in England as a method for indicating steady lines of turbulence transition on airfoils. The China clay technique, reported by Richards and Burnstall (1945), was developed to overcome some difficulties of similar, but toxic, chemical method used by Gray (1944). The method relies on the differential rate of evaporation in laminar and turbulent flow of a liquid absorbed on a solid spread over a test surface. In practice a solid layer of China clay is sprayed on the surface to form a smooth white absorbent film. When a liquid of the same index of refraction is sprayed over the film it becomes transparent and subsequent evaporation of the absorbed liquid then exposes the white China clay. When subjected to an external flow, the visualization technique is simply an indicator of relative rates of evaporation in the boundary layer adjacent to the surface. Since China clay has a low index of refraction ($n = 1.56$), there are a number of liquids of nearly the same refractive index which may be used. This enables the experimentalist to choose a liquid with an evaporation rate suitable for a particular application. Some nontoxic liquids with indices of refraction compatible with China clay, listed in order of increasing evaporation rate, are · nitrobenzene, ortho-nitrotoluene, ethyl-silicate, chloronaphthalene, and iso-safrole. For comparison, iso safrole has a relative rate of evaporation 7.6 times that of nitrobenzene.

Gregory, et al (1955) were evidently the first to apply China clay to rotating flow. They investigated the flow induced along the surface of a flat circular plate rotating at constant speed below a quiescent fluid. In this system fluid in the Ekman boundary layer is centrifuged radially outward in a spiral pattern and replaced by the uniform vertical drift of the quiescent fluid above the rotating plate. Their observations with China clay applied to the surface of the plate indicated two critical radii separating three flow regimes. Within the inner radius the slow rate of evaporation indicated a purely laminar, stable flow. Beyond the outer radius the high rate of evaporation revealed a wholly turbulent region. In the intervening region equi-angular spirals were recorded in the China clay surface. Their theoretical analysis for this flow indicated two forms of instability, one composed of spiral vortices which are stationary with respect to the rotating plate and the other composed of rapidly rotating spiral vortices. The China clay method revealed the stationary vortices which were made visible by nonuniform evaporation across an instability wavelength. Although the method was originally developed to discern transition to turbulence, this investigation showed that it could be used to detect more delicate differences in evaporation rate associated with the secondary flow produced by laminar instability.

3.2.2 Smoke

We use the terminology "smoke" in a generic sense to denote various types of fume-like visualizations for gaseous flow. These include smoke, oil mists, kerosene sprays, and chemical reactions producing visible fumes. Conventional smoke wires and smoke generators used for continuous visualization in wind tunnel flows may also be used in nonrecirculatory rotating flows. Titanium tetrachloride and tin tetrachloride release dense smoke when brought in contact with damp air. A discussion of titanium tetrachloride visualization in air is given by Freymuth (1989). A fine kerosene mist may be generated using a Preston-Sweeting mist generator and operational details of this system are reported by Maltby and Keating (1962) and Warpinski, et al (1972). A comprehensive discussion of smoke visualization methods is given by Mueller (1983).

Kaye and Elgar (1958) introduced smoke into the air flow between concentric rotating cylinders to observe the formation process of laminar Taylor vortices and their evolution to turbulent vortex cells with increasing Taylor number. Kreith, et al (1962) visually examined the flow over a rotating cone by introducing smoke from a kerosene smoke generator into the boundary layer from a point directly below the vertex of the cone. They observed that the boundary layer flow curls up into spiral vortices similar to those observed by Gregory, et al (1955) for flow over a flat rotating disk. So (1967) injected smoke to visualize low speed spiraling air flow through a conical diffuser. He pointed out that the smoke technique is not suitable for the determination of flow patterns at high Reynolds numbers when the flow in the diffuser becomes turbulent. Smoke is often used in laboratory simulations of atmospheric vortices because the swirling flow created in the test apparatus is generally exhausted to the atmosphere. Wan and Chang (1972) injected smoke near the top plate of a such an apparatus to visualize vertically recirculating cells in their spiral vortex flow. Smoke trapped in the cells allowed the investigators to distinguish between a central core recirculation cell and an annular recirculation cell at different flow conditions. In a similar experiment Curch and Snow (1979) observed spiral waves circumscribing a vortex breakdown by injecting smoke into the Ekman boundary layer feeding the vortex.

In experiments on circular shear flow instability, Rabaud and Couder (1983) placed droplets of ammonia solution and hydrochloric acid on the bottom of their enclosed test cell before rotating the cylindrical plates. Subsequent rotation of the disks mixed the fumes and created a dense ammonium chloride smoke which remained in suspension. Excellent photographs of unstable vortex patterns observed in the cavity were obtained with lateral illumination from a white light source.

3.2.3 Helium Bubbles

Another method of visualizing air flow is to inject soap bubbles into the stream. Bubbles are particularly ideal tracers since their size and buoyancy can be controlled. A bubble filled with air will slowly sink due to the weight of the soap film. Air-Helium mixtures injected into bubbles of controlled size can be adjusted to render the bubble neutrally buoyant. Practical use of this method of visualization in wind tunnel flows can be traced back to Redon and Vinsonneau (1936). Helium bubble generators are now commercially available. In these systems a console meters the helium, the bubble film solution, and air to control bubble size, mean specific weight, and bubble generation rate. Because soap bubbles typically reflect only 5% of incident light, care must be taken with lighting and photography. The goal is to shine as much light as possible on the bubbles and yet keep the background dark. Further details on the helium bubble flow visualization technique are given in Mueller (1983).

The first application of this visualization to rotating flows is evidently that of Owen, et al (1961). They injected helium bubbles into jet-driven vortex flow contained in a cylindrical tube. Mullen and Maxworthy (1977) injected neutrally buoyant helium bubbles into a laboratory flow simulating dust devil vortices. Planform and elevation photography of bubble streaks verified the azimuthal symmetry of the vortex flow. The visualization method helped identify both one-cell and two-cell vortex breakdown flow regimes.

3.2.4 Soap Films

Couder (1984) performed some experiments of grid turbulence on a soap film. In this experiment the soap film stretched across a rectangular wire frame is the two-dimensional fluid medium. He found that the flow lines produced by towing a small cylinder along the film could be visualized readily by reflection of monochromatic light from the film surface. The observed interference fringes are due to local nonuniformities in film thickness induced by motion on the liquid film.

Rabaud and Couder (1983) made use of a soap film for flow visualization of instabilities created in a circular shear layer. The circular shear layer in air is created by the differential rotation between inner and outer sections of the boundaries forming a cylindrical cavity. The soap film, comprised of a 0.5% solution of commercial liquid soap, is stretched across the horizontal cylindrical mid-plane of the test cavity where the flow is two-dimensional. A 5% addition of glycerol produced stable films and the experiment was conducted in a humid atmosphere to minimize evaporation. They observed that the horizontal film was flat, there being negligible curvature due to its own weight. The soap film was easily set into motion by air flow parallel to its surface. During the experiment the film slowly thins down at its center by a drainage mechanism due to local centrifugal forces in the rotating film. This causes undesirable circular interference fringes in the center of the test cell, but they do not interfere with the distinct instability flow patterns produced at the radius of differential rotation between the inner and outer halves of the test cavity. The vortices appear as concentric elliptical fringes on the film owing to the local thinning produced by centrifugal forces inside each rotating vortex. Further application of the visualization method to this flow system has been reported by Chomaz, et al (1988).

Extremely sharp definition of the morphology of the unstable flow is observed through the transparent walls of the test cavity using soap film visualization, and one can readily measure azimuthal mode numbers and propagation speeds with the aid of a video recorder mounted through the eyepiece of a rotoscope. A comparison of the unstable flow patterns with those observed using smoke visualization shows nearly identical results. As pointed out by Rabaud and Couder (1983) there are two basic concerns which must be borne in mind when using soap film visualization. First, it is probably only practical in two-dimensional flow where, under proper conditions of film thickness, type, and composition, the film exactly follows an air flow parallel to its surface. If the air flow that interacts with the film is three-dimensional, capillary forces in the film will oppose deformation in the film plane and impose zero normal velocity on its surface. In this case the film will be dragged by the air in regions where the parallel velocity is largest and return currents will appear elsewhere in order to maintain a divergence-free flow of the film in its own plane, in which case unambiguous interpretation of the observed fringes will not be possible. Second, although the introduction of the film does not disturb the flow patterns, it does produce a slight decrease in the effective kinematic viscosity of the fluid (air) in the cell. Thus control parameters based on viscosity, such as the critical Reynolds number for onset of instability, will be slightly in error if determined by soap film visualization without appropriate corrections.

3.3 Optical Viewing Systems

Special problems arise when trying to view flow patterns in a rotating system. In most geophysical flow simulations the entire apparatus is placed on a rotating table and one is interested in observing fluid motion *relative* to the rotating system. Furthermore, flow patterns are often identified by observation in a coordinate section of the flow. For example, horizontal sections are the most useful viewing planes in the rotating laterally heated annulus experiment. Meridional section views of toroidal and spiral instabilities formed in the gap between concentric rotating cylinders or spheres are particularly informative. All-around (360⁰) visualization of Taylor-Couette instabilities is likewise highly desirable. In some cases one is interested in tracking a flow feature which is not stationary with respect to the rotating apparatus or the laboratory reference frame. Various solutions to these optical viewing problems are discussed below.

A continuous stationary image of a rotating flow observed from a stationary reference frame may be obtained with a rotating Dove prism. Mechanical rotation of the prism along its optical axis causes a stationary image to rotate at twice the prism's rotation rate. Thus an image rotating at angular speed Ω may be "derotated" by viewing the image through a Dove prism counter-rotating along its optical axis at angular speed $\Omega/2$. This device, first introduced by Thoma (see Fischer, 1931), has come to be known as a rotoscope. Careful alignment of the rotating prism along its optical axis is necessary to prevent nutation of the derotated image (Sullivan, 1972). Hide (1958) and Fowlis and Hide (1965) coupled a rotoscope mounted overhead of a rotating heated annulus experiment directly to the turn-table to observe free-surface azimuthal waves. Many photographs of these free-surface flow patterns are presented in Hide's (1958) paper. Hide and Titman (1967) employed a rotoscope to observe the non-axisymmetric instability of detached Stewartson layers in a rotating fluids experiment. Rabaud and Couder (1983) adjusted the rotation rate of a rotoscope focused on a circular shear flow to stop the motion of rapidly-rotating vortical instabilities to facilitate viewing and measurement.

A planar light sheet projected through a fluid suspension of reflecting particles yields good visualization when viewed at right angles to the plane of illumination. Various methods have been used to produce light sheets. Greenspan (1969) projected light through a vertical slit to illuminate the meridional plane of a cylinder spinning-up from rest in order to visualize the advancing shear front. To illuminate an axial section of Taylor vortex flow, Benjamin (1978) and Benjamin and Mullin (1981) inserted two razor-blades in the slide-holder of a slide-projector so that light only passed through the gap between the sharp edges to form a narrow beam slit. Fitzjarrald (1982) projected a laser beam along the central axis of a rotating heated annulus onto a differentially rotating 45⁰ mirror. Light reflecting radially from the rotating mirror then swept out circles illuminating a thin horizontal plane of fluid motion. Bornstein and Escudier (1984) illuminated fluorescein dye through a meridional section of a vortex breakdown with laser light reflected from a rapidly oscillating mirror. Lourenco and Krothapalli (1987) diverged a focused laser beam through a cylindrical lens to produce a narrow light sheet for laser speckle photography.

There is often the need to observe fluid motion 360° around the circumference of an experiment. This is especially true in annulus experiments which exhibit flow dislocations that disrupt regular modes of instability (Weidman and Mehrdadtehranfar, 1985). Coles (1965) solved this problem in an elegant manner for all-around visualization of Taylor vortex flow. He wrapped photographic film around the circumference of the outer cylinder and exposed it with a Xenon light source placed along the central axis inside the annulus. The light flash was first diffused through horizontal baffles before traversing radially across a fluid suspension of aluminum flakes and onto the film strip. All-around photographs of doubly periodic wavy vortex flow produced in this manner are presented in Coles' (1965) paper. Donnelly, et al (1980) identified dislocations in Taylor vortex flow by taking a photograph through mirrors arranged to give three contiguous 120° views of the cylinder. Although the flow patterns are distorted by the curved cylinder walls, the dislocations are clearly evident. Gorman and Swinney (1982) placed two mirrors at right angles symmetrically behind a Taylor-Couette apparatus to obtain a five-sectioned all-around view of wavy vortex flow with a single camera snapshot. Care must be taken in interpreting these photographs since two of the five views are reversed.

Onji, et al (1986) resolved the problem of all-around visualization in a Taylor-vortex experiment by fitting a mirror of conical section around the outer cylinder at the base of the apparatus. The curved mirror is inclined 45° to the central axis so that an overhead view of toroidal vortex instabilites are revealed as concentric circles. The number of vortices viewed in this manner depends on the instability wavelength and the height of the conical mirror. Couder has also used a conical mirror for all-around visualization of Taylor vortex flow (Chomaz, private communication).

Isolated features in rotating flows have been tracked by various means. In experiments performed in a long channel of density stratified fluid rotating at constant angular speed, Maxworthy (1983) followed a solitary internal Kelvin wave with a motorized camera. A 45° mirror attached to the trolley supporting the overhead camera permitted simultaneous photography of both the plan view and side view of the propagating dyed wave front. King, et al (1984) followed azimuthally propagating wavy vortex structures by mounting a television camera on a separate table rotating coaxially around their Taylor-Couette apparatus. The table speed was adjusted until the wave pattern was stationary when viewed on the television monitor. They found this method of measuring the wave speed very useful because defects in the flow pattern made interpretation of separately measured power spectra difficult.

3.4 Interferometer, Shadowgraph, and Schilieren Systems

Optical techniques sensitive to changes in the magnitude or gradients of refraction index include differential interferometers, shadowgraphs, and Schilieren measurement systems. The operational details of each system are well-known and have been reviewed by a number of authors. Two particularly useful references are Goldstein (1983) and Merzkirch (1987). In what follows we simply outline the basic features of each optical system and indicate what measurements can be made. The fundamental difference between the imaging techniques is that interferometers are sensitive to changes in the optical path length, shadowgraphs respond to the first derivative of the index of refraction normal to the light beam and Schilieren optics provide a measure of variations in the second derivative of the index of refraction normal to the light beam. In each system light rays projected through a fluid test section are imaged on a screen and the resulting image represents the *integrated* optical distortion to which that system is sensitive. Since the density ρ of a homogeneous transparent medium is related to the index of refraction n through the Lorenz-Lorentz relation,

$$\rho = \text{const.} \left[\frac{n^2 - 1}{n^2 + 2} \right] , \tag{1}$$

the optical imaging techniques respond to density variations in the fluid test volume. In low speed rotating gas flows the pressure field often may be assumed constant. In this case the density field can be directly related to the temperature field through the ideal gas equation of state and the imaging techniques are then responsive to temperature variations in the fluid volume. Similarly, in rotating homogeneous liquid flows with small density gradients, the density is directly proportional to the temperature through the coefficient of thermal expansion, so again the imaging techniques respond to thermal nonuniformities in the fluid volume.

Because the imaged fields produced by the three systems are integrated measurements along the path of the light beam, these optical methods are best-suited for measurement in two-dimensional flow fields for which there is no change in refractive index along the illumination path. However, this does not preclude their use for three-dimensional flow visualization. In fact, the complicated geometrical optics analysis necessary to compute the light refraction path through the fluid for the Schilieren and shadowgraph techniques generally precludes their use for quantitative measurement in all but the simplest two-dimensional flows. The interferometer is most amenable to direct measurement of density or temperature since the light image does not depend on the refraction of light through the test cell but only on differences in optical path length.

Holographic interferometry and tomography have not been used in rotating flows, although the latter shows great promise for determining arbitrary three-dimensional distributions of refractive index variations. Lauterborn and Vogel (1984) review the application of these optical techniques for measurement and visualization of fluid flows.

Schmitt and Lambert (1979) used a horizontal shadowgraph setup developed by Shirtcliffe and Turner (1970) to observe the development of salt-finger instabilities produced at the interface of a sugar-salt solution undergoing uniform rotation. Direct measurements of the thickness of the fingering layer were obtained from photographs of the shadowgraph images taken in the rotating reference frame. Busse and Heikes (1980) observed the evolution of convective instabilities in a rotating thin fluid layer heated from below with shadowgraph imaging. In this application the shadowgraph responds to temperature variations in the convecting fluid and light and dark regions can be associated with hot rising and cold descending fluid, respectively. Density profiles for thermal cellular convection of nitrogen gas in rotating rectangular boxes were measured by Buhler and Oertel (1982) with a differential interferometer. The interferometer could be made sensitive to horizontal or vertical density gradients by simple rotation the prism optics, and steady convection photographs produced in each manner are presented in their paper.

Hart, et al (1986) employed both color Schilieren and shadowgraph photography in a spacelab experiment on thermal fluid convection in a rotating differentially-heated hemispherical shell. The reason for conducting the experiments in space was to have an environment relatively free of background gravity so that a purely radial gravitational field could be produced electrostatically in a dielectric test fluid. With the hemisphere rotating, a nonrotating optical train focused light from a grating onto the reflecting inner sphere. When the incident rays are compared with the reflected rays one obtains various measures of the depth-averaged temperature field in the fluid shell. Several photographs of the complex flow structures found with and without imposed north-south heat flux gradients are presented in their report.

3.5 Naturally Visualized Flows

In some experiments static or dynamic flow patterns may be observed naturally without the need to introduce a foreign visualizing agent. For example, interfacial boundaries between water and air or between two immiscible liquids are often naturally visible owing to innate differences in color or refractive index between the two fluids. These and other examples of naturally visualized flows are given below.

It is well known that some non-Newtonian liquids will climb up the surface of a rotating rod immersed vertically in the liquid bath. Joseph, et al (1973) recorded naturally visible free-surface profiles of climbing liquids by photographing the steady shapes at grazing incidence across the surface of the liquid bath. Measurements were obtained from slides projected to give very enlarged surface profiles. Values of the height and radius accurate to ±0.1 mm were easily obtained using this simple observation technique. Joseph, et al (1984b) made direct measurements of climb heights for various non-Newtonian liquids by direct observation of the visible contact line on the surface of the rod. Using a microscope mounted on a vertical vernier scale, they reported measurements repeatable to within ±0.05 mm. Tieu, et al (1984) measured the interfacial shape between immiscible oil additives in a vertical cylinder undergoing angular oscillations by taking high-speed photographs of the naturally visible interface.

Standing waves on hollow-core vortices produced by gaseous cavitation and/or injection of air in rotating water flows are easily visualized because of the refractive index discontinuity at the air-water interface. Keller and Escudier (1980) photographed the steady interfacial varicose wave patterns in order to determine the wavelength dependence on the hollow-core diameter. Manuel, et al (1987) produced hollow-core vortices by injecting air through the tip of a blade placed at angle of attack to the axial flow of water in a cylindrical tube. Travelling solitary waves and hydraulic jumps were observed on the interface of the air cavity. Measurements of the phase speed of the solitary waves and of the change in diameter across the hydraulic jumps were made with the aid of high speed photography. In a somewhat similar situation Benjamin and Barnard (1964) observed the motion of an air cavity travelling along the central axis of a rapidly rotating horizontal pipe initially full of water. Diffuse backlighting provided sufficient contrast for high-speed photography of the naturally visible cavity boundary.

In studies on the centrifugal separation of a mixture, particle suspensions inherently in the fluid provide a means for visualization. In this industrial process, the axis of rotation is vertical and the particles separate out radially in a nearly horizontal planes because the centrifugal acceleration is much greater than the gravitation acceleration. Schaflinger, et al (1986), Schaflinger (1987), and Schaflinger and Stibi (1987) have observed the separation fronts of monodisperse suspensions of tiny glass beads or polystryrene spheres during the centrifugation process by strobing a light beam axially through the suspension. With radial partitions, the separation fronts in a horizontal plane are not cylindrical because the Coriolis forces acting on the particles become significant at high rotation speeds. These separation fronts are readily observed by the marked contrast in axial illumination fore and aft of the front separating the mixture ahead from clear liquid behind.

Hakim and Schowalter (1980) observed the deformation of a silicone oil drop suspended in castor oil and sheared by eccentrically rotating parallel disks. The interfacial boundary of the 1 mm diameter drops were easily viewed through a camera-microscope system owing to the difference in the indices of refraction of the two liquids. Measurement of the modal shapes of acoustically levitated, oscillating rotating drops have been obtained by Annamalai, et al (1985). The naturally visible boundary of the silicone oil/carbon tetrachloride drops rapidly vibrating in air were recorded on cine film and later analyzed for measurement of drop oscillation frequencies and amplitude decay constants. Krumdieck and Weidman (1985) have made measurements of the centrifugal instability of capillary rivulets formed in the annular groove of a rotating disk. The spontaneous departure of liquid from the groove at a critical angular velocity is readily detected by eye. Observation of multiple equilibrium states of water naturally visible in an eccentrically rotating glass U-tube have been reported by Denardo (1988). The motion of the rotating tube was stopped with a synchronized strobe flash and height measurement of the liquid surface was made with a telescope mounted on a vernier traverse.

Two other types of naturally visible flows have been observed in laboratory simulations of atmospheric vortices which include sources of buoyancy. Turner and Lilly (1963) introduced carbonated water into a rotating cylinder as a means of producing buoyancy by changes of phase in the fluid medium. The addition of small nuclei (eg., common salt or Alkaseltzer pellets) along the axis of rotation produces vigorous convection in the form of a strong vortex containing tiny carbon dioxide gas bubbles which make visible the very flow they create. Luguovtsov (1982) generated a tornado-like vortex by rotating a vertical hollow cylinder just above the surface of a heated water bath. He observed and photographed fog condensation in the columnar vortex which made its spiral motion readily visible.

4. PRESSURE MEASUREMENT

Flush-mounted wall pressure taps, Pitot probes and yaw probes commonly employed in wind tunnel testing have been used to some degree in rotating flows. Differential pressures between a probe and a reference pressure is measured with a differential liquid manometer or other pressure transducer. Wall taps give the local static pressure and yaw probes determine flow direction by balancing pressures from two or more static pressure taps separated $60°$ apart on the surface of a sphere pointed into the oncoming stream. Pitot probes aligned with the free stream measure the total pressure. Pitot-static probes buck the total pressure with the local static pressure to yield the dynamic pressure $\rho U^2/2$ from which one can calculate the velocity U if the fluid density ρ is known. Pitot-static probes for velocity measurement have been largely replaced by hot-wires and hot-films, which themselves have been superseded by laser Doppler velocimetry. Pitot-static and yaw probes represent a major intrusion in rotating flows and should be used with care. Wall pressure taps, on the other hand, do not interfere with the flow. A complete discussion of the principles of operation of these probes and their use with differential liquid manometers is given by Pope (1958).

Owing to their restricted time response, liquid manometers give only time-averaged or slowly-varying pressure measurements. Measurement of fluctuating pressures can be obtained using condenser microphones, piezoelectric crystals or strain gage transducers. Condenser microphones are capacitive devices whose capacitance varies with changes in plate separation caused by an applied external pressure. These transducers are fragile and the fact that the dielectric constant of the capacitor is determined by the humidity and temperature of air between the plates generally rules out their use in liquids. Piezoelectric crystals generate an electric field when mechanically deformed. In transducers the crystals are usually plated on opposite surfaces and the electric charge produced by deformation of one surface relative to the other provides a measure of the applied force over the sensor area, and hence the local pressure. The impedance of the transducer is basically capacitive, but reversible in contrast to condenser microphones. The third common type of pressure transducer consists of a flexible diaphragm whose deformation is measured by a strain gage attached to its surface. Fundamentals on the operation, use, and frequency response of these transducers are given in Beckwith, et al (1982) and Blake (1983). A comprehensive review of pressure measurement techniques has been reported by Soloukhin, et al (1981). A report on state-of-the-art dynamic wall pressure measurement techniques is given by Leehey (1989).

Taylor (1935) reported what may be the earliest use of pressure probes in a rotating flow. He constructed tiny pitot pressure tubes to make pressure surveys across the gap of supercritical circular Couette flow in an effort to determine the radial distribution of azimuthal velocity. Taylor observed strong probe interference (see §2) but was able to extrapolate measurements from two different probe diameters to zero probe diameter pressures. Gregory, et al (1955) used a simple form of an acoustical stethoscope to detect changes in behavior of air flow above a rotating disk. They listened through one end of a short length of 0.25 inch diameter plastic tubing with the other end held in the boundary layer about 0.01 inch above the rotating surface. A radial traverse from the inner laminar flow to the outer turbulent flow was made and they reported that one could detect critical radii associated with flow changes as accurately as with the China clay technique (§3.2.1). No audible noise was heard in the laminar region at small radii, a fairly definite pitch was heard in the middle region where laminar spiral waves developed and a roar typical of turbulent flow was produced at large radii. Kreith, et al (1962) also detected boundary layer transition on rotating cones using a stethoscope.

Pitot-static probes, wall pressure taps and yaw probes were used in an investigation by So (1967) of swirling flow through a 6° conical diffuser. Both axial and azimuthal velocity profiles were measured for swirl angles in the range 0-70°. Regions of reversed flow were observed at high swirl angles and care was taken to correctly determine the flow direction and measure velocities in those regions. However, similar measurements by Hummel (1965) have shown that intruding probes must have disturbed the flow to some degree, especially in and near regions of reversed flow. Nevertheless, So's pressure measurements were instrumental in elucidating five fundamental flow regimes, including two types of vortex breakdown phenomena.

Many laboratory experiments simulating intense atmospheric vortices have incorporated pitot-static probes to measure axial and swirl velocity distributions. In a critical review of this work Maxworthy (1982) contended that such measurements in rotating flow should be regarded with suspicion owing to probe interference. Static pressure measurements in the ground plane below a vortex flow, on the other hand, can provide useful signatures of the presence of a vortex breakdown near the wall (Maxworthy, 1972). Comprehensive measurements of static pressure fields beneath tornado-like vortices produced in the laboratory have been reported by Snow (1982). He has shown that the pressure fields can be interpreted to indicate the evolution from one-cell to two-cell tornado flows and suggested use of the data for better interpretation of barographic records of real tornadoes.

An innovative use of pressure measurements was reported by Aldridge and Toomre (1969) in their investigation of the axisymmetric inertial oscillations of a fluid contained in a rotating sphere. The sphere was mounted on a rotating table and inertial waves were excited via small forced sinusoidal oscillation of the sphere coparallel with the axis of steady rotation. Relative to the rotating system, a small hollow tube was inserted along the axis of sphere rotation to its center. Amplitude and phase measurements of the disturbance pressure differences between probe tip and fluid free surface at the north pole were obtained with an inductive differential pressure transducer. The measurements provided one of the first definitive verifications of theory predicting modes of inertial oscillation in a rotating flow. The same dynamic pressure measurement technique was used by Beardsley (1970) to study inertial waves in a rotating cone.

Whitehead and Porter (1977) observed unexpected azimuthally propagating free-surface waves in an experiment on the critical withdrawal of a rotating fluid. Time traces of the travelling waves at different rotation speeds were obtained from a miniature pressure transducer installed at a fixed point beneath the free surface. These signals substantiated the existence of the azimuthal waves and further showed that the frequency of oscillation was more than twice that of the background rotation. In a another experiment on the steady flow between separate basins through a connecting channel, Whitehead (1986) constructed a differential manometer on the rotating table to measure the difference in fluid levels between the two reservoirs. Care was taken to mount the manometer U-tube symmetrically across the rotational axis to balance centrifugal effects distorting the manometer fluid interfaces.

Recently H. Snyder (private communication) reported using a miniature condenser microphone to detect the onset of turbulence in the flow produced by a disk rotating inside a cylindrical housing in studies associated with the development of laser disks for information storage. The microphone was placed inside the stationary housing near the outer rim of the rotating disk. The distinct roar which became audible at a critical rotation speed was interpreted to be due to the onset of turbulence on the disk.

5. TEMPERATURE PROBES

In the following sections specific temperature probes which have found use in rotating flows are discussed. These include resistance wire thermometers, thermocouples and thermistors. In §5.3.2 and §5.3.3 discussions on the use of thermistors for near simultaneous measurement of temperature and speed, and of temperature and velocity, are included.

5.1 Resistance Thermometers

The resistance thermometer works on the principle that the resistance of a length of wire depends on its temperature. Platinum wire is often used because of its high temperature coefficient of resistance. The wire elements may be coiled around an insulator rod or used in straight sections. Bare wire has a faster response time, but in certain environments it may be necessary to use insulated wire. The measuring circuit is usually a Wheatstone bridge operated in an unbalanced condition so that it may be calibrated to give a direct reading of the wire temperature. Compared to a thermistor, even platinum wire has a very low temperature coefficient and hence the components of the galvanometer circuit must be of high quality for good temperature resolution. With care a resolution of 0.0001 °C may be obtained although in typical laboratory applications a resolution of 0.005 °C can be expected.

An interesting application of resistance thermometry to rotating flow has been reported by Mullen and Maxworthy (1977) in a laboratory simulation of dust devil vorticies. They constructed several resistance thermometers out of tungsten wire and simultaneously swept the vertical array of wires horizontally through the vortex center at a rate much faster than the drift rate of the vortex. The instantaneous voltages from the thermometers were recorded on a multichannel tape recorder for future processing which initially consisted of A/D transfer to a computer, smoothing, shifting, and averaging data from repeated sweeps through the vortex. This left a data set at each height in the form of an integrated temperature versus time which was then converted to an integrated temperature at equally spaced radial intervals from the measured sweep rate of the resistance thermometer array. In the final step the integrated temperature data was unfolded vial numerical inversion of an Abel integral to obtain radial temperature profiles. These profiles corroborated the existence of one-cell and two-cell vortical structures observed with helium bubble flow visualizations (see §3.2.3).

5.2 Thermocouples

A thermocouple is formed by the junctions of two dissimilar metals formed at their opposite ends. If the junctions are at different temperatures an electromotive force is produced. Thermocouples made of copper-constantan, chromel-constantan and iron-constantan are commonly used. Copper-constantan is preferred in the laboratory because of the reliability of the junctions and the nearly linear voltage-temperature relation in the range 10-30 °C. Furthermore, the 40 μV/°C sensitivity of copper-constantan junctions is stable to within 1%. A resolution of 0.01 °C can be obtained with accurate voltmeters. Further details of thermocouple operation are given in Beckwith, et al (1982).

Thermocouples have been used extensively to obtain temperature profiles in thermally active rotating flows. The thermocouple junctions are usually mounted on the tip of a small hypodermic needle and traversed through the flow with a micrometer screw. Fultz and Kaiser (1971) have documented thermocouple probe interference with the fluid thermal field in a laterally-heated rotating annulus (see §2). Hide (1958) used microthermocouple probes to obtain time series temperature measurements at ten different levels in a rotating heated annulus to exhibit the vertical structure of baroclinic instability waves. The electric signals were transferred to the laboratory through a set of rotating mercury slip rings. In a similar experiment Bowden and Eden (1965) mapped out isotherms in the main body of the fluid for different rotation speeds using an array of miniature thermocouple probes. Haas and Nissan (1961) obtained radial temperature profiles for both subcritical and supercritical radially heated Taylor vortex flow using a copper-constantan thermocouple junction mounted on the tip of a hypodermic needle. Abell and Hudson (1975), in their study of centrifugally driven free convection in a rotating rectangular cavity heated from above and cooled from below, measured centerline temperature profiles with a series of thermocouples strung across the channel. Hide, et al (1977) scanned an array of thermocouples spaced at equal intervals around the perimeter of a circular ring positioned at mid-height in a laterally-heated rotating annulus to obtain wavenumber spectra, amplitude and phase of the azimuthal wavenumber components, and vacillation drift rates. Probe interference effects in this experiment have been discussed in §2. Simmers and Coney (1979) measured radial temperature profiles in Taylor-Couette flow with axial throughflow by traversing a fine thermocouple probe across the annular gap.

White and Koschmieder (1981a) made novel use of a pair of copper-constantan thermocouples in a rotating heated annulus experiment. The thermocouples were positioned at the top and bottom level of the cooling circuit for the outer copper cylinder bounding the fluid annulus. The difference signal between the thermocouples provided a qualitative measure of the local heat flux into the annulus along the vertical line connecting the thermocouples. Moreover, because of the high conductivity through the copper cylinder wall, the sensors were responsive to temperature variations induced by the slowly-propagating azimuthal waves *inside* the annulus. Thus a measure of the vertically averaged wave motion in the annulus was obtained without introducing probes into the flow. Example traces of amplitude vacillations recorded by this thermocouple setup are presented in White and Koschmieder (1981b).

5.3 Thermistors

5.3.1 Temperature Measurement

The word thermistor derives its name from *therm*ally sensitive res*istor*. Thermistors, first introduced in the early 1930s, are made of solid semiconducting materials whose resistance decreases in the range 3-6% per degree centigrade. Even at the low end of thermistor sensitivity the temperature coefficient of resistance is over 100 times that of thermocouples. Becker, et al (1946) have reviewed a wide variety of thermistor applications. The high sensitivity combined with their small size, durability, and long-time stability make thermistors ideally suited for temperature measurement in fluid systems. Like any other resistance thermometer, the thermistor is operated in a resistance bridge, and care must be taken to keep the current sufficiently small so that it produces no appreciable heating. Since thermistors are readily made to operate at resistances substantially higher than metallic resistance thermometers or thermocouples, lead resistance is not ordinarily a problem. This permits greater flexibility in the use of thermistors since they can be located remotely from their measuring circuit and still maintain good measurement precision.

A thermistor placed in a simple Wheatstone bridge with a standard galvanometer can easily detect a temperature change of 0.0005 °C. The semiconductor materials are glass coated to minimize noise and are manufactured in a variety of shapes and sizes. Flush mounted wall thermistors and small isolated beads are most often used for temperature measurement in fluid flows. Recent miniaturization has sufficiently reduced the time constant to make them useful in low frequency time-dependent flows, particularly in high thermal conductivity liquid flow. Typical 1/e time constants for 0.015 inch diameter beads are 1.0 sec in still air and 0.1 sec in still water and faster time responses are obtained in moving fluids.

Thermistors have been used individually or *en masse*. For example, Lambert and Snyder (1966) mounted a triangular array of three bead thermistors flush with the inner wall in a rotating heated annulus experiment with weak axial flow. With this configuration it was possible not only to detect the onset of instability, but also measure the drift rate, amplitude and wavenumber of the disturbances. Fowlis and Pfeffer (1969) mounted a staggered array of 48 small bead thermistors in a rotating heated annulus to sample 8 different phases of a four-wave amplitude vacillation at 6 different radii. The beads were suspended axially throughout the fluid volume by the fine wire leads emanating from opposite sides. Significant probe interference was observed. They noticed a long-period modulation of the amplitude vacillation that was not observed in the absence of the thermistor array. In an extension of this work, Pfeffer, et al (1970) used a single thermistor to detect transitions between steady waves and amplitude vacillation. Davies and Walin (1977) measured vertical temperature profiles in a laterally-heated rotating annulus with a single moveable thermistor mounted on the end of a hypodermic needle. Pfeffer, et al (1980a) measured the time-dependent behavior of amplitude vacillation, structural vacillation and geostrophic turbulence in a laterally-heated rotating annulus experiment using an array of 2016 thermistor bead sensors strung vertically through the working fluid.

5.3.2 Simultaneous Speed and Temperature Measurement

As with hot-wire and hot-film probes, thermistors become sensitive to fluid speed when heated. The theory of operation of these "hot-thermistors" was described by Rasmussen (1962) who determined temperature and speed sensitivities as a function of thermistor overheat. At low overheat the thermistor is sensitive to temperature only, while at sufficiently high overheat it is sensitive to both temperature and speed. At the same time Lumley (1962) reported distinct advantages of thermistors over platinum wire when used as a constant-temperature anemometer in liquid flow. He also reported preliminary frequency response measurements for both bead thermistors and thin films operating in the constant-current and constant-temperature modes. Beardsley (1975) used heated thermistors as a null-velocity sensor to detect the propagation of velocity nodes in a rotating sliced-cylinder model of wind-driven ocean circulation.

Fowlis (1970) conceived a system for near-simultaneous determination of both temperature and speed using thermistors. In a study of baroclinic instability in a laterally-heated rotating annulus, Fowlis, et al (1974) strategically located an array of 98 bead thermistors throughout the test volume and the speed and temperature measurement system is described here. Using computer control, a low overheat signal is sent out to all thermistors and at time t_1 a scan is made of offset bridge voltages. These voltages are indicative of the temperature field only. Next, a high overheat signal is sent out to all thermistors and at time t_2 a scan is made of the offset voltages now sensitive to both temperature and speed. A final scan at low overheat is then made at time t_3. After each scan the voltages are digitized and recorded on magnetic tape. The scan and recording times (0.8 sec) are short compared to the equilibration times for a change in overheat (5.0 sec). The temperature fields determined at times t_1 and t_3 are linearly interpolated to time t_2. With the temperatures known at time t_2, the fluid speeds at each thermistor location can then be calculated from calibration curves. In this manner information gathered during the 14.6 sec temperature-velocity-temperature scanning period is folded into a near-simultaneous speed and temperature field since the 0.8 sec thermistor array sampling period at time t_2 is small compared to the 196 sec period of amplitude vacillation under investigation.

Fowlis, et al (1974) then spatially interpolated the speed and temperature fields derived from the 98 thermistors to construct isotachs and isotherms in a horizontal plane where the fluid motion is nearly two-dimensional. Photographs of the flow visualized by aluminum flakes illuminated in the same horizontal plane were taken at times corresponding to the temperature and speed measurements. The isotherms exhibited good phase correlation with the amplitude of wave vacillation but the isotachs did not correlate well, demonstrating the need to construct streamlines from velocity measurements rather than isotachs from speed measurements. A very important contribution from this study was the assessment of probe interference. Insertion of the probes substantially reduced the amplitude of vacillation. Moreover, the vacillation period with probes (196 sec) was considerably longer than without probes (150 sec). The authors argued that these probe effects were similar to what one would observe by increasing the fluid viscosity in a probe-free flow. This conjecture was supported by measurements of the spin-up time of the fluid to small changes in rotational speed with and without probes. In a given experiment the measured spin-up time was 44 sec without probes and 26 sec with probes. Using linearized spin-up theory, Fowlis, et al (1974) determined that effect of probe drag was to increase the effective kinematic viscosity of the liquid by a factor of three over that of a probe-free flow.

Large arrays of thermistor probes have been used by Pfeffer, et al (1980b) to report measurements of radial eddy heat fluxes, eddy temperature variances and radial gradients and variances of the azimuthally averaged temperature in the laterally-heated rotating annulus experiment over a broad range of dimensionless parameters. Buzyna, et al (1984) studied transition to geostrophic turbulence in the rotating heated annulus experiment. In both studies 2016 thermistors were axially strung throughout the fluid volume in order to capture higher order temporal and spatial frequencies. Although no discussion of probe interference was given for this large thermistor array, Buzyna (private communication) reports that since the annulus was substantially larger than that used for studies with 98 thermistors, the ratio of the total wire drag to the Ekman layer drag was comparable in each experiment. The study by Fowlis, et al (1974) suggests that the qualitative features of the measurements may be representative of an unperturbed flow, but that the reported Taylor numbers need to be modified to represent the effective increase in kinematic viscosity due to the presence of the probes.

5.3.3 Simultaneous Velocity and Temperature Measurement

Fowlis, et al (1972) proposed a technique for the simultaneous measurement of *velocity* and temperature in two-dimensional flow using three satellite thermistors clustered about a central heated thermistor. The satellite thermistors operate at low overheat to measure the local temperatures, and are spaced 120° apart forming an equilateral triangle in the plane of fluid motion. The heated thermistor, located at the center of the triangle, measures flow speed using the average temperature of the surrounding sensors. Flow direction is determined through the detection of the central bead's thermal wake by the satellite beads. Kung, et al (1987) have recently built, optimized and calibrated a four thermistor anemometer patterned after the system described by Fowlis, et al (1972). Consideration of thermal wake sensitivity and flow blocking led to a center-to-center distance between central and satellite thermistors of about two bead diameters. In each anemometer configuration six wires were strung vertically through the fluid volume: one for each of the four sensors, one return lead for the heater thermistor and one return lead which could be switched to each of the three satellite sensors. The return leads, though not needed in theory, were used to help stabilize the support system. In a temperature-velocity-temperature scan sequence, the estimated measurement accuracies for temperature, speed, and flow direction are reported to be ±0.01 ºC, ±0.5 mm/sec, and ±5º, respectively.

Without a doubt Kung, et al (1987) hold the record for the maximum number of wires traversing a given experiment. They constructed a network of 360 thermistor anemometer clusters and 572 single thermistors for the rotating heated annulus experiment. This represents a total of 2012 thermistors with an associated 2732 wires traversing the fluid volume. Isotherm, velocity fields and streamline contours were constructed by spatial interpolation of the temperature and velocity measurements. There is no discussion of probable flow distortion in this paper.

6. VELOCITY MEASUREMENT SYSTEMS

In this section various experimental techniques for measuring velocity are presented. Time series velocity data are obtained through the use of hot-wire, hot-film, and laser Doppler anemometry systems. Full field measurement of fluid velocity in a plane at discrete times can be obtained through the use of streakline and particle displacement photography or laser speckle velocimetry. The use of hot-thermistor arrays for simultaneous measurement of temperature and speed, or temperature and velocity has been discussed in §5.3.2 and §5.3.3, respectively.

6.1 Streakline and Particle Displacement Photography

A simple method for obtaining estimates of a velocity field in both steady and unsteady flow is to track isolated particles moving along their respective pathlines at small intervals of time. The technique lends itself to flows that are quasi-two-dimensional. Short duration time-lapse photography of the motion of particles moving in the plane will result in small streaks from which the velocity field can be measured if the directions of particle motion are known *a priori*. This often has been referred to as "streakline photography," a confusing misnomer since it is actually the photograph of short sections of particle *pathlines*. Another way to use pathline information is to illuminate a plane of motion with a strobed light source and record the particle positions on film with the shutter held open. In this particle displacement photography the velocity field may be deduced at successive time intervals from a single photographic record. Various methods for producing light sheets for illumination purposes are described in §3.3. In this section we consider the use of streakline photography for quantitative flow measurements. Examples of streakline photography for flow visualization purposes have been presented in §3.1.2.

A detailed analysis of particle tracer methods for obtaining velocity fields from photographic data is given by Sommerscales (1981). Errors in the measurement of the trace length of either a streak or a particle displacement include: (a) random errors in the measuring device locating the absolute position of particles on developed film, (b) errors in the definition of the ends of a streak, (c) emulsion shrinkage in the development of the film, (d) errors in producing enlargement photographs for the purposes of measurement, (e) uncertainty in the timed interval of particle displacement and (f) errors in the measurement of the magnification of the camera optics. It is estimated that the total contribution of the random errors probably does not exceed 5% for laminar flow. Investigations by Agui and Jimenez (1987) and Rignot and Spedding (1988) emphasize the importance of errors incurred in interpolating measured velocities taken from a set of irregularly spaced data points to form a velocity field on a regular grid.

6.1.1 Manual Data Reduction

Faller (1963) used particle displacement photography following the motion of miniature floating light bulbs to compute flow circulation in a rotating source-sink flow. In experiments on Ekman layer flow over a regular array of hills in a rotating tank Faller and Mooney (1971) tracked the motion of polyethylene rods weighted at the bottom and tapered at the top so as to float vertically. Particle displacement photographs of three successive positions of the floats were used to manually extract the surface stress of the liquid flowing over the bottom topography. In the rotating sliced-cylinder experiment Beardsley and Robbins (1975) overlayed numerical computations of flow streamlines on streakline photographs obtained at mid-depth to make a convincing corroboration between experiment and theory. In the same geometry, Beardsley (1975) used streak photography to record at mid-depth the elliptical Lagrangian paths described by suspended particles in response to an oscillatory shear stress imposed at the upper fluid surface. Flierl, et al (1983) tracked the motion of paper pellets on a water surface to observe the free surface manifestation of internally generated modons in a rotating tank. Velocity fields were obtained from particle displacements recorded by an overhead movie camera. Boyer, et al (1984a) simultaneously released approximately 500 surface floats on a 10 cm grid pattern downstream of the a vertical cylinder in rotating channel flow to measure two-dimensional velocity fields in the unsteady cylinder wake. Both absolute and relative velocities of the cyclones and anticyclones shed from the cylinder were obtained by manual data reduction of particle displacement photographs. Li, et al (1986) measured velocity fields in a horizontal plane of a laterally-heated annulus. Illumination for particle displacement photography was obtained by a rotating laser beam to sweep out circular paths at mid-depth. The laser beam was interrupted for one rotation cycle in a series of cycles so a missing dot in a series of dots indicated the direction of particle motion. The position of approximately 450 dot pairs were manually located and the coordinates logged into a computer with the aid of a digitizer. Streamline patterns derived from these velocity fields were then computed. The 6 sets of 24 streamline fields reported required the manual identification of 129,600 individual dots.

6.1.2 Automated Data Reduction

Manual reduction of streakline or particle displacement photographs have provided measurement of velocity, surface stress, relative vorticity and Lagrangian particle displacement. However the extensive labor involved to obtain full-field maps as exemplified by the work of Li, et al (1986) points out the need for automated data reduction. Advances in this direction for laser speckle velocimetry are reported in §6.5. Here only particle streakline and particle displacement photography are considered. Elkins, et al (1977) have reported on the development of automatic and semi-automatic film reading techniques to process stereoscopic trace particle records of turbulent flow fields. Although still in the evaluation stage, this system has produced three-dimensional stereoscopic views of the instantaneous velocity pattern in turbulent pipe flow from which one-dimensional velocity distributions or two-dimensional velocity fields may be extracted. An automated image processing technique to derive velocity field data from particle-tracked photographs was reported by Jonas and Kent (1979). In this scheme particles from two successive photographs were identified and stored on a computer. Use was made of an approximate analytical function describing the expected modal content of the velocity field. Iteration back and forth between photographs permitted computation of an increasing number of undetermined coefficients in the analytical function. In application to flow in a horizontal plane of a laterally-heated annulus, instantaneous trace particle positions were located with an image-analyzing computer. Negatives were interrogated with a special Vidicon scanner with a 720-line by 890-point resolution. For the flow 20 azimuthal and 3 radial mode numbers were taken and convergence for the complete analysis took about 20 sec on a mainframe computer. Endpoints of some of the streaks had to be extracted by hand due to imperfect detection by the automatic scanner. Nonetheless, this measurement technique shows great promise for flow fields whose modal structure can be determined *a priori*. Read and Hide (1984) applied the automated particle measurement technique to obtain velocities of nonaxisymmetric flow in a laterally-heated rotating annulus simulation of Jupiter's Great Red Spot. Unfortunately no discussion was presented on the assumed analytical form for the velocity field in their experiment, although the authors had clear streakline photographs available from which a reasonable approximation to the modal flow structure could be made.

Gharib, et al (1985) used automated particle tracking to follow the motion of luminescent particles on a free surface water flow (see §3.1.4). Here the particles were excited with two 20 μs light pulses separated in time by 250 ms. Each spot on the photographic negative was located by a Vidicon camera and image digitizer with the process controlled by a host computer. Detection of the isolated spots was then followed by invoking a trace-following-algorithm which covers each spot with eight surrounding masks to test for the direction of movement as detected by the continuous phosphorescent particle trace. Successive maskings of points along the particle trace lead to the end point illuminated by the second flash. This process was continued until the entire photograph had been surveyed. In instances of trace crossings or overlaps, the computer software would halt and ask for the operator's assistance. Gharib (1989) has reported on recent progress for automating this particle tracing technique.

Ruiz, et al (1986) used image processing techniques to scan digitized photographs of streaks made by magnesium powder in an axial plane of Czochralski flow. They located the Cartesian coordinates of the streak's center of gravity as well as its length and angle of inclination. The results were logged onto magnetic tape for subsequent computer analysis of the velocity field as presented in their paper. Unfortunately, the authors did not explicitly state whether the scanning was performed manually or automatically. Recently, Rignot and Spedding (1988) have devised an automated image processing and grid interpolation technique for measuring velocity fields from streakline photographs. This study includes a careful analysis of errors in velocity measurement from the streakline data and interpolation of that irregularly spaced data set to produce a velocity field and its first spatial derivatives on a regular grid. The automated streakline measurement system was successfully applied to several two-dimensional laboratory vortex flow experiments. A floppy disk is included in this report with machine-readable programs for implementation of the data processing techniques for calculation of vorticity, circulation and streamfunction fields from the velocity data.

6.2 Circulation Meters

In axisymmetric swirling flows paddle blades may be constructed to respond to the local circulation. Truxillo and Hussey (1969) built a simple circulation meter for experiments on spin-up from rest in a vertical open-topped cylinder filled with liquid. They mounted two small flat plates 4 X 7.5 mm^2 at the ends of a tiny horizontal rod. A vertical support rod was fixed normal to the horizontal rod at its midpoint to form a T-shaped paddle wheel. This assembly was then submerged along the central cylinder axis into the liquid and suspended from a quartz torsion fiber. With the cylinder impulsively rotated to a fixed angular velocity, the paddle wheel measured the delay time for the radially-propagating shear front to reach the paddle blades.

McEwan (1982) constructed three circulation meters rotating about a common shaft to measure the average circulation in a rotating source-sink flow designed to simulate a tropical disturbance. The smallest 5.2 cm diameter meter had small rectangular blades affixed to the ends of its support rod while the larger 13.8 cm and 26.0 cm diameter meters had 8 mm diameter spheres at the end points of their rod supports. The central pivot shaft was vibrated acoustically to overcome rotational bearing friction. Through calibration in steady swirl flow, McEwan determined that the circulation meters reliably measured the mean azimuthal relative motion at their respective radii without bias. Circulation measurements from the meters showed that fluid angular momentum was almost completely conserved in this source-sink vortex simulation at low rotation rates.

6.3 Hot-Wire Anemometers

The hot-wire is a thermal transducer used to measure fluid velocity. An electric current is passed through a fine wire exposed to fluid flow. Changes in flow velocity produce changes in heat transfer from the wire which in turn effects a change in wire resistance. When the hot-wire is placed in a simple bridge circuit, measurement of the imbalance voltage across the bridge produces an analog signal proportional to the instantaneous velocity in the fluid stream. Wire elements are commonly made of tungsten, platinum, or platinum-iridium (80% Pt, 20% Ir) because of their good physical properties and availability in small diameters. Typically a 5 μm diameter wire is soldered across the tips of needle supports located 2 mm apart. The needle prongs are affixed to a cylindrical housing with electrical leadouts. The housing is made as small as possible to minimize probe interference and yet maintain rigidity when placed in a fluid stream.

According to Freymuth (1983) the idea of using a hot-wire for velocity measurement was originally proposed by Oberbeck (1895). Since then the concept was reinvented several times, but Riabouchinsky (1909) was the first to build and use a hot-wire anemometer. The underlying theory of its thermal heat balance was first reported by King (1914). Assuming a tiny wire placed normal to a fluid stream is cooled by forced convection only, King derived the functional relation between the flow velocity and the heat transfer from the wire. In nondimensional form "King's law" may be written

$$Nu = A + B \, (Re)^n , \tag{2}$$

where $Nu = hd/k$ is the Nusselt number and $Re = Ud/\nu$ is the Reynolds number. Here h is the heat transfer coefficient, k is the fluid thermal conductivity, ν is the fluid kinematic viscosity, U is the flow velocity and d is the wire diameter. In general the dimensionless constants A and B and the exponent n are a function of fluid Prandtl number. Since typically $n \cong 1/2$, the hot-wire is most sensitive to velocity at low speeds. A limit on the validity of (2) is imposed at low velocity where neglected buoyancy effects become important. For standard hot-wire probes used in air the velocity limit is of order 5 cm/sec. A hot-wire with large length-to-diameter ratio operating at low overheat indeed satisfies (1) very accurately. Since the joule heating in the wire is proportional to the Nusselt number, a form of equation (2) suitable for direct calibration is given by

$$E^2 = C + D \, U^n , \tag{3}$$

where E is the voltage across the wire and C and D are dimensional constants to be determined with the exponent n by direct calibration.

At moderately high wire overheat Betchov (1948) showed that King's law must be modified to include a response proportional to the square of the temperature difference between wire and fluid. In this same paper Betchov analyzed the effect of conduction to the prong supports. Collis and Williams (1959) also reported a simple modification to King's law for high wire overheat. Champagne, et al (1967a) measured actual temperature distributions across hot-wires positioned normal and also inclined to a uniform flow.

Hot-wires may be operated in the constant-current or constant-temperature mode. The *constant-current* mode maintains uniform current in the wire and variations in wire resistance due to forced convection are detected as voltage fluctuations across the wire. The main disadvantage of the constant-current system is that its frequency response is limited by the thermal inertia of the wire. In the *constant-temperature* mode introduced by Morris (1912), a feedback loop is incorporated to continuously compensate for the thermal inertia of the wire. The feedback circuit is designed to maintain constant wire resistance and thus temperature. The result is a considerably higher frequency response essential for the measurement of turbulent velocity fluctuations.

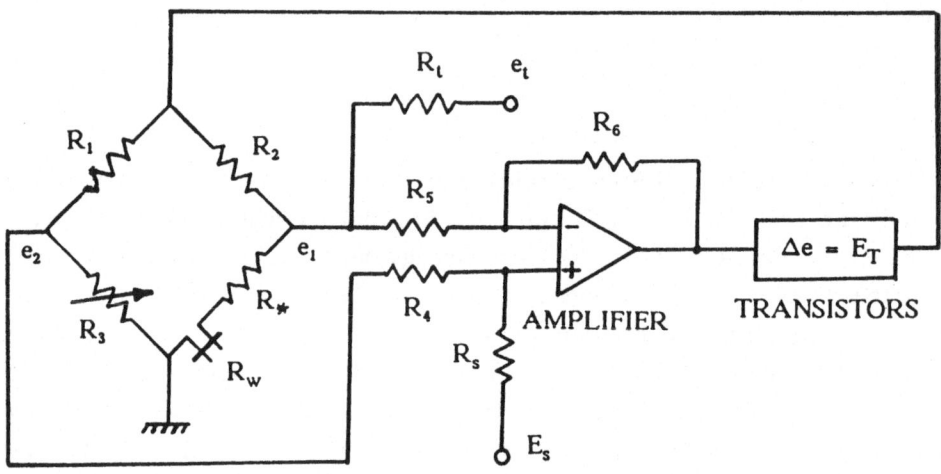

Figure 1. Simplified diagram of a constant-temperature anemometer circuit.

A simplified diagram of a constant-temperature circuit is presented in Figure 1. The hot-wire is represented by the resistance R_w and the transistors providing power in the feedback loop are denoted schematically by the voltage drop ΔE. To obtain measurements a wire overheat is selected and the bridge is balanced at mean flow conditions. With flow variations the voltage imbalance $(e_2 - e_1)$ is magnified across the operational amplifier with gain R_6/R_5 and fedback to the top of the bridge to maintain constant wire resistance. Staritz (1960) has shown that an electronic square wave signal e_t injected through resistance R_t to artificially excite the hot-wire can be used to determine the system response on an oscilloscope.

Optimum frequency response then may be obtained by trimming the d-c offset voltage E_s to the operational amplifier.

A complete analysis of the dynamic response of a constant-temperature circuit must include its interaction with the hot-wire probe. The electrical circuit can be analyzed to a degree of accuracy commensurate with the need. Freymuth (1968) considered the effects of electronic noise on the signal-to-noise ratio for hot-wires operating in both the constant-current and constant-temperature mode. Davis (1970) showed the effect of probe inductance on the dynamic response of a constant-temperature hot-wire. Perry and Morrison (1971) pointed out the sometimes crucial effects of amplifier offset voltage, bridge imbalance, and finite amplifier gain on hot-wire stability and frequency response. Weidman and Browand (1975) made a complete circuit and probe analysis for the simple second-order constant-temperature anemometer depicted in Figure 1. They compared the predicted response over a wide frequency range with that measured from a sine wave injected at R_t. Freymuth (1977a) developed a linear theory of electronic square-wave and sine-wave testing for a third-order system applicable to circuits with two adjustable controls for optimizing the frequency response. Freymuth (1977b) then showed that the optimization given by the linear analysis also provides optimal response for large-amplitude velocity fluctuations.

Multi-wire probes have been configured to obtain local measurement of multiple velocity components, vorticity, and Reynolds stress. The wires mounted at the tip of a common probe holder are fed to matched constant-temperature circuits. Two wires in an X-configuration are sufficient for making simultaneous measurements of all three velocity components (Fabris, 1978). Turbulent shear stress measurements using inclined hot-wires have been reported by Champagne, et al (1967b). A single inclined wire rotated about its axis can be used to measure mean Reynolds shear stress, but normal shear stress measurement requires the use of crossed wires (Perry, 1982). Sophisticated multi-wire probes have been built for special purposes. A case in point is the miniature 9-wire probe constructed by Balint, et al (1987) designed to simultaneously measure all three components of velocity and vorticity. Recent progress on the use of multi-wire probes for vorticity measurement is reported by Foss and Wallace (1989).

Many treatises on hot-wire anemometry are available. Details on the use of hot-wires for turbulence measurements have been reported by Hinze (1959), Corrsin (1963) and Bradshaw (1971). Sandborn (1972), Perry (1982) and Fingerson and Freymuth (1983) emphasize the fundamental principles of hot-wire operation. An excellent review of hot-wire anemometry has been given by Compte-Bellot (1976). Freymuth (1983) has presented an historical survey of the field.

Hot-wires may not significantly affect rotating flows in conditions where there is a substantial radial or axial velocity component to convect the probe wake "downstream" and away from the measurement station. Van Atta (1966) reported careful measurements of spiral turbulence in a large Taylor-Couette facility. Hot-wire probes were used to map out the laminar-turbulent interface of the spiral zone and to measure its rate of rotation. Weak probe interference effects in that facility were reported by Coles and Van Atta (1966) for subcritical circular Couette flow (see §2). Coles (1965) reported hot-wire measurements used to determine the azimuthal speed of wavy disturbances riding on Taylor vortex cells. Mellor, et al (1968) used a constant-current anemometer to measure radial and tangential velocity profiles in cellular flow between horizontal concentric rotating disks. Possible probe interference in this recirculating flow was not discussed.

Hot-wire probes were used extensively by Green (1968) and Ingram (1971) to identify inertial modes of oscillation in spiraling source-sink flow in an annulus. Tatro and Mollo-Christensen (1967) and Caldwell and Van Atta (1970) placed hot-wires in spiraling boundary layer flow to detect Type I and Type II Ekman layer instabilities. Cerasoli (1975), after having performed a careful study of free shear layer instability induced by probes in rotating source-sink flows similar to those used by the above-cited investigators, concluded that the inertial eigenmode measurements of Green (1968) and Ingram (1971) were all rod-induced instabilities. In addition Cerasoli showed that the measurements of Type II Ekman waves reported by Tatro and Mollo-Christensen (1967) were also probe-induced instabilities. On the other hand, Cerasoli concluded that the hot-wire measurements of Caldwell and Van Atta (1970) were not tainted by probe induced instabilities.

Ibbetson and Tritton (1975) used a rotating constant-current hot-wire probe to measure the decay of grid generated turbulence in a rotating annulus. The arm supporting the hot-wire was rotated at sufficiently low speed relative to the fluid in the annulus to preclude a turbulent wake. Nevertheless, the background noise of the rotating hot-wire was observed in the hot-wire signal. At low levels of turbulence intensity, care was taken to correct for this probe-induced noise by subtracting the mean-square noise from the mean-square turbulence signal under the reasonable assumption that the signal and noise were uncorrelated.

Simmers and Coney (1979) made hot-wire measurements of stable and unstable laminar circular Couette flow with superposed axial throughflow. They found, under certain assumptions validated by probe calibration, that all three velocity components could be obtained from just two measurements from a single probe mounted normal to the outer cylinder and free to rotate about its axis. Measured velocity profiles with the inner cylinder rotating compared favorably with known laminar flow solutions except that the axial velocity distributions were skewed towards the inner wall. A careful analysis showed that this was due to self-interference of the probe and not an artifact of the velocity decomposition technique. The low axial velocities near the inner wall were insufficient to sweep the wake of the probe downstream of the measurement volume during one revolution of the fluid particles. In a follow-on study Simmers and Coney (1980) used the same velocity decomposition technique to determine the effect of surface heat transfer on the measured velocity distributions. Wan and Coney (1980, 1982) employed X-wire probes to measure power spectra, turbulence intensities and auto- and cross-correlations in a study of mode transition in adiabatic spiral-vortex flow.

6.4 Hot-film Anemometers

Hot-wire measurement in low-speed liquid flows has met with only marginal success due to their fragile nature and fouling by lint and dirt. The need for turbulence measurements in liquid flows and high-speed gas flows led to the development of a wedge shaped hot-film probe introduced by Ling (1955). These probes are made by fusing a thin platinum film to the surface of a quartz substrate. Three fundamental probe configurations for velocity measurement have come into existence: film plated around a tiny cylindrical rod, film plated on the tip of a 30° wedge and a ring of film plated on the surface of a cone upstream of its apex.

As with hot-wires, hot-films may be operated in either the constant-current or constant-temperature mode. Ling and Hubbard (1956) used a constant-temperature circuit to assess the probe's frequency response. The frequency response is considerably lower than that of a hot-wire due to conduction to the substrate. Because of the high heat transfer in liquid flows, more current is necessary to maintain a given overheat for hot-films operating in water. This also mandates higher power transistors in the feedback loop for constant-temperature operation. In practice the resistance overheat in water is limited to about 8% owing to the formation of bubbles on the surface of the heated film.

The thermal analysis for hot-film probes is rendered difficult because heat from the film is transferred both to the liquid by forced convection and to the underlying substrate by conduction. Neglecting axial heat conduction through the substrate, Lowell and Patton (1955) analyzed the heat transfer response of a cylindrical hot-film subject to sinusoidal surface temperature perturbations. Ling (1960) reported empirical heat transfer correlations similar to King's law, equation (2), for the steady operation of cylindrical and wedge probes. Bellhouse and Schultz (1967) made a one-dimensional analysis for a wedge film probe and showed the complicated nature of the thermal feedback through the substrate. In an investigation of the turbulence measuring characteristics of hot-film probes, Bankoff and Rosler (1962) reported satisfactory results for turbulence intensities measured in a water jet with wedge and cylindrical hot-film probes. However, turbulence measurements made by the two probes in an air jet were low by a factor of two or three. Kidron (1966) made direct measurements of the constant-current response of cylindrical hot-films using a novel technique. With a hot-film mounted in a steady wind tunnel flow, he projected high-frequency (microwave) electromagnetic energy across the test section through an X-band waveguide. At microwave frequencies the electromagnetic energy is applied to the surface of the hot-film, thereby simulating the action of heat transfer between the film and the air stream. The energy source was modulated over a wide range of frequencies and the voltage response across the film operating at constant current was measured. Kidron verified Lowell and Patton's (1955) prediction of a 10 dB roll-off at high frequencies. Rodriguez, et al (1970) compared measurements from wedge, cylindrical and conical hot-film probes in liquids and found that the conical probe gave turbulent intensities consistently 5-15% lower than the other probes.

A probe and circuit analysis for a cylindrical hot-film probe operating at constant-temperature has been given by Weidman and Browand (1975). They combined the thermal response of a film plated on a cylindrical substrate with the electrical response of the constant-temperature feedback circuit given in Figure 1. Measurements of the system frequency response to sinusoidal perturbations were carried out over the range 1 Hz - 100 KHz. They reported excellent agreement between experiment and theory, as long as the thermal time-constant for the probe was obtained by direct measurement in lieu of using the theoretical value calculated from the manufacturer's specifications. A similar probe and circuit frequency analysis was made by Freymuth (1980) for noncylindrical hot-film anemometers using the Bellhouse-Schultz heat transfer model.

Cylindrical hot-film probes can be fabricated so that the upper half of the cylindrical film coating is electrically insulated from the lower half. In practice each half of the film is plated around 170° of the circumference of the cylinder leaving 10° gaps between the films. The advantage of this "split-film" configuration is that the outputs of each half film can be independently operated in separate constant-temperature circuits. Then, since the heat transfer to each half is sensitive to the the angle of the incident velocity vector, the split-film probe can be calibrated to simultaneously measure both streamwise and transverse velocity components. Calibrations by Browand and Weidman (1976) have shown that split-films operated at the same overheat in matched constant-temperature circuits give the total velocity accurate to ±2% and the flow angle to ±1°. However, the use of these probes for precise quantitative measurement has been questioned by Fuller (1974) and Herzog and Lumley (1978). Indeed, Ho (1982) has shown there are problems due to thermal cross talk between the two halves of the split-film. He demonstrated both static and dynamic thermal interference between the output signals when each half of the split-film was operated at a different overheat. This points out the need to match the overheat ratios, but even then the problem of thermal cross talk due to asymmetric heat transfer at low frequencies is not well understood.

Utilization of hot-films for measurement in rotating flows has not been extensive. Wan and Coney (1982) placed a hot-film probe at mid gap in spiral Couette flow to obtain power spectra and autocorrelograms. They noted that visual detection with suspended aluminum flakes of the highly-supercritical flow transitions was extremely difficult owing to the gradual appearance of higher frequency instability modes. The autocorrelograms, on the other hand, provided a clear indication of flow transitions. In another experiment Tritton (1985) mounted a free shear layer water channel on a rotating table. Measurements of the mean velocity profile, turbulence intensity profiles, and Reynolds stress profiles were obtained by traversing a split-film probe across the rotating shear layer. Barcilon, et al (1979) placed two hot-film sensors 30° apart in a Taylor vortex flow and used cross-correlations to determine the speed of instability waves propagating past the probes. With the double probe setup they were able to determine power spectra as a function of wavenumber and hence distinguish between Taylor and Gortler vortex instabilities.

Reviews encompassing both hot-film and hot-wire operation are given by Blackwelder (1981) and Fingerson and Freymuth (1983). Hot-films mounted flush with flow boundaries can be used to determine the local skin friction. This application of hot-film probes is discussed in §7.2.

6.5 Laser Doppler Velocimetry

Yeh and Cummins (1964) published a short note that revolutionized velocity measurement techniques in fluid dynamics. They showed how to utilize the Doppler shifted frequency of laser light scattered from particles suspended in a moving fluid to determine the speed of those particles. In their optical setup a laser beam was split into to beams, one of which was focussed on a small volume within the fluid seeded with micron-size scattering particles. The second (reference) beam was optically routed around the experiment and back into alignment with radiation scattered from the fluid measurement volume. The aligned beams were then heterodyned on the face of a photomultiplier tube. The Doppler shifted frequency $\Delta \nu$ in the heterodyned signal is related to the flow velocity V through the equation

$$\Delta \nu = \left[\frac{n}{\lambda_0}\right] V \cdot (k_s - k_i) \ , \tag{4}$$

where n is the index of refraction of the test liquid, k_i is the unit vector along the incident beam, k_s is the unit vector in the direction of scattered radiation and λ_0 is the laser wavelength in air. Yeh and Cummins (1964) demonstrated this nonintrusive measurement technique using a 5 mw He-Ne laser to measure very slow velocities (0.007 cm/sec) in Poiseuille flow.

Modes of Operation

The laser Doppler velocimeter (LDV) has two common modes of operation: the *reference beam* mode and the *dual scattering* mode. The arrangement described above is an example of the reference beam mode, but is somewhat cumbersome because the reference beam is routed around the experiment and has to be carefully realigned with the radiation scattered from the measurement volume. A simpler variation of the reference beam mode is shown in Figure 2.

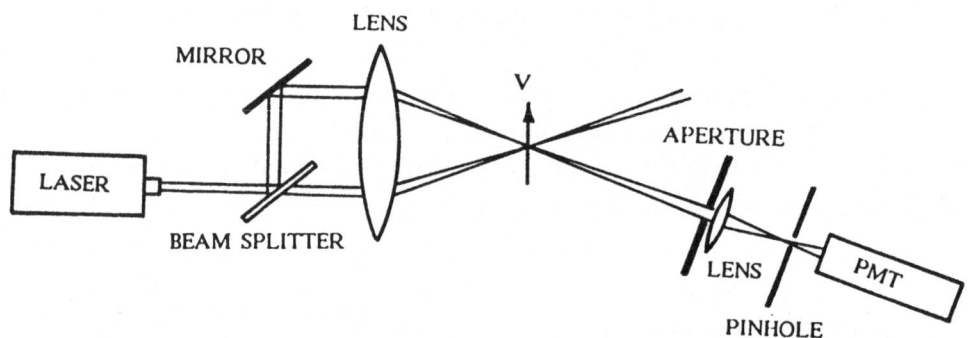

Figure 2. The reference beam mode.

Here the *sending optics* splits the laser beam into an intense scattering beam and a weak reference beam. Both beams are focussed at the same point in the fluid and the *receiving optics*, aligned along the direction of the weak reference beam, includes radiation scattered through an angle θ from the intense beam. The aperture, lens and pinhole insure that the photomultiplier only detects radiation emanating from the focal volume. Since the reference beam is not Doppler-shifted, the heterodyning takes place on the face of the photomultiplier tube (PMT).

The dual scattering mode of LDV operation seems to have been first introduced by Bourquin and Shigemoto (1968), but was independently reported by Mazumder and Wankum (1969) and Penny (1969). The principle of operation is intrinsically different from the reference beam mode in that the heterodyning is done in the focal volume formed by two crossed beams in the fluid. The optical setup is sketched in Figure 3.

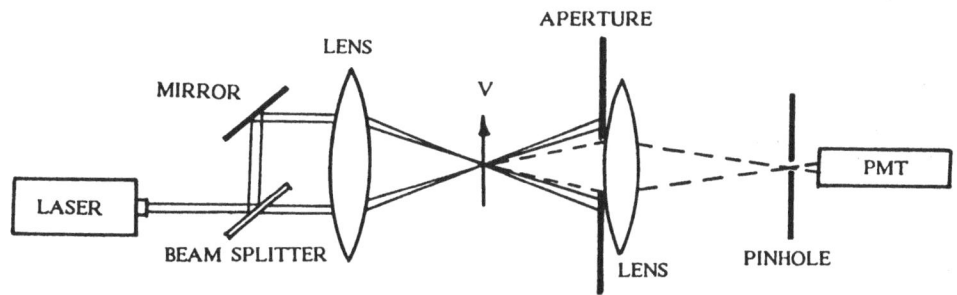

Figure 3. The dual scattering mode.

The sending optics is similar to that of the reference beam mode except that now beams of equal intensity intersect to produce a pattern of plane parallel fringes within the focal volume. The fringes are formed from the constructive and destructive interference of the spatially coherent beams. A particle traversing normal to the fringes produces a modulated oscillatory signal with frequency $\Delta\nu$ directly proportional to the velocity of the particle. The magnitude of this velocity,

$$V = \left[\frac{\lambda_0}{2\,\sin(\theta/2)}\right]\Delta\nu, \tag{5}$$

is independent of the direction in which the scattered light is collected. The dual scatter mode has reduced spectral broadening and a significantly higher signal-to-noise ratio when compared to the reference beam mode. Also, there are no stringent requirements for alignment of the receiving optics which are generally directed along the bisector of the crossed beams to realize maximum signal intensity by forward scattering. The configurations sketched in Figures 2 and 3 above are of the *forward scattering* type. In some experiments the receiving optics must be placed on the same side as the sending optics, an arrangement denoted as the *back-scattering* mode of operation. Clearly the forward scattering mode is preferred because the weak signal of the back-scattering mode necessitates higher laser power and/or more sensitive photomultiplier tubes. In both modes of operation the design should be configured to produce equal path lengths of the split laser beams up to the point of phase mixing, either within the probe volume or on the face of a PMT. Foreman (1967) has shown that substantial reduced signal power results if the path length mismatch is a significant fraction of the laser's optical cavity length, particularly in gas lasers operating with a large number of axial modes.

There has been a long history of development of LDV systems to their present day commercially available modular form. We summarize some of the important improvements in parallel with a review of applications to rotating fluids. The nonintrusive LDV is particularly well-suited to investigations of unsteady flow inside closed cylinders for which the streamlines formed by the O(1) velocity field are circular. The first use of LDV in rotating flow is credited to Bien and Penner (1970). They obtained steady and unsteady flow measurements inside a stationary cylinder driven by a rotating endplate using the forward scattering, reference beam mode with a 1 mw He-Ne laser. The laser beams passed through the transparent walls of the cylinder and were focussed at different radial positions by the adjustment of mirrors. The signal from the PMT was fed into a frequency meter for instantaneous measurement in unsteady flow, and to a spectrum analyzer for time-averaged measurement in steady flow. Follow-on studies in the same cylindrical apparatus have been reported by Bien and Penner (1971) and Jerskey and Penner (1973).

Watkins and Hussey (1973) used the reference beam mode to test theoretical predictions for spin-up from rest of fluid contained in an impulsively accelerated cylinder. They measured the fluid spin-up times at four radial positions in cylinders of different aspect ratios. Weidman (1976) also performed detailed measurements of the nonlinear spin-up process in a cylinder uniformly accelerated from rest and further considered the problem of fluid spin-down at uniform cylinder deceleration. An LDV system with self-aligning optics patterned after the original idea of Brayton (1969) was constructed. The experimental setup is sketched in Figure 4. The main component of this system is the optical flat, a piece of glass ground with opposing faces parallel and flat to within a fraction of a wavelength of the He-Ne laser. When placed in line and tilted with respect to the axis of a laser beam, the optical flat automatically splits the incident light into two perfectly parallel beams whose separation distance can be adjusted by the angle of tilt. A silver deposition on the flat surface at the point of second internal reflection reduces power loss in that beam. The parallel beams are then self-focussed at a common point in the fluid with a high quality acromat lens mounted on a motor-driven traverse. Of course the self-focussing feature in this application requires uniformity and concentricity of the rotating transparent plastic cylindrical wall. The receiving optics were setup in the dual scatter mode.

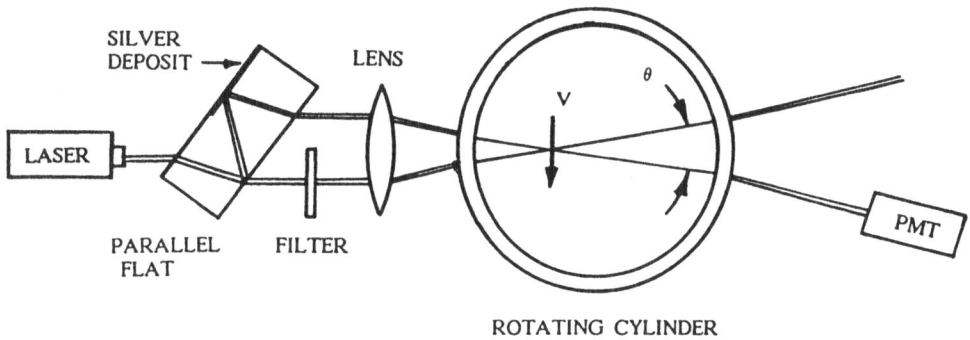

Figure 4. Self-focussing optics.

Time-Dependent Measurements

In the above cited experiments, unsteady motions were tracked manually by following the Doppler-shifted frequency on a frequency meter or a spectrum analyzer. This method of data acquisition is clearly impractical for turbulent flow and even for rapidly varying laminar flow. The two types of signal processing commonly employed to record time-dependent flow are the frequency tracker and the frequency counter. In the tracker-type signal processor, the Doppler frequency from a photomultiplier is fed to an electronic phase-locked loop. Within its tracking range, typically ±30% of a center frequency, the phase-locked loop automatically adjusts the voltage of its voltage-controlled oscillator to yield a frequency identical to the input signal. This provides a frequency-to-voltage conversion with analog voltage signal proportional to the instantaneous velocity. The counter-type signal processor gives the instantaneous velocity of individual particles by measuring the time between zero-crossings of a fixed number of Doppler cycles. Both methods are sensitive to signal drop-out, that is, to short periods of measurement time for which no particles are detected in the measurement volume. The type of system preferred depends in part on the strength of particle seeding and the dynamic range of the unsteady flow.

Frequency Shifting

A major drawback of the early LDV systems was that simple heterodyning techniques yielded the magnitude of the beat frequency but not its direction. Hence the measurement of oscillatory flow about zero mean, or the velocity field in regions of separated flow encompassing reversed streamlines, could not be accommodated. Denison and Stevenson (1970) overcame this problem by shifting the frequency of the reference beam with a rotating diffraction grating. The positive offset frequency was selected so that all Doppler-shifted frequencies encounted in their experiment were positive. An excellent comparison between theory and experiment for Poiseuille flow driven by a sinusoidal pressure gradient was obtained with this system. Frequency shifting is now commonly effected with a Bragg cell.

When using the dual scatter mode it is often advantageous to shift the frequency of both beams with two separate Bragg cells. This gives improved tracking capability and still permits resolution of low and reversed velocities. Faler and Leibovich (1978) used this technique to make a careful investigation of the mean recirculating flow inside a vortex breakdown bubble. Pfister, et al (1983) used double Bragg cells in the measurement of velocity autocorrelation functions in supercritical Taylor-Couette flow. It is shown how the velocity offset is implemented by following the recent work of Heinrichs, et al (1988). In this experiment designed to measure axial velocities over a vortex wavelength in supercritical Taylor-Couette flow, equal power incident beams (cf. Figure 3) are passed through separate Bragg cells driven at 40.0 and 40.1 MHz, respectively. Since one beam is shifted 100 KHz relative to the other, the fringe pattern formed in the beam intersection region is constantly translating. A particle travelling at velocity V normal to the fringes then scatters light with intensity modulated at a frequency of 100 KHz + V/Δx, where Δx is the fringe spacing (4.9 μm for the experimental setup). A measurement of the deviation of the modulation frequency from 100 KHz is then proportional to the local axial velocity of the scattering particles. The output of their phase-locked frequency tracker was stable to ±3 Hz which corresponds to a measuring accuracy of ±0.015 mm/sec. This allowed very accurate resolution of the axial velocity in a Taylor cell which had a peak-to-peak amplitude of nearly 15 cm/sec over one vortex wavelength.

Substantial advances in the understanding of rotating flows have been made with laser Doppler velocimetry. For example, Warn-Varnas, et al (1978) mounted an LDV system on a rotating table to resolve time and space scales associated with inertial modes in a rotating fluid. Fenstermacher, et al (1979) used a back-scatter mode while L'vov and Predtechensky (1979) incorporated a forward-scatter mode to test competing theories on the sequence of bifurcations leading to chaos in Taylor vortex flow. Garg and Leibovich (1979) extended their earlier LDV measurements of vortex breakdown into the turbulent flow regime. Escudier, et al (1982) were able to classify three distinct types of vortex breakdown phenomena according to the wake-like or jet-like axial velocity profiles measured with LDV. Hopfinger, et al (1982) used LDV to measure the mean flow of organized vortex motion generated by grid turbulence in a rotating tank. Szeri, et al (1983) reported careful LDV measurements of the basic flow states between closely spaced co-rotating and counter-rotating disks with and without axial throughflow. Aitta, et al (1985) made LDV measurements of tricritical phenomena in Taylor-Couette flow and found agreement with Landau theory applied to that system. Sirivat, et al (1988) have recently made measurements of non-Newtonian flow between rotating disks using a reference beam back-scatter mode of operation. In this investigation scattered light from the measuring volume was picked up by a PMT through the same optical components which comprised the the sending optics.

Multiple Velocity Components

Multiple velocity components at the same point in a fluid flow may be measured using several single component LDV systems. The problem lies in identifying from which one-component system the scattered light is emanating. Beam identification may be effected by labeling each pair of beams with a distinct optical property. The possible choices are to use different color beams, beams of different frequencies, or beams with different polarizations. At least three separately labeled beams are needed to measure two velocity components and three-component LDV systems present special problems. A discussion of multi-component LDV operation is given by Adrian (1983).

The only application of a multi-component LDV system to the study of rotating flow known to this author is found in the doctoral thesis of Ronnenberg (1977). He constructed a three-component laser anemometer operating in the reference beam mode to obtain measurements of fluid motion in a stationary cylinder driven by a rotating end plate. Since the flow was steady, the Doppler-shifted frequencies heterodyned on three separate photodetectors were recorded successively in time. Axisymmetric streamlines were then computed from the measured velocity field in an axial plane. Simultaneous velocity measurement at separate locations in a rotating flow have been reported by Savas (1987) in his study of Bodewadt instability. Using a color separator to generate blue and green light beams from an Argon ion laser, two independent forward scattering LDV setups focused beams at different radii in the Bodewadt boundary layer. Simultaneous measurement of identical horizontal velocity components at neighboring radial positions helped interpret certain features of this new time-dependent instability composed of concentric circular waves observed in the Ekman layer on the endwall of a rotating cylinder impulsively brought to rest.

Correlation and Structure Functions

Improved accuracy and additional information on a velocity field may be obtained by measurement of velocity *correlation* functions. Auto- and cross-correlation functions can provide time scales and amplitudes of velocity fluctuation as well as their first- and second-order statistics. Original correlation function measurements were made by recording LDV data on tape for subsequent computer processing. Vehrenkamp, et al (1979a) reported a frequency shifted, cross-beam LDV setup for direct measurement of the velocity autocorrelation function using the Erdmann-Gellert (1978) rate correlation technique. A double cross-beam LDV system gives the cross-correlation function with no greater measurement difficulty. A comparison between conventional, correlation, and rate correlation LDV techniques for a simple time-periodic flow has made by Vehrenkamp, et al (1979b). With sufficiently high scattering particle density, the cross-beam rate correlation method was found to be superior in signal-to-noise and in the rejection of errors due to inhomogeneous particle concentration.

The long integration time required to construct the correlation functions generally limits on-line measurements to low-speed hydrodynamic flows. Efforts to alleviate this restriction are in progress. Schulz-DuBois and Rehberg (1981) have demonstrated the advantages of computing the *structure* function (Yaglom, 1952) which contains the same information as the correlation function. Both auto- and cross-structure functions were computed from real experimental data and compared to their correlation function counterparts. It was found that convergence to approximately the same accuracy could be obtained with only 200 terms for the structure function compared to 20,000 terms for the correlation function. In a separate effort, Brown and Jones (1983) have reported a new burst-correlation LDV system, so called because it processes in real time the autocorrelation function produced by a single particle during its passage through the measurement volume.

The development of these correlation and structure function measurement methods permit greater detail in measurement of complicated flows exhibiting secondary bifurcations. Pfister, et al (1983) have produced excellent measurements of modulated jet and wavy mode Taylor vortex flow. Mullin, et al (1983) measured the presence of side bands on the primary instability frequency produced by a secondary bifurcation in a nonstandard Taylor vortex flow characterized by discrete symmetries. Brandstater, et al (1983) and Brandstater and Swinney (1987) used a pulse correlator to measure the variation of attractor dimension in supercritical Taylor vortex flow. A codimension-2 bifurcation in Taylor-Couette flow with rotating end plates has been documented by Mullin, et al (1987) using a microprocessor-based correlator.

Several references on the principles of laser Doppler velocimetry are available. For an introduction the reader is referred to Durst, et al (1981). Other references include Penner and Jerskey (1973) and Adrian (1983). The use of LDV for turbulence measurement is reviewed by Buchhave, et al (1979). A special report on the use of scanning laser anemometry systems for measurement of separated flows is given by Simpson (1989).

6.6 Laser-Speckle Velocimetry

A grainy appearance is observed when coherent light is diffusely reflected from the surface of an object if the surface roughness is large compared to the wavelength of illumination. This *speckle pattern* (Dainty, 1975) is caused by constructive and destructive interference of the scattered light. Speckle patterns can also be produced in liquids by light reflected from a particle suspension. The diameter d of a photographed speckle illuminated by coherent light of wavelength λ_0 can be estimated from the Rayleigh resolution criterion,

$$d = 1.2 \, \lambda_0 \, f \left[\frac{m + 1}{m} \right], \tag{6}$$

where f is the f-number and m is the magnification of the camera optics. Typical speckle diameters photographed using standard laser illumination and camera optics are of order 2-10 μm. Such speckle patterns can be faithfully recorded with high resolution photographic film.

Grousson and Mallick (1977), Barker and Fourney (1977), and Dudderar and Simpkins (1977) independently described how speckle patterns may be used to measure fluid velocity in a plane. Each pair of investigators chose to demonstrate the measurement technique by measuring the parabolic velocity profile in Poiseuille flow in a glass tube. The fluid is seeded with small particles and an axial plane is illuminated with a sheet of laser light. A double exposed photographic record of the movement of the speckle pattern in the plane is obtained using a double pulsed laser. If the fluid motion between laser pulses is large enough to cause a shift of at least one speckle diameter, an optically interrogated exposed speckle photograph, or specklegram, will produce a regular pattern of fringes similar to Young's fringes. The direction of speckle movement is normal to the fringe pattern and its displacement l may be determined from the equation

$$l = \left[\frac{\lambda}{m \, \sin\alpha} \right] , \qquad\qquad (7)$$

where α is the angular spacing of the fringes and λ is the wavelength of the interrogating beam. The velocity V is then given by $V = l/\Delta t$, where Δt is the time between exposures. Barker and Fourney (1977) seeded their flow with a colloidal suspension of milk, Grousson and Mallick (1977) added 0.5 μm polystyrene spheres, and Dudderar and Simpkins (1977) used white latex paint. Simpkins and Dudderar (1978) subsequently applied the laser speckle velocimetry (LSV) technique to an investigation of unsteady, two-dimensional Rayleigh-Benard convection. They reported that the dynamic range for their experiment was limited to particle displacements in the range $d \leq l \leq 20d$, the upper limit being set by the loss of correlation between individual speckles. Since the time interval O(10 μs) between speckle exposures was much smaller than the characteristic time of the unsteady motion, information recorded on a specklegram provided a nearly instantaneous measurement of the velocity field. In each of these early investigations the specklegram was interrogated point-by-point with coherent light from a laser beam.

A variation of laser speckle velocimetry was introduced by Meynart (1980). In this application several 20 ms time exposures were taken at one second intervals on a single photographic plate. The resulting photograph does not exhibit a true speckle pattern but rather a well-resolved series of points. In this case, interrogation of a small area of film using coherent plane waves gives rise to multiple-beam Young fringes. Meynart (1980) points out that these fringes are much sharper than the fringes on a conventional specklegram. Thus the point-by-point measurement of the fringe spacing and angular orientation can be obtained with increased precision. Meynart also reported a second method of analysis giving at once a two-dimensional map of the velocity field. In this setup, the entire photograph is interrogated by coherent light and spatial filtering is accomplished with the aid of Fourier transform lenses. The end result is a filtered image of the flow field in a plane exhibiting equal-velocity fringes of known magnitude. Beautiful velocity fringe patterns (isotachs) of Rayleigh Benard convection are exhibited in his paper. This mode of analysis is very attractive since it provides a simple and rapid full-field picture without the necessity of scanning many individual sections of a photograph. Because Meynart's method depends on particle tracking, it has come to be known

(Adrian and Yao, 1983) as particle image displacement velocimetry (PIDV).

Both LSV and PIDV are presently under intense development because of their full-field velocity measurement capability. Meynart (1983) was the first to apply speckle velocimetry to a gas flow. Adrian and Yao (1985) performed a study of the optimum particle size for LSV and PIDV in air and water. An optimization study of camera optics and film resolution parameters was reported by Lourenco and Krothapalli (1987). Further details on the PIDV technique are given in the review paper by Krothapalli (1989).

It appears that Fitzjarrald (1982) is the only investigator to have used PIDV in a rotating fluid experiment. He recognized that it was well-suited to the laterally-heated rotating annulus experiment wherein the flow is nearly two-dimensional in level planes. A horizontal section of the flow was strobed by a laser beam reflecting off a 45° mirror mounted on the end of a vertical rotating shaft coincident with the central axis of the annulus. Multiple exposures of the aluminum particle suspension during successive laser beam sweeps were taken with overhead photography. The negatives were interrogated point-by-point with a narrow laser beam using an automated scanning system which permitted fringe patterns analysis to proceed at the rate of approximately two points per minute of operator time. A comparison of velocity vectors derived from PIDV generally exhibited good alignment with the series of dots describing the direction of fluid motion on the original photographs. However, some of the veclocity magnitudes were in error due the difficulty of trying to count indistinct fringes in some parts of the flow field. Fitzjarrald pointed out that this was due in part to the alignment of particles with the stream surfaces: near the outer cylinder wall where there is upwelling, the tilted particles reflected significantly less light. Another difficulty encountered was insufficient dynamic range to record velocities over the entire spatial field. The dynamic range is limited by the smallest fringe that can be measured (largest spot separation on the film) and by the slowest speed for which the spots to run together on the film. Fitzjarrald computed the mean kinetic energy and the eddy kinetic energy from the available data and compared results with those reported by Pfeffer, et al (1974) using an *in situ* array of 98 hot-thermistor beads. He noted that the phase of the mean and eddy kinetic energies was nearly the same as observed by measurement with the *in situ* probes, but the magnitudes found with the PIDV method were significantly higher. This comparison offers further evidence of probe interference effects encountered in large thermistor arrays (see §5.3).

7. SHEAR STRESS PROBES

We consider three basic shear stress probes which have been used in rotating flows, namely, Preston tubes, hot-films and hot-thermistors. An up-to-date report on the measurement of mean and fluctuating wall shear stress in nonrotating flow using hot-wire probes is given by Haritonidis (1989).

7.1 Preston Tubes

Local wall shear stress in turbulent flow may be determined by measurement of the pressure difference ΔP between a miniature Pitot probe placed at the wall and a nearby static pressure tap. The miniature Pitot probe placed at the wall has come to be known as a Preston tube. A typically 2 mm diameter probe aligned into the free stream measures the total pressure very close to the wall. Based on previous investigations, Preston (1953) argued that the velocity variation close to a wall in a turbulent boundary layer is a universal function that depends only on the local wall shear stress according to the relation

$$\left[\frac{\Delta P \ d^2}{\rho \ \nu^2}\right] = K \ \left[\frac{\Delta P}{\tau_w}\right] , \tag{8}$$

where d is the inside diameter of the tube, τ_w is the local wall shear stress, and ρ and ν are the fluid density and kinematic viscosity, respectively. A universal calibration for the constants K has been reported by Head and Ram (1971). Preston tubes necessitate accurate machining and positioning to cause minimum flow disturbance. Bellhouse and Schultz (1966) report that slight errors in positioning a Preston tube on the surface of a 3 inch diameter cylinder can cause 30% skin friction measurement errors. Less difficulties are encountered on flat wall surfaces. Hanratty and Campbell (1983) have reviewed the operation of the Preston tube and other wall shear stress probes.

Preston tubes have been used to some extent in rotating flows. Moore (1967) measured local values of wall shear stress using a Preston tube in rotating channel flow with air as the working fluid. In a similar experiment Johnston, et al (1972) made Preston tube wall stress measurements in rotating water channel flow. Strong asymmetries in the wall shear stress on opposite channel walls have been measured in both cases. A discussion of the use of Preston tubes for shear stress measurement in rotating flows is given in the appendix of Johnston, et al (1972).

7.2 Hot-Films

Various flush-mounted heated elements have been used to measure skin friction in boundary layer flow. Early analysis this problem can be traced to Leveque (1928) who considered heat transfer from the film into a fluid flow without heat loss to the solid boundary surrounding the film. Leveque's result relating the Nusselt number $Nu = QL/k(T_w - T_\infty)$ to a nondimensional wall shear stress parameter $N_w = CL^2 \tau_w / k\nu$ is given as

$$Nu = 0.807 \, (N_w)^{1/3}, \qquad\qquad (9)$$

where Q is the local heat-transfer rate per unit film area, L is the length of the hot-film in the flow direction, T_w is the local wall temperature, T_∞ is the freestream fluid temperature, τ_w is the wall shear stress, and C, k and ν are the fluid heat capacity, thermal conductivity and kinematic viscosity, respectively. In some pioneering work Fage and Falkner (1931) experimented with heated foil elements. Liepmann and Skinner (1954) rederived the fundamental 1/3 power shear stress law using dimensional analysis. In practice there is a constant heat loss to the film substrate at a given operating condition which can be taken into account by adding a constant to equation (9). In terms of the voltage drop V across the film, the heat transfer balance given by (9) may be written

$$V^2 = A \, (\tau_w)^{1/3} + B , \qquad\qquad (10)$$

which provides a suitable form for direct calibration of the shear stress probe. Bellhouse and Schultz (1966) and Spence and Brown (1968) extended previous analyses to include the effect of pressure gradient on the operation of the heated sensor. In both these investigations the authors reported the development of tiny surface-mounted probes which could be installed and removed at will. This led the way to plug-in hot-films now commercially available. A simple calibration procedure to determine the constants A and B was reported by Brown and Davey (1971). In this setup the probe is installed flush with a disk rotating above a stationary plate. Theoretical results for the flow in the gap in this differentially rotating disk system are well-known. Variation of the radial location of the probe on the rotating disk, the disk-plate separation distance and the angular rotation speed provide a wide range of skin friction calibration points.

An example application of shear stress probes in rotating flow is given by Caldwell, et al (1972) who operated a hot-film sensor at constant-temperature to measure skin friction in a turbulent Ekman layer. The elongated rectangular shape of the sensing element had a strong sensitivity to the flow direction which permitted accurate measurement (±2°) of the local direction of wall shear stress in addition to its amplitude. In spiral laminar flow between concentric differentially-rotating cylinders, Coney and Simmers (1979) detected three shear stress regimes with the aid of a hot-film probe mounted flush with the outer cylinder wall. Other references to the use of heated-film sensors in Taylor vortex flow are given in their paper. Sdougos, et al (1984) used a flush mounted heated-film sensor to measure the radial shear stress distribution on the flat boundary beneath a shallow rotating cone. This cone-and-plate device is often used as a fluid viscometer by measuring of the torque required to rotate the cone against the retardation of the viscous fluid that fills the cavity between cone and plate.

7.3 Hot-Thermistors

Like the hot-film, heated thermistors embedded flush in a wall respond to the normal gradient of velocity parallel to the wall. Niler (1965) analyzed this problem for laminar Couette flow between parallel plates under the assumption of a dissipationless fluid and negligible forced convection. He modeled the thermistor as a spot generating an axisymmetric Gaussian distribution of heat in zero flow. Asymptotic solutions to the heat equation were reported in the limits of small and large Peclet number. At low Peclet number Niler found that the hot-thermistor responds to the entire velocity profile of the impinging fluid, an undesirable result from the point of view of an experimental test probe. For asymptotically large values of the Peclet number, however, the hot-thermistor responds to the local velocity gradient at the wall. In this limit the response is the same as that for a heated surface film sensor, namely the power dissipated is proportional to the 1/3 power of the velocity gradient or wall shear stress. Niler derived this result without apparent knowledge of the similar result obtained earlier by Leveque (1928) and Liepmann and Skinner (1954).

Lambert, et al (1965) devised an experiment to test Niler's high Peclet number result in high Prandtl number circular Couette flow. Operating the thermistor in a constant-temperature circuit, they found that indeed the probe is sensitive to the 1/3 power of the local velocity gradient as long as the thermistor is operated below the region of voltage saturation. These investigators pointed out that glass-coated hot-thermistors are relatively noise-free probes compared to some heated-film elements they fabricated. In most practical applications the time response of the wall shear sensor is about 2 Hz. Although this prohibitively low response precludes the use of the hot-thermistor for the study of transition in classical boundary layer flow, it is ideally suited for investigations of the stability of high Prandtl number circular Couette flow where the drift rates of instability waves are low. Snyder and Lambert (1966) reported on the simultaneous operation of a triangular array of three hot-thermistors mounted flush with an inner rotating cylinder to measure wavelengths and phase speeds in a detailed study of finite-amplitude Taylor vortex flow.

8. HEAT TRANSFER MEASUREMENT

One-dimensional heat flux measurements in a bounded fluid system are generally reported in terms of the Nusselt number Nu which is the ratio of the total heat flux Q transferred by conduction and convection to that transferred by conduction only. Thus the Nusselt number may be written

$$Nu = \left[\frac{QL}{k\Delta T} \right], \tag{11}$$

where k is the fluid thermal conductivity, ΔT is the differential temperature between opposing boundaries and L is the relevant length scale normal to the boundaries. Thus measurements of both the net heat flux Q and the temperature contrast ΔT are required. In some instances heat transfer has been determined from the measurement of local temperature gradients near a flow boundary. We divide the presentation into three basic experiments which have received particular attention: Taylor-Couette flow with radial heating, the laterally-heated rotating annulus, and bounded rotating fluids driven by vertical temperature gradients.

The understanding of the dissipation of heat in the air gap between rotor and stator in rotating electrical machinery has motivated a considerable number of experimental heat transfer investigations. The simplest analog to this practical problem is circular Couette flow. A preliminary experimental heat transfer study of this system was reported by Kaye and Elgar (1958). A heated nichrome ribbon wound around the inner rotating cylinder provided the uniform source of heat which was removed through a water cooling jacket encasing the outer cylinder. An axial flow of air was introduced to simulate forced convection cooling for electric motors. The net heat transfer across the gap was taken as the difference between the rate of electrical heating and the rate of heat transfer by radiation. Temperatures on the inner and outer cylindrical surfaces were measured with thermocouples. The Nusselt number was determined for a range of Taylor numbers and axial Reynolds numbers. Bjorklund and Kays (1959) continued these measurements in air and Haas and Nissan (1960) reported measurements in liquids. In the latter experiment, the net radial heat transferred across the gap was measured by monitoring the power input to an electrically heated bath inside the inner cylinder after thermal equilibrium was attained. The Nusselt number was calculated using average temperatures obtained from 13 or more thermocouples mounted along the surface of each cylinder bounding the liquid. Simmers and Coney (1979) obtained measurements of radial temperature distributions across the gap of a Taylor-Couette facility with combined rotational and axial flow using a miniature thermocouple. The local heat transfer was computed from the slope of the temperature curve near the outer cylinder, and results were reported for a range of Taylor numbers for conditions of developing and fully-developed axial flow.

Hide (1958) made some early estimates of radial heat transfer in the laterally heated rotating annulus experiment. For the purpose of determining the heat transferred across the outer cylindrical boundary, he measured the heat rise $\rho C Q \Delta T$ in the water circuit cooling the outer cylinder. Knowing the density ρ and specific heat C of the coolant, it was only necessary to determine volume flux Q and temperature rise ΔT of the water in the cooling circuit. The temperature rise in the cooling circuit and the temperature difference across the annulus were measured with miniature thermocouple probes. Hide's results showed for the first time that the radial heat transfer was independent of fluid depth in the annulus. More precise measurements designed to minimize heat losses from the top and bottom surfaces were carried out by Bowden and Eden (1965) using similar measurement techniques. A sharp break in the Nusselt number curve as a function of rotation rate was observed at the transition between symmetric flow and the onset of steady waves. Buzyna, et al (1974) measured temperature and velocity perturbations around a mid-radius ring in the rotating heated annulus using an array of 98 thermistors for concomitant measurement of temperature and speed. Simultaneous streakline photography was employed to record flow direction in order to convert speed into velocity. The zonally-averaged radial eddy heat flux $T'v'$ across the central "latitudinal circle" was computed as a function of time. Here T' and v' are the perturbation temperature and radial velocity, respectively. The eddy heat flux was observed to rise from a low value to a maximum and fall back to the low value again in phase with the azimuthal instability wave being studied in their experiment. This rather sophisticated heat flux measurement would not have been possible without the large thermistor array.

Rossby (1969) conducted careful heat transfer measurements for rotating Rayleigh-Benard convection. In this experiment a layer of fluid is *heated from below* so that, in the absence of rotation, heat is transferred by conduction through motionless fluid until a critical Rayleigh number is reached, at which point natural convection sets in. A careful determination of the vertical heat flux was obtained by measuring the temperature drop across uniform thickness epoxy layers sandwiched between a copper plate bounding the fluid and an aluminum plate opposite. The temperature drops were calibrated with prescribed heat fluxes prior to an experiment. Heat transfer measurements were made over 5 decades of Taylor number and 4 decades of Rayleigh number for water at fixed Prandtl number. A complete discussion of this measurement technique and associated error sources is given in an appendix of Rossby's paper.

Abell and Hudson (1975) determined the vertical heat transfer across a rotating fluid layer in a rectangular cavity *heated from above*. In this stably stratified configuration, convection is triggered by the coupling of the vertical density gradient with the orthogonal centrifugal acceleration. The upper aluminum plate was electrically heated, the bottom plate cooled by a constant temperature water circulation system, and the sidewalls were insulated. The total heat supplied to the fluid system was calculated from the resistance heating. Heat lost to the surroundings conducted down through the vertical sidewalls and radially out the insulation was determined by first making heat transfer measurements across the test chamber with the fluid replaced by a solid piece of plastic with known thermal conductivity. The heat loss so determined at different power settings was then subtracted from the total measured heat input to an active fluid experiment. The temperature contrast across the horizontal surfaces was measured from an array of copper-constantan thermocouples embedded in the bounding aluminum plates. Hudson, et al (1978) extended these experiments to measure axial heat transfer across a fluid contained in a cylindrical cavity rotating about its central axis. Following the procedure of Rossby (1969), the heat flux across the top and bottom plates was determined from thermistor measurements of the temperature difference across calibrated thermal resistances consisting of epoxy layers sandwiched between aluminum disks.

9. MASS TRANSFER MEASUREMENT

Mass transfer in rotating flows has been measured by various means. The simplest technique is direct weighing of the mass sample before and after a mass transfer experiment. Kreith, et al (1959) measured the rate of mass transfer of napthalene cast into a flat circular recess in a rotating disk by weighing the disk before and after test runs up to 10,000 rpm. In a similar experiment Hansford and Litt (1968) measured the mass transfer of benzoic acid and β-naphthol as a function of plate rotation speed. Care must be taken in these experiments to account for mass loss from the disk due to natural convection during the time interval between weighing and mounting, both before and after a test run.

Another technique for determining mass transfer is to measure the concentration of material diffused from a solid surface into a fluid. Flower, et al (1969) used two variations of this technique to measure the rates of mass transfer from the surface of a stationary outer cylinder in a circular Couette apparatus with axial flow and rotating inner cylinder. In the first method mercury vapor was transferred from a solid surface of amalgamated silver and its concentration in the air downstream was determined by ultraviolet absorption measurements. Experiments of both longitudinal and circumferential diffusion into the air were performed. In the second method long sections of falling liquid films were produced on the wall of the stationary outer cylinder which in some cases was lined with filter paper. The mass transfer of both water and n-heptane was determined by thermal conductivity measurements in the air downstream of the liquid film surfaces. Control of the axial flow between the cylinders and the rotation rate of the inner cylinder gave mass transfer measurements over a range of Reynolds and Taylor numbers.

9.1 Electrochemical Technique

The above measurement techniques yield only time-averaged mass transfer over large surface areas. The electrochemical technique offers a method of making rapid time-dependent mass transfer measurements over large or small surface areas. The axisymmetric flow generated by a rotating disk beneath a quiescent electrolyte provides a "uniformly accessible" mass transfer surface, i.e., mass transfer occurs normal to the disk and not laterally. The principle of operation of the electrochemical method is as follows. An electrochemical reaction is carried out at the surface of an electrode mounted flush with the wall. The voltage on the electrode is kept sufficiently high that the reaction rate is so fast that it is mass-transfer controlled. At the same time the voltage must not be so great that side reactions occur. Under this condition of current-limiting diffusion, the mass transfer coefficient K is directly related to the limiting current I_{lim} produced at the surface of the electrode through the equation

$$I_{lim} = n F C_{\infty} K ,$$ (12)

where n is the valence of the metal ions, F is the Faraday constant, and C_{∞} is the ion concentration of the electrolyte solution far from the active surface. Levich

(1942) worked out the mass transfer relationship for the case when the electrode is the entire surface of a rotating disk. Kabanov and Siver (1948) were the first to use a rotating disk electrode (RDE) to verify the Levich theory. Eisenberg, et al (1954) used the electrochemical technique to make mass transfer measurements in a diffusion-controlled reaction of potassium ferricyanide and ferrocyanide in sodium hydroxide. These measurements were performed in an annulus with the rotating inner nickel cylinder serving as the electrode. The symmetry of this geometry also guarantees that the mass transfer at the cylindrical wall is uniformly accessible. In high rate metal dissolution studies, the material in a RDE will recede into its support and disrupt the fluid flow characteristics there. To overcome this deficiency Chin (1971) developed a rotating (hemi-) spherical probe (RSE). The advantage of this probe is that metal lost in dissolution gradually changes the radius of the hemisphere but does not noticeably disturb the flow around it.

The local mass transfer on both the RDE and RSE increases with radius owing to the locally increasing relative velocity between probe and electrolyte. Therefore, these probes can provide only an average mass transfer measurement over the entire active surface. A more local measurement of mass transfer can be made with disks having only a narrow annular ring as the active electrode. This case was also analyzed by Levich (1962), but experiments by Avdeyeva (Levich, 1962) gave results 20-30% lower than that predicted by the Levich theory.

Truly local mass transfer rate measurement using very small diameter probes mounted flush with a rotating disk and the theory for these point probes have been reported by Chin and Litt (1972a). They found that for a point electrode of diameter d at radius r on a disk rotating at angular velocity ω, the nondimensional equation governing the mass transfer from the electrode to the electrolyte in a diffusion-controlled reaction is given by

$$Sh = 0.780 \ \epsilon^{1/6} \ R^{1/2} \ Sc^{1/3}, \tag{13}$$

where $Sh = Kd/D$ is the Sherwood number, $\epsilon = d/r$ is a geometrical parameter, $R = r\omega d/\nu$ is the rotational Reynolds number and $Sc = \nu/D$ is the Schmidt number relating the kinematic viscosity ν to the mass diffusion coefficient D. Chin and Litt performed an experiment using the diffusion-controlled reaction of potassium ferricyanide and ferrocyanide in an aqueous solution of potassium chloride. Measurements obtained with a 0.038 cm diameter electrode placed at various radial positions flush with the rotating plate permitted wide variation of ϵ and the change in angular speed covered a wide range of R. Substituting carboxymethylcellulose for potassium chloride in the electrolyte solution permitted a four-fold change in Sc. Using equation (12) to determine the local mass transfer coefficient K, equation (13) was verified in regions of laminar flow. At high rotational Reynolds number a break-point was observed in the mass transfer coefficient indicating the onset of transition to turbulence.

In a follow-on study Chin and Litt (1972b) showed how the point electrode could be used as a sensitive experimental tool for studying fluid instability adjacent to a rotating disk. They argued that since the limiting current measures the instantaneous rate of mass transfer to the electrodes, time-dependent fluctuations of the current should respond to flow instabilities. They showed, in fact, that stationary Type I Ekman waves and their evolution to turbulence could be detected over a wider range of Reynolds number with the mass transfer electrode than had previously been possible with other measurement systems. The electrochemical technique has also been used in studies of diffusion-limited heterogeneous catalytic reactions. For example, using a rotating disk electrode White, et al (1974) measured the hydrogenation of α-methylstyrene and White and Litt (1975) measured the hydrogenation of phenylacetylene over palladium.

Kataoka (1975) employed the electrochemical method to make mass transfer measurements in Taylor vortex flow between concentric cylinders with the outer cylinder stationary and the inner one rotating. Assuming the validity of Reynolds analogy between heat and mass transfer for the O(1) Prandtl number fluids used in the experiment, Kataoka inferred supercritical heat transfer rates from the mass transfer measurements. Further details on the electrochemical mass transfer measurement technique are reported by Mizushina (1971).

9.2 Electrochemiluminescence Technique

The use of electrochemiluminescence for flow visualization has been described in §3.1.3. The electrolytic solution is composed of luminol dissolved in water with hydrogen peroxide as the active electrolyte. As with the electrochemical technique described in §9.1, the electrolyte cell is operated at limiting current conditions to insure that the active electrolyte at the surface of the anode is negligible compared to the free-stream concentration. For this condition Springer (1964) showed that the light intensity emitted as a blue glow on the model electrode is proportional to the local mass transfer rate of the active electrolyte to the anode surface. A rotating water channel produced by the solid body rotation of an annular chamber filled with the luminol solution was constructed by Colello and Springer (1966) for measurement purposes. A photomultiplier light-measuring unit was focussed on small areas of the test anode suspended into the rotating channel flow. Light intensity measurements in three completely different experiments gave results consistent with theoretical mass transfer results. The experiments afforded three independent checks on the calibration constant $K = I/W$ where I is the light intensity and W is the mass transfer rate to the surface, both related to the same unit surface area. Apart from these exploratory experiments, this relatively powerful measurement technique apparently has not been used for mass transfer measurement in other rotating flow systems.

9.3 Optical Absorption Method

A simple technique for mass transfer measurement in turbulent Taylor vortex flow has been reported by Tam and Swinney (1988). The optical absorption of laser light was employed to determine an effective diffusion coefficient in liquid flow. Methylene blue dye was injected into the annular gap at mid-height from an orifice on the inner rotating cylinder. Time series measurements of the optical absorption of laser beams directed through the gap onto photodiodes at two axial positions below the injection point were digitized and recorded on a computer. As the dye diffused through the turbulent Taylor cells down the annulus, the intensities of the transmitted beams decreased. A simple one-dimensional diffusion model permitted calculation of an effective axial diffusion coefficient for experiments conducted over a wide range of supercritical Taylor numbers.

10. TORQUE AND FORCE MEASUREMENT

Torque measurements date back to the earliest investigations of rotating flow because of their practical implications on the resistance of rotating machinery. We review in turn torque and force measurements and show that while torque measurements are relatively commonplace, direct measurement of forces in a rotating fluid has only recently been accomplished.

10.1 Torque Measurement

In an experimental study on the efficiency of turbine water-wheels and centrifugal pumps, Thompson (1855) reported the mechanical work consumed through the action of friction for a disk rotated inside a closed cylindrical cavity filled with water. The details of the measurement technique were only revealed 30 years later when Unwin (1885) repeated Thompson's experiments. A hollow cylinder closed at the bottom and filled with water was placed in a concentric cylindrical water bath, capped with a top lid, and suspended by three fine wires attached to its outer rim at 120° intervals. The flat test disk was attached to a shaft mounted through the vertical axis of the water-filled cylinder. The shaft was driven at constant speed by a Rider's hot-air 1/2 hp engine --- this represented a marked improvement over Thompson's hand-cranked shaft. A silk thread wound around the rim of the cylinder ran over a pulley to a weighing pan. With the inner disk rotating at constant speed, the outer cylinder would twist in the direction of disk rotation and the torque on the cylinder was determined by adding weights to zero the position of the cylinder relative to a stationary indicator.

This torsional suspension torque measurement technique was employed with various modifications in several other early experimental investigations. Using, for the first time, a concentric rotating cylinder apparatus, Mallock (1889) measured the torque on a stationary inner cylinder with the outer one rotating. In this case the inner cylinder was suspended from a single torsion wire made of either brass or nickel. The experiments were performed by driving the outer cylinder at uniform speed and recording the angle through which the inner cylinder turned when it came to rest. The deflection angle was read from divisions marked around the perimeter of a circular disk mounted rigidly to the stationary cylinder. The calibration of the torque necessary to rotate the torsion wire through a given angle was made with an elaborate system of threads and weights carefully designed to provide only horizontal forces tangent to the perimeter of the circular disk. No external damping was provided so and hence the cylinder was observed to oscillate in both its sheared and unsheared states with a period of about 15 seconds. In a subsequent series of experiments Mallock (1896) suspended the outer cylinder from three torsion wires and rotated the inner cylinder.

Couette (1890) performed an almost identical experiment to that reported by Mallock (1889) in that he suspended the inner cylinder on a single torsion wire and rotated the outer cylinder. However, in Couette's experiment the torque was measured by winding thread wound around a horizontal pulley attached to the stationary cylinder and drawing it out over a vertical pulley down to a weighing pan. With the outer cylinder rotating, weights added to the pan brought the inner cylinder back to its undeflected position, thereby providing a direct measurement of the torque. As in Mallock's (1889) experiment, no external damping was provided and the suspended cylinder oscillated with a 20 second period. Couette used theoretical results for fluid shear acting on the cylindrical surface to determine fluid viscosity.

Taylor (1936) measured the stopping force at fixed radius on an inner cylinder with the outer one rotating and *vice versa*. In contrast to previous torsion experiments, Taylor's cylinders were supported in rotary bearings. The measurement technique used was identical to that of Couette (1890) except that during an experiment weights were placed in the scale pan (i) until the cylinder was just able to move in the direction of the driving cylinder and (ii) until it was just able to move in the opposite direction. The difference between the two measurements was due to the static friction of the cylinders against their supports, so the fluid friction torque was taken to be the average of the two measurements.

More recently, Donnelly (1958) carefully repeated the earlier torque measurements in the concentric cylinder apparatus. Having reviewed the difficulties encountered by Mallock (1896) and taken note of Taylor's (1923) comments on Mallock's apparatus, Donnelly set up a fused quartz rod and fiber torsion system with optical rotation measurement, magnetic damping, precision alignment, and employed a well-regulated constant temperature bath surrounding the concentric cylinders. The system was calibrated using standard viscosity oils. The consistency of measurements obtained through the use of different fluids of calibrated viscosity attest to the care with which the experiment was conducted. These experimental results probably remain the most precise fluid torque measurements made to date.

Wan and Coney (1982) employed an inductive torque transducer in their investigation of highly supercritical spiral vortex flow in a Taylor-Couette apparatus with axial throughflow. The transducer was installed between the driving motor and the driven inner cylinder to monitor torque required to maintain steady rotation. Measurements were made at constant axial Reynolds number with increasing Taylor number above the onset of spiral vortex instability. They observed the interesting feature that the torque slowly decreased with increasing Taylor number as the spiral vortices were observed (aluminum flake suspension visualization) to break down into chaotic laminar waves. At higher Taylor numbers the torque increased dramatically as the flow evolved into finer scales of turbulence.

Cone-and-plate viscometers are generally equipped with a torque meter for direct measurement of the torque required to rotate the cone against the viscous fluid that fills the cavity between cone and plate. In a fundamental study of this device for low angle cones, Sdougos, et al (1984) made extensive torque measurements as a function of rotational Reynolds number using a commercially available torque indicator for their Brookfield cone-and-plate viscometer.

10.2 Force Measurement

The measurement of the drag force on spheres translating *vertically* along the axis of a rotating fluid can be obtained by employing a standard technique used to measure drag in Stokes flow. In Stokes flow experiments (see, for example, Lasso and Weidman, 1986) an object is released from rest in a viscous fluid, allowed to come to terminal velocity, and its time of flight between two accurately marked positions is recorded. The latter measurement gives the steady translation velocity and the drag is measured by differencing the object's weight and its buoyancy force. Huner and Hussey (1977) have shown that the best way to reduce accumulated errors is to submerge the object in a sample of the test fluid and directly measure the net gravitational force using an electronic balance. Maxworthy (1965, 1968) employed this classical technique in the following manner. He constructed a set of buoyant spheres, released them from the bottom of a nonrotating tank, and measured their terminal ascent rate. The rise rate and sphere diameter d give the sphere Reynolds number, and the drag D at that Reynolds number can be determined from previous nonrotating experimental results. The spheres then were released in a rotating tank and a smaller ascent rate U was measured owing to the increased dissipation in the sphere Ekman layers. Since the drag in the rotating liquid is still D, a drag coefficient based on D, U, d and the fluid properties can be formulated. As with nonrotating fluid experiments, care must be taken to account for container blockage effects which depend on the ratio of sphere diameter to the diameter of the cylindrical test facility. Maxworthy (1968) determined the drag in a fluid of infinite extent by performing measurements in cylinders of increasing radius and extrapolating results to zero sphere blockage.

The measurement of forces on a body moving *horizontally* in a rotating fluid poses a special challenge to the experimentalist. Mason (1975) introduced an ingenious experiment from which such forces can be inferred. In the experiment a spherical test object is tethered to the end of a 0.005 cm diameter copper wire suspended from a rigid support mounted on a rotating turntable 1 m above a rectangular tank of water. The axis of the pendulum wire is carefully aligned with the axis of rotation. Initially the sphere is drawn to one side of the tank and held in place. After the fluid is spun up to solid body rotation, the object is released and executes a spiraling motion towards the central axis. Successive points along the trajectory were recorded photographically at short time intervals with the aid of stroboscopic illumination. Both the normal and tangential forces acting on the sphere were determined from measurements of three successive positions on the trajectory by balancing forces in the horizontal plane. These include the horizontally projected tension force in the wire, the centripetal force and the Coriolis force. There are several inherent limitations of this measurement technique. The experiment is restricted to objects of prolate spheroidal shape so that the "flow" sees the same cross section at every point along its trajectory. Even oblate spheroids were unacceptable since they wobbled back and forth as they moved through the fluid and this jerky motion precluded accurate measurement. Furthermore, eddies shed off the test objects induce a time-dependent drag and the measurement method averages this effect to some degree. Cusps in the spiral trajectories, due to inertial overshoot of the test object, were observed at low tank rotation speeds and reliable data could not be obtained for these cases. This inertial effect restricted the experiments to Rossby numbers $Ro > 10^{-2}$, approximately. In spite of these limitations, the novel measurement technique permitted a relatively simple determination of horizontal forces on objects moving in a rotating fluid with a reported accuracy of about ±20%. These very small forces, typically on the order of dynes, would be extremely difficult to measure directly in a rotating flow. In a subsequent study, Mason (1977) used the same pendulum technique to measure forces on a sphere moving horizontally in a salt-stratified rotating fluid.

It appears that the only *direct* force measurements made on an object moving horizontally in a rotating fluid are those reported by Chabert d'Hieres, et al (1988). These experiments were performed in the large 14 m diameter rotating table at the Institut de Mechanique in Grenoble, France. In this experiment a hollow plastic cylinder 1.05 m high and 0.5 m diameter was suspended vertically in the fluid from a universal joint on a radial overhead carriage riding with the rotating table. Relative motion was obtained by driving the carriage circumferentially in either a prograde or retrograde direction. The unsteady forces generated by asymmetric vortex shedding from the cylinder were detected by the displacements of four ring transducers symmetrically located between the inner wall of the cylinder and its central support axle. Ring distortion measured by inductive probes within the rings was sufficiently small (2 mm) to produce negligible tilt of the cylinder from its vertical axis. With this configuration forces as small as 0.05 N could be resolved. Further details of the system design and measurement error analysis may be found in Chabert d'Hieres, et al (1987). During an experimental run 1000 discrete force measurements were recorded. This allowed a time resolution of 120 msec at the highest speeds and 1 sec at the lowest speed, sufficient to resolve the Strouhal frequencies encountered in the experiment.

11. DENSITY GRADIENT FORMATION AND MEASUREMENT

Of major interest in geophysical fluid dynamics is the rotating flow of density stratified. Density stratification permits baroclinic production of vorticity not possible in homogeneous fluid flow. Density gradients in the oceans and atmosphere are inevitably continuous, although sharp gradients are often associated with the upper thermocline layer of lakes and oceans and also with thermal inversions in the atmosphere. The formation of controlled density gradients in rotating laboratory flow offer some challenge to the experimentalist.

11.1 Stratification Techniques

Immiscible two-layer laboratory models of geophysical flow provide the simplest form of fluid stratification. Hart (1972, 1973, 1985) has made extensive use of this system for the study of baroclinic instability in a rotating cylindrical tank driven by a differentially rotating upper lid. Immiscible fluids with weak interfacial tension minimize undesirable surface tension restoring forces. The benefit of finite surface tension is that the fluids remain separated during spin-up. Self-diffusion and mixing of miscible liquids would be accelerated by advection from the side walls during spin-up, particularly if the density contrast is weak or the spin-up process is rapid. Hart used silicone oil of fixed density in the upper layer and water-alcohol solutions in the lower layer. The density contrast across the layers can be readily adjusted by changing the water-alcohol mixture ratio. Also, density measurements can be made in the laboratory prior to layer formation in the test facility, and hence standard measuring devices such as hydrometers, optical refractometers or densitometers can be employed.

Continuous stratification is generally produced by forming a vertical concentration gradient of salt or sugar in water. In the absence of external mixing, a desired stratification profile will remain unaltered for a considerable period of time because of the slow vertical diffusion of salt (1 cm/hr) and especially sugar (1 cm/day). Weidman and Johnson (1982) have shown that nearly hyperbolic tangent density profiles can be obtained by the natural diffusion of two liquid layers of differing concentrations carefully placed one on top of the other. The characteristic vertical length scale of the profile increases with time so that a range of profiles are available to the patient experimentalist. Maxworthy (1983) used the following procedure in a rotating experiment to form a hyperbolic tangent density profile for an investigation of solitary internal Kelvin waves. The lower heavy layer was initially placed in the tank and brought to solid body rotation. Liquid forming the lighter upper layer was slowly metered from the laboratory, through a rotating gland to a pipe rotating with the experiment and into a porous container floating on the surface of the lower layer. The upper layer fluid then gently spread out over the surface of the heavier fluid without appreciable mixing.

Linear density stratification is readily produced by the technique described by Oster (1965). Two identical tanks placed well above the experiment are interconnected at their bottom level by a pipe and shut-off valve. We label one tank the reservoir and the other the mixing tank which is outfitted with a motor-driven stirrer that keeps the liquid within at uniform density. An outlet pipe from the mixing tank runs to the experiment. The problem of generating a uniform density gradient between salty water of density 1.04 g/cc and fresh water of density 1.00 g/cc is considered for purposes of illustration. The reservoir is filled with the fresh water to a height determined by half the volume of the experimental tank and the mixing tank is filled to the same height with the 1.04 g/cc salty water. At an initial instant he motor is turned on and all valves are opened. The outlet pipe runs to a porous float that rises with the fluid level in the test apparatus. As gravity drains the mixing tank, fresh water enters from the reservoir to equilibrate the free-surface levels. The density in the mixing tank, and therefore that to the experiment, gradually decreases until the tanks are empty at which point the density is that of fresh water. The experiment has then acquired a linear density stratification ranging from 1.04 g/cc at the bottom to 1.00 g/cc at the top.

Such a stratification can be easily formed in a stationary tank, but if the fluid is subjected to any appreciable rotation it will mix and even undergo instability to form discrete layers of uniform density (Baker, 1971). The problem is solved (Buzyna and Veronis, 1971) by filling the tank while rotating by the "slip-under method" used in the preparation of density gradient columns (Fortuin, 1960). In this case the mixing tank in the Oster method is filled with fresh water and the reservoir is filled with salty water. The test apparatus is filled from the bottom through a rotating gland and the initial fresh water layer rises as it is replaced by denser, saltier water from below. This method avoids the need for a free surface float, but usually incorporates some system to disperse the liquid uniformly over the base of the tank so it will fill uniformly without appreciable mixing. Whitehead (1980) buried the outlet manifold in a layer of crushed rock spread over the the bottom of a rotating tank. Weidman and Browand (private communication) seeped the stratifying liquid up through a large matrix of small holes drilled through a false bottom in a rotating test facility.

As long as the tank levels decrease at the same rate and the liquid in the mixing tank is well agitated, the Oster or Fortuin method of producing linear density gradients is fool-proof. Nevertheless, the investigator is inevitably desirous of some form of verification. Buzyna and Veronis (1971), in their study of the spin-up of a linearly stratified fluid, introduced calibrated density floats into the fluid and measured their equilibrium heights with a cathetometer while the tank was rotating. Mason (1977) also produced a linearly stratified fluid by bottom-filling for measurements of sphere drag in a rotating system. He injected small amounts of dye in the filling pipes at the same moment fluid samples were taken. Plotting the observed elevations of the resulting dye sheets in the rotating tank versus the measured densities of the fluid samples gave verification of stratification linearity to within a few percent. Linden (1977) produced a linear salt-stratified fluid layer in a rotating annulus by filling from below through a sponge diffuser. Small 2 ml samples of the fluid were withdrawn at three fluid levels via stainless steel tubes attached to syringes for subsequent density measurement with a refractometer. Whitehead (1980), in an experiment on the selective withdrawal of rotating stratified fluid, measured salinity with an optical refractometer at equal depth intervals. Griffiths and Linden (1981) found that a constant vertical density gradient in the deep central portion of a rotating fluid could be easily produced by bringing two layers of salt solution into solid body rotation and then mixing vertically with a horizontal grid. Boyer and Chen (1987) used the Oster method to produce a linear salt stratification in a rotating fluid for experiments simulating large-scale atmospheric flow over mountain ranges.

Thermal stratification provides an alternative means of generating statically stable density gradients. Gershenfeld, et al (1981) constructed a rotating flume with uniformly flowing, linearly stratified water by this means. In brief, a rectangular channel was mounted on a rotating table and fluid at the entrance to the channel was pumped through flow straighteners and past a vertical array of horizontal rod heaters. Each heater rod in the rotating flume was connected to a separate variable voltage supply in the laboratory through a set of slip rings. Again the goal was linear stratification, and individual voltage adjustments helped maintain the desired profile at each flowrate. *In situ* measurement of the temperature was made with a thermistor traversed vertically through the flow. The facility was used to test predictions on selective withdrawal of fluid from a rotating stratified water current. Density profiles were readily calculated from the temperature distributions from known values of the coefficient of thermal expansion. Thermal stratification was also employed by Davies and Rahm (1982) to determine the effects of density-profile shape on the vertical attenuation of Taylor columns produced by the horizontal translation of a sphere through a rotating fluid.

Vertical density gradients weakly dependent on radius are generated in the laterally-heated rotating annulus experiment. Read and Hide (1984) observed isolated baroclinic eddies near the transition to nonaxisymmetric flow which had certain features in common with Jupiter's Great Red Spot. The stratification is a manifestation of the flow in that it depends on the details of the mechanical and thermal boundary conditions, and therefore cannot be independently controlled. However, the rotating heated annulus does provide an environment which may be exploited for certain studies in the flow of a rotating stratified fluid.

11.2 Conductivity Probe

Continuous measurement of time-dependent density fluctuations may be made in an isothermal system with a conductivity probe. The use of platinized-electrode conductivity cells in an a-c circuit to determine the resistivity of electrolyte solutions dates back to Kohlrausch in 1875 (Glasstone, 1942). These and other investigations focussed on techniques to precisely measure the conductivity of uniform solutions in large electrolytic cells. Prausnitz and Wilhelm (1956) constructed a conductivity probe with two small electrodes in close proximity in an attempt to make localized measurements of salt concentration fluctuations in a density-stratified flow. Their goal was to employ the conductivity probe in lieu of a hot-wire for making turbulence measurements in liquids. Lamb, et al (1960) improved on the idea by making one probe significantly larger than the other so that the signal would be dominated by the solution resistivity very close to the smallest electrode. Gibson and Schwarz (1963) pointed out the the influence of the larger electrode on the cell resistance of a two-electrode probe may be entirely eliminated by allowing its area and separation from the smaller electrode to become large. This single-electrode system has become the standard for conductivity probe operation. The sensor consists of a tiny metallic cathode bead fixed at the drawn tip of a long slender glass cylinder enclosing a platinum wire filament to transmit the carrier signal. The tip of the probe is anodized in a solution of platinum chloride. This produces a spongy layer of platinum-black which serves to increase the capacitive impedance and helps to minimize drift (Amen, 1985). The single-electrode conductivity probe is operated in one arm of an a-c Wheatstone bridge circuit. Fluctuations in the probe impedance due to variations in solution conductivity cause a carrier-suppressed, amplitude-modulated unbalance voltage across the bridge which may be demodulated to obtain a signal proportional to electrolyte concentration in an isothermal fluid. Weidman and Johnson (1982) reported strong nonlinearity in the density-voltage calibration curve at low salt concentrations. For densities in the limited range 1.01-1.04 g/cc, Koop (1976) found a parabolic calibration between voltage and density quite acceptable.

As noted by Maxworthy and Browand (1975), there is no satisfactory theory for the operation of the conductivity probe although contributions have been made by Robinson (1968) and Schanne, et al (1968). Since the complicated electric field in the vicinity of the probe has not been adequately described, one must resort to some sort of estimate of the probe's spatial resolution. Gibson and Schwarz (1963) suggested that the effective cell volume over which the density is averaged is roughly 10 probe tip diameters, but experimentally observed that hydrodynamic interference of the probe tip has a marked effect on the probe measurement volume.

The reader is referred to Koop (1976) and Amen (1985) for additional information concerning the operation and calibration of conductivity probes.

In spite of their popularity for measurement of wave propagation in stationary stratified systems and density fluctuations in stratified channel flow, the only application of a conductivity probe in a rotating stratified system known to this author is that of Renouard, et al (1987). They monitored the disturbance height of internal Kelvin waves propagating in a rotating miscible two-fluid system by placing a conductivity probe in a feedback circuit to match a local density in the pycnocline with a reference density outside the experiment. The propagating internal wave distorts the pycnoclines at the fixed probe location and causes an imbalance in the conductivity bridge. This signal is then compared with the reference density signal and the difference voltage drives the probe back to its original pycnocline. In this unique application the conductivity probe serves as a wave follower. The authors report a precision of ±0.1 mm on the interfacial height measurement and a response time of about 0.5 sec.

12. INTERFACIAL HEIGHT GAGES

Free-surface and liquid-liquid height detection systems of varying degrees of complexity have been developed according to need. Relatively simple systems often suffice for measurement of stationary and slowly-varying interfacial position. For example, Whitehead (1974), in a study of rotating hydraulics, measured the level of a lower fluid layer in a two-fluid system using a depth micrometer connected to an ohmmeter through the conducting fluid. The depth gage was driven down by a small motor until it made contact with the lower layer as observed by a deflection of the ohmmeter, at which point a measurement was taken. In another experiment, Whitehead (1986) constructed a differential manometer to monitor the height difference between two rotating fluid basins (see §4). Direct photography often can provide accurate measurement of interfacial height profiles in naturally visible rotating flows (see §5). In the following sections more sophisticated systems required to monitor the time-dependent movement of fluid interfaces are described.

12.1 Feedback Systems

Goller and Ranov (1968) made measurements of free-surface height profiles for water contained in a cylinder rotating along its vertical axis. Initially the fluid level is horizontal and after an abrupt acceleration of the cylinder to constant angular speed, the free surface continuously deforms towards its final steady-state paraboloidal shape. An electrohydraulic servo system was constructed to follow the free surface level. An electric current was fed through a fine wire probe submerged just below the water surface to an electrode on the bottom of the tank. In this position the probe resistance was balanced on an electrical bridge. When the probe moved relative to the water surface an imbalance signal was generated, amplified and fed to a servo-valve that hydraulically repositioned the probe to eliminate the imbalance. The response of the system was sufficiently rapid to accurately follow the liquid surface profile in a three-second sweep over the five inch radius of the tank. A pen attached to the opposite end of the piston on which the probe was mounted directly traced out free surface shapes at selected times during the spin-up period.

Renouard, et al (1987) monitored the interfacial height of internal Kelvin waves propagating on the interface of a miscible two-fluid system in a rotating channel using a conductivity probe. The feedback operation of this interface follower is described in §11.2.

12.2 Capacitive Level Sensor

Height measurements at an interface can be accurately measured by a capacitive sensor originally reported by McGoldrick (1970) for free-surface wave measurement in nonrotating flow and modified by Hart (1972) for use at a liquid-liquid interface in rotating flow. For the latter case, the technique consists of stretching a thin wire on the order of 0.002 cm through the interface formed by two liquids, one of which is an insulator and the other weakly conducting. Variations in interfacial height are reflected in changes in impedance between the wire sensor and another electrode placed in the conducting fluid. The gage is driven by a constant amplitude high-frequency voltage source and reactance changes measured across an a-c bridge, rectified and amplified are extremely sensitive to interfacial height fluctuations. Hart reports that excursions less than 0.01 mm can be detected and that the frequency response is typically 10 Hz. In studies of rotating baroclinic flow with application to geophysical fluid dynamics, one is usually interested in nearly equal densities and viscosities with negligible surface tension and this coupled with appropriate fluid conductivities restricts the choice of working fluids. For experiments of this type, Hart (1972) used 2 centistokes Dow Corning silicone oil for the upper layer and a mixture of 40% distilled water, 55% methanol and 5% isopropyl alcohol in the lower layer, to which is added a few grams of salt to enchance electrical conductivity. Owing to the high sensitivity of the sensor, it is imperative that the fluids be kept scrupulously clean; any dust or lint trapped on the interface will inevitably hit the sensor wire in a closed rotating flow system and cause noisy signals. Other than the problem of fluid selection, the capacitive level sensor is a very attractive measurement system because of its simplicity and high sensitivity.

Hart (1972, 1973) has used the capacitance level sensor to study large-amplitude baroclinic waves and associated instabilities. Farmer, Hart and Weidman (1982) employed the capacitance gage to produce the first experimental evidence of quasiperiodic flow on a torus and study its breakdown to chaos in this baroclinic flow system. Well-defined Poincare sections and return maps characterizing the attractor flow were readily constructed from a single time series of interfacial amplitude, a tribute to the accuracy of the miniature tidal gage. Hart (1985) made further use of the level sensor in laboratory experiments of baroclinic chaos on an f-.plane. Recently, Ohlsen (1988) modified the circuitry to obtain simultaneous measurement from three wires driven by separate carrier frequencies in a β-plane study of nonlinear baroclinic instability.

12.3 Optical Polarimeter

It is well known that when a beam of plane polarized light passes through a solution of an optically active substance, its plane of polarization is rotated through an angle which depends on the path length in the solution, the concentrations of the solution, and the molecular properties of the solute (Jenkins and White, 1957). Lambert and Davey (1974) constructed a direct-reading polarimeter for density contrast measurement in nonrotating stratified fluids. In this application the path length of a polarized beam through the fluid is fixed and the polarization rotation becomes a measure of the average density across the beam path. Hart and Kittleman (1986) used this same polarimetry principle to develop a new instrument for recording the height field of a liquid-liquid or liquid-air interface. In this application the rotation of a light beam directed vertically through a uniformly optically active liquid layer is directly proportional to the distance traversed. The interfacial height between two liquid layers can also be measured by making one layer optically active and the other optically inactive. D-limonene (or d-carvene), a liquid distilled from orange peels, was found to have high rotational sensitivity for polarized light. In a 10 cm liquid layer the rotation angle ranged from 80° at a wavelength of 700 nm to 220° at 400 nm. Hart and Kittleman demonstrated full-field capability of the system in a rotating fluid apparatus designed to generate baroclinic waves at the interface between two 13 cm deep liquid layers, with one optically active and the other optically inactive. White light transmitted through a diffuser and linear-polarizer beneath the tank was projected across the liquid layers, through an analyzer, and onto a color camera. Color photographs taken from the TV monitor exhibited the 2 cm range of interfacial height distribution in a color map. Analysis of the photographs and comparison with similar experiments incorporating a capacitive level sensor (§12.2) indicated that the method can be used to detect depth fluctuations on the order of 1 mm in the visible range.

It is clear from the rotational sensitivity curve for D-limonene that the highest resolution can be obtained by working with monochromatic deep blue light. To exploit this high sensitivity Hart and Kittleman (1986) developed a digital polarimeter, the operation of which is described below. Circularly polarized light from a He-Cd laser (442.0 nm) is transmitted through a rotating (30 Hz) hollow shaft containing a quarter-wave plate that converts the circularly polarized beam to one with a linear polarization aligned along the optical shaft axis. The rotating polarized beam is passed through the liquids, through an analyzer and focussed onto a photodiode. The 60 Hz signal from the diode is notch-filtered and zero-crossings of the resultant oscillations are used to trigger a counter whose clock is obtained from a high resolution encoder monitoring the rotation of the hollow shaft. It was reported that the digital polarimeter gave a local measurement of interface height to better than ±0.05 mm. Moreover, the measurement system has the advantage of being relatively insensitive to laser beam amplitude fluctuations.

The polarimetry measurement technique is in some respects superior to Hart's (1972) capacitive level sensor. Besides being non-intrusive, Hart and Kittleman (1986) report that the measurement signal is relatively unaffected by dirt on the liquid-liquid interface, providing the dirt does not appreciably alter the polarization of the light beam. However, the capacitive level sensor is somewhat more accurate than the polarized beam measurement system.

13. SPECIALIZED FLOW STABILITY DETECTORS

13.1 The Ion Technique

Reif and Meyer (1960) have used ion transport measurements to study the properties of liquid helium II. Donnelly and Reif (1962) attempted to extend the method to study the motion of cryogenics. For this purpose they constructed a pair of coaxial cylinders with the inner cylinder coated with radioactive polonium to produce ions in liquid nitrogen or liquid helium and an electric field applied across the cylinders. With the inner cylinder rotating above the critical speed for onset of Taylor vortices, it was believed that axial variations in the radial motion associated with the instability would spatially modulate the transport of ions to the outer cylinder. Donnelly accidently observed that if carbon tetrachloride at room temperature were substituted for the cryogenic fluid, the collected current on the outer cylinder showed variations of the type described above but did not require a radioactive source. This discovery led to the development of a powerful measurement tool presently known as the "ion technique" and was first employed by Donnelly, et al (1962) to study modulated rotating Couette flow.

The theory underlying the ion technique pivots on an understanding of ion transport from the electrical double layer on an inner cylinder across the charged gap to the electrical double layer on the outer cylinder. Donnelly and Tanner (1965) analyzed the electrically conducting fluid equations whereby ions are transported in the fluid by convection, diffusion and conduction. To simulate Taylor cells, it was assumed that the fluid possessed a weak secondary radial motion of the form $A \cos\kappa z$, where A is the equilibrium amplitude and κ is the wavenumber in the axial z-direction. Approximate solution to the transport equations under the assumption of a purely radial electric field show that the ion current I_{ion} collected on the outer cylinder is given by

$$I_{ion} = -QA\left[\frac{\delta_0}{d}\right]\cos\kappa z - I_0 , \qquad (14)$$

where Q is the charge on the cylinders, d is the gap width, $\delta_0 \ll d$ is a characteristic electrical layer thickness, and I_0 is the standing current. This result supports all experimentally observed trends, namely that (i) the current at a collector mounted on the outer cylinder wall is unaffected by the laminar axial and azimuthal flow, (ii) the change in collector current is linearly proportional to the magnitude of the radial component of velocity, and (iii) the change in current is independent of the standing current under conditions where the fluid conductivity may be considered constant.

The high sensitivity of the ion technique to variations in radial fluid motion in the vicinity of the electrical double layer at a cylinder boundary make it ideally suited for studies of supercritical Taylor-Couette flow. To see how the ion technique is set up in practice, we describe Donnelly's (1965) second application of the method to the study of transition from Couette flow to Taylor vortex flow. Two brass cylinders, plated with a thin layer of gold in order to reduce stray electrochemical potentials, are insulated from each other through the use of plastic bearings. The outer cylinder is grounded and the inner cylinder is connected to a constant voltage source. The annulus is filled with reagent grade carbon tetrachloride which absorbs impurities from the annular cell and reaches an equilibrium concentration in about 24 hours. The ion current on the order of 10^{-12} amps is collected on a thin ring mounted flush with the outer cylinder wall and insulated from it above and below by thin plastic washers. It is important that the electric field between the cylinders be directed radially and this is accomplished with a vibrating reed eletrometer. The output of the electrometer is fed to an integrator circuit with an adjustable RC time constant and the integrated signal is monitored on a strip chart recorder. In the experiment with the inner wall rotating, the collector current was only weakly dependent on cylinder rotation speed below the onset of instability because of the absence of radial flow. Above the onset the current was very sensitive to the rotation speed, increasing (decreasing) if the ring collector was located at a radial inflow (outflow) boundary of the Taylor cells. A slow vertical displacement of the outer cylinder at supercritical conditions exhibited a periodic signal consistent with the axial modulation of the radial velocity in Taylor vortex flow.

The ion technique has proven to be an effective tool for testing theoretical predictions for finite amplitude supercritical Taylor vortex flow (Donnelly and Schwarz, 1965). Current collected around a circular ring embedded in the outer cylinder is sensitive to the *average* of radial velocity variations around the ring. This configuration is a good detector for axisymmetric flow perturbations but would be ineffective for nonaxisymmetric disturbances. Walden and Donnelly (1979) have instrumented a cylinder with 1500 ion collectors 1 mm in diameter in order to obtain *local* measurements in unstable circular Couette flow. The investigators computed power spectral density curves using time series measurements from the point ion collectors to map out newly observed highly-supercritical Taylor vortex flow states.

13.2 Light Transmission Probes

The discussion in §3.1.2 contained many examples of how particle flake suspensions can provide global visualization information on steady and unsteady rotating flows, particularly with regard to instability. In this and the following section we describe how light beams transmitted through and reflected from these suspensions may be used to interrogate the dynamics of rotating flows.

Coles (1965) was apparantly the first to use transmitted light through an aluminum flake suspension to obtain quantitative information concerning unsteady Taylor vortex flow. He placed an incandescent lamp at the central axis inside the annulus and monitored the intensity of light transmitted radially through the fluid and onto a photomultiplier located outside the annulus. The optical setup insured that light detected by the phototube emanated from a region in the fluid only 0.5 mm in diameter. Satisfactory time signals of wavy vortex flow were obtained with a pigment concentration of 0.0005 g/cc for an optical path length through the fluid of about 7 mm. Ohji (1987) used this same technique to study modulated wavy vortex flow. He obtained two-point azimuthal correlations of the flow by monitoring the intensity of forward-scattered light produced by two laser beams projected radially through the annulus onto separate photodiodes. Tam and Swinney (1988) determined an effective diffusion coefficient for mass transport in turbulent Taylor vortex flow using optical absorption of laser light (see §9.2). Methylene blue dye injected into the annular gap at mid-height from an inner rotating cylinder diffused axially through the turbulent vorticies. Comparison of the optical absorption of two laser beams directed through the gap onto photodiodes at different axial positions below the injection point were used to compute an effective axial diffusion coefficient.

13.3 Light Reflection Sensors

Measurements of unsteady supercritical Taylor vortex flow using laser light reflected from Kalliroscope particle suspensions began to appear at the beginning of this decade. Park and Donnelly (1981) reflected laser light from a rotating mirror to draw a line along the entire vertical column of a Taylor-Couette apparatus as the Reynolds number was slowly increased. Light reflected from suspended guanine platelets was recorded by a photomultiplier located at the focal point of a telescope. In this manner the time evolution of all vortex cells as they first appeared were recorded. This simple but elegant measuring technique showed that the onset Reynolds number for Taylor vortex cells as measured at mid-cylinder height was virtually independent of cylinder aspect ratio. In another Taylor vortex experiment Ahlers and Cannell (1983) suddenly increased the speed of an inner rotating cylinder supporting stable Couette flow to slightly supercritical conditions to observe the propagation of Taylor cells from the endwalls of the annulus. Monitoring laser light reflected from scattering particles at a fixed point some distance away from the endwalls, the authors were able to detect the propagating wave front and thereby measure its average propagation speed. Park, et al (1983) recorded reflected laser light on a 256-channel optical scanning device to make precise measurements of the axially modulated position of the wavy inflow boundary between a pair of counter-rotating Taylor vorticies.

Baxter and Andereck (1986) obtained power spectra of Taylor instabilities by recording the intensity of reflected laser light on a 1024-pixel charge-coupled device.

Also in Taylor-vortex flow Zhang and Swinney (1985), Andereck, et al (1986) and Ross and Hussain (1987) obtained power spectra from long time records of reflected laser light sensed by a photodiode. The former authors documented three new types of nonpropagating oscillatory Taylor vortex modes using this relatively simple data acquisition technique. Walsh, et al (1987) used an infrared beam source and photocell sensor to measure reflected light as a means of detecting the onset of instability in modulated Taylor vortex flow. Kohuth and Neitzel (1988) obtained power-spectra histories by recording reflected laser beam intensities on a Reticon photodiode array in experiments on the stability of impulsively-initiated circular Couette flow.

14. PROBE INTERFERENCE REVISITED

The advent of laser Doppler anemometry provided a non-intrusive method for measuring velocity at a point. By and large, the focussing of crossed laser beams at a point in the flow does not induce measurable disturbances except in certain delicate experiments. A case in point is the stability study of circular Couette flow reported by Cooper, et al (1985). In this experiment a laser Doppler system was installed to monitor the axial (vertical) velocity at a point in the gap while the inner cylinder, initially rotating at constant speed in a stable flow regime, was ramped to higher speeds in a prescribed manner. Measurements showed that the time to onset of instability was markedly affected by the laser heating at the measurement volume. The investigators reasoned that since the initial flow was circular, a ring of fluid passing continuously through the laser measurement volume became preheated and the attendant decrease in fluid density was sufficient to cause local natural convection, thereby inducing vertical fluid motion prior to the onset of instability. Estimates of the local heating rate of the silicone oil ring traced out at the focal point of the laser gave 0.01 °F/min for their experimental conditions.

The remaining two interference problems have to do with disturbances induced by flow visualization techniques. In a simple, yet profound study of electrolytic visualization techniques using thin wires to produce timelines normal to the flow upstream of a bluff body, Taneda, et al (1974) pointed out a fundamental drawback: the dye sheet emanating from the wire does not connect with the body, but leaves an unmarked region adjacent to the body surface. This defect in flow visualization was observed for both the hydrogen bubble and the electrolytic precipitation techniques, but was not observed with aluminum particle suspensions. After conducting a number of tests it was concluded that the formation of a layer into which visualizing agents could not penetrate must be due to a hydrodynamic effect.

For example, it was observed that the thickness of the "invisible layer" increased when the free-stream velocity, the wire diameter or the body size increased, or when the separation distance between wire and body decreased. It was concluded that since the width of the wake of the tracer-generating wire was much smaller than the size of the obstacle whose flow was being visualized, the pressure distribution around the obstacle must have been governed by the main flow. Therefore, since the velocity U_w at the centerline of the wake of the wire is smaller than than of the free-stream velocity around the obstacle, the kinetic energy in the wake is insufficient to surmount the pressure gradient in the vicinity of the forward stagnation point. Using a cylindrical body for calculation purposes Taneda, et al (1974) computed the upstream position at which the flow markers bend away from the body as the radius r_0 at which the total pressure in the wire wake equals the static pressure induced by the cylinder along its stagnation streamline. It was determined that

$$r_0 = \frac{d}{2} \left[1 - \left[1 - \sqrt{U_w/U_0} \right] \right]^{1/2} , \tag{15}$$

where U_0 is the free-stream velocity and d is the cylinder diameter. Thus the thickness δ of the invisible region in the neighborhood of the stagnation point is

$$\delta = r_0 - \frac{d}{2} , \tag{16}$$

in agreement with all the observed trends in the experimental observations. This important disturbance effect will most certainly be evident in dye sheet visualizations around bluff bodies in rotating flow, but with probable asymmetries in the thickness of the invisible layer due to Coriolis effects.

In a final example of interference effects, Dominguez-Lerma, et al (1985) documented a problem directly related to the introduction of a flow visualizing agent. It was observed that the spatial distribution of Taylor vortices in a vertically-mounted circular Couette system was significantly affected by guanine platelets (Kalliroscope). An initially uniform wavelength of Taylor vortices evolved into a condition where the vortices in the upper half of the cylindrical apparatus were as much as 10% wider than those in the lower half. Subsequent experiments showed that this nonuniformity in wavenumber distribution increased with increasing concentration of the Kalliroscope suspension. Furthermore, at later times a time-dependent flow emerged, characterized by an aperiodic vertical sloshing of the Taylor cells. These effects were traced to a nonuniform distribution of the Kalliroscope flakes inside the rotating cells. It was reported that the flakes slowly accumulated in the core of the vortices leaving a reduced concentration in the vicinity of the separation streamlines. Since this anomalous behavior was not observed when the experimental apparatus was operated horizontally, it was concluded that the undesired effects resulted from a coupling of the periodic concentration distribution of the flakes to the gravity field. It is clear that this subtle interaction had a profound disturbance effect on supercritical Taylor vortex flow.

15. CONCLUDING REMARKS

It is clear that the art of making measurements in laboratory rotating flows is continuously evolving through both technological advances and innovative thinking. We summarize the present stage of development in the following concluding remarks.

Flow visualization plays a crucial role in rotating flows, particularly smoke, dye and particle flake suspensions which often do not disturb the fluid motion in a fundamental manner. Radiation-induced visualization techniques have proven eminently successful in the form of laser induced fluorescence and are under development in the areas of particle luminescence and photochromism. The exploitation of conical mirrors for all-around circumferential viewing offers an example of innovative development. The novel use of soap films for visualization of two-dimensional air flow is found to be particularly effective. With some exceptions, index-of-refraction sensitive techniques such as the shadowgraph and the interferometer have not received much attention in rotating flows, yet holographic interferometry and tomography show promise for visualization of fluid motion in three dimensions.

Velocity measurement with Pitot-static probes, hot-wires and hot-films have generally given way to non-intrusive laser Doppler velocimetry. Point measurement in time-varying rotating flows is being pushed to the limit of technology by the quest for simultaneous measurement of multiple velocity components and the acquisition of structure functions. Laser speckle velocimetry and especially particle image displacement velocimetry are under intense development owing to their full-field capability in a plane of two-dimensional flow. These measurement tehcniques pivot to a large degree on the evolution of computer algorithms and image processing capabilities for automated scanning and analysis of data recorded on photographic film.

High frequency response heated films are more desirable than Preston tubes for surface shear stress measurement in situations for which the heat transfer/shear stress correlation is well understood. Heated thermistors have also been employed for the purpose of shear stress measurement, but are restricted to low frequency unsteady flows at high Prandtl number.

Thermally driven rotating flows offer a special challenge to the experimentalist. Non-disturbing temperature probes are not available for measurement of temperature interior to a rotating fluid flow, although hot-films and thermistors may be used at fluid boundaries. The need to measure thermal fields or simultaneous temperature and velocity fields in rotating geophysical flows has motivated the introduction of large thermistor arrays into the fluid volume. All available evidence shows that these arrays alter the velocity and temperature fields under scrutiny. Furthermore, unstable flows are more strongly affected than stable flows, and the very nature of an instability can be altered by the presence of a thermistor array. Hence there is no guarantee that a supercritical flow bifurcation sequence observed with a large array of thermistors strung throughout the fluid volume will be the same as that observed in its absence. The intent here is not to criticize the substantial progress made in the understanding of geophysical flows through the development of simultaneous velocity and temperature measurement systems incorporating thermistor arrays. However, it is imperative from the measurement point of view to bear in mind that confined thermally active rotating flows are disturbed both quantitatively and sometimes qualitatively by the insertion of numerous thermistors into the fluid volume.

The use of lasers in probing rotating flows has been very successful. Much can be learned from light intensity measurements of laser transmission or reflection. These simple systems have been employed to detect the onset of instability, measure various characteristics of an instability flow, deduce diffusion coefficients and measure the propagation speed of moving wave fronts. Specifically, laser power absorption and reflection allows one to measure instability frequencies, power spectra, auto- and cross-correlation coefficients, phase plots, and Poincare sections and return maps for chaotic flow systems. The recent development of an optical polarimeter for interfacial height detection is yet another example of the innovative application of lasers for interrogating fluid systems.

The most important item to bear in mind when setting up an experimental apparatus and attendant measurement systems for understanding a particular rotating flow, especially when investigating flow instabilities, is the omnipresent spectre of probe disturbance. The pronounced interference effects noted in §14 by the seemingly innocuous use of laser Doppler velocimetry and particle flake suspension visualization offer a warning to the unwary experimentalist.

ACKNOWLEDGMENTS

The author takes this opportunity to express his gratitude to the more than one hundred active researchers who responded to a call for publications relevant to the subject of this review. From them I learned several new features about rotating fluid measurement techniques and I hope they in turn will benefit from a reading of this article. The author also appreciates the editorial suggestions provided by his colleagues Mohamed Gal-El-Hak and George Buzyna, his friend Susan McCann and his mother Ora Weidman. The writing of this article was supported in part by an ONR/NOAA grant under Contract Number N00014-86K-0728.

REFERENCES

Abell, S. and Hudson, J. L. (1975) An experimental study of centrifugally driven free convection in a rectangular cavity. *Int. J. Heat Mass Trans.*, **18**, 1415-1423.

Adrian, R. J. (1983) Laser velocimetry. *Fluid Mechanics Measurements*, ed. R. J. Goldstein, 155-244 (Hemisphere Publishing, Washington).

Adrian, R. J. and Yao, C. S. (1983) Development of pulsed laser velocimetry measurement of fluid flow. *Proceedings, Eighth Biennial Symposium on Turbulence*, eds. G. Patterson and J. L. Zakin (University of Missouri, Rolla).

Adrian, R. J. and Yao, C. S. (1985) Pulsed laser technique application to liquid and gaseous flows and the scattering power of seed materials. *Appl. Opt.*, **24**, 44-52.

Agui, J. C. and Jimenez, J. (1988) On the performance of particle tracking. *J. Fluid Mech.*, **185**, 447-468.

Ahlborn, F. (1902) On the mechanism of hydrodynamic drag. (German), *Abhandl. Gebiete Naturwiss.*, **17**, Hamburg. See also Tietjens, O. G. (1934) *Applied Hydro- and Aeromechanics* (Dover, New York, 1957).

Ahlers, G. and Cannell, D. S. (1983) Vortex-front propagation in rotating Couette-Taylor flow. *Phys. Rev. Lett.*, **50**, 1583-1586.

Aitta, A., Ahlers, G. and Cannell, D. S. (1985) Tricritical phenomena in rotating Couette-Taylor flow. *Phys. Rev. Lett.*, **54**, 673-676.

Aldridge, K. D. and Toomre, A. (1969) Axisymmetric inertial oscillations of a fluid in a rotating spherical container. *J. Fluid Mech.*, **37**, 307-323.

Amen, R. L. (1985) The decay of grid generated turbulence in a two layer stratified fluid. PhD thesis, University of Southern California, Los Angeles, CA.

Andereck, D. D., Dickman, R. and Swinney, H. L. (1983) New flows in a circular Couette system with co-rotating cylinders. *Phys. Fluids*, **26**, 1395-1401.

Andereck, C. D., Liu, S. S. and Swinney, H. L. (1986) Flow regimes in a circular Couette system with independently rotating cylinders. *J. Fluid Mech.*, **164**, 155-183.

Annamalai, P., Trinh, E. and Wang, T. G. (1985) Experimental study of the oscillations of a rotating drop. *J. Fluid Mech.*, **158**, 317-327.

Baker, D. J. (1966) A technique for the precise measurement of small fluid velocities. *J. Fluid Mech.*, **26**, 573-575.

Baker, D. J. (1967) Shear layers in a rotating fluid. *J. Fluid Mech.*. **29**, 165-175.

Baker, D. J., Jr. (1968) Demonstrations of fluid flow in a rotating system. II: The "spin-up" problem. *Amer. J. Phys.*, **36**, 980-986.

Baker, D. J., Jr. and Robinson, A. R. (1969) A laboratory model for the general oceanic circulation. *Phil. Trans. Roy. Soc. Lond.*, **A265**, 533-566.

Baker, D. J. (1971) Density gradients in a rotating stratified fluid: experimental evidence for a new instability. *Science*, **172**, 1029-1031.

Balint, J.-L., Vukoslavcevic, P., and Wallace, J. M. (1987) A study of the vortical structure of the turbulent boundary layer. *Advances in Turbulence*, eds. G. Compte-Bellot and J. Mathieu, 456-464 (Springer-Verlag, Berlin).

Bankoff, S. G. and Rosler, R. S. (1962) Constant-temperature hot-film anemometer as a tool in liquid turbulence measurement. *Rev. Sci. Instrum.*, **33**, 1209-1212.

Barcilon, A., Brindley, J., Lessen, M. and Mobbs, F. R. (1979) Marginal instability in Taylor-Couette flows at a very high Taylor number. *J. Fluid Mech.*, **94**, 453-463.

Barker, D. B. and Fourney, M. E. (1977) Measuring fluid velocities with speckle patterns. *Opt. Lett.*, **1**, 135-137.

Bar-Yoseph, P., Roesner, K. G. and Seelig, S. (1986) Flow in an eccentrical spherical gap. *Proc. 6th Workshop on Gases in Strong Rotation*, Tokyo, Japan, 1-9.

Baxter, G. W. and Andereck, C. D. (1986) Formation of dynamical domains in a circular Couette system. *Phys. Rev. Lett.*, **57**, 3046-3049.

Beardsley, R. C. (1969) A laboratory model of the wind-driven ocean circulation. *J. Fluid Mech.*, **38**, 255-271.

Beardsley, R. C. (1970) An experimental study of inertial waves in a closed cone. *Stud. Appl. Math.*, **49**, 187-196.

Beardsley, R. C. and Robbins, K. (1975) The 'sliced-cylinder' laboratory model of the wind-driven ocean circulation. Part 1. Steady forcing and topographic Rossby wave instability. *J. Fluid Mech.*, **69**, 27-40.

Beardsley, R. C. (1975) The 'sliced-cylinder' laboratory model of the wind-driven ocean circulation. Part 2. Oscillatory forcing and Rossby wave resonance. *J. Fluid Mech.*, **69**, 41-64.

Becker, J. A., Green, C. B. and Pearson, G. L. (1946) Properties and uses of thermistors - Thermally sensitive resistors. *Trans. Amer. Inst. Elect. Engr.*, **65**, 711-725.

Beckwith, T. G., Buck, N. L. and Marangoni, R. D. (1982) *Mechanical Measurements*. (Addison-Wesley, Reading, MA).

Bellhouse, B. J. and Schultz, D. L. (1966) Determination of mean and dynamic skin friction, separation and transition in low-speed flow with a thin-film heated element. *J. Fluid Mech.*, **24**, 379-400.

Bellhouse, B. J. and Schultz, D. L. (1967) The determination of fluctuating velocity in air with heated thin film gauges. *J. Fluid Mech.*, **29**, 289-295.

Benjamin, T. B. (1978) Bifurcation phenomena in steady flows of a viscous fluid. II. Experiments. *Proc. Roy. Soc. Lond.*, **A359**, 27-43.

Benjamin, T. B. and Barnard, B. J. S. (1964) A study of the motion of a cavity in a rotating liquid. *J. Fluid Mech.*, **19**, 193-209.

Benjamin, T. B. and Mullin, T. (1981) Anomalous modes in the Taylor experiment. *Proc. Roy. Soc. Lond.*, **A377**, 221-249.

Benjamin, T. B. and Mullin T. (1982) Notes on the multiplicity of flows in the Taylor experiment. *J. Fluid Mech.*, **121**, 219-230.

Betchov, R. (1948) L'influence de la conduction thermique sur les anemometres a fils chaud. *Proc. K. Ned. Akad. Wet.*, **51**, 721-730.

Bien, F. and Penner, S. S. (1970) Velocity profiles in steady and unsteady rotating flows for a finite cylindrical geometry. *Phys. Fluids*, **13**, 1665-1671.

Bien, F. and Penner, S. S. (1971) Spin-up and spin-down of rotating flows in finite cylindrical containers. *Phys. Fluids*, **14**, 1305-1308.

Bjorklund, I. S. and Kays, W. M. (1959) Heat transfer between concentric rotating cylinders. *J. Heat Trans.*, **81**, 175-186.

Blackwelder, R. F. (1981) Hot-wire and hot-film anemometers. *Methods of Experimental Physics: Fluid Dynamics - Part A*, **18**, ed. R. J. Enrich, 259-314 (Academic Press, New York).

Blake, W. K. (1983) Differential pressure measurement. *Fluid Mechanics Measurements*, ed. R. L. Goldstein, 61-97 (Hemisphere, Washington).

Bornstein, J. and Escudier, M. P. (1984) LDA easurements within a vortex-breakdown bubble. *Laser Anemometry in Fluid Mechanics*, 253-263 (Ladoan, Lisbon).

Bourquin, K. R. and Shigemoto, F. H. (1968) Investigation of air-flow velocity by laser backscatter. *NASA Tech. Note*, No. D-4453.

Bowden, M. and Eden, H. F. (1965) Thermal convection in a rotating fluid annulus: Temperature, heat flow and flow field observations in the upper symmetric regime. *J. Atm. Sci.*, **22**, 185-195.

Boyer, D. (1971) Rotating flow over long shallow ridges. *Geophys. Fluid Dyn.*, **2**, 165-184.

Boyer, D. L. and Guala, J. R. (1972) Model of the antarctic circumpolar current in the vicinity of the MacQuarie ridge. *Antarctic Res. Ser.; Antarctic Oceanology II: The Australian-New Zeland Sector*, **19**, ed. D. E. Hayes, 79-93.

Boyer, D. L. and Davies, P. A. (1982) Flow past a circular cylinder on a β-plane. *Phil. Trans. Roy. Soc. Lond.*, **A306**, 533-556.

Boyer, D. L., Kmetz, M., Smathers, L., Chabert d'Hieres, G. and Didelle, H. (1984a) Rotating open channel flow past right circular cylinders. *Geophys. Astrophys. Fluid Dyn.*, **30**, 271-304.

Boyer, D. L., Davies, P. A. and Holland, W. R. (1984b) Rotating flow past disks and cylindrical depressions. *J. Fluid Mech.*, **141**, 67-95.

Boyer, D. L. and Biolley, F. M. (1986) Linearly stratified, rotating flow over long ridges in a channel. *Phil. Trans. Roy. Soc. Lond.*, **A318**, 411-440.

Boyer, D. L., Davies, P. A., Holland, W. R.. Biolley, F. and Honji, H. (1987a) Stratified rotating flow over and around isolated three-dimensional topography. *Phil. Trans. Roy. Soc. Lond.*, **A322**, 213-241.

Boyer, D. L., Chen, R. and Davies, P. A. (1987b) Some laboratory models of flow past the Alpine/Pyrenees mountain complex. *Meteorol. Atmos. Phys.*, **36**, 187-200.

Bradshaw, P. (1971) *An introduction to turbulence and its measurement.* (Pergamon Press, New York).

Brandstater, A., Swift, J., Swinney, H. L., Wolf, A., Farmer, J. D., Jen, E., and Crutchfield, J. P. (1983) Low-dimensional chaos in a hydrodynamical system. *Phys. Rev. Lett.*, **51**, 1442-1445.

Brandstater, A., and Swinney, H. L. (1987) Strange attractors in weakly turbulent Couette-Taylor flow. *Phys. Rev..* **35A**, 2207-2220.

Brayton, D. B. (1969) A simple, laser, Doppler shift, velocimeter with self-aligning optics. *AEDC Tech. Rep.*, No.70-45.

Browand, F. K. and Weidman, P. D. (1976) Large scales in the developing mixing layer. *J. Fluid Mech.*, **76**, 127-144.

Brown, G. L. and Davey, R. F. (1971) The calibration of hot films for skin friction measurement. *Rev. Sci. Instrum.*, **42**, 1729-1731.

Brown, R. G. W. and Jones, R. (1983) Burst-correlation laser Doppler velocimetry. *Opt. Lett.*, **8**, 449-451.

Buchhave, P. George, W. K. Jr., and Lumley, J. L. (1979) The measurement of turbulence with the laser-Doppler anemometer. *Ann. Rev. Fluid Mech.*, **11**, 443-503.

Buhler, K. (1983) Instabilitaten spiralformiger Stromungen zwischen konzentrischen Kuglen. *ZAMM*, **63**, T235-T239.

Buhler, K. (1986) Stromungsmechanische Instabilitaten im Kugelspalt. *Stromunnsmechanik und Stromungsmaschinen*, **38**, 11-24.

Buhler, K. and Oertel, H. (1982) Thermal cellular convection in rotating rectangular boxes. *J. Fluid Mech.*, **114**, 261-282.

Burkhalter, J. E. and Koschmieder, E. L. (1973) Steady supercritical Taylor vortex flow. *J. Fluid Mech.*, **58**, 547-560.

Busemann, A. (1971) Compressible flow in the thirties. *Ann. Rev. Fluid Mech.*, **3**, 1-12.

Busse, F. H. and Heikes, K. E. (1980) Convection in a rotating layer: A simple case of turbulence. *Science*, **208**, 173-175.

Buzyna, G. and Veronis, G. (1971) Spin-up of a stratified fluid: theory and experiment. *J. Fluid Mech.*, **50** 579-608.

Buzyna, G., Pfeffer, R. L. and Kung, R. (1984) Transition to geostrophic turbulence in a rotating differentially heated annulus. *J. Fluid Mech.*, **145**, 377-403.

Caldwell, D. R. and Van Atta, C. W. (1970) Characteristics of Ekman boundary layer instabilities. *J. Fluid Mech.*, **44**, 79-95.

Caldwell, D. R., Van Atta, C. W. and Helland, K. N. (1972) A laboratory study of the turbulent Ekman layer. *Geophys. Fluid Dyn.*, **3**, 125-160.

Cerasoli, C. P. (1975) Free shear layer instability due to probes in rotating source-sink flows. *J. Fluid Mech.*, **72**, 559-586.

Chabert d'Hieres, G., Davies, P. A. and Didelle, H. (1987) A laboratory study of the lift forces on a moving solid obstacle in a rotating fluid. *Coriolis Laboratory Technical Report*, Institut de Mechanique, Grenoble, France.

Chabert d'Hieres, G., Davies, P. A. and Didelle, H. (1988) A laboratory study of the lift forces on a moving solid obstacle in a rotating fluid. *Dyn. Atmos. Oceans*, under review.

Champagne, F. H., Sleicher, C. A. and Wehrmann, O. H. (1967a) Turbulence measurements with inclined hot-wires. Part 1. *J. Fluid Mech.*, 28, 153-176.

Champagne, F. H., Sleicher, C. A. and Chao, J. L. (1967b) Turbulence measurements with inclined hot-wires. Part 2. *J. Fluid Mech.*, 28, 177-182.

Chin, D.-T. (1971) An experimental study of mass transfer on a rotating spherical electrode. *J. Electrochem. Soc.*, 118, 1764-1769.

Chin, D.-T. and Litt, M. (1972a) Mass transfer to point electrodes on the surface of a rotating disk. *J. Electrochem. Soc.*, 119, 1338-1343.

Chin, D.-T. and Litt, M. (1972b) An electrochemical study of flow instability on a rotating disk. *J. Fluid Mech.*, 54, 613-625.

Chomaz, J. M., Rabaud, M., Basdevant, C. and Couder, Y. (1988) Experimental and numerical investigation of a forced circular shear layer. *J. Fluid Mech.*, 187, 115-140.

Church, C. R. and Snow, J. T. (1979) The dynamics of natural tornadoes as inferred from laboratory simulations. *J. Rech. Atmos.*, 13, 111-133.

Clayton, B. R. and Massey, B. S. (1967) Flow visualization in water: a review of techniques. *J. Sci. Instrum.*, 44, 2-11.

Clutter, E. W. and Smith, A. M. O. (1961) Techniques of flow visualization using water as the working medium. *Aero. Engrg.*, 20, 24-27, 75-76.

Coles, D. (1965) Transition in circular Couette flow. *J. Fluid Mech.*, 21, 385-425.

Coles, D. and Van Atta, C. W. (1966) Measured distortion of laminar circular Couette flow by end effects. *J. Fluid Mech.*, **25**, 513-521.

Colello, R. G. and Springer, G. S. (1966) Mass-transfer measurements with the technique of electrochemiluminescence. *Int. J. Heat Mass Trans.*, **9**, 1391-1399.

Collis, D. C. and Williams, M. J. (1959) Two-dimensional convection from heating wires at low Reynolds numbers. *J. Fluid Mech.*, **6**, 357-389.

Compte-Bellot, G. (1976) Hot-wire anemometry. *Ann. Rev. Fluid Mech.*, **8**, 209-231.

Coney, J. E. R. and Simmers, D. A. (1979) A study of fully-developed, laminar, axial flow and Taylor vortex flow by means of shear stress measurements. *J. Mech. Engrg. Sci.*, **21**, 19-24.

Cooper, E. R., Jankowski, D. F., Neitzel, G. P. and Squire, T. H. (1985) Experiments on the onset of instability in unsteady circular Couette flow. *J. Fluid Mech.*, **161**, 97-115.

Corrsin, S. (1963) Turbulence: experimental methods. *Handbuch der Physik*, **VIII/2**, eds. S. Flugge and C. Truesdell, 524-590 (Springer-Verlag, Berlin).

Couder, Y. (1984) Two-dimensional grid turbulence in a thin liquid film. *J. Physique. Lett.*, **45**, L353-L360.

Couette, M. M. (1890) Etudes sur le frottement des liquides. *Annal. Chimie Phys.*, **21**, 433-510.

Dainty, J. C. (1975) *Laser Speckle and Related Phenomena.* (Springer, Berlin).

Davies, P. A. and Rahm, L. (1982) The interaction between topography and a nonlinearly stratified rotating fluid. *Phys. Fluids*, **25**, 1931-1934.

Davies, P. A. and Walin, G. (1977) Some further experiments with a heated rotating annulus having semi-conducting walls. *Tellus*, **29**, 161-170.

Davis, M. R. (1970) The dynamic response of constant resistance anemometers. *J. Phys. E: Sci. Instrum.*, **3**, 15-20.

Davis, W. and Fox, R. W. (1967) An evaluation of the hydrogen bubble technique for the quantitative determination of fluid velocities within clear tubes. *J. Basic Engr.*, **89**, 771-781.

Denardo, B. (1988) Equilibrium states of a rotating U-Tube. *American J. Sci.*, under review.

Denison, E. B. and Stevenson, W. H. (1970) Oscillatory flow measurements with a directionally sensitive laser velocimeter. *Rev. Sci. Instrum.*, **41**, 1475-1478.

Dewey, F. C. Jr. (1976) Qualitative and quantitative flow field visualization utilizing laser-induced flourescence. *Applications of nonintrusive instrumentation in fluid flow research*, AGARD-CP-193, Paper No. 17.

Dickenson, S. C. and Long, R. R. (1983) Oscillating-grid turbulence including effects of rotation. *J. Fluid Mech.*, **126**, 315-333.

Dimotakis, P. E., Miake-Lye, R. C. and Papantoniou, D. A. (1983) Structure and dynamics of round turbulent jets. *Phys. Fluids*, **26**, 3185-3192.

Dominguez-Lerma, M. A., Ahlers, G. and Cannell, D. S. (1985) Effects of "Kalliroscope" flow visualization on rotating Couette-Taylor flow. *Phys. Fluids*, **28**, 1204-1206.

Dominguez-Lerma, M. A., Cannell, D. S. and Ahlers, G. (1986) Eckhaus boundary and wave-number selection in rotating Couette-Taylor flow. *Phys. Rev.*, **A34**, 4956-4970.

Donnelly, R. J. (1958) Experiments on the stability of viscous flow between rotating cylinders. 1. Torque measurements. *Proc. Roy. Soc. Lond.*, **A286**, 312-325.

Donnelly, R. J. and Reif, F. (1962) Study of hydrodynamic stability with ion current. *Bull. Amer. Phys. Soc.*, **7**, 371.

Donnelly, R. J., Reif, F. and Suhl, H. (1962) Enchancement of hydrodynamic stability by modulation. *Phys. Rev. Lett.*. **9**, 363-365.

Donnelly, R. J. (1965) Experiments on the stability of a viscous flow between rotating cylinders. IV. The ion technique. *Proc. Roy. Soc. Lond.*, **A283**, 509-519.

Donnelly, R. J. and Tanner, D. J. (1965) Experiments on the stability of a viscous flow between rotating cylinders. V. The theory of the ion technique. *Proc. Roy. Soc. Lond.*, **A283**, 520-530.

Donnelly, R. J. and Schwarz, K. W. (1965) Experiments on the stability of a viscous flow between rotating cylinders. VI. Finite amplitude experiments. *Proc. Roy. Soc. Lond.*, **A283**, 531-556.

Donnelly, R. J., Park, K., Shaw, R. and Walden, R. W. (1980) Early nonperiodic transitions in Couette flow. *Phys. Rev. Lett.*, **44**, 987-989.

Douglas, H. A., Hide, R. and Mason, P. J. (1972) An investigation of the structure of baroclinic waves using three-level streak photography. *Quart. J. Roy. Met. Soc.*, **98**, 247-263.

Dudderar, T. D. and Simpkins, P. G. (1977) Laser speckle photography in a fluid medium. *Nature*, **270**, 45-47.

Durst, F., Melling, A., and Whitelaw, J. H. (1981) *Principles and practice of laser-Doppler anemometry.* (Academic Press, London, New York)

Eichorn, R. (1961) Flow visualization and velocity measurement in natural convection with the tellurium dye method. *J. Heat Trans.*, **83**, 379-381.

Eisenberg, M., Tobias, C. W. and Wilke, C. R. (1954) Ionic mass transfer and concentration polarization at rotating electrodes. *J. Electrochem. Soc.*, **101**, 306-319.

Elkins, R. E., Jackman, G. R., Johnson, R. R., Lindgren, E. R. and Yoo, J. K. (1977) Evaluation of steroscopic trace particle records of turbulent flow fields. *Rev. Sci. Instrum.*, **48**, 738-746.

Erdmann, J. C. and Gellert, R. P. (1978) Recurrence rate correlation in scattered light intensity. *J. Opt. Soc. Am.*, **68**, 787-795.

Escudier, M. P. (1982) Vortex breakdown and the criterion for its occurrence. *Topics in Atmospheric and Oceanographic Sciences: Intense Atmospheric Vortices*, eds. L. Bengtsson and J. Lighthill, 247-257 (Springer-Verlag, Berlin).

Escudier, M. P. (1983) Vortex breakdown in the absence of an endwall boundary layer. *Exp. Fluids*, 1, 193-194.

Escudier, M. P. (1984) Observations of the flow produced in a cylindrical container by a rotating endwall. *Exp. Fluids*, 2, 189-196.

Escudier, M. P. and Zehnder, N. (1982) Vortex-flow regimes. *J. Fluid Mech.*, 115, 105-121.

Escudier, M. P., Bornstein, J. and Zehnder, N. (1980) Observations and LDA measurements of confined turbulent vortex flow. *J. Fluid Mech.*, 98, 49-63.

Escudier, M. P., Bornstein, J. and Maxworthy, T. (1982) The dynamics of confined vortices. *Proc. Roy. Soc. Lond.*, A382, 335-360.

Euteneuer, G.-A. (1969) Storwellenlangen-Messung bei Langswirbeln in laminaren Grenzschlichten an konkav gekrummten Wanden. *Acta Mech.*, 7, 161-168.

Euteneuer, G.-A. (1972) Eie entwicklung von Langswirbeln in zeitlich anwachsenden Grenzschlichten an konkaven Wanden. *Acta. Mech.*, 13, 125-223.

Euteneuer, G.-A. and Reimann, J. (1971) Der Mechanismus der Sichtbarkeit von Gortler-Taylor-Wirbeln in Flussigkeiten mittels feingemahlenem Pulver. *Acta Mech.*, 12, 89-97.

Fabris, G. (1978) Probe and method for simultaneous measurement of "true" instantaneous temperature and three velocity components in turbulent flow. *Rev. Sci. Instrum.*, 49, 654-664.

Fage, A. and Falkner, V. M. (1931) On the relation between heat transfer and surface friction for laminar flow. *Aero. Res. Counc., Lond., Rept. & Mem.*, No. 1408.

Faler, J. H. and Leibovich, S. (1978) An experimental map of the internal structure of a vortex breakdown. *J. Fluid Mech..* 86, 313-335.

Faller, A. J. (1960) Further examples of stationary planetary flow patterns in bounded basins. *Tellus*, 12, 159-171.

Faller, A. J. (1963) An experimental study of the instability of the laminar Ekman boundary layer. *J. Fluid Mech.*, 15, 560-576.

Faller, A. J. and Kaylor, R. E. (1966) Investigations of stability and transition in rotating boundary layers. *Dyn. Fluids Plasma*, 309-329 (Academic Press, New York).

Faller, A. J. and Mooney, K. A. (1971) The Ekman boundary-layer stress due to flow over a regular array of hills. *Bound. Layer Meteor.*, 2, 83-107.

Faller, A. J. and Porter, D. L. (1976) A note on eastern boundary currents in a laboratory analogue of the ocean circulation. *Tellus*, 28, 88-89.

Farmer, D., Hart, J. and Weidman, P. (1982) A phase space analysis of baroclinic flow. *Phys. Lett.*, 91A, 22-24.

Fenstermacher, P. R., Swinney, H. L. and Gollub, J. P. (1979) Dynamical instabilities and the transition to chaotic Taylor vortex flow. *J. Fluid Mech.*, 94, 103-128.

Fingerson, L. M. and Freymuth, P. (1983) Thermal Anemometers. *Fluid Mechanics Measurements*, ed. R. J. Goldstein, 99-154 (Hemisphere Publishing, Washington).

Firing, E. and Beardsley, R. C. (1976) The behavior of a barotropic eddy on a β-plane. *J. Phys. Ocean.*, 6, 57-65.

Fischer, K. (1931) *Mitt. Hydraul. Inst. Munch.*, 4, 7.

Fitzjarrald, D. E. (1982) An investigation of wave-amplitude vacillation using a light-speckle velocity measuring technique. *J. Phys. E: Sci. Instrum.* 15, 911-915.

Flierl, G. R., Stern M. E. and Whitehead, J. A., Jr. (1983) The physical significance of modons: Laboratory experiments and general integral constraints. *Dyn. Atm. Oceans*, 7, 233-263.

Flower, J. R., MacLeod, N., and Shahbenderian, A. P. (1969) The radial transfer of mass and momentum in an axial fluid stream between coaxial rotating cylinders - I Experimental measurements. *Chem. Engrg. Sci.*, 24, 637-650.

Focke, W. W. and Knibbe, P. G. (1986) Flow visualization in parallel-plate ducts with corrugated walls. *J. Fluid Mech.*, **165**, 73-77.

Foreman, J. W. Jr. (1967) Optical path-length difference effects in photomixing with multimode gas laser radiation. *Appl. Opt.*, **6**, 821-826.

Fortuin, J. M. H. (1960) Theory and application of two supplementary methods of constructing density gradient columns. *J. Polymer Sci.*, **44**, 505-515.

Foss, J. F. and Wallace, J. M. (1989) The measurement of vorticity in transitional and fully developed turbulent flows. *Lecture Notes in Engineering: Advances in Fluid Mechanics Measurements*, ed M. Gad-el-Hak (Springer-Verlag, New York).

Fowlis, W. W. (1970) Techniques and apparatus for the fast and accurate measurement of fluid temperature and flow speed fields. *Rev. Sci. Instrum.*, **41**, 570-576.

Fowlis, W. W. (1979) Remote optical techniques for liquid flow and temperature measurement for Spacelab experiments. *Opt. Engrg.*, **18**, 281-286.

Fowlis, W. W. and Hide, R. (1965) Thermal convection in a rotating annulus of liquid: Effect of viscosity on the transition between axisymmetric and non-axisymmetric flow regimes. *J. Atm. Sci.*, **22**, 541-558.

Fowlis, W. W. and Pfeffer, R. L. (1969) Characteristics of amplitude vacillation in a rotating, differentially heated fluid determined by a multi-probe technique. *J. Atm. Sci.*, **26**, 100-108.

Fowlis, W. W., Buckley, J. D. and Ruppert, J. W. (1972) The measurement of flow direction, flow speed, and temperature in a liquid using a miniaturized array of thermistor beads. *Geophysical Fluid Dynamics Institute Tech. Rep.* No. 37, Florida State University, Tallahassee, FL.

Fowlis, W. W., Pfeffer, R. L., Buzyna, G., Buckley, J. C. and Ruppert, J. (1974) The measurement of time-dependent fluid temperature and flow speed fields: Techniques, apparatus and results. *Flow - Its Measurement and Control in Science and Industry*, **1**, ed. R. E. Wendt, Jr., 613-622 (Instrument Society of America, Pittsburg).

Frantisak, F., Palade de Iribarne, A., Smith, J. W. and Hummel, R. L. (1969) Nondisturbing tracer technique for quantative measurements in turbulent flow. *Ind. Eng. Chem. Fundamentals*, **8**, 160-167.

Freymuth, P. (1968) Noise in hot-wire anemometers. *Rev. Sci. Instrum.*, **39**, 530-536.

Freymuth, P. (1977a) Frequency response and electronic testing for constant-temperature hot-wire anemometers. *J. Phys. E: Sci. Instrum.*, **10**, 705-710.

Freymuth, P. (1977b) Further investigation of the nonlinear theory for constant-temperature hot-wire aneometers. *J. Phys. E: Sci. Instrum.*, **10**, 710-713.

Freymuth, P. (1980) Sine-wave testing of non-cylindrical hot-film anemometers according to the Bellhouse-Schultz model. *J. Phys. E: Sci. Instrum.*, **13**, 98-102.

Freymuth, P. (1983) History of thermal aneometry. *Handbook of Fluids in Motion*, eds. N. P. Cheremisinoff and R. Gupta (Ann Arbor Science, Ann Arbor)

Freymuth, P. (1989) Air flow visualization using titanium tetrachloride. *Lecture Notes in Engineering: Advances in Fluid Mechanics Measurements*, ed. M. Gad-el-Hak (Springer-Verlag, New York).

Fuller, W. R. (1974) Calibration of a split-film sensor. MS thesis, University of Southern California, Los Angeles, CA.

Fultz, D. (1949) A preliminary report on experiments with thermally produced lateral mixing in a rotating hemispherical shell of liquid. *J. Meteor.*, **6**, 17-33.

Fultz, D. and Kaiser, J. A. C. (1971) The disturbing effects of probes in meteorological fluid-model experiments. *J. Atm. Sci.*, **28**, 1153-1164.

Garg, A. K. and Leibovich, S. (1979) Spectral characteristics of vortex breakdown flowfields. *Phys. Fluids*, **22**, 2053-2064.

Geller, E. W. (1955) An electrochemical method of visualizing the boundary layer. *J. Aeronaut. Sci.*, **22**, 869-870.

Gershenfeld, N., Frazel, R. E. and Whitehead, J. A. Jr. (1981) Rotating flume with uniformly flowing linear stratified water. *Rev. Sci. Instrum.*, 52, 1556-1559.

Gharib, M. and Willert, C. (1989) Particle tracing: Revisited. *Lecture Notes in Engineering: Advances in Fluid Mechanics Measurements*, ed. M. Gad-el-Hak (Springer-Verlag, New York).

Gharib, M., Hernan, M. A., Yavrouian, A. H. and Sarohia, V. (1985) Flow velocity measurement by image processing of optically activated tracers. *AIAA 23rd Aerospace Sciences Meeting*, No. 85-0172.

Gibson, C. H. and Schwarz, W. H. (1963) Detection of conductivity fluctuations in a turbulent flow field. *J. Fluid Mech.*, 16, 357-364

Glasstone, S. (1942) *An Introduction to Electrochemistry*, (Van Nostrand, New York).

Goldsmith, H. L. and Mason, S. G. (1962) Particle motions in sheared suspensions. XII. The spin and rotation of disks. *J. Fluid Mech.*, 12, 88-96.

Goldstein, R. J. (1983) Optical systems for flow measurement: Shadowgraph, Schlieren, and interferometric techniques. *Fluid Mechanics Measurements*, ed. R. L. Goldstein, 377-422 (Hemisphere, Washington).

Goller, H. and Ranov, T. (1968) Unsteady rotating flow in a cylinder with a free surface. *J. Basic Engrg.*, 90, 445-454.

Gorman, M. and Swinney, H. L. (1982) Spatial and temporal characteristics of modulated waves in the circular Couette system. *J. Fluid Mech.*, 117, 123-142.

Gray, W. E. (1944) A chemical method of indicating transition in the boundary layer. *Roy. Aircraft Estab. Tech. Note*, No. 1466.

Green, A. (1968) An experimental study of the interactions between Ekman layers and an annular vortex. PhD thesis, Massachusetts Institute of Technology, Boston, MA.

Greenspan, H. P. (1969) *The Theory of Rotating Fluids*. (Cambridge University Press, Cambridge).

Gregory, N., Stuart, J. T. and Walker, W. S. (1955) On the stability of three-dimensional boundary layers with application to the flow due to a rotating disk. *Phil. Trans. Roy. Soc. Lond.*, **A248**, 155-199.

Griffiths, R. W. and Linden, P. F. (1981) The stability of vortices in a rotating, stratified fluid. *J. Fluid Mech.*, **105**, 283-316.

Griffiths, R. W. and Hopfinger, E. J. (1987) Coalescing of geostrophic vortices. *J. Fluid Mech.*. **178**, 73-97.

Grousson, R. and Mallick, S. (1977) Study of flow patterns in a fluid by scattered laser light. *Appl. Optics*, **16**, 2334-2336.

Haas, F. C. and Nissan, A. H. (1961) Experimental heat transfer characteristics of a liquid in Couette motion and with Taylor vorticies. *Proc. Roy. Soc. Lond.*, **A261**, 215-226.

Hakimi, F. S. and Schowalter, W. R. (1980) The effects of shear and vorticity on deformation of a drop. *J. Fluid Mech.*, **98**, 635-645.

Hanratty, T. J. and Campbell, J. A. (1983) Measurement of wall shear stress. *Fluid Mechanics Measurements*, ed. R. L. Goldstein, 559-614 (Hemisphere, Washington).

Hansford, G. S. and Litt, M. (1968) Mass transport from a rotating disk into power-law liquids. *Chem. Engrg. Sci.*, **23**, 849-864.

Haritonidis, J. H. (1989) Measurements of mean and fluctuating wall shear stress. *Lecture Notes in Engineering: Advances in Fluid Mechanics Measurements*, ed. M. Gad-el-Hak (Springer-Verlag, New York).

Hart, J. E. (1971) Instability and secondary motion in a rotating channel flow. *J. Fluid Mech.*, **45**, 341-351.

Hart, J. E. (1972) A laboratory study of baroclinic instability. *Geophys. Fluid Dyn.*, **3**, 181-209.

Hart, J. E. (1973) On the behavior of large-amplitude baroclinic waves. *J. Atmos. Sci.*, **30**, 1017-1034.

Hart, J. E. (1985) A laboratory study of baroclinic chaos on the f-plane. *Tellus*, 37A, 286-296.

Hart, J. E. and Kittleman, S. (1986) A method for measuring interfacial wave fields in the laboratory. *Geophys. Astrophys. Fluid Dyn.*, 36, 179-185.

Hart, J. E., Toomre, J., Deane, A. E., Hurlburt, N. E., Glatzmaier, G. A., Fichtl, G. H., Leslie, F., Fowlis, W. W. and Gilman, P. A. (1986) Laboratory experiments on planetary and stellar convection performed on Spacelab 3. *Science*, 234, 61-64.

Harvey, E. N. (1940) *Living Light*. (Princeton University Press, Princeton).

Hathaway, D. H. and Fowlis, W. W. (1986) Flow regimes in a shallow rotating cylindrical annulus with temperature gradients imposed on the horizontal boundaries. *J. Fluid Mech.*, 172, 401-418.

Head, M. R. and Ram, V. V. (1971) Simplified presentation of Preston tube calibration. *Aeronaut. Q.*, 22, 295-300.

Heikes, K. E. and Maxworthy, T. (1982) Observations of inertial waves in a homogeneous rotating fluid. *J. Fluid Mech.*, 125, 319-345.

Heinrichs, R. M., Cannell, D. S., Ahlers, G. and Jefferson, M. (1988) Experimental test of the perturbation expansion for the Taylor instability at various wavevectors. *Phys. Fluids*, 31, 250-255.

Herzog, S. and Lumley, J. L. (1979) Determination of large eddy structures in the viscous sublayer: A progress report. *Proceedings of the Dynamic Flow Conference*, 869-336, P. O. Box 121, DK-2740 Skolunde, Denmark.

Hide, R. (1958) An experimental study of thermal convection in a rotating liquid. *Proc. Roy. Soc. Lond.*, A250, 441-477.

Hide, R. and Titman, C. W. (1967) Detached shear layers in a rotating fluid. *J. Fluid Mech.*, 29, 39-60.

Hide, R., Ibbetson, A. and Lighthill, M. J. (1968) On slow transverse flow past obstacles in a rapidly rotating fluid. *J. Fluid Mech.*, 32, 251-272.

Hide, R., Mason, P. J. and Plumb, R. A. (1977) Thermal convection in a rotating fluid subject to a horizontal temperature gradient: Spatial and temporal characteristics of fully developed baroclinic waves. *J. Atm. Sci..* 34 930-950.

Hinze, J. O. (1959) *Turbulence, an introduction to its mechanism and theory.* (McGraw Hill, New York).

Ho, C.-M. (1982) Response of a split film probe under electrical perturbations. *Rev. Sci. Instrum.*, **58**, 1240-1245.

Hopfinger, E. J., Browand, F. K. and Gagne, Y. (1982) Turbulence and waves in a rotating tank. *J. Fluid Mech.*, **125**, 505-534.

Howland, B., Pitts, W. H. and Gesteland, R. C. (1962) Use of electrochemiluminescence for visualizing fields of flow. *Tech. Rep.* 404, Research Laboratory of Electronics, M.I.T.

Hsueh, Y. and Legeckis, R. (1973) Western intensification in a rotating water tunnel. *Geophys. Fluid Dyn.*, **5**, 333-358.

Hudson, J. L., Tang, D, and Abell, S. (1978) Experiments on centrifugally driven thermal convection in a rotating cylinder. *J. Fluid Mech.*, **86**, 147-159.

Hummel, D. (1965) Untersuchungen uber das Aufplatzen der Wirbel an schlanken Deltaflugeln. *Z. Flugwiss.*, **13**, 158-168.

Huner, B. and Hussey, R. G. (1977) Cylinder drag at low Reynolds number. *Phys. Fluids*, **20**, 1211-1218.

Ibbetson, A. (1967) Some laboratory experiments on Rossby waves in a rotating annulus. *Tellus*, **19**, 81-87.

Ibbetson, A. and Tritton, D. J. (1975) Experiments on turbulence in a rotating fluid. *J. Fluid Mech.*, **68**, 639-672.

Imaichi, K. and Ohmi, K. (1983) Numerical processing of flow-visualization pictures - measurement of two-dimensional vortex flow. *J. Fluid Mech.*, **129**, 283-311.

Ingram, G. R. (1971) Experiments in a rotating source-sink annulus. PhD thesis, Massachusetts Intsitute of Technology, Cambridge, MA.

Iribarne, A., Frantisak, F., Hummel, R. L. and Smith, J. W. (1972) An experimental study of instabilities and other flow properties of a laminar pipe jet. *Amer. Inst. Chem. Engr. J.*, 18, 689-698.

Jeffrey, G. B. (1922) The motion of ellipsoidal particles immersed in a viscous fluid. *Proc. Roy. Soc. Lond.*, A102, 161-179.

Jeong, K. and Park, K. (1987) Observation of a very-low-frequency oscillation in a Taylor-Couette flow. *Phys. Rev.*, A35, 4854-4855.

Jerskey, T. and Penner, S. S. (1973) Velocity profiles in steady and unsteady rotating flows for a finite cylindrical geometry. *Phys. Fluids*, 16, 769-774.

Johnston, J. P., Halleen, R. M. and Lezius, D. K. (1972) Effects of spanwise rotation on the structure of two-dimensional fully developed turbulent channel flow. *J. Fluid Mech.*, 56 533-557.

Jonas, P. R. and Kent, P. M. (1979) Two-dimensional velocity measurement by automatic analysis of trace particle motion. *J. Phys. E: Sci. Instrum.*, 12, 604-609.

Joseph, D. D., Beavers, G. S. and Fosdick, R. L. (1973) The free surface on a liquid between cylinders rotating at different speeds. Part II. *Arch. Rat. Mech. Anal.*, 49, 381-401.

Joseph, D. D., Nguyen, K. and Beavers, G. S. (1984a) Non-uniqueness and stability of the configuration of flow of immiscible fluids with different viscosities. *J. Fluid Mech.*, 141, 319-345.

Joseph, D. D., Beavers, G. S., Cers, A., Dewald, C., Hoger, A. and Than, P. T. (1984b) Climbing constants for various liquids. *J. Rheology*, 28, 325-345.

Kabanov, G. N. and Siver, Y. G. (1948) *Zhur. Fiz. Khim.*, 22, 53. See also Levich, V. G., *Physicochemical Hydrodynamics.* (Printice Hall, New Jersey, 1962).

Kaiser, J. A. C. (1969) Rotating deep annulus convection. I. Thermal properties of the upper symmetric regime. *Tellus*, 21, 789-805.

Karlsson, S. K. F. and Snyder, H. A. (1965) Observations on a thermally induced instability between rotating cylinders. *Annals Phys.*, **31**, 314-324.

Kataoka, K. (1975) Heat-transfer in a Taylor vortex flow. *J. Chem. Engrg. Japan*, **8**, 271-276.

Kaye, J. and Elgar, E. C. (1958) Modes of adiabatic and diabatic fluid flow in an annulus with an inner rotating cylinder. *A.S.M.E. Trans.*, **80**, 753-765.

Keller, J. J. and Escudier, M. P. (1980) Theory and observations of waves on hollow-core vortices. *J. Fluid Mech.*, **99**, 495-511.

Kidron, I. (1966) Measurement of the transfer function of hot-wire and hot-film turbulence transducers. *IEEE Trans. Instrum. Meas.*, **15**, 76-81.

King, G. P., Li, Y., Lee, W., Swinney, H. L. and Marcus, P. S. (1984) Wave speeds in wavy Taylor-vortex flow. *J. Fluid Mech.*, **141**, 365-390.

King, L. V. (1914) On the convection of heat from small cylinders in a stream of fluid: Determination of the convection constants of small platinum wires, with applications to hot-wire anemometry. *Proc. Roy. Soc. Lond.*, **90**, 563-570.

Kohuth, K. R. and Neitzel, G. P. (1988) Experiments on the stability of an impulsively-initiated circular Couette flow. *Exp. Fluids*, to appear.

Koop, G. C. (1976) Instability and turbulence in a stratified shear layer. PhD thesis, University of Southern California, Los Angeles, CA.

Koschmieder, E. L. (1972) Convection in a rotating laterally-heated annulus. *J. Fluid Mech.*, **51**, 637-656.

Koschmieder, E. L. (1979) Turbulent Taylor vortex flow. *J. Fluid Mech.*, **93**, 515-527.

Koschmieder, E. L. and Lewis, E. R. (1986) Hadley circulations on a nonuniformly heated rotating plate. *J. Atm. Sci.*, **43**, 2514-2526.

Kung, R. K., Buzyna, G. and Pfeffer, R. L. (1987) Velocity and temperature measurement with thermistor anemometers in a thermally stratified rotating fluid. *J. Phys. E: Sci. Instrum.*, **20** 461-467.

Kreith, F., Taylor, J. H. and Chong, J. P. (1959) Heat and mass transfer from a rotating disk. *J. Heat Transfer*, **81**, 95-105.

Kreith, F., Ellis, D. and Giesing, J. (1962) Boundary layer and transition characteristics of a rotating cone. *A.S.M.E.*, Paper No. 62-WA-105 (American Society of Mechanical Engineers, New York).

Krothapalli, A. and Smith, C. A. (1989) Particle image displacement velocimetry. *Lecture Notes in Engineering: Advances in Fluid Mechanics Measurements*, ed. Gad-el-Hak (Springer-Verlag, New York)

Krumdieck, S. and Weidman, P. D. (1988) The shape and stability of rotating capillary rivulets. *Bull. Amer. Phys. Soc.*, **30**, 1732.

Lamb, D. E., Manning, F. S. and Wilhelm, R. H. (1960) Measurement of concentration fluctuations with an electrical conductivity probe. *Amer. Inst. Chem. Eng. J.*, **6**, 682-685.

Lambert, R. B., Snyder, H. A. and Karlsson, S. K. F. (1965) Hot thermistor anemometer for finite amplitude stability measurements. *Rev. Sci. Instr.*, **36**, 924-928.

Lambert, R. B. and Snyder, H. A. (1966) Experiments on the effect of horizontal shear and change of aspect ratio on convective flow in a rotating annulus. *J. Geophys. Res.*, **71**, 5225-5234.

Lambert, R. B. and Davey, M. (1974) Continuously direct-reading polarimeter for density contrast measurements in optically active solutions. *Rev. Sci. Instrum.*, **45**, 1531-1536.

Larsen, J. and Rosner, K. G. (1982) Optical flow-velocity measurement in irregularly shaped cavities. *Recent Contributions to Fluid Mechanics*, ed. W. Haase (Springer-Verlag, Berlin).

Lasso, I. and Weidman, P. D. (1986) Stokes drag on hollow cylinders and conglomerates. *Phys. Fluids*, 29, 3921-3934.

Lauterborn, W. and Vogel, A. (1984) Modern optical techniques in fluid mechanics. *Ann. Rev. Fluid Mech.*, 16, 223-244.

Leveque, M. A. (1928) Transmission de chaleur par convection. *Ann. des Mines*, 13, 201-299.

Liepmann, H. W. and Skinner, G. T. (1954) *NACA Tech. Note*, No. 3268.

Linden, P. F. (1977) The flow of a stratified fluid in a rotating annulus. *J. Fluid Mech.*, 79, 435-447.

Leehey, P. (1989) Dynamic wall pressure measurements. *Lecture Notes in Engineering: Advances in Fluid Mechanics Measurements*, ed. M. Gad-el-Hak (Springer-Verlag, New York).

Levich, V. G. (1942) *Acta Physicochim. U.R.S.S.*, 17, 257. See also Levich, V. G., *Physicochemical Hydrodynamics*. (Printice Hall, New Jersey, 1962).

Levich, V. G. (1962) *Physicochemical Hydrodynamics*. (Printice Hall, New Jersey).

Li, G.-Q., Kung, R. and Pfeffer, R. L. (1986) An experimental study of baroclinic flows with and without two-wave bottom topography. *J. Atm. Sci.*, 43, 2585-2599.

Ling, S. C. (1955) Measurements of flow characteristics by the hot-film technique. PhD thesis, State University of Iowa, Ames, Iowa.

Ling, S. C. (1960) Heat-transfer characteristics of hot-film sensing element used in flow measurement. *J. Basic Engrg.*, 82, 629-634.

Ling, S. C. and Hubbard, P. G. (1956) The hot-film: A new device for fluid mechanics research. *J. Aeronaut. Sci.*, 23, 890-891.

Lorenzen, A. (1985) Anomalous modes and finite length effects in Taylor Couette flow. *Max-Planck-Instutut Fur Stromungsforschung*, Bericht 102, Gottingen.

Lourenco, L. and Krothapalli, A. (1987) The role of photographic parameters in laser speckle or particle image displacement velocimetry. *Exp. Fluids*, 5, 29-32.

Lowell, H. H. and Patton, N. (1955) *NACA Tech. Note*, No. 3415.

Luguovtsov, B. A. (1982) Laboratory models of tornado-like vortices. *Topics in Atmospheric and Oceanographic Sciences: Intense Atmospheric Vortices*, eds. L. Bengtsson and J. Lighthill, 299-312 (Springer-Verlag, Berlin).

Ludwieg, H. (1964) Experimentelle Nachprufung der Stabilitatstheorien fur reibungsfreie Stromungen mit schraubenlinienformigne Stromlinien. *Z. Flugwiss.*, 12, 304-309.

Lumley, J. L. (1962) The constant temperature hot-thermistor. *ASME Symposium Proceedings: Measurements in Unsteady Flow.* (ASME, New York).

L'vov, V. S. and Predtechensky, A. A. (1979) On Landau and "stochastic attractor" pictures in the problem of transition to turbulence. *Institute of Automation and Electrometry, Siberian Branch, USSR Academy of Science*, Preprint No. 111.

Malkus, W. V. R. (1968) Precession of the Earth as a couse of geomagnetism. *Science*, 160, 259-264.

Mallock, A. (1889) Determination of the viscosity of water. *Proc. Roy. Soc. Lond.*, A45, 126-132.

Mallock, A. (1896) Experiments on Fluid Viscosity. *Phil Trans. Roy. Soc. Lond.*, A187, 41-56.

Maltby, R. L. and Keating, R. F. A. (1962) Smoke techniques for use in low speed wind tunnels. *AGARDograph*, No. 70, 87-109.

Manuel, F., Crespo, A. and Castro, F. (1987) Wave and cavity propagation along a tip vortex interface. *Physico-Chemical Hydrodynamics*, 9, 611-630.

Mason, P. J. (1975) Forces on bodies moving transversely through a rotating fluid. *J. Fluid Mech.*, 71, 577-599.

Mason, P. J. (1977) Forces on spheres moving horizontally in a rotating stratified fluid. *Geophys. Astrophys. Fluid Dyn.*, **8**, 137-154.

Matisse, P. and Gorman, M. (1984) Neutrally buoyant anisotropic particles for flow visualization. *Phys. Fluids*, **27**, 759-760.

Maxworthy, T. (1965) An experimental determination of the slow motion of a sphere in a rotating viscous fluid. *J. Fluid Mech.*, **23**, 373-384.

Maxworthy, T. (1967) The flow creating a concentration of vorticity over a stationary plate. *Jet Prop. Lab. Space Prog. Sum.*, **IV**, 243-250.

Maxworthy, T. (1968) The observed motion of a sphere through a short, rotating cylinder of fluid. *J. Fluid Mech.*, **31**, 543-655.

Maxworthy, T. (1970) The flow created by a sphere moving along the axis of a rotating, slightly-viscous fluid. *J. Fluid Mech.*, **40**, 453-479.

Maxworthy, T. (1972) On the structure of concentrated, columnar vortices. *Astron. Acta.*, **17**, 363-374.

Maxworthy, T. (1982) The laboratory modelling of atmospheric vortices: A critical review. *Topics in Atmospheric and Oceanographic Sciences: Intense Atmospheric Vortices*, eds. L. Bengtsson and J. Lighthill, 229-246 (Springer-Verlag, Berlin).

Maxworthy, T. (1983) Experiments on solitary internal Kelvin waves. *J. Fluid Mech.*, **129**, 365-383.

Maxworthy, T. and Browand F. K. B. (1974) Experiments in rotating and stratified flows: Oceanographic application. *Ann. Rev. Fluid Mech.*, **7**, 273-305.

Maxworthy, T., Hopfinger, E. J. and Redekopp, L. G. (1985) Wave motions on vortex cores. *J. Fluid Mech.*, **151**, 141-165.

Mazumder, M. K. and Wankum, D. L. (1969) SNR and spectral broadening in turbulence structure measurement using CW laser. *IEEE J. Quant. Electr.*, **5**, 316-318.

McEwan, A. D. (1982) Convection and mixing at high Rossby numbers in rotating systems. *Topics in Atmospheric and Oceanographic Sciences: Intense Atmospheric Vortices*, eds. L. Bengtsson and J. Lighthill, 271-283 (Springer-Verlag, Berlin).

McGoldrick, L. F. (1970) An experiment on second-order capillary gravity resonant wave interactions. *J. Fluid Mech.*, **40**, 251-271.

Merzkirch, W. (1987) *Flow Visualization.* (Academic Press, New York).

Mellor, G. L., Chapple, P. J. and Stokes, V. K. (1968) On the flow between a rotating and a stationary disk. *J. Fluid Mech.*, **31**, 95-112.

Meynart, R. (1980) Equal velocity fringes in a Rayleigh-Benard flow by a speckle method. *Appl. Optics*, **19**, 1385-1386.

Meynart, R. (1983) Speckle velocity study of vortex pairing in a low-Re unexcited jet. *Phys. Fluids*, **26**, 2074-2079.

Mizushina, T. (1971) *Advances in Heat Transfer*, **7**, p. 87 (Academic Press, New York).

Moore, J. (1967) *Gas Turbine Lab, M.I.T. Rep.*, No. 89.

Morris, J. T. (1912) The electrical measurement of wind velocity. *Engineering*, **94**, 892-894.

Mory, M., Stern, M. E. and Griffiths, R. W. (1987) Coherent baroclinic eddies on a sloping bottom. *J. Fluid Mech.*, **183**, 45-62.

Mullen, J. B. and Maxworthy, T. (1977) A laboratory study of dust devil vorticies. *Dyn. Atm. Oceans* **1**, 181-214.

Mueller, T. J. (1983) Flow visualization by direct injection. *Fluid Mechanics Measurements*, ed. R. L. Goldstein, 307-372 (Hemisphere, Washington).

Mullin, T. (1982) Mutations of steady cellular flows in the Taylor experiment. *J. Fluid Mech.*, **121**, 207-218.

Mullin, T., Lorenzen, A. and Pfister, G. (1983) Transition to turbulence in a non-standard rotating flow. *Phys. Lett.*, **96A**, 236-238.

Mullin, T., Tavener, S. J. and Cliffe, K. A. (1987) A codimension-2 bifurcation in Taylor-Couette flow with rotating ends. TP No. 1231, Theorteical Physics Division, Harwell Laboratory, Oxon, England.

Nakatani, N. Fujiwara, K. Matsumoto, M. and Yamada, T. (1971) Measurement of flow velocity distribution by luminescence. *Japan J. Appl. Phys.*, **10**, 1748-1749.

Nakatani, N. Fujiwara, K. Matsumoto, M. and Yamada, T. (1975) Measurement of velocity distributions by pulse luminescence method. *J. Phys. E: Sci. Instrum.*, **8**, 1042-1056.

Narimousa, S. and Maxworthy, T. (1985) Two-layer model of shear-driven coastal upwelling in the presence of bottom topography. *J. Fluid Mech.*, **159**, 503-531.

Niler, P. P. (1965) Performance of a thermistor anemometer in constant density shear flow. *Rev. Sci. Instr.*, **36**, 921-924.

Oberbeck, A. (1895) Uber die Abkuehlende Wirkung von Lufstroemen. *Annalen Physik Chemie*, **56**, 397-411.

Ohji, M., Shionoya, S. and Amagai, K. (1986) Mode selection in the transition of circular Couette flow. *Proceedings of the 3rd Asian Congress on Fluid Mechanics*, 34-37, Tokyo, Japan.

Ohji, M. (1987) Structure of modulated wavy vortical flows in the circular Couette system. *IUTAM Symposium on Fundamental Aspects of Vortex Motion*, 92-95, Tokyo, Japan.

Ohlsen, D. (1988) Nonlinear baroclinic instability on the beta-plane. PhD thesis, University of Colorado, Boulder, CO.

Oster, G. (1965) Density gradients. *Sci. Amer.*, **213**, 70-76.

Owen, F. S., Hale, R. W., Johnson, B. V. and Travers, A. (1961) Experimental investigation of characteristics of confined jet-driven vortex flows. *United Aircraft Res. Lab. Rep.*, No. R-2494-2, AD-328.

Park, K. and Donnelly, R. J. (1981) Study of the transition to Taylor vortex flow. *Phys. Rev.*, **A24**, 2277-2279.

Park, K., Crawford, G. L. and Donnelly, R. J. (1983) Characteristic lengths in the wavy vortex state of Taylor-Couette flow. *Phys. Rev. Lett..* 51, 1352-1354.

Penner, S. S. and Jerskey, T. (1973) Use of lasers for local measurement of velocity components, species densities, and temperatures. *Ann. Rev. Fluid Mech.*, 5, 9-30.

Penney, C. M. (1969) Differential doppler velocity measurements. *IEEE J. Quant. Electr.*, **318**, 318.

Perry, A. E. (1982) *Hot Wire Anemometry.* (Clarendon Press, Oxford).

Perry, A. E. and Morrison, G. L. (1971) A study of the constant-temperature hot-wire anemometer. *J. Fluid Mech.*, 47, 577-599.

Pfeffer, R. L., Fowlis, W. W., Fein, J. and Buckley, J. (1970) Experimental determination of the transition between the symmetrical and wave regimes in a rotating differentially heated annulus of fluid. *Rev. Pure Appl. Geophys.*, **81**, 263-271.

Pfeffer, R. L., Buzyna, G. and Kung, R. (1980a) Time-dependent modes of behavior of thermally driven rotating fluids. *J. Atmos. Sci.*, 37, 2129-2149.

Pfeffer, R. L., Buzyna, G. and Kung, R. (1980b) Relationships among eddy fluxes of heat, eddy temperature variances and basic-state temperature parameters in thermally dirven rotating fluids. *J. Atmos. Sci.*, 37, 2577-2599.

Pfister, G., Gerdts, U., Lorenzen, A. and Schatzel, K. (1983) Hardware and software implementation of on-line velocity correlation measurements in oscillatory and turbulent rotational Couette flow. *Photon Correlation Techniques in Fluid Mechancis*, ed. Schultz-duBois (Springer-Verlag, Berlin).

Plateau, J. (1863) Experimental and theoretical researches on the figures of equilibrium of a liquid mass withdrawn from the action of gravity. *Annual Report of the Board of Regeants of the Smithsonian Institution*, (Government Printing Office, Washington, D. C.).

Pope, A. (1958) *Wind-tunnel testing.* (Wiley, New York).

Popovich, A. T. and Hummel, R. L. (1967) A new method for non-disturbing turbulent flow measurements very close to the wall. *Chem. Eng. Sci.*, 22, 21-25.

Prandtl, L. (1904) Uber Slussigkeitsbewegung bei sehr kleiner Reibung. *Proceedings 3rd International Mathematics Congress*, 484-491 (Heidelberg, Germany).

Prandtl, L. (1937) Betrachtungen zur Mechanik der freien Atmosphare. *Abhandlungen der Gesellschaft der Wissenchaften zu Gottingen, Mathematisch-physikalische Klasse, III*, Folge, Heft 18, 75-84. See also *Ludwig Prandtl Gesammelte Abhandlungen*, 1100-1108 (Springer-Verlag, Berlin, 1961).

Prausnitz, J. M. and Wilhelm, R. H. (1956) Turbulent concentration fluctuations through electrical conductivity measurements. *J. Sci. Instrum.*, 26, 941-943.

Preston, J. H. (1953) The determination of turbulent skin friction by means of Pitot tubes. *J. Roy. Aero. Soc.*, 58, 109-121.

Pritchard, W. G. (1969) The motion generated by a body moving along the axis of a uniformly rotating fluid. *J. Fluid Mech.*, 39, 443-464.

Rabaud, M. and Couder, Y. (1983) A shear-flow instability in a circular geometry. *J. Fluid Mech.* 136, 291-319.

Rasmussen, R. A. (1962) Application of thermistors to measurements in moving fluids. *Rev. Sci. Instr.*, 33, 38-42.

Read, P. L. and Hide, R. (1984) An isolated baroclinic eddy as a laboratory analogue of the Great Red Spot on Jupiter. *Nature*, 308, 45-48.

Redon, M. H. and Vinsonneau, M. F. (1936) Etude de l'ecoulement de l'air autour d'une maquette. *Aeronautique*, 18, 60-66.

Reif, F. and Meyer, L. (1960) Study of superfluidity in liquid He by ion motion. *Phys. Rev.*, 119, 1164-1173.

Renouard, D. P., Chabert D'Hieres, G. and Zhang, X. (1987) An experimental study of strongly nonlinear waves in a rotating system. *J. Fluid Mech.*, 177, 381-294.

Riabouchinsky, D. (1909) Appareil pour l'etude du frottement de l'air contre un plan. *Bull. Inst. Aerodyn. Doutchino*, 2, 115-120.

Richards, E. J. and Burstall, F. H. (1945) The "China Clay" method of indicating transition. *Rep. Memor. Aero. Res. Coun., Lond.*, No. 2126.

Rignot, E. J. M. and Spedding, G. R. (1988) Performance analysis of automated image processing and grid interpolation techniques for fluid flows. *Univ. Southern. Cal. Aero. Engr. Rep.*, No. 143.

Robinson, D. A. (1968) The electrical properties of metal microelectrodes. *Proc. IEEE*, 56, 1065-1071.

Rodriguez, J. M., Patterson, G. K. and Zakin, J. L. (1970) *J. Hydronautics*, 4, 16-21.

Ronnenberg, B. (1977) Ein selfstjustierendes 3-Komponenten-Laserdoppler-anemometer nach dem Vergleichsstrahlverfahern, angewandt fur Untersuchungen in einer stationaren zylindersymmetrischen Drehstromung mit einem Ruckstromgebiet. *Max-Planck-Institute Fur Stromungsforschung*, ISSN 0436-1199.

Ross, M. P. and Hussain A. K. M. F. (1987) Effects of cylinder length on transition to doubly periodic Taylor-Couette flow. *Phys. Fluids*, 30, 607-609.

Rossby, H. T. (1969) A study of Benard convection with and without rotation. *J. Fluid Mech.*, 36, 309-335.

Ruiz, X., Massons, F. D., Aguilo, M. and Gali, S. (1986) Image processing of Czochralski bulk flow. *J. Crystal Growth*, 79, 92-95.

Roesner, K. G. (1988) Zur Wirbelbidund in rotierenden fluessigkeiten. *ZAMM*, 68, to appear.

Sambuco, E. and Whitehead, J. A. (1976) Hydraulic control by a wide weir in a rotating fluid. *J. Fluid Mech.*, 73, 521-528.

Sandborn, V. A. (1972) *Resistance temperature transducers.* (Metrology Press, Fort Collins).

Sarpkaya, T. (1971) On stationary and travelling vortex breakdowns. *J. Fluid Mech.,* **45**, 545-559.

Savas, O. (1983) Circular waves on a stationary disk in rotating flow. *Phys. Fluids,* **26**, 3445-3448.

Savas, O. (1985) On flow visualization using reflective flakes. *J. Fluid Mech.,* **152**, 235-248.

Savas, O. (1987) Stability of Bodewadt flow. *J. Fluid Mech.,* **183**, 77-94.

Schaflinger, U. (1987) Enhanced centrifugal separation with finite Rossby numbers in cylinders with compartment walls. *Chem. Engrg. Sci.,* **42**, 1197-1205.

Schaflinger, U., Koppl, A. and Filipezak, G. (1986) Sedimentation in cylindrical centrifuges with compartments. *Ing. Arch.,* **56**, 321-331.

Schaflinger, U. and Stibi, H. (1987) On centrifugal separation of suspensions in cylindrical vessels. *Acta Mech.,* **67**, 163-181.

Schanne, O. F., Lavallee, M. Laprade, R. and Gagne, S. (1968) Electrical properties of glass microelectrodes. *Proc. IEEE,* **56**, 1072-1082.

Schmitt, R. W. and Lambert, R. B. (1979) The effects of rotation on salt fingers. *J. Fluid Mech.,* **90**, 449-463.

Schultz-Grunow, F. and Hein, H. (1956) Beitrag zur Couettestromung. *Z. Flugwiss.,* **4**, 28-30.

Schwarz, K. W., Springett, B. E. and Donnelly, R. J. (1964) Modes of instability in spiral flow between rotating cylinders. *J. Fluid Mech.,* **20**, 281-289.

Schraub, F. A., Kline, S. J., Henry, J., Runstadler, P. W., Jr. and Littell, A. (1965) Use of hydrogen bubbles for quantitative determination of time-dependent velocity fields in low-speed water flows. *J. Basic Engrg.,* **87**, 429-444.

Sdougos, H. P., Bussolari, S. R. and Dewey, C. F. (1984) Secondary flow and turbulence in a cone-and-plate device. *J. Fluid Mech.*, **138**, 379-404.

Shulz-DuBois, E. O. and Rehberg, I. (1981) Structure function in lieu of correlation function. *Appl. Phys.*, **24**, 323-329.

Shirtcliffe, T. G. L. and Turner, J. S. (1970) Observations of the cell structure of salt fingers. *J. Fluid Mech.*, **41**, 707-720.

Simmers, D. A. and Coney, J. E. R. (1979) The experimental determination of velocity distribution in annular flow. *Int. J. Heat and Fluid Flow*, **1**, 177-184.

Simmers, D. A. and Coney, J. E. R. (1980) Velocity distributions in Taylor vortex flow with imposed laminar axial flow and isothermal surface heat transfer. *Int. J. Heat and Fluid Flow*, **2**, 85-91.

Simpkins, P. G. and Dudderar, T. D. (1978) Laser speckle measurements of transient Benard convection. *J. Fluid Mech.*, **89**, 665-671.

Simpson, R. L. (1989) Scanning laser anemometry and other measurement techniques for separated flows. *Lecture Notes in Engineering: Advances in Fluid Mechanics Measurements*, ed. M. Gad-el-Hak (Springer-Verlag, New York).

Sirivat, A., Rajagopal, K. R. and Szeri, A. Z. (1988) An experimental investigation of the flow of non-Newtonian fluids between rotating disks. *J. Fluid Mech.*, **186**, 243-256.

Snow, J. T. (1982) Pressure fields beneath tornado-like vortices. *Topics in Atmospheric and Oceanographic Sciences: Intense Atmospheric Vortices*, eds. L. Bengtsson and J. Lighthill, 259-270 (Springer-Verlag Berlin).

Snyder, H. A. and Karlsson, S. K. F. (1964) Experiments on the stability of Couette motion with a radial thermal gradient. *Phys. Fluids*, **7**, 1696-1706.

Snyder, H. A. and Lambert, R. B. (1966) Harmonic generation in Taylor vorticies between rotating cylinders. *J. Fluid Mech.*, **26**, 545-562.

So, K. L. (1967) Vortex phenomena in a conical diffuser. *AIAA J.*, **5**, 1072-1078.

Soloukhin, R. I., Curtis, C. W. and Emrich, R. J. (1981) Measurement of Pressure. *Methods of Experimental Physics: Fluid Dynamics - Part B*, 18, ed. R. J. Emrich, 499-610 (Academic Press, New York).

Sommeria, J., Meyers, S. D. and Swinney, H. L. (1989) Experiments on vortices and Rossby waves in eastward and westward jets. *Nonlinear Topics in Ocean Physics*, ed. A. Osborne (North Holland, Amsterdam).

Sommerscales, E. F. C. (1981) Measurement of velocity: Tracer methods. *Methods of Experimental Physics: Fluid Dynamics - Part A*, 18, ed. R. J. Emrich, 1-240 (Academic Press, New York).

Spence, D. A. and Brown, G. L. (1968) Heat transfer to a quadratic shear profile. *J. Fluid Mech.*, 33, 753-773.

Springer, G. S. (1964) Use of electrochemiluminescence in the measurement of mass transfer rates. *Rev. Sci. Instrum.*, 35, 1277-1280.

Staritz, R. F. (1960) Die Elektronische Messung der Stroemungsgeschwindligkeit und der Turbulenz, *VDI Zeitschrift*, 102, 94-97.

Stern, M. E., Whitehead, J. A. and Hua, B.-L. (1982) The intrusion of a density current along the coast of a rotating fluid. *J. Fluid Mech.*, 123, 237-265.

Stommel, H., Arons, A. B., Faller, A. J. (1958) Some examples of stationary planetary flow patterns in bounded basins. *Tellus*, 10, 179-187.

Sullivan, D. L. (1972) Alignment of rotational prisms. *Appl. Opt.*, 11, 2028-2032.

Szeri, A. Z., Schneider, S. J., Labbe, F. and Kaufman, H. N. (1983) Flow between rotating disks. Part I. Basic flow. *J. Fluid Mech.*, 134, 103-131.

Tagg, R., Cammack, L., Croonquist, A. and Wang, T. G. (1980) Rotating liquid drops: Plateau's experiment revisited. *Jet Prop. Lab. Pub.*, No. 80-66.

Takeuchi, D. I. and Jankowski, D. F. (1981) A numerical and experimental investigation of the stability of spiral Poiseuille flow. *J. Fluid Mech.*, 102, 101-126.

Tam, W. Y. and Swinney, H. L. (1988) Mass transport in turbulent Couette-Taylor flow. Submitted to *Phys. Rev. A.*

Taneda, S. and Honji, H. (1971) Unsteady flow past a flat plate normal to the direction of motion. *J. Phys. Soc. Japan*, **30**, 262-272.

Taneda, S., Honji, H. and Tatsuno, A. (1974) The behaviour of tracer particles in flow visualization by electrolysis of water. *J. Phys. Soc. Japan*, **37**, 784-788.

Tatro, P. R. and Mollo-Christensen, E. (1967) Experiments on Ekman layer instability. *J. Fluid Mech.*, **28**, 531-543.

Taylor, G. I. (1917) Motion of solids in fluids when the flow is not irrotational. *Proc. Roy. Soc. Lond.*, **A93**, 99-113.

Taylor, G. I. (1922) The motion of a sphere in a rotating liquid. *Proc. Roy. Soc. Lond.*, **A102**, 180-189.

Taylor, G. I. (1923a) Experiments on the motion of solid bodies in rotating fluids. *Proc. Roy. Soc. Lond.*, **A104**, 213-218.

Taylor, G. I. (1923b) Stability of a viscous liquid contained between two rotating cylinders, *Phil. Trans. Roy. Soc. Lond.*, **A223**, 289-343.

Taylor, G. I. (1935) Distribution of velocity and temperature between concentric rotating cylinders. *Proc. Roy. Soc. Lond.*, **A151**, 494-512.

Taylor, G. I. (1936) Fluid friction between rotating cylinders. I - Torque measurements. *Proc. Roy. Soc. Lond.*, **A157**, 546-564.

Thompson, J. (1855) Report made to the President and Council of the Royal Society, of experiments on the friction of discs revolving in water. *Proc. Roy. Soc. Lond.*, **7**, 509-511.

Tietjens, O. G. (1934) *Applied Hydro- and Aeromechanics.* (Dover, New York, 1957).

Tieu, H. A., Joseph, D. D. and Beavers, G. S. (1984) Interfacial shapes between two superimposed rotating simple fluids. *J. Fluid Mech.*, **145**, 11-70.

Tritton, D. J. (1985) Experiments on turbulence in geophysical fluid dynamics. I. Turbulence in rotating fluids. *Proc. International School of Physics, "Enrico Fermi" Course LXXXVIII: Turbulence and predictability in geophysical fluid dynamics and climate dynamics*, 172-192 (Soc. Stat. Fis., North Holland)

Truxillo, S. G. and Hussey, R. G. (1969) Delay times in fluid spin-up; Contrast to liquid helium. *Phys. Rev. Lett.*, **22**, 509-510.

Turner, J. S. and Lilly, D. K. (1963) The carbonated-water tornado vortex. *J. Atmos. Sci.*, **20**, 468-471.

Unwin, W. C. (1885) Experiments on the friction of disks rotated in fluid. *Inst. Civ. Engin. Proc.*, **80**, 221-230.

Van Atta, C. (1966) Exploratory measurements in spiral turbulence. *J. Fluid Mech.*, **25**, 495-512.

Vehrenkamp, R., Schatzel, K., Pfister, G. and Schultz-DuBois, E. O. (1979a) Direct measurement of velocity correlation functions using the Erdmann-Gellert rate correlation technique. *J. Phys. E: Sci. Instrum.*, **12**, 119-125.

Vehrenkamp, R., Schatzel, K., Pfister, G., Fedders, B. S. and Schultz-DuBois, E. O. (1979b) A comparison between analog LDA, photon correlation LDA and rate correlation techniques. *Physica Scripta*, **19**, 379-382.

Versteegen, P. L. and Jankowski, D. F. (1969) Experiments on the stability of viscous flow between eccentric rotating cylinders. *Phys. Fluids*, **12**, 1138-1143.

Vogel, H. U. (1971) Uber den stetigen Anschluss reibungsfreier Stromungen an Stromungsfelder met Newtonscher Reibung. *ZAMM*, **51**, T177-T179.

Walden, R. W. and Donnelly, R. J. (1979) Reemergent order of chaotic circular Couette flow. *Phys. Rev. Lett.*, **42**, 301-304.

Walsh, T. J., Wagner, W. T. and Donnelly, R. J. (1987) Stability of modulated Couette flow. *Phys. Rev. Lett.*, **58**, 2543-2546.

Wan, C. A. and Chang, C. C. (1972) Measurement of the velocity field in a simulated tornado-like vortex using a three-dimensional velocity probe. *J. Atmos. Sci.*, **29**, 116-127.

Wan, C. C. and Coney, J. E. R. (1980) Transition modes in adiabatic spiral vortex flow in narrow and wide annular gaps. *Int. J. Heat and Fluid Flow*, **2**, 131-138.

Wan, C. C. and Coney, J. E. R. (1982) An investigation of adiabatic spiral vortex flow in wide annular gaps by visualisation and digital analysis. *Int. J. Heat and Fluid Flow*, **3**, 39-44.

Wang, T. G., Trinh, E. H., Croonquist, A. P. and Elleman, D. D. (1986) Shapes of rotating free drops: Spacelab experimental results. *Phys. Rev. Lett.*, **56**, 452-455.

Warn-Varnas, A., Fowlis, W. W., Piacsek, S. and Lee, S. M. (1978) Numerical solutions and laser-Doppler measurements of spin-up. *J. Fluid Mech.*. **85**, 609-639.

Warpinski, N. R., Nagib, H. M. and Lavan, Z. (1972) Experimental investigation of recirculating cells in laminar coaxial jets. *AIAA J.*, **10**, 1204-1210.

Watkins, W. B. and Hussey, R. G. (1973) Spin-up from rest: Limitations of the Wedemeyer model. *Phys. Fluids*, **16**, 1530-1531 (1973).

Weidman, P. D. (1976) On the spin-up and spin-down of a rotating fluid. Part 2. Measurements and stability. *J. Fluid Mech.*, **77**, 709-735.

Weidman, P. D. and Browand, F. K. (1975) Analysis of a simple circuit for constant temperature anemometry. *J. Phys. E: Sci. Instrum.*, **8**, 553-560.

Weidman, P. D. and Johnson, M. (1982) Experiments on leapfrogging internal solitary waves. *J. Fluid Mech.*, **122**, 195-213.

Weidman, P. D. and Mehrdadtehranfar, G. (1985) Instability of natural convection in a tall vertical annulus. *Phys. Fluids*, **28**, 776-787.

Werle, H. (1960) Etude effectuee a la cuve a huile et au tunnel hydrodynamique a visualisation de l'O.N.E.R.A. *Rech. Aeronaut.*, **79**, 9-26.

Werle, H. (1973) Hydrodynamic flow visualization. *Ann. Rev. Fluid Mech.*, 5, 361-383.

White, D. E., Litt, M. and Heymach, G. J. III. (1974) Diffusion-limited heterogeneous catalytic reactions on a rotating disk. I. Hydrogenation of α-methylstyrene. *Ind. Eng. Chem., Fundam.*, 13, 143-150.

White, D. E. and Litt, M. (1975) Diffusion-limited heterogeneous catalytic reactions on a rotating disk. II. Hydrogenation of phenylacetylene over palladium. *Ind. Eng. Chem., Fundam.*, 14, 183-190.

White, H. D. and Koschmieder, E. L. (1981a) Convection in a rotating, laterally heated annulus. The wavenumber transition. *Geophys. Astrophys. Fluid Dynamics*, 18, 279-299.

White, H. D. and Koschmieder, E. L. (1981b) Convection in a rotating, laterally heated annulus. Pattern velocities and amplitude oscillations. *Geophys. Astrophys. Fluid Dynamics*, 18, 301-320.

Whitehead, J. A. Jr. (1980) Selective withdrawal from a rotating stratified fluid. *Dyn. Atmos. Oceans*, 5, 507-515.

Whitehead, J. A. Jr. (1985) A laboratory study of gyres and uplift near the straight of Gibraltar. *J. Geophys. Res.*, 90, 7045-7060.

Whitehead, J. A. (1986) Flow of a homogeneous rotating fluid through straights. *Geophys. Astrophys. Fluid Dyn.*, 36, 187-205.

Whitehead, J. A. and Chapman, D. C. (1986) Laboratory observations of a gravity current on a sloping bottom: the generation of shelf waves. *J. Fluid Mech.*, 172, 373-399.

Whitehead, J. A., Jr. and Gershenfeld, N. (1981) Selective withdrawal from a rotating stratified current with applications to OTEC. *Ocean Engrg.*, 8, 507-515.

Whitehead, J. A. and Porter, D. L. (1977) Axisymmetric critical withdrawal of a rotating fluid. *Dyn. Atm. Oceans*, 2, 1-18.

Whitehead, J. A., Leetmaa, A. and Knox, R. A. (1974) Rotating hydraulics of strait and sill flows. *Geophys. Fluid Dyn.*, 6, 101-125.

Wilcke, J. C. (1785) Forsok til Uplysning om Luft-hvirflar och Sky-drag. *K. Vet. Acad. nya Hand.*, 6, 290-307.

Wimmer, M. (1976) Experiments on a viscous fluid flow between concentric rotating spheres. *J. Fluid Mech.*, 78 317-335.

Wimmer, M. (1978) Die zahe Stromung im Spalt zwischen einer rotierenden Scheibe und einem ruhenden Gehause. *ZAMM*, 58, T350-T353.

Wimmer, M. (1981) Experiments on the stability of viscous flow between two concentric rotating spheres. *J. Fluid Mech.*, 103, 117-131.

Wimmer, M. (1983) Die viskose Stromung zwischen rotierenden Kegelflachem. *ZAMM*, 63, T299-T301.

Wortmann, F. X. (1953) Eine Methode zur Beobachtung un Messung von Wasserstromungen mit Tellur. *Z. Angew. Phys.*, 5, 201-206.

Yaglom, A. M. (1962) *An introduction to the theory of stationary random functions.* English edition: (Prentice Hall, Englewood Cliffs).

Yeh, Y. and Cummins, H. Z. (1964) Localized fluid flow measurements with an He-Ne laser spectrometer. *Appl. Phys. Lett.*, 4, 176-178.

Zhang, L.-H. and Swinney, H. L. (1985) Nonpropagating oscillatory modes in Couette-Taylor flow. *Phys. Rev.*, A31, 1006-1009.

EXPERIMENTS IN DRAG-REDUCING POLYMER FLOWS

E.W. Hendricks, J.V. Lawler[*],

M.P. Horne, R.A. Handler and J.D. Swearingen

Naval Research Laboratory, Washington, D.C. 20375-5000

1. Introduction

Small amounts of high molecular weight polymers (a few parts per million by weight) dissolved in a solvent can reduce the turbulent flow frictional resistance from that of the pure solvent by as much as 90 percent. Since the viscosity of a polymer/solvent mix is slightly increased over that of the solvent alone, the fact that viscous drag is decreased is somewhat surprising. The study of these drag-reducing, non-Newtonian fluid flows received initial impetus from the study of Toms (1948). Since this work, the examination of the turbulent flow properties of dilute solutions of long chain high-molecular weight polymers has generated many hundreds of publications, of which a majority are experimental investigations. This preponderance of experimental studies is no doubt due to the difficult theoretical nature of the problem which must address not only the turbulent kinematics but also the interaction of a viscoelastic fluid with the turbulence. Several excellent reviews of the available experiments and possible explanations of this phenomenon have been published since Toms original work: Lumley (1969) & (1973), Hoyt (1972), Virk (1975), Little et al. (1975), and Berman (1978). Theoretical discussions of the drag reduction phenomenon can be found in Lumley (1973), Berman (1978) and a recent work by Ryskin (1987a). The text that follows includes a review of experimental methods utilized by both rheologists and fluid dynamicists in the examination of turbulent flows of drag-reducing polymer solutions. In addition, we review some of the major experimental findings and make recommendations for future research. Finally, we present some new measurements of wall pressure in a turbulent non-Newtonian channel flow.

2. Rheological Characterization of Drag-Reducing Fluids

In this section, we discuss the measurements of the rheological properties of dilute polymer solutions as they relate to turbulent flow drag reduction. The flow of a viscoelastic polymer solution is markedly different in many ways from a Newtonian flow. These differences are due to the long chain structure of polymers. The flow properties of viscoelastic fluids are often rate-dependent. For example, once a threshold strain rate has been exceeded the shape of the

---

[*] Current address: Hoechst Celanese, Research Division, Summit, New Jersey 07901

polymer molecule becomes anisotropic and partially aligns with the flow. This anisotropy produces stresses along the backbone of the aligned molecule. In shearing flows, components of the anisotropic stresses in a viscoelastic fluid that act in a direction normal to the deformation gradient are, not surprisingly, called "normal" stresses. The large number of atoms in a polymer molecule allow large deformations. Thus, the time required for a polymer to diffuse back to equilibrium is many orders of magnitude greater than the corresponding relaxation times of Newtonian fluids. Polymer solutions have relaxation times from milliseconds to seconds. Finally, the stresses in a polymer solution are a function of the previous deformations of the fluid, since these deformations have altered the current conformation of the polymers. This dependence on past events is termed "memory".

2.1. Constitutive Equations

Constitutive equations are mathematical expressions which model the stress level and distribution in a polymer solution or melt (see recent books by Tanner 1985, Bird *et al.* 1987a & 1987b, and Larson 1988). Two examples of typical constitutive equations are those attributed to Maxwell and Oldroyd. The Maxwell constitutive equation represents a viscoelastic fluid as two beads connected by a spring in series with a dashpot. The relaxation time is equal to the ratio of the viscosity of the dashpot to the modulus of the spring. To incorporate a solvent viscosity into a molecular model, another dashpot in parallel with the dashpot/spring elements can be added. The resultant constitutive equation is called the Oldroyd-B model.

The validity of the many proposed constitutive equations as accurate models for dilute polymer solutions in a turbulent flow remains largely untested. The state of the polymers in turbulent flows is so poorly understood that many of the assumptions invoked to develop a specific constitutive equation cannot be tested. Since several constitutive equations exist and turbulent flow is extremely complex, these equations have historically been tested in laminar shear and extensional flows. The rationale of this approach is that measurements of stresses, velocities, orientations, or geometry-specific flow parameters can be conducted in these simpler flows for the independent determination of the constants and functions which appear in a constitutive equation. As we will show, the values for the constants arrived at with this approach vary greatly from study to study. The best determinations are within a factor of two or three and the worst can differ by several orders of magnitude.

2.2. Shear Viscosity and Normal Stresses

Shear viscosity (hereafter referred to as simply "viscosity") is the easiest rheological property of dilute solutions to measure accurately. That is not to say that the measurement of viscosity is easy, merely that it is easier to measure viscosity than any other rheological property of a dilute polymer solution. Most standard techniques for measuring non-Newtonian viscosities cannot be used with dilute aqueous solutions due to the low viscosity of water. Non-Newtonian entrance effects (Bagley 1957) and adsorption-entanglement layers (Hikmet *et al.* 1985) that develop in capillary rheometers, which are the instruments most often used to characterize dilute solutions, lead to inaccuracies in the measurement. However, it can be reliably stated that the viscosities of aqueous dilute solutions of polymers (less than 100 *ppm*) are less than twice the viscosity of water.

The viscosities of dilute polymer solutions are weak functions of shear rate. The combination of a relative viscosity (ratio of solution viscosity to solvent viscosity) close to unity and a low solvent viscosity makes determining this shear rate dependence difficult with current rheometers. Hoyt & Fabula (1964) have measured no change in viscosity for polyethylene oxide solutions (PEO, Union Carbide WSR-301) ranging in concentration from 10 to 100 *ppm* up to a shear rate of 500,000 s^{-1}. Belokon *et al.* (1973) have reported that aqueous solutions of PEO (WSR-301) only begin to depend on shear rate above a concentration of 200 *ppm*. Solutions with concentrations less than 100 *ppm* are the critical fluids for testing drag reduction theories since these solutions are the most efficient drag reducers. Hoyt (1971) finds that the minimum concentration required to reach the maximum drag reduction asymptote is only 20 to 30 *ppm* for an aqueous PEO (WSR-301) solution, with a 2 to 4 *ppm* solution producing one-half of the maximum reduction.

The lack of significant shear thinning (the reduction of viscosity with increasing shear rate) of the strongly drag-reducing solutions cited above demonstrates that shear-thinning is not the principal mechanism for drag reduction. Further evidence is reported by Hoyt & Fabula (1964). They have measured no reduction in turbulent drag for a carboxy vinyl polymer (Carbopol 941) solution, which is shear-thinning but inelastic.

Finally, very few measurements of the normal stresses of these fluids have been reported. The problem, as in the viscosity measurements, is the dependence of these stresses in dilute aqueous solutions on the shear rate. The best measurement method for dilute solutions is a jet-thrust technique. The measurement of the thrust produced by a jet flowing out of a capillary can be used to predict the normal stresses. Davies *et al.* (1977/1978) find that the normal stress values arrived at by this technique are high by as much as a factor of two. The magnitude of the normal stresses is related to the relative viscosity of the solution. Consequently, at higher concentrations (that is, high relative viscosities), the normal stresses can be separated from the shear stresses and other sources of uncertainty in standard rheological measurements and reliable measurements can be obtained.

2.3. Elongational Properties

Elongational viscosity is defined as the constant of proportionality between the difference of the first and second normal stresses and the elongation rate. The first normal stress acts along the direction of elongation, and the second normal stress acts normal to the first. The modifications of turbulent flow, cavitation, and other flow instabilities by polymer solutions are most likely the result of the high elongational viscosity of dilute polymer solutions, as shown by Ting & Hunston (1977). On a unit concentration basis, the magnitude of the effect increases with decreasing concentration. From this observation, Ting & Hunston postulate that the large elongational viscosity is a polymer-solvent and not polymer-polymer interaction because there are ten billion to one trillion solvent molecules for each polymer molecule in these dilute solutions. It is unfortunate that accurate, quantitative measurements of elongational viscosity remain unavailable. The attempts to obtain an estimate of the elongational viscosity are discussed below.

As a result of the high elongational viscosity of dilute polymer solutions (compared to Newtonian), an excess pressure is required to force these solutions through an orifice. To obtain an order of magnitude estimate of the elongational viscosity, Fruman & Barigah (1982) utilize data from an orifice flow (Figure 1). The limitation of this technique is that it assumes that the elongational viscosity is a constant and independent of strain or strain rate. These analyses do indicate that the elongational viscosity is several orders of magnitude larger than

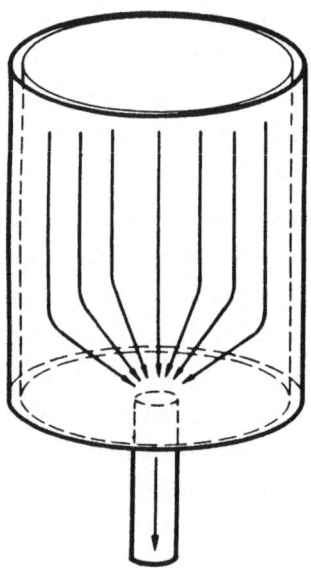

FIGURE 1. Orifice flow geometry

the shear viscosity. Hasegawa *et al.* (1988) have measured the axial velocity field above and below an orifice while simultaneously measuring the pressure drop across the orifice. While the scatter in their measured pressure data is large, they find that a pressure calculated from the measured velocity field using a Maxwell fluid model modified to allow a deformation-rate dependent relaxation time roughly correlates the data. A Second Order Fluid (SOF) model was also used but was found to be inferior. The SOF model is derived by truncating to second order a Taylor series expansion of the stress as a function of velocity gradient. This model is able to predict normal forces, but does not include stress as a function of past deformations (no memory).

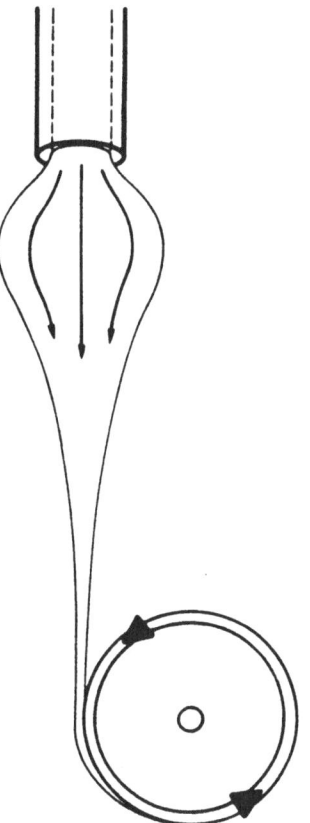

FIGURE 2. Fiber-spinning geometry.

Fiber spinning and a ductless syphon have been employed to measure the elongational viscosity of dilute solutions. The fiber-spinning and ductless-syphon geometries are depicted in Figures 2 and 3, respectively. By assuming an absence of radial gradients and measuring the force along the fluid, one can calculate the elongational stresses and elongational viscosities from photographs taken of the profile of the fluid. The durations of the elongations in fiber-spinning/ductless-syphon experiments are too short to produce steady-state stresses in the solutions, although this assumption is routinely made in the analyses. As with orifice flow, the measured elongational viscosities of solutions are found to be about three orders of magnitude larger than the shear viscosities (Weinberger & Goddard 1974 and Usui & Sano 1981).

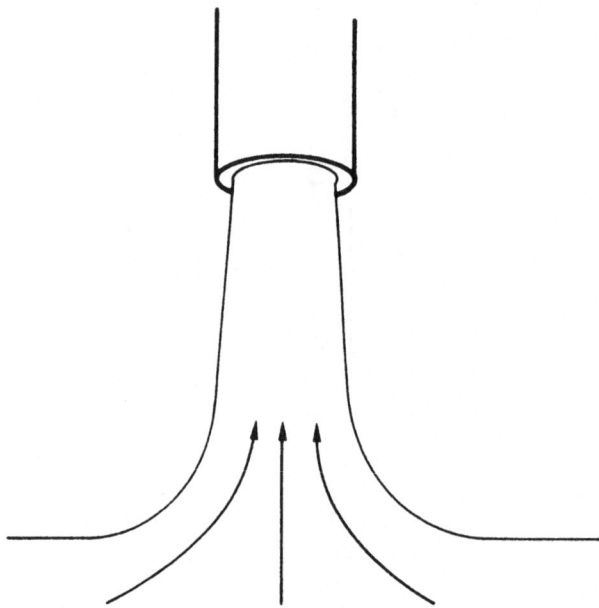

FIGURE 3. Ductless syphon geometry.

Attempts have been made to minimize the errors in the measurement of elongational properties. Baid & Metzner (1977) have calculated the response of various fluid models to the time-dependent strain history occurring along the length of a fiber to simulate more accurately the stress history. Matthys (1987) has shown that radial gradients do exist and that the axial velocity on the outer surface is greater than at the core. Becraft & Metzner (1988) discuss methods to reduce other sources of errors in making measurements during a fiber-spinning experiment.

FIGURE 4. Crossed-slots geometry.

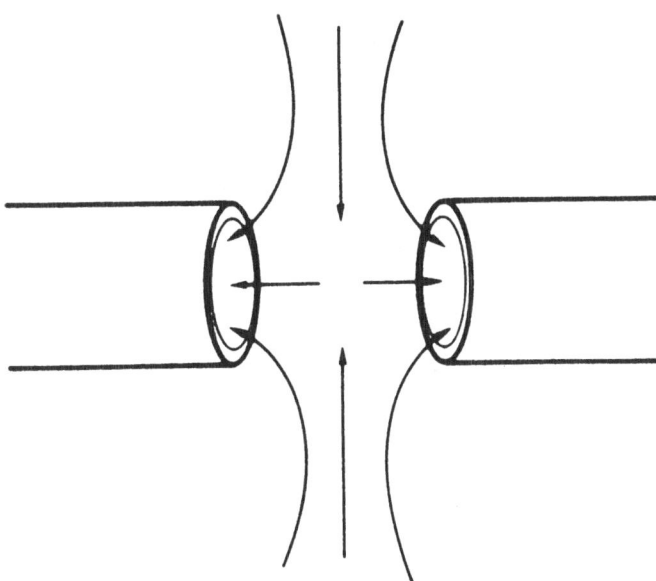

FIGURE 5. Opposed-jets geometry.

An important tool in the measurement of rheological properties is flow birefringence. The refractive index of a flowing polymer solution is anisotropic because the polymers deform into anisotropic conformations. These conformations produce bulk anisotropic stresses which can be determined if the level of the birefringence is linearly related to the stress level for the fluid (Lodge 1955, Philippoff 1956, and Brodnyan *et al.* 1957). Thus, the stress in a solution can be determined by optically measuring the level of the birefringence. Similarly, the orientation of the stress is obtained by measuring the birefringence at two different angles of polarization. Dandridge *et al.* (1979) have measured the birefringence of an aqueous solution of polyethylene oxide (PEO) over a concentration range of 50 to 1350 *ppm* in a shear flow with a shear rate greater than 2000 s^{-1}. The birefringence of polymer solutions has been measured in elongational flow with the crossed-slots and opposed-jets geometries. Schematic diagrams of these geometries are shown in Figures 4 and 5. In these experiments the extension rate at which the polymers begin to uncoil and hence to show strong birefringence can be determined. In the small regions of these flows undergoing elongational deformations, once uncoiling begins the polymers extend completely if the residence time is sufficiently long (DeGennes 1974). This complete extension produces the strong birefringence in a small region of the flow domain. Recently, Wiest & Bird (1988) have shown that complete polymer extension does not suddenly occur at a critical extension rate. If a polymer is modeled as a chain of bead-spring dumbells, Wiest & Bird find that dumbells along the chain extend first, and then the chain as a whole begins to uncoil.

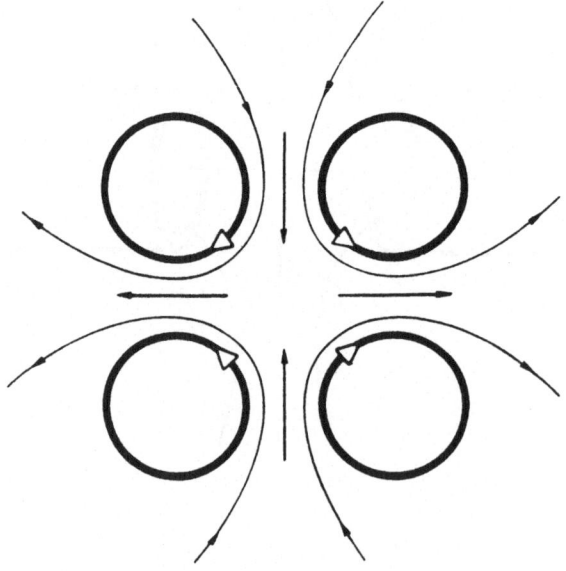

FIGURE 6. Four-roll mill geometry.

The four-roll mill, as depicted in Figure 6, is a good geometry in which to examine the effect of elongational deformations on polymer solutions. The deformation history with counter-rotating rollers in a Newtonian fluid is close to an ideal hyperbolic flow (Crowley *et al.* 1976). The principal limitation of this device is that high elongation rates cannot be obtained at a low Reynolds number. The stresses in a dilute solution of polystyrene (PS, 50-100 *ppm*) in a viscous solvent are reported by Fuller & Leal (1980) for this geometry. The birefringence levels observed during elongation are three to four orders of magnitude higher than those observed previously in shear flows at comparable deformation rates. Fuller & Leal are able to model closely the birefringence and stresses of this PS solution with a dumbbell model which included a nonlinear hydrodynamic friction, an internal viscosity, and a molecular weight distribution.

Most constitutive equations predict high values for elongational viscosity. Chakraborty & Metzner (1986) have been able to simulate qualitatively the orifice flow experiments of James & Saringer (1982) by calculating the stresses in the flow with an Oldroyd-B fluid model. Ryskin (1987b) has simulated these same experiments with his "Yo-Yo" model, which is a bead-rod representation of a polymer molecule that begins to uncoil at the midpoint of the backbone with each end of the chain remaining coiled. Further uncoiling lengthens the elongated portion of the chain between the coiled ends. He has also used this model to predict some aspects of turbulent drag reduction (Ryskin 1987a). These interesting predictions are not readily generalized to multi-dimensional flows or flows containing both shear and elongational deformations. Baid & Metzner (1977) have been able to model combined shear and elongational flows with the Oldroyd-B model, but only by selecting a different set of fluid model parameters for each type of flow.

2.4. Relaxation Times From Shear Measurements

The relaxation time of a polymer solution represents the time scale over which deformed polymer conformations return to equilibrium conformations. Baid & Metzner (1977) have measured the viscosity and normal stresses for a 100 *ppm* polyacrylamide (PAC, Dow Separan AP-30) in a glycerin/water solution. They have estimated the Maxwell relaxation time (normal stress divided by twice the shear stress and shear rate) to be about 50 *ms*. Darby & Chang (1984) have measured the shear rate at which the viscosity begins to decrease (shear thin) for PAC (AP-30) in water. The reciprocal of this critical shear rate is another estimate of the relaxation time for a viscoelastic fluid. Darby & Chang's estimate is about 10 to 20 *s* for a 100 *ppm* solution. Thus, these two investigations disagree on the value of the relaxation time by 3 orders of magnitude.

The relaxation time, estimated from linear viscoelastic theory, is proportional to the viscosity of the solvent if one neglects the change in the intrinsic viscosity due to a change in the solvent (Rouse 1953). Therefore, the differences in relaxation times discussed above may be even greater if measured in the same solvent. The ambiguity and uncertainty of shear measurements in dilute solutions makes determining relaxation times and other constitutive equation parameters from shear flow measurements inadvisable.

2.5. Relaxation Times Based On Elongational Flow Measurements

To obtain estimates for relaxation times, it is assumed that the critical elongation rate obtained by one of the various methods discussed above is equal to one-half of the reciprocal of the relaxation time of the polymer molecules. Farrell *et al.* (1980) find that the relaxation time and its dependence on molecular weight for polystyrene solutions agree with a bead-spring molecular model with hydrodynamic interaction for the motion of non-free draining molecules in a solvent, called the Zimm model (Zimm 1956). Non-free draining means that solvent molecules inside the strands of the polymer molecule move with the polymer. James & Saringer (1982) measure a relaxation time in their orifice experiments for a PEO solution. James & Saringer's measured relaxation time is different from the relaxation time predicted using the Zimm model by a factor of three. This discrepancy could be due to the fact that water is a very good solvent for PEO while Zimm's model more accurately describes polymers in poor solvents (Bird *et al.* 1987b). Another possibility for error is the uncertainty in determining the elongation rates that occur in this geometry. Elongation rates above an orifice have been measured by Hasegawa & Iwaida (1984). They find that for aqueous solutions of 100 and 200 *ppm* PEO the product of the elongation rate just above the orifice and the relaxation time (estimated from jet-thrust experiments) is roughly equal to one half of the value when vortices begin to form upstream. More experiments are needed before it can be determined if the formation of these vortices can be used to infer values for rheological properties.

2.6. Mixed Flows

Turbulent flows are "mixed" flows in that they contain combinations of shear and elongational deformations. Recent analysis of the numerical simulations of wall-bounded Newtonian turbulence by Lawler *et al.* (1987) finds that the shear component of deformation near the wall is large enough to enhance the resistance to elongation (increase elongational viscosity) for a large class of fluid models. Recent experimental results for flow through an orifice by James and his co-workers (1987) demonstrate that preshearing of the fluid before the fluid enters an orifice has a profound effect on the excess pressure drop. One explanation is that a deformed polymer molecule is much more rapidly elongated than a molecule in an equilibrium configuration. Another explanation is that networks form during the shearing, and these structures require higher stress levels to deform in an elongational flow than do individual molecules. The effect of preshearing might be predicted by an Oldroyd-B or "Yo-Yo" model if this new mixed flow were simulated. However, the analyses by Chakraborty & Metzner (1986) and Ryskin

(1987b) must be expanded to at least two-dimensional flow problems if preshearing or more realistic kinematics are to be modelled.

The aforementioned four-roll mill is an excellent geometry in which to study mixed flows. A wide range of deformations, between pure elongation and simple shear, can be generated by varying the ratio of the angular velocities of the rollers. As discussed above, Fuller & Leal (1980) predicted the birefringence of a PS solution in pure elongation. They obtained equally accurate predictions of the birefringence in mixed flows with their model. Dunlap & Leal (1987) extend the measurements in the four-roll mill to a wider range of mixed flows. They then calculate the birefringence of several molecular models to determine which models are able to predict the flow behavior in these mixed flows. Dunlap & Leal determined that the model of Phan-Thien *et al.* (1984), which includes conformation-dependent, anisotropic friction and strain-inefficient rotation, fit the behavior of the PS polymer solution better than molecular models without these features. This work by Dunlap & Leal is an excellent example of how mixed flow experiments can and should be used to test and develop constitutive equations.

2.7. Simulation of Complex Flows

The direct numerical simulation of low-Reynolds number Newtonian turbulent flow is now possible with the advent of supercomputers. Currently, channel flows at Reynolds numbers of up to 3300, based on centerline velocity and channel half width, have been simulated. To predict turbulent drag-reducing flows, one must incorporate a non-Newtonian constitutive equation into the simulations. However, the determination of the polymer properties necessary to form an accurate constitutive equation has not been successful. A research approach which takes advantage of the comparison of experiments to large-scale direct numerical simulations is required. The increasing power of computers makes feasible the simulation of continuum fluid models in complex geometries. Comparison of experiments with these calculations may be a better method of selecting fluid models and determining the parameters in these models than trying to measure rheological properties of these fluids directly.

A good candidate for a coupled simulation/experimental approach might be a vortex flow. Balakirshnan & Gordon (1971) have observed experimentally that a vortex does not form during the draining (through an orifice) of a tank filled with a dilute polymer solution. Schematic diagrams of the differences between the flows of a Newtonian fluid and a polymer solution are shown in Figure 7. Later, these researchers (Gordon & Balakirshnan 1972) showed that the minimum concentration for vortex inhibition is close to the concentration needed to achieve maximum drag reduction in a pipe. A third study (Chiou & Gordon 1976) contains measurements of the axial velocity above the orifice as a function of position. They found that this velocity is reduced by an order of magnitude compared with the flow of a Newtonian fluid.

FIGURE 7. Tank draining through an orifice geometry. A) A Newtonian fluid forms a central vortex extending down to the orifice. B) A polymer solution develops a much weaker vortex which does not extend down to the orifice.

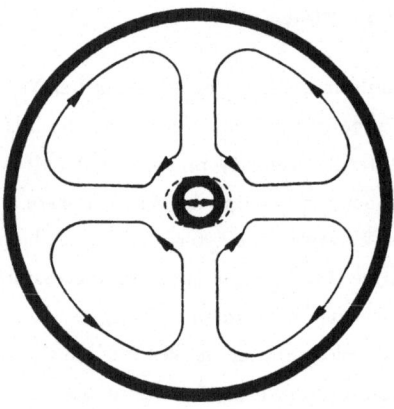

FIGURE 8. Vibrating rod geometry.

Another possible model geometry for simulations is the flow around a vibrating rod inside a larger cylinder (see Figure 8). Chang & Schowalter (1974) present experiments which show that the vortices which form in a polymer solution are rotating in a direction opposite those which form in a Newtonian fluid. Small vortices (not shown on the figure) occur near the vibrating cylinder in a Newtonian fluid. Perturbation calculations by Chang & Schowalter (1979) show that in an Oldroyd-B fluid model these small vortices grow as the elasticity of the fluid is increased. Eventually, these vortices, which are rotating in the opposite direction compared to rotation of the larger vortices in a Newtonian fluid, fill the gap between the cylinders, and the original large vortices disappear. Thus, the fluid model has been able to predict a macroscopic change in the flow field. The relaxation time of the polymer solutions can be estimated by fitting the flow behavior to the numerical simulations. Sufficient rheological data are unavailable to obtain an independent estimate.

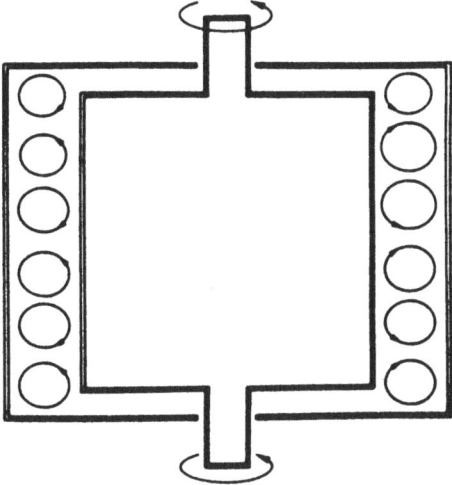

FIGURE 9. Concentric cylinder geometry.

The onset of Taylor vortices between concentric rotating cylinders (Figure 9) is another good choice for the comparison of the flows of dilute solutions and fluid models. The vortices are generated by a centrifugal instability in the flow which Taylor (1923) has accurately predicted for a Newtonian fluid using small perturbation theory analysis. Denn and his co-workers (Ginn & Denn 1969, Denn & Roisman 1969, Denn et al. 1971, and Sun & Denn 1972) have compared experiments with polymer solutions and stability calculations for a SOF fluid model to determine the rheological properties of the solutions. For some of the polymer solutions, these rheological properties agree with properties measured by other more standard rheological measurements. Comparisons for other polymer solutions do not show agreement. Jones et al. (1973) discuss the sensitivity of the estimated solution properties to errors in the stability experiments. Despite the large number of experiments to measure the onset of Taylor vortices in dilute polymer solutions, only Beard et al. (1966) and Deutsch & Phillips (1977) have calculated the onset of Taylor vortex flow for a fluid model with memory. Belokon et al. (1973) find that the onset Reynolds number is not affected by the very dilute drag reducing solutions. Therefore, the other aspects of the secondary flow which are altered by the presence of the polymers (for example, the radial and axial velocities) must be measured to test the predictive capabilities of fluid models. These perturbation velocities can be measured accurately with a laser-Doppler velocimeter (LDV) as shown by Gollub & Swinney (1975) and Lawler (1986).

2.8. Dimensions of Polymer Molecules and Entanglements

Large aspect ratios, large included volumes, and formation of aggregates are three aspects of polymer solutions which greatly magnify the effect that polymers have on the flow of the solution. If the polymers in a solution are elongated, the volume fraction based on their circumscribed volume can be quite large. The ratio of elongated to coiled volume fraction is approximately proportional to the number of links in the chain to the 3/2 power. In their four-roll mill geometry, Dunlap & Leal (1987) find that the volume fraction of a 100 ppm PS solution reaches about 5000 before the laminar velocity field is effected. Wolff (1980) reports the mean radius of gyration of isolated PEO (WSR-301) molecules in water as 49 nm and the radius of aggregated PEO as 340 nm. The radius of gyration is the root-mean-square distance from each atom in a polymer to the center of mass of the polymer (or aggregate). Wolff also reports that the weight-average molecular weight of this PEO is 2.5×10^6 for isolated molecules and 1.2×10^8 for aggregated molecules. If the polydispersity (ratio of weight-average to number-average molecular weights) is assumed to be two, the average distance between the centers of mass of neighboring polymers in a solution can be estimated. This distance between polymers in a dispersed 30 ppm solution is about 460 nm, and in an aggregated solution of the same concentration the distance is 1680 nm. Therefore, the volume contained inside the radii of gyration is 0.7 % of the total volume in a solution of isolated polymers and 5.0 % in an aggregated solution. For comparison, a face-centered cubic arrangement of hard spheres of equal radii has a maximum volume concentration of 74 %. Clearly, interactions between PEO molecules even at a weight concentration of 0.003 % are important, especially since the radius of gyration is a statistical quantity and the strands of the polymer can extend well beyond this distance. Kowalik et al. (1987) increased polymer-polymer interaction by increasing the intermolecular forces in a polymer solution and found that higher polymer-polymer interaction increases drag reduction. Georgelos & Torkelson (1988) report that the critical shear rate in an orifice flow experiment (the shear rate at which the pressure drop into the orifice increases dramatically) is a function of the age of the PEO solutions. The critical shear rate increases as a solution ages, which indicates that the aggregates are disentangling with time. Different experiments may indicate different degrees of entanglement for the same solution depending on the time allowed for disentanglement to occur. The state of entanglements also depends on previous flow history. For example, experiments limited to pure elongational flows examine conformations which are different than the conformations found in mixed flows. Drag reduction occurs at very low weight fraction but the conformation, volume fraction, and degree of entanglement of the polymers in turbulence are unknown.

2.9. Polymer Degradation

The breaking of polymer chains does not cause any of the phenomena associated with viscoelasticity or drag reduction, but degradation of the polymer solution must be of concern when conducting drag reduction and solution characterization experiments. Polymers in elongational flow fields break near the center of the backbone of the polymer which provides strong evidence that the polymers are unraveled during elongational flows (see Odell & Keller 1986 and Merrill & Horn 1984). Breaking in the center greatly reduces the average molecular weight, and the higher molecular weight polymers break first, so degradation greatly changes the viscoelasticity of polymer solutions. Also, chemical degradation can ruin polymer solutions. For example, the polyethylene oxide backbone is attacked by both ferric ions and chlorine (see Hoyt 1980).

2.10. Summary

In summary, the elongational viscosity appears to be the rheological property responsible for the drag-reducing characteristics of dilute polymer solutions. Unfortunately, elongational viscosity is also the most difficult property of a dilute solution to measure accurately. In fact, with the exception of the viscosity, attempts to determine correct values of any of the rheological properties of dilute polymer solutions used in predictive models have been unsuccessful or of limited value. Additionally, molecular entanglements and polymer degradation must be considered in any non-Newtonian flow experiment.

Finally, recommendations for future research directions include an approach which examines flows which contain mixtures of shear and elongation, as demonstrated in, for example Dunlap & Leal (1987). Coordination of numerical simulation and experimental investigations holds promise to determine the efficacy of candidate constitutive equations to predict changes in turbulence due to the presence of polymer additives.

3. Turbulence Measurements in Drag-Reducing Fluids

Drag reduction has been experimentally examined by fluid dynamicists using many techniques. Early studies concentrated on pressure drop/flow rate experiments which showed the magnitude and scope of the phenomenon. These relatively simple early experiments established that the flow must be turbulent for drag reduction to occur and that there exists an ultimate limit of drag reduction in pipes (Virk 1975). In the following section we summarize subsequent attempts to delineate the interactions between non-Newtonian fluids and turbulent flow fields.

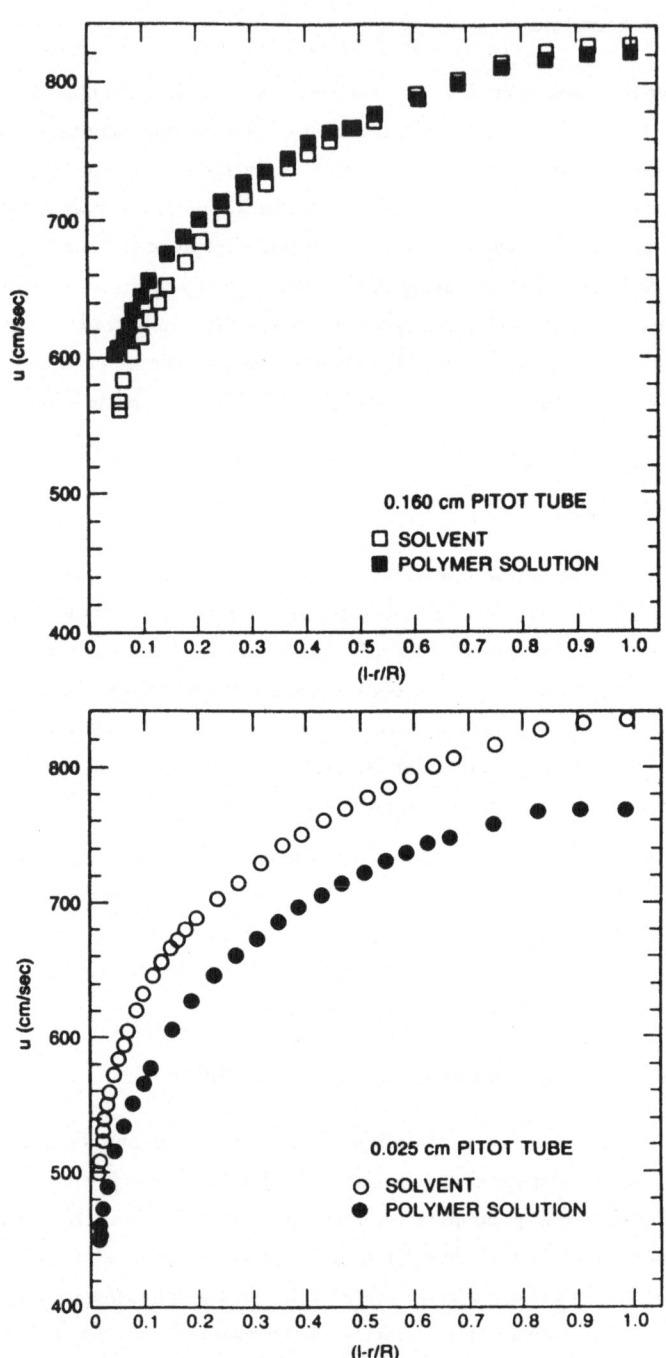

FIGURE 10. Streamwise velocity profiles of PEO solution in a 3.21 cm i.d. pipe with two pitot tubes of different sizes. From the data of Smith *et al.* (1967).

3.1. Velocity Field Measurements

The velocity measurement techniques used to probe polymer/turbulence interactions can be divided into five categories: Pitot probes, heated element probes, optical techniques, laser Doppler velocimeters, and flow visualization.

Pitot probes have been used extensively: Wells(1965), Elata *et al.* (1966), Spangler (1969), Patterson & Florez (1969), Nicodemo *et al.* (1969), Virk *et al.* (1970), Ernst (1966), Giles (1968), Goren & Norbury (1967), and Tomita (1970). However, the data obtained by pressure probes are contaminated by viscoelastic effects (Metzner & Astarita 1967 and Smith *et al.* 1967). Pitot tube measurements in polymer solutions can contain large and varying errors because of the non-linear stress/strain relationship and the influence of the normal stresses in polymer solutions. Variables such as probe size, shear rate, polymer type, and concentration all affect the measurements. The data of Smith *et al.* (1967) for two different sized pitot tubes in identical polymer solution flows are shown in Figure 10. Note the large disagreement between the two probes. Corrections are not usually attempted because of the complexity and

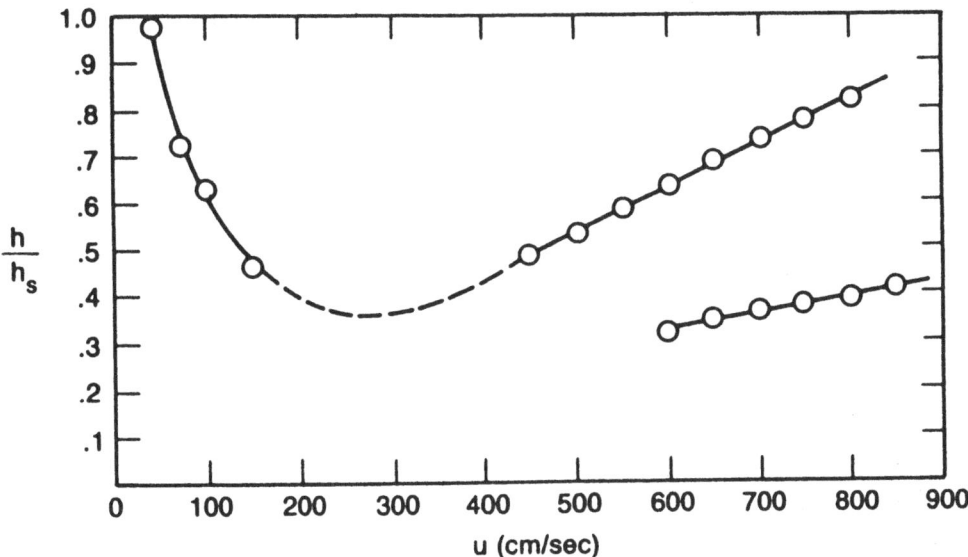

FIGURE 11. Heat transfer coefficient ratio as a function of velocity for a .005 cm cylindrical hot-film probe in a polymer pipe flow. The heat transfer coefficient of the probe in the solvent is h_s. From the data of Smith *et al.* (1967).

large number of unknowns involved. If pitot tubes are used in a low-speed dilute solution flow far from the wall, the velocity information obtained can be considered reliable. However, the lack of frequency response and the inability of these pressure probes to provide accurate information near the wall limit their utility in turbulence research.

Hot-film and hot-wire sensors have also been utilized for turbulence measurements in non-Newtonian flows: Virk et al. (1967), Fabula (1966) and Patterson & Florez (1969). These probes depend upon the heat transfer characteristics of the medium, and as such, are difficult to calibrate in dilute polymer solutions. Metzner & Astarita (1967) and Friehe & Schwartz (1969) have found calibrations of heated element sensors difficult to obtain and particularly sensitive to drift in the calibration due to contaminants. Figure 11, from the data of Smith et al., shows the variation in heat transfer of a hot-film probe in polymer solutions. Their data indicates that heated element probes can only be relied upon to give accurate measurements in low-speed flows. They show that the Reynolds number, based on the probe sensor diameter, at which the first drop in the heat transfer occurs is 50, which in Newtonian flows corresponds to the formation of the Karman vortex street. Further, Smith et al. indicate that an unstable region exists where the heat transfer can change by a factor of three. Once again, much care must be taken in the interpretation of heated element probe data.

Some novel measurement techniques have been employed with varying degrees of success. Wells et al. (1968) utilized a piezo-electric crystal pressure probe to measure turbulence intensity and frequency spectra. Seyer & Metzner (1969) have studied turbulence intensities in a pipe using bubble tracing. Fortuna & Hanratty (1972) used an electro-chemical technique to measure the gradient of fluctuating velocity at the wall.

The advent of the laser Doppler velocimeter (LDV) has made reliable turbulence measurements in a variety of non-Newtonian flow fields possible. The LDV has quickly become the instrument of choice for probing non-Newtonian flows. LDV measurements of mean and fluctuating velocities in a turbulent non-Newtonian flow are reliable because the probe is non-invasive and is not affected by the viscoelastic properties of the fluid. LDV measurements of streamwise velocity profiles and turbulence intensities of non-Newtonian flow in a pipe have been conducted by Chung & Graebel (1972), Goldstein et al. (1969), Mizushina & Usui (1977), McComb & Rabie (1982). Non-Newtonian turbulence in a duct has been studied with an LDV by Kumor & Sylvester (1973), Logan (1972), Rudd (1972), Reischman & Tiederman (1975). Streamwise velocity profiles show an essentially unchanged (from Newtonian) viscous sub-layer and the existence of a logarithmic region. The most significant change in the mean profiles takes place in the buffer layer region as demonstrated in Figure 12, where the quantities have been non-dimensionalized with the wall variables consisting of the friction velocity u_τ and the kinematic viscosity ν. In all studies the velocity profiles in the non-Newtonian flows show enlarged buffer layers. Additionally, measurements of the root-mean-square (RMS) profile of

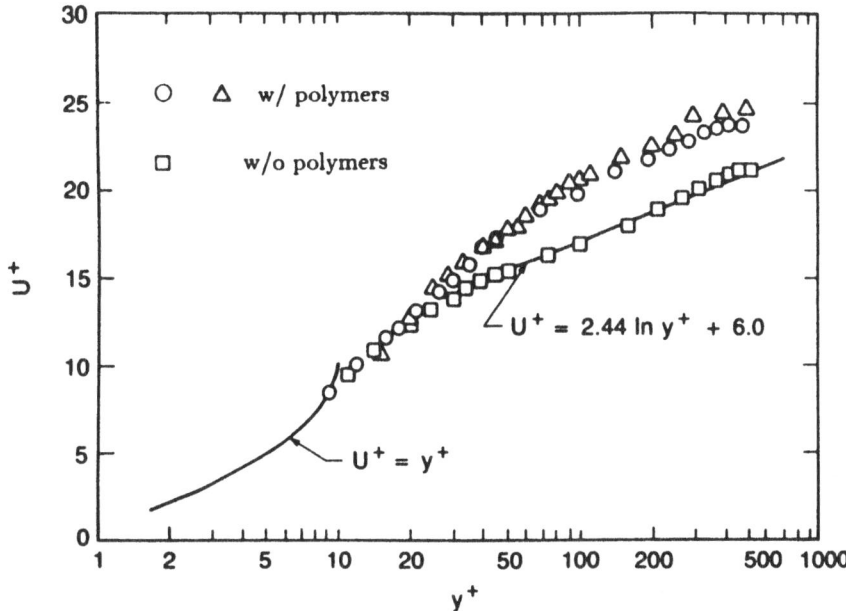

FIGURE 12. Mean streamwise velocity profile scaled with inner variables in a channel flow with polymers. From the data of Luchik & Tiederman (1988).

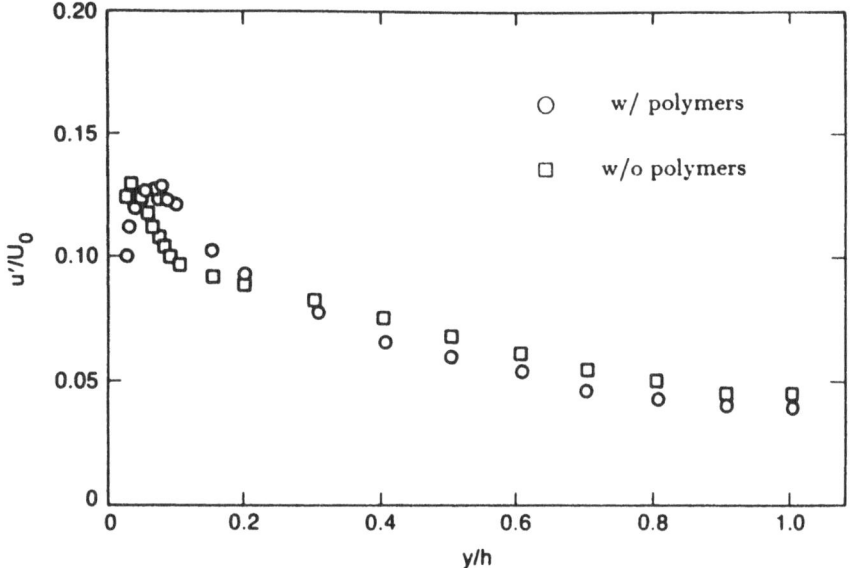

FIGURE 13. RMS streamwise velocity fluctuation profile scaled with outer variables for a channel flow with polymers. From the data of Luchik & Tiederman (1988).

the streamwise velocity fluctuations exhibit an increase in the peak value, as demonstrated in the work of Luchik & Tiederman (1988) shown in Figure 13. Here the velocity and normal coordinates have been non-dimensionalized with the outer variables, centerline velocity U_0 and channel half-width h, to allow direct comparison of the absolute magnitudes. This peak is broader in extent and located farther from the wall in drag reducing flows than in solvent flows.

Figure 14 shows recent measurements of the fluctuating normal velocity component made by Luchik & Tiederman (1988). This data and the data of Willmarth *et al.* (1987) confirm the earlier findings of Logan (1972), showing a suppression of the RMS value of the normal component of velocity. The data of Luchik & Tiederman and Willmarth *et al.* have some unresolved differences which may be facility-related. However, the trend of the Reynolds stress is consistent in both studies. The peak in Reynolds stress in non-Newtonian flows is lower and broader than in Newtonian flows as demonstrated in Figure 15.

Flow visualization techniques have been used to detect changes in coherent features in the turbulence due to polymers. Wells & Spangler (1967), Wu & Tulin (1972), and McComb & Rabie (1982) demonstrate that polymers need to be contained in the wall region to be effective drag reducers. Therefore, much effort has been devoted to discovering the changes to wall layer structures. Eckelmann *et al.* (1972), Donohue *et al.* (1972), Achia & Thompson (1977) and Oldaker & Tiederman (1977) report that the addition of polymers increases the dimensionless spanwise spacing of low-speed streaks. Oldaker & Tiederman make the additional observation that the non-Newtonian streaks are much longer than their Newtonian counterparts.

Tiederman, Luchik & Bogard (1985) used a water channel flow with a polyacrylamide solution injected at the wall and examined the flow in close proximity to the injector. They found, through flow visualization, that polymers contained solely within the viscous sublayer do not alter the streak spacing or bursting rates from values for Newtonian flows and exhibit no drag reduction. However, with the polymers mixed in the buffer layer, they find the spanwise spacing of the streaks increases and the bursting rate decreases. The decrease in the bursting rate is larger than one would expect from the change in streak spacing.

In their most recent work using a two-component LDV, Luchik & Tiederman (1988), find that for a well mixed aqueous solution of 2 *ppm* polyacrylamide, there is a decrease in the bursting rate. However, in contrast to their earlier results, the decrease in bursting is in direct proportion to the increase in the average streak spacing. At first this might seem to be a disagreement with earlier work but if, as they point out, one considers that the early measurements were made in a higher polymer concentration (20-50 *ppm*) wall layer that had not developed fully (i.e., near the injection point) it raises the possibility that there might be some damping of the large scale structures in this type of flow that might not be present in

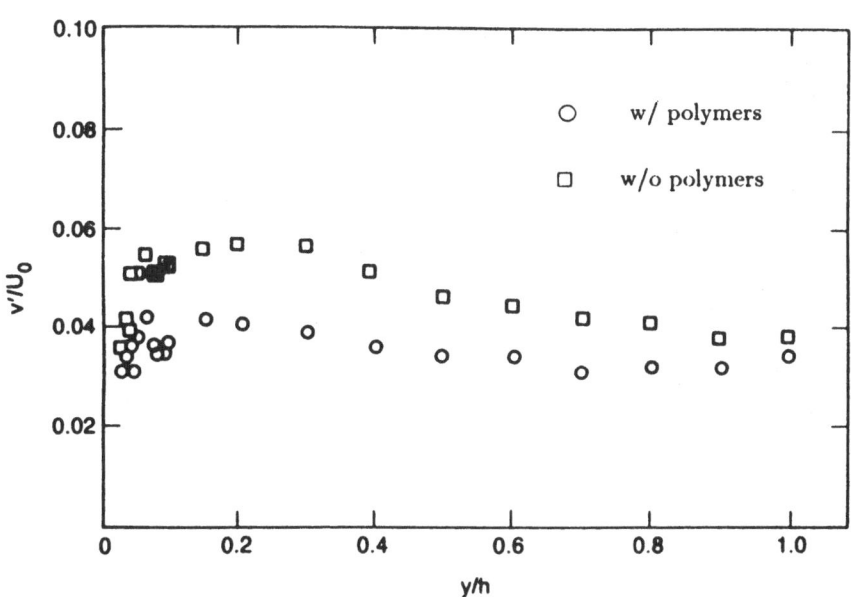

FIGURE 14. RMS normal velocity profile scaled with outer variables for a channel flow with polymers. From the data of Luchik & Tiederman (1988).

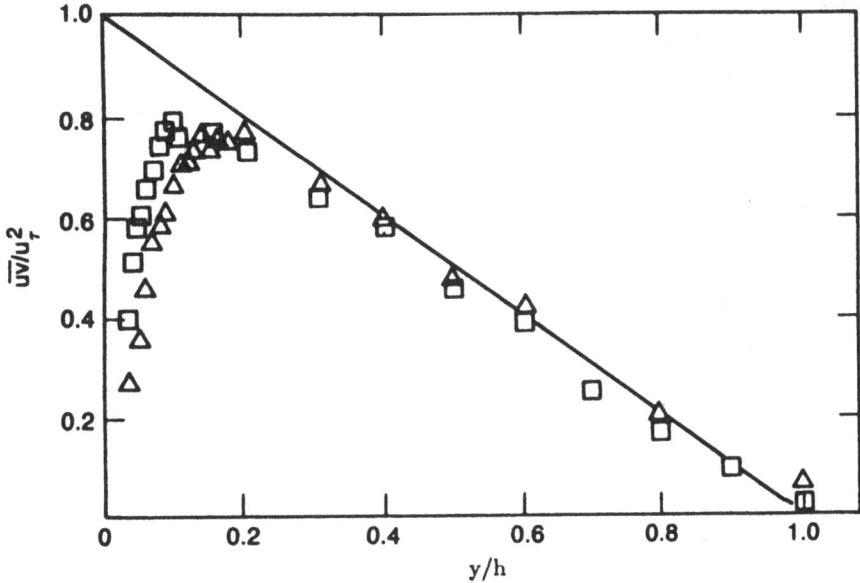

FIGURE 15. Reynolds stress profile scaled with inner variables for a channel flow with polymers. From the data of Luchik & Tiederman (1988).

their current experiments. Going a step further by applying conditional sampling techniques to the data for the well mixed fully-developed flow, Luchik & Tiederman have shown that lower threshold Reynolds stress producing motions are damped while the higher threshold motions are not.

In review, we have shown that quantitative information about the velocity field of non-Newtonian flows can be obtained reliably with the advent of the LDV. Further, we have shown that flow visualization can and has provided important insight into polymer/turbulence interaction. Examination of turbulent velocity data with an emphasis on turbulent structure (Luchik & Tiederman) is clearly a very promising research direction.

3.2. Wall Pressure Measurements

Wall pressure structure is a topic which has received relatively little attention by the non-Newtonian turbulence community. Considerable interest in determining the nature of the wall pressure field in turbulent shear flows exists for several reasons. First, wall pressure fluctuations induce unwanted structural vibrations. These vibrations may, in turn, radiate acoustic energy to the far field. Also, pressure fluctuations have been thought of as a "footprint" of the

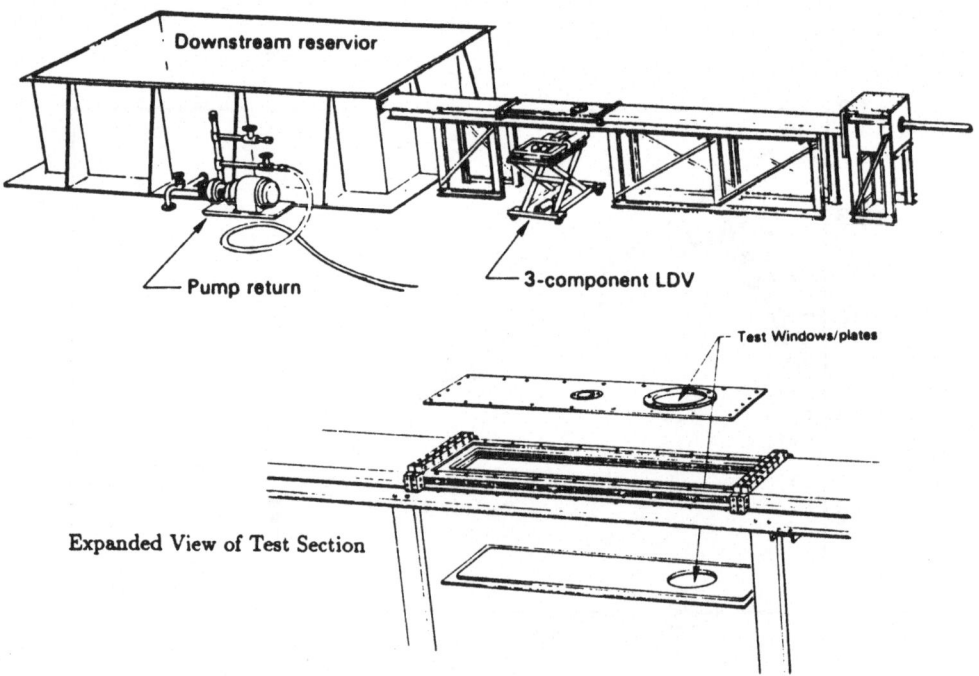

FIGURE 16. 18:1 aspect ratio blowdown channel flow facility.

turbulence. The wall pressure field may contain considerable information about coherent events (such as bursting) occurring in the energetic regions of the turbulent boundary layer. For these reasons, it is of interest to study not only the wall pressure fluctuations but also the sources of these fluctuations and means that can be employed to modify the turbulent pressure energy spectrum. Experiments by Greshilov *et al.* (1975) have shown clear evidence that dilute polymer solutions can significantly reduce the high frequency content of the wall pressure field. In contrast to the experiments of Greshilov *et al.*, in which a low aspect ratio (3.5:1) channel was used, measurements presented here were performed in a high aspect ratio (18:1) channel containing a fully-developed turbulent flow.

The wall pressure measurements were made in the 457 mm × 25 mm (18:1) channel. A drawing of the test section is shown in Figure 16. The measurement location was on the centerline of the channel and 4.1 m downstream from the beginning of the developing turbulent boundary layer. The ratio of the downstream distance of the measurement point to the half-height of the channel, x/h, is 320. This value of x/h is within the range for which fully-developed turbulent channel flow can be expected, as demonstrated by Hussain & Reynolds (1975). The flow channel is a blow-down type facility and can attain bulk flow speeds of up to 2 m/sec.

A pressure transducer array of 4 probes (ENDEVCO Model 8514-10) was flush mounted along the centerline of the channel and aligned with the flow direction. The nominal sensitivity of the probes is approximately -227 dB re 1.0 $\frac{V}{\mu Pa}$ and the frequency response is flat out to 140 Khz. The active area of the transducer is 0.5 mm (Galib & Zandina 1984), giving a channel half-height to sensor diameter ratio, h/d, of 25:1 and a viscous scale, $d^+ = \frac{du_\tau}{\nu}$, in the range 20-40. The signals were low-pass filtered to prevent aliasing, then digitized and stored on magnetic tape for post-processing. An FFT algorithm, utilizing 2048 time-domain data points, was used to compute the power spectral density (PSD), coherence, and phase between each pair of transducers within the array. Eight hundred realizations of the PSD were averaged to give a statistical error of approximately 3.5 %.

Mean velocity profiles for the channel flow, obtained from LDV measurements, with and without polymers are shown in Figure 17. The Reynolds number, based on mass average velocity and channel half-width, was 20,000. Note the presence of an enlarged buffer layer and a shift in the logarithmic region. Figure 18 shows the spectral density of the wall pressure fluctuations scaled on the outer variables, ρ (density) and $\bar{U}$ (average velocity in the channel), versus the nondimensional frequency $\frac{\omega h}{\bar{U}}$, where ω is the radian frequency and h is the channel half width. The effect of a dilute polymer solution (a concentration of 50 ppm Union Carbide WSR-301 PEO in water) is clearly to reduce the fluctuation energy significantly for frequencies above $\frac{\omega h}{U} = 10$. The noise represented by the large peaks in the spectrum below $\frac{\omega h}{U} = 1$ is facility related. In Figure 19, the difference between these two spectra is presented. Note that this substantial reduction in fluctuation energy is qualitatively consistent with the

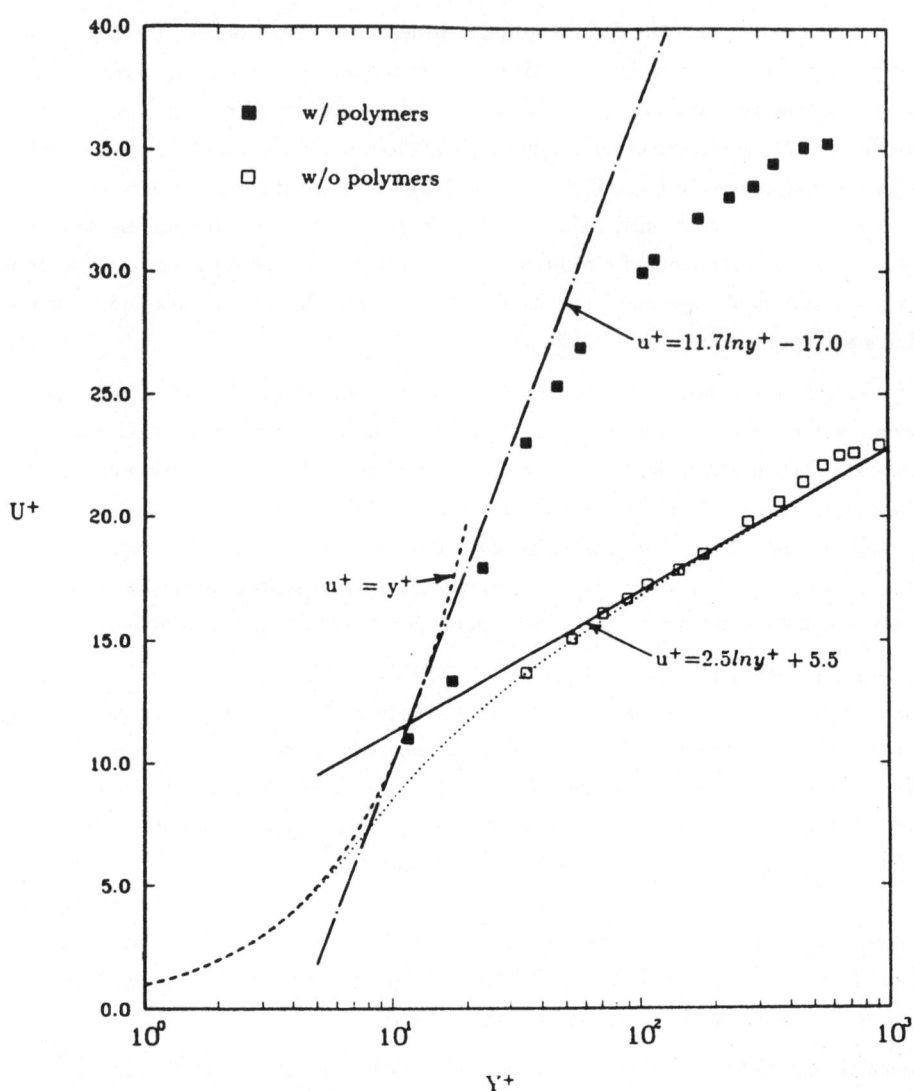

FIGURE 17. Mean streamwise velocity profile scaled with inner variables for a 50 ppm PEO solution channel flow with Re=20,000.

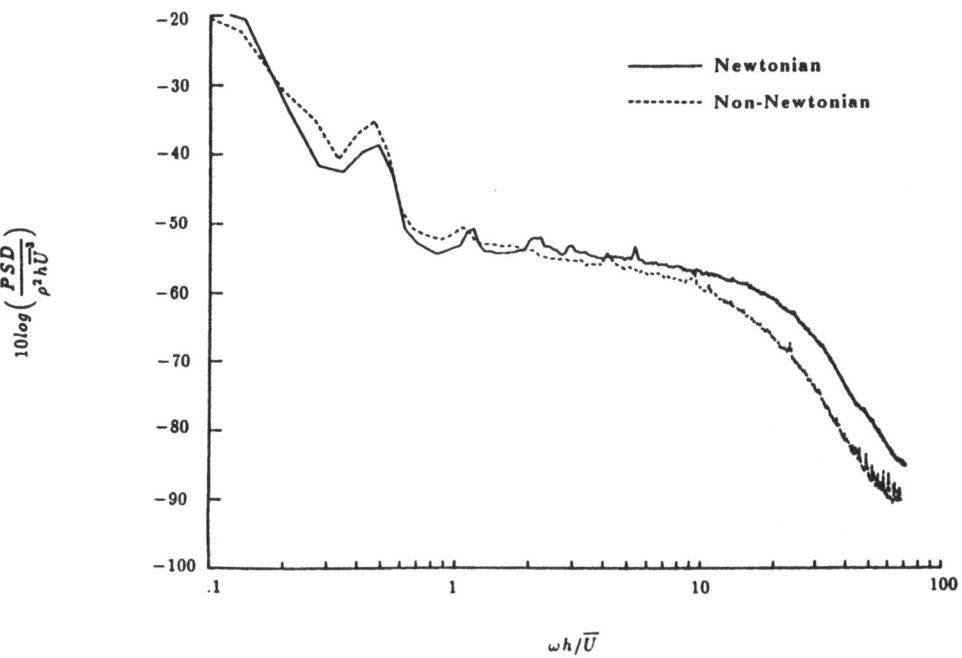

FIGURE 18. Power spectral density vs. frequency, scaled with outer variables.

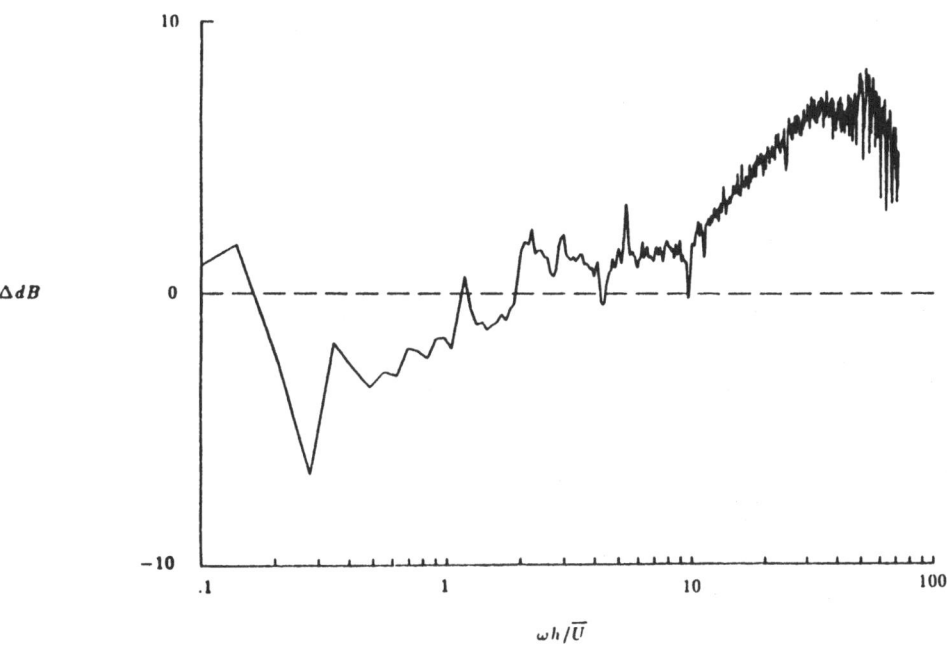

FIGURE 19. Power spectral density deficit (PSDD) vs. frequency, outer variable scaling.

results of Greshilov *et al.*. Quantitatively, however, their experiments (at the same polymer concentration) indicate fluctuation reductions of 15 to 20 *db* while we observe only a maximum reduction of 6 *db*. We suspect that significant differences in the aspect ratios of the channels and the Reynolds numbers of the two experiments (20,000 versus 40,000) contributed to this difference, but the cause for the discrepancy is not understood at the present time.

The coherence of the pressure fluctuations has also been measured by performing a narrow band correlation between signals from two transducers separated by 2.54 *mm* (center to center) in the streamwise direction as seen in Figure 20. The effect of the polymer solution is to increase the streamwise coherence dramatically over the entire range of frequencies. If we assume $\frac{U}{\omega}$ is proportional to a characteristic streamwise wavelength, then one interpretation of this result is that the polymer increases the "lifetime" of a characteristic turbulent eddy. That is, a characteristic structure of wavelength $\frac{U}{\omega}$ maintains its integrity for a greater distance as the structure convects downstream. This result is qualitatively consistent with the flow visualization studies of Oldaker & Tiederman (1977) in which a streamwise elongation of the low-speed streaks in the wall region was noted. The results of Greshilov *et al.*, however, appear to be inconsistent with our results. They actually observe a slight decrease in streamwise coherence. Again, we cannot explain such differences, but the lack of proper development of coherent structures near the wall in their low aspect ratio channel may be significant.

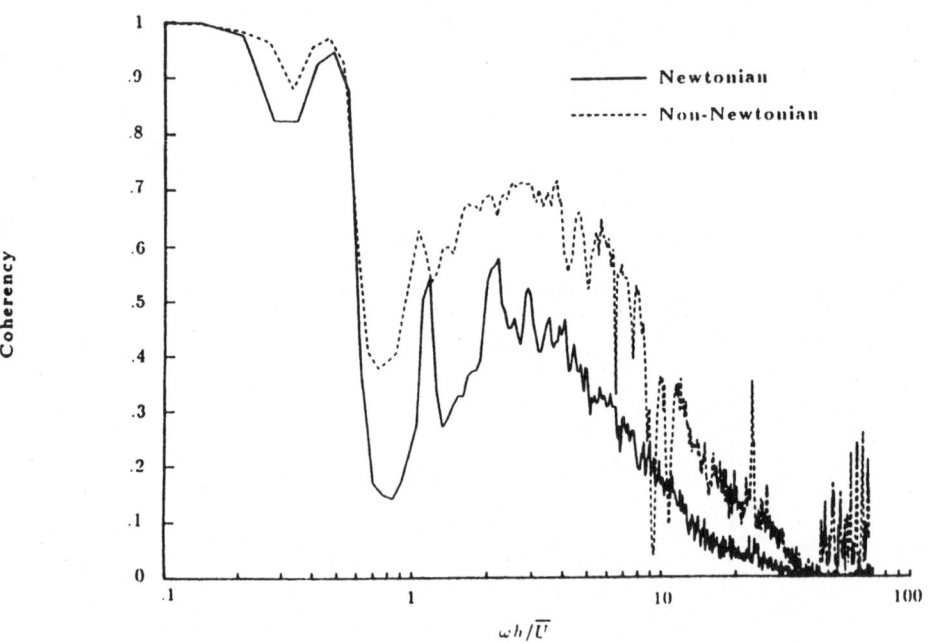

FIGURE 20. Coherency vs. frequency, outer variable scaling.

These preliminary results have indicated the existence of some inconsistences with previous investigations on the effect of polymers on wall pressure fluctuation statistics. However, the present results are in qualitative agreement with flow visualization studies of the flow structures in the wall region of turbulent boundary layers. Simultaneous flow visualization and two-point measurements of the velocity are required to determine the sources of the wall pressure field.

4. Concluding Remarks

We have reviewed the many techniques used to study the flow of drag-reducing non-Newtonian fluids in turbulence. We have evaluated the techniques and made some recommendations. Attempts to measure the properties of viscoelastic fluids that are relevant to drag reduction in turbulent flows have been largely unsuccessful. In particular, little progress has been made towards developing tests of the applicability of constitutive relations to turbulent flows. The most crucial property is the elongational viscosity. The high ratio of this property to the shear viscosity of dilute polymer solutions appears to produce drag reduction.

Currently no predictive model exists that can reproduce experimental turbulence measurements, although there does exist some capability to predict the amount of drag reduction attainable in a given geometry. In order to improve predictive capability a research approach that involves a joint experimental/simulation effort for a mixed flow (shear and elongational) is recommended. Finally, it is crucial in any experiment involving long chain polymers and turbulence to consider the possible effects of degradation, molecular entanglements and the initial condition of the polymer (pre-sheared).

For turbulence measurements in dilute polymer flows, it was shown that the LDV and flow visualization are effective and reliable experimental tools. Many observations on the polymer/turbulence synergy have been reported, but much information remains to be discovered. Research aimed at understanding the interaction between polymers and the structure of turbulence along with measurements of wall pressure will be valuable in identifying the mechanisms of polymer drag reduction. We should also emphasize that along with the new experimental efforts, it should be possible to simulate non-Newtonian turbulence. The ability to include a non-Newtonian constitutive equation in a direct numerical simulation of turbulence is currently under development at NRL and other institutions. It is our opinion that although this work is in its infancy, direct numerical simulation of turbulent non-Newtonian flows coupled with experiments holds great promise for opening new vistas in the investigation of drag reduction physics.

Acknowledgments

The authors gratefully acknowledge the help of Kimberly Gardiner in preparation of the original draft of this work. JVL thanks the Office of Naval Technology for their assistance through a postdoctoral fellowship and JDS similarly thanks the National Research Council for their support through an NRC-NRL Research Associateship. This research was sponsored by the Naval Research Laboratory and by the Office of Naval Research Code 1132 through Dr. Michael Reischman.

References

Achia, B.U. & Thompson, D.W. 1977 Structure of the turbulent boundary layer in drag-reducing pipe flow. *J. Fluid Mech.* **81**, 439.

Bagley, E.B. 1957 End correction in the capillary flow of polyethylene. *J. Appl. Phys.* **28**, 624.

Baid, K.M. & Metzner, A.B. 1977 Rheological properties of dilute polymer solutions determined in extensional and shearing experiments. *Trans. Soc. Rheol.* **21**, 237.

Balakirshnan, C. & Gordon, R.J. 1971 New viscoelastic phenomenon and turbulent drag reduction. *Nature* **231**, 177.

Beard, D.W., Davies, M.H. & Walters, K. 1966 The stability of elastico-viscous flow between rotating cylinders. Part 3. Overstability in viscous and Maxwell fluids. *J. Fluid Mech.* **24**, 321.

Becraft, M.L. & Metzner, A.B. 1988 Bourdon tube effects in the fiber spinning apparatus. *J. Rheol.* **32**, 243.

Belokon, V.S., Kalashnikov, V.N, Kudin, A.M. & Vlasov, S.A. 1973 Rheological properties of polymers reducing drag friction. *Prog. Heat Mass Trans.* **5**, 233.

Berman, N.S. 1978 Drag reduction by polymers. *Ann. Rev. Fluid Mech.* **10**, 47.

Bird, R.B., Armstrong, R.C., & Hassager, O. 1987a *Dynamics of Polymeric Liquids, 2nd Ed., Vol.1: Fluid Mechanics.* Wiley, New York. 647 p.

Bird, R.B., Curtiss, C.F., Armstrong, R.C., & Hassager, O. 1987b *Dynamics of Polymeric Liquids, 2nd Ed., Vol.2: Kinetic Theory.* Wiley, New York. 437 p.

Brodnyan, J.G., Gaskins, F.H. & Philippoff, W. 1957 On normal stresses, flow curves, flow birefringence, and normal stresses of polyisobutylene solutions. Part II. Experimental. *Trans. Soc. Rheol.* **1**, 109.

Chakraborty, A.K. & Metzner, A.B. 1986 Sink flows of viscoelastic fluids. *J. Rheol.* **30**, 29.

Chang, C. & Schowalter, W.R. 1974 Flow near an oscillating cylinder in dilute viscoelastic fluid. *Nature* **252** 686. & Errata. 1975 *Nature* **253**, 572.

Chang, C. & Schowalter, W.R. 1979 Secondary flow in the neighborhood of a cylinder oscillating in a viscoelastic fluid. *J. N.-N. Fluid Mech.* **6**, 47.

Chiou, C.S. & Gordon, R.J. 1976 Vortex inhibition: Velocity profile measurements. *AIChE J.* **22**, 947.

Chung, J.S. & Graebel, W.P. 1972 Laser anemometer measurements of turbulence in non-Newtonian pipe flows. *Phys. Fluids* **15**, 546.

Crowley, D.G., Frank, F.C., Mackley, M.R & Stephenson, R.G. 1976 Localized flow birefringence of polyethylene oxide solutions in a four roll mill. *J. Poly. Sci.: Poly. Phys. Ed.* **14**, 1111.

Dandridge, A., Meeten, G.H., Layec-Raphalen, M.N. & Wolff, C. 1979 Flow birefringence of dilute solutions of polyethyleneoxide of high molecular weight at high rates. *Rheol. Acta* **18**, 275.

Darby, R. & Chang, H.D. 1984 Generalized correlation for friction loss in drag reducing polymer solutions. *AIChE J.* **30**, 274.

Davies, J.M., Hutton, J.F., & Walters, K. 1977/78 A critical re-appraisal of the jet-thrust technique for normal stresses, with particular reference to axial velocity and stress rearrangement at the exit plane. *J. N.-N. Fluid Mech.* **3**, 141-160.

DeGennes, P.G. 1974 Coil-stretch transition of dilute flexible polymers under ultrahigh velocity gradients. *J. Chem. Phys.* **60**, 5030.

Denn, M.M. & Roisman, J.J. 1969 Rotational stability and measurement of normal stress functions in dilute polymer solutions. *AIChE J.* **15**, 454.

Denn, M.M., Sun, Z.-S. & Rushton, B.D. 1971 Torque reduction in finite amplitude secondary flows of dilute polymer solutions. *Trans. Soc. Rheol.* **15**, 415.

Deutsch, S. & Phillips, W.M. 1977 The use of the Taylor-Couette stability problem to validate a constitutive equation for blood. *Biorheol.* **14**, 253.

Donohue, G.L., Tiederman, W.G. & Reischman, M.M. 1972 Flow visualization of the near-wall region in a drag reducing channel flow. *J. Fluid Mech.* **56**, 559.

Dunlap, P.N. & Leal, L.G. 1987 Dilute polystyrene solutions in extensional flows: birefringence and flow modification. *J. N.-N. Fluid Mech.* **23**, 5.

Eckelmann, L.D., Fortuna, G., & Hanratty, T.J. 1972 Drag reduction and the wavelength of flow-oriented wall eddies. *Nature* **236**, 94.

Elata, C., Lehrer, J. & Kahanovitz, A. 1966 Turbulent shear flow of polymer solutions. *Israel J. Tech.* **4**, 87.

Ernst, W.D. 1966 Investigation of the turbulent shear flow of dilute aqueous cmc solutions. *AIChE J.* **12**, 581.

Fabula, A.G. 1966 An experimental study of grid turbulence in dilute high-polymer solutions. *Proceedings of the 6th Symposium on Navy Hydrodynamics*, Washington ONR-ACR-136. 39.

Farrell, C.J., Keller, A., Miles, M.J. & Pope, D.P. 1980 Conformational relaxation time in polymer solutions by elongational flow experiments: 1. Determination of extensional relaxation time and its molecular weight dependence. *Polymer* **21**, 1292.

Fortuna, G. & Hanratty, T.J. 1972 The influence of drag-reducing polymers on turbulence in the viscous sublayer. *J. Fluid Mech.* **53**, 575.

Friehe, C.A. & Schwartz, W.H. 1969 The use of pitot static tubes and hot-film anemometers in dilute polymer solutions. in: C.S. Wells (editor), *Viscous Drag Reduction*, Plenum Press, New York., 281.

Fruman, D.H. & Barigah, M. 1982 Rheological interpretation of pressure anomalies of aqueous dilute polymer solutions (ADPS) in orifice flow. *Rheol. Acta* **21**, 556.

Fuller, G.G. & Leal, L.G. 1980 Flow birefringence of dilute solutions in two-dimensional flows. *Rheol. Acta* **19**, 580.

Galib, T.A. & Zandina A. 1984 Turbulent pressure fluctuations with conventional piezoelectric and miniature piezoresistive transducers. *JASA Suppl.* **1**, 76.

Georgelos, P.N. & Torkelson, J.M. 1988 The role of solution structure in apparent thickening behavior of dilute PEO/water systems. *J. N.-N. Fluid Mech.* **27**, 191-204.

Ginn, R.F. & Denn, M.M. 1969 Rotational stability in viscoelastic liquids: Theory. *AIChE J.* **15**, 450.

Giles, W.B. 1968 Similarity laws of friction-reduced flows. *J. Hydronautics* **2**, 34.

Goldstein, R.J., Adrian, R.J. & Kreid, D.K. 1969 Turbulent and transition pipe flow of dilute aqueous polymer solutions. *Ind. Eng. Chem. Fund.* **8**, 498.

Gollub, J.P. & Swinney, M.L. 1975 Onset of turbulence in a rotating fluid. *Phys. Rev. Lett.* **35**, 927.

Gordon, R.J. & Balakirshnan, C. 1972 Vortex inhibition: a new viscoelastic effect with importance in drag reduction and polymer characterization. *J. Appl. Poly. Sci.* **16**, 1629.

Goren, Y., & Norbury, J.F. 1967 Turbulent flow of dilute aqueous polymer solutions. *Trans. ASME: J. Basic Engg.* **89**, 814.

Greshilov, E.M., Evutshenko, A.V. & Lyamshev, L.M. 1975 Hydrodynamic noise and the Toms effect. *Sov. Phys. Acoust.* **21**, 247.

Hasegawa, T. & Iwaida, T. 1984 Experiments on elongational flow of dilute polymer solutions. Part II: Velocity field for the flow through small aperatures. *J. N.-N. Fluid Mech.* **15**,279-307.

Hasegawa, T., Fukutomi, K., & Narumi, T. 1988 Experimental estimation of elongational stresses of dilute polymer solutions and a related examination of some constitutive equations. *J. N.-N. Fluid Mech.* **27**, 133-151.

Hikmet, R.A.M., Narh, K.A., Barham, P.J. & Keller, A. 1985 Adsorption- entanglement layers in flowing high-molecular weight polymer solutions. *Prog. Coll. Poly. Sci.* **71**, 32.

Hoyt, J.W. & Fabula, A.G. 1964 The effect of additives on fluid friction. *5th Symposium on Naval Hydrodynamics*, Bergen, Norway, ONR ACR-112, 947.

Hoyt, J.W. 1971 Drag-reduction effectiveness of polymer solutions in the turbulent-flow rheometer: a catalog. *Poly. Lett.* **9**, 851.

Hoyt, J.W. 1972 The effects of additives in fluid friction. *Trans. ASME: J. Basic Engg.* **94**, 258.

Hoyt, J.W. 1980 Effect of ferric ions on drag reduction effectiveness of polyacrylamide. *Poly. Sci. Engg.* **20** , 493.

Hussain, A.K.M.F. & Reynolds, W.C. 1975 Measurements in fully developed turbulent channel flow. *Trans. ASME: J. Fluids Engg.* **97**, 568.

James, D.F., McLean, B.D. & Saringer, J.H. 1987 Presheared extensional flow of dilute polymer solutions. *J. Rheol.* **31**, 453.

James, D.F. & Saringer, J.H. 1982 Flow of dilute polymer solutions through converging channels. *J. N.-N. Fluid Mech.* **11**, 317.

Jones, W.M., Davies, D.M., & Thomas, M.C. 1973 Taylor vortices and the evaluation of material constants: a critical assessment. *J. Fluid Mech.* **60**, 19.

Kowalik, R.M., Duvdevani, I., Peiffer, D.G., Lundberg, R.D., Kitano, K. & Schulz, D.N. 1987 Enhanced drag reduction via interpolymer associations. *J. N.-N. Fluid Mech.* **24**, 1.

Kumor, S.M. & Sylvester, N.D. 1973 Effects of a drag-reducing polymer on the turbulent boundary layer. Drag reduction. *AIChE Symp. Ser.* **69**, 1.

Larson, R. 1988 *Constitutive Equations for Polymer Melts and Solutions.* Butterworth, Stoneham, Massachusetts. 304 p.

Lawler, J.V. 1986 Laser-Doppler velocimetry of viscoelastic flow between eccentric rotating cylinders. Ph.D. Thesis. MIT, Cambridge, Massachusetts.

Lawler, J.V., Handler, R.A. Hendricks, E.W. & Leighton, R.I. 1987 Transient normal stresses in turbulent flows. Paper F5. *59th Ann. Meet. Soc. Rheol.*, Atlanta, Georgia (in preparation).

Little, R.C., Hansen, R.J., Hunston, D.L., Kim, O., Patterson, R.L. & Ting, R.Y. 1975 The drag-reduction phenomenon. Observed characteristics, improved agents, and proposed mechanisms. *Ind. Eng. Chem. Fund.* **14**, 283.

Lodge, A.S. 1955 Variation of flow birefringence with stress. *Nature* **176**, 838.

Logan, S.E. 1972 Laser velocimeter measurement of Reynolds stress and turbulence in dilute polymer solutions. PhD. Thesis, California Institute of Technology.

Luchik, T.S. & Tiederman, W. G. 1988 Turbulent structure in low-concentration channel flows. *J. Fluid Mech.* (in press).

Lumley, J.L. 1969 Drag reduction by additives. *Ann. Rev. Fluid Mech.* **1**, 367.

Lumley, J.L. 1973 Drag reduction in turbulent flow by polymer additives. *J. Polym. Sci: Macromol. Rev.* **7**, 263.

Matthys, E.F. 1987 Laser-induced photochromic flow visualization: measurement of velocities and deformation rates. Paper E14. *59th Annual Meeting of the Society of Rheology,* Atlanta, Georgia, October 1987.

McComb, W.D. & Rabie, L.H. 1982 Local drag reduction due to injection of polymer solutions into turbulent flow in a pipe. Pt I: Dependence on local polymer concentration; Pt II: Laser-Doppler measurements of turbulent structure. *AIChE J.* **28**, 547.

Merrill, E.W. & Horn, A.F. 1984 Scission of macromolecules in dilute solution: Extensional and turbulent flows. *Poly. Comm.* **25**, 144.

Metzner, A.B., & Astarita, G. 1967 External flow of viscoelastic materials: fluid property restrictions on the use of velocity-sensitive probes. *AIChE J.* **13**, 550.

Mizushina, T. & Usui, H. 1977 Reduction of eddy diffusion for momentum and heat in viscoelastic fluid flow in circular tube. *Phys. Fluids* **20**, S100.

Nicodemo, L., Acierno, D. & Astarita, G. 1969 Velocity profiles in turbulent pipe flow of drag-reducing liquids. *Chem. Engg. Sci.* **24**, 1241.

Odell, J.A. & Keller, A. 1986 Flow-induced chain fracture of isolated linear macromolecules in solution. *J. Poly. Sci.: Poly. Phys. Ed.*, **24** , 1889.

Oldaker, D.K. & Tiederman, W.G. 1977 Spatial structure of the viscous sublayer in drag-reducing channel flows. *Phys. Fluids* **20**, S133.

Patterson, G.K. & Florez, G.L. 1969 Velocity profiles during drag reduction. in: C.S. Wells (editor), *Viscous Drag Reduction*, Plenum Press, New York., 223.

Phan-Thien, N., Manero, O., & Leal, L.G. 1984 A study of conformation-dependent friction in a dumbbell model for dilute solutions. *Rheol. Acta* **23**, 151.

Philippoff, W. 1956 Flow-birefringence and stress. *Nature* **178** , 811.

Reischman, M.M. & Tiederman, W.G. 1975 Laser-Doppler anemometer measurements in drag-reducing channel flows. *J. Fluid Mech.* **70**, 369.

Rouse, P.E. 1953 A theory of the linear viscoelastic properties of dilute solutions of coiling polymers. *J. Chem. Phys.* **21**, 1272.

Rudd, M.J. 1972 Velocity measurements made with a laser Dopplermeter on the turbulent pipe flow of a dilute polymer solution. *J. Fluid Mech.* **51**, 673.

Ryskin, G. 1987a Turbulent drag reduction by polymers: A quantitative theory. *Phys. Rev. Lett.* **59**, 2059.

Ryskin, G. 1987b Calculation of the effect of polymer additive in a converging flow. *J. Fluid Mech.* **178**, 423.

Seyer, F.A. & Metzner, A.B. 1969 Turbulence phenomena in drag reducing solutions. *AIChE J.* **15**, 427.

Smith, K.A., Merrill, E.W., Mickley, H.S. & Virk, P.S. 1967 Anamalous pitot tube and hot film measurements in dilute polymer solutions. *Chem. Engg. Sci.* **22**, 619.

Spangler, J.G. 1969 Studies of viscous drag reduction with polymers including turbulence measurements and roughness effects. in: C.S. Wells (editor), *Viscous Drag Reduction*, Plenum Press, New York., 131.

Sun, Z. & Denn, M.M. 1972 Stability of rotational Couette flow of polymer solutions. *AIChE J.* **18**, 1010.

Tanner, R.I. 1985 *Engineering Rheology*. Oxford University, New York. 451 p.

Taylor, G.I. 1923 Stability of a viscous liquid contained between two rotating cylinders. *Phil. Trans. Roy. Soc. London A* **223**, 289.

Tiederman, W.G., Luchik, T.S. & Bogard, D.G. 1985 Wall-layer structure and drag reduction. *J. Fluid Mech.* **156**, 419.

Ting, R.Y. & Hunston, D.L. 1977 Polymeric additives as flow regulators. *Ind. Eng. Chem. Prod. Res. Dev.* **16**, 129.

Tomita, Y. 1970 Pipe flows of dilute aqueous polymer solutions. *Bull. JSME* **13**, 926.

Toms, B.A. 1948 Some observations on the flow of linear polymer solutions through straight tubes at large Reynolds numbers. *Proc. 1st Intl. Congr. Rheol.* **2**, 135.

Usui, H. & Sano, Y. 1981 Elongational flow of dilute drag reducing fluids in a falling jet. *Phys. Fluids* **24**, 214.

Virk, P.S. 1975 Drag reduction fundamentals. *AIChE J.* **21**, 625.

Virk, P.S., Merrill, E.W., Mickley, H.S., Smith, K.A. & Mollo-Christensen, E.L. 1967 The Toms phenomenon: Turbulent pipe flow of dilute polymer solutions. *J. Fluid Mech.* **30**, 305.

Virk, P.S., Mickley, H.S. & Smith, K.A. 1970 The ultimate asymptote and mean flow structure in Toms phenomenon. *Trans. ASME: J. Appl. Mech.* **37**, 488.

Weinberger, C.B. & Goddard, J.D. 1974 Extensional flow behavior of polymer solutions and particle suspensions in a spinning motion. *Intl. J. Multiphase Flow,* **1**, 465.

Wells, C.S. 1965 Anomalous flow of non-Newtonian fluids. *AIAA J.* **3**, 1800.

Wells, C.S., Harkness, J. & Meyer, W.A. 1968 Turbulence measurements in pipe flow of a drag-reducing non-Newtonian flow. *AIAA J.* **6**, 250.

Wells, C.S. & Spangler, J.G. 1967 Injection of a drag-reducing fluid into turbulent pipe flow of Newtonian fluid. *Phys. Fluids* **10**, 1890.

Wiest, J.M. & Bird, R.B. 1988 On coil-stretch transitions in dilute polymer solutions. *University of Wisconsin-Madison, Rheology Research Center Rep.* 116.

Willmarth, W.W., Wei, T. & Lee, C.O. 1987 Laser anemometer measurements of Reynolds stress in a turbulent channel flow with drag reducing polymer additives. *Phys. Fluids* **30**, 933.

Wolff, C. 1980 On the real molecular weight of polyethylene oxide of high molecular weight in water. *Can. J. Chem. Eng.* **58**, 634-636.

Wu, J. & Tulin, M.P. 1972 Drag reduction by ejection of additive solutions into pure-water boundary layer. *Trans. ASME: J. Basic Engg.* **94**, 749.

Zimm, B.H. 1956 Dynamics of polymer molecules in dilute solution: viscoelasticity, flow birefringence and dielectric loss. *J. Chem. Phys.* **24**, 269.

NUCLEAR AEROSOL MEASUREMENT TECHNIQUES

Patrick F. Dunn
Department of Aerospace and Mechanical Engineering
University of Notre Dame
Notre Dame, Indiana   46556

and

Vincent J. Novick
Engineering Division

and

Barbara J. Schlenger
Reactor Analysis & Safety Division
Argonne National Laboratory
Argonne, Illinois   60439

Abstract

In this chapter the considerations and techniques used for aerosol measurement in nuclear reactor test facilities are presented. High pressure, temperature, levels of radioactivity and changing chemical composition during this type of experiment impose a number of constraints on the aerosol sampling system. The mechanisms of aerosol formation, transport and deposition within the system are described. The general equations of aerosol transport and deposition are given in detail and are applied for typical reactor aerosol sampling system conditions. The various techniques that could be employed for such systems also are reviewed. Finally, examples of three systems that have been used successfully in large-scale reactor accident experiments are discussed in detail.

Nomenclature

B   Boltzman's constant (1.38E-16 erg/K)

C   Cunningham slip correction factor = $1 + (\lambda/D_p)\ 2.514 + 0.80\exp\left(\dfrac{-0.55D_p}{\lambda}\right)$

C_p Gas specific heat at constant pressure

d_w Wire diameter

D   Diffusion coefficient = $BT\ \tau/m$

D_p Particle diameter

D_m Diameter of average mass

D_T Tube diameter

F   Fraction of aerosol lost or collection efficiency

g   Gravitational acceleration $(9.8\ \text{m/s}^2)$

h   STEP channel height

h'   STEP distance between wires

I   Light intensity

k   Coagulation coefficient

K_a Gas thermal conductivity

K_p Particle thermal conductivity

Kn   Knudsen number = $2\lambda/D_p$

L   Tube length, path length

m   Particle mass

N   Particle concentration

Pe   Peclet number = $Pr \cdot Re$

Pr   Prandtl number = $C_p\eta/K_a$

Q   Volumetric flow rate

Q_{ext} Extinction coefficient

R   Interception parameter = D_p/d_w

R_T Tube radius

R_b Radius of bend

Re   Reynolds number = $\rho V D_T/\eta$

Stk   Stokes number (for STEP wire) = $\rho D_p{}^2 CV/18\eta d_w$

STK   Stokes number (for pipe flow) = $4QC\rho D_p{}^2/9\pi\eta D_T{}^3$

T   Temperature

V   Velocity

V_a Bulk gas velocity in a tube

V'   Gas streamline velocity

V_f Terminal settling velocity

V_{th} Thermophoretic velocity

W   Nozzle diameter

δ   Skin depth of laminar sublayer

ΔR Radial displacement

∇T Temperature gradient

λ   Mean free path of gas molecules

φ   Angle of bend in radians

η   Gas absolute viscosity

ρ   Particle density

ρ_g Fluid or gas density

τ   Relaxation time $= mC/3\pi D_p \eta$

θ   Tube inclination angle in degrees

Introduction

This chapter concerns the methods used to characterize the size, number concentration, radioactivity and chemical composition of nuclear aerosols. From a fluid-mechanical viewpoint, these aerosols can be considered to be a two-phase flow of particles suspended in a flowing gas. In most nuclear aerosol experiments, however, the concentration of particles is not sufficiently high enough to alter the fluid-mechanical behavior of the gas. Yet, on the other hand, the properties of the gas, such as its mean free path, viscosity, and velocity, fundamentally affect many of the ongoing aerosol processes, such as diffusion, impaction and settling. Therefore, it is the fluid mechanical properties of the gas to a large extent that determine the manner in which these particles are to be collected and analyzed.

In the event of a severe accident in a nuclear reactor, a significant portion of the radioactive material available for release to the environment is postulated to be in aerosol form. It is necessary that the nature of these aerosols be evaluated in order for the nuclear community to develop and validate models appropriate for nuclear reactor safety, siting and population exposure analyses. This includes the experimental determination of the amount and timing of their release and their physical characteristics. The latter encompasses their particle size distribution, number concentration, mass concentration, shape, chemical composition and level of radioactivity. This information is essential in determining the transport of material out of the reactor plenum and the deposition and transport through the reactor piping and into the containment building. In order to correctly interpret the aerosol measurements, local thermal hydraulic conditions, including pressure, temperature, flow velocity and gas composition must be known. Data is obtained from experiments that range in size and complexity from bench-scale setups to large tests that are conducted in research reactors.

The focus of this chapter is on aerosol measurement considerations and techniques that are applicable to the large facilities where reactor accident conditions are simulated most closely. Although these considerations and techniques apply directly to the sampling of nuclear reactor accident aerosols, they also relate to the sampling of aerosols formed in other high-temperature and/or high-pressure environments, such as in high-temperature combustion, fires, explosions, etc. The unique features of the reactor aerosol sampling environment are described. The requirements and methodology involved in developing an appropriate aerosol measurement system are presented. This includes the aspects of both sampling and particle measurement. This chapter does not provide an extensive review of the many approaches that have been taken to sample nuclear aerosols. Rather, it provides examples of several successful systems that have been used.

The foremost consideration in developing an approach to sampling aerosols is first to identify the possible physicochemical nature of the aerosols and their environment. For the case of nuclear aerosols, the nature of the aerosol particles and the gas in which they are suspended depends to a large degree on the type of nuclear reactor being considered. Light water reactors (LWRs), which are cooled with water, are of particular interest, mainly because of their prevalence. Although much of what is contained in this chapter relates directly to aerosol measurements in simulated LWR accident environments, the information is generally applicable to aerosol measurements in other types of reactors.

The accident scenario leading to the formation of aerosols in LWRs can be summarized as follows. The accident is initiated when heat removal from the reactor core is substantially impaired by failure of one or more components in the reactor coolant system. The resulting temperature increase causes the uranium fuel and its cladding to melt. In addition, the cladding can be oxidized by the steam, yielding hydrogen. These processes allow for the release of radioactive fission products from the fuel pellets that are contained within the cladding. Portions of the fuel, cladding, fission products, control rods and structural material in both vapor and aerosol form are carried by the steam and hydrogen from the core to other parts of the reactor and into the containment building from which they may be released to the environment. If the accident progresses far enough, additional aerosol release will result from the interaction of the molten core and the concrete structures below it. However, this chapter deals primarily with aerosol measurements related to the first phase of the accident.

Conditions in experimental facilities where the accident environment is simulated presents the aerosol experimenter with formidible sampling and measurement problems. The pressures can range from 0.2 to 8 MPa, and may vary with time. Temperatures are also high, up to 2500 K, and vary with time and location. The steam-hydrogen gas mixture is chemically reactive and radioactive, primarily due to the fission products that are present in the mixture in both gaseous and aerosol form. The transient nature of the accident results in continuously changing aerosol characteristics. The presence of vapors that can condense onto surfaces or onto existing aerosols rather than directly forming aerosols further complicates the situation.

Ideally, the aerosol sampling and measurement systems should be designed to obtain sufficient information about the aerosols to yield a complete description of their composition, formation, growth, transport and deposition as a function of time and location within an experimental assembly. Such information then would be applied to the understanding of the broader problem of aerosol behavior within an actual reactor undergoing an accident.

Determination of the specific experimental approach to aerosol sampling and measurement depends upon several factors. These include the thermal hydraulic conditions (pressure, temperature and velocity) of the aerosol carrier gas, the anticipated aerosol characteristics such as size distribution, number concentration, chemical composition and level of radioactivity, and the space within and access to the test capsule for instrumentation. These factors primarily dictate the proportion of real-time versus post-test measurements to be made. Most often, the aerosol measurements are made post-test on the basis of collected samples. This is primarily because more information can be obtained from post-test measurements and because most real-time aerosol measurement systems are relatively complex and are not well suited to operation in severe environments. A majority of measurements are made on samples that are extracted from the flow using a sampling probe. Care must be taken to ensure that the sample is representative and that the aerosol size distribution and composition did not change in transit to the collection device.

In the following sections, specific information related to aerosol sampling and measurement techniques is presented. Examples of several methods are included. These topics are preceeded by a brief description of relevant aerosol mechanisms along with some basic definitions related to aerosol properties.

1.0 Aerosol Mechanisms

The first step in developing an effective approach to sampling and measurement is to identify the mechanisms by which the aerosols are formed, transported and deposited within the reactor containment system.

1.1 Aerosol Formation and Growth

Aerosols can form during a reactor accident by several mechanisms. These include homogeneous nucleation of a vapor, heterogeneous nucleation of vapor onto an existing aerosol (condensation), and fragmentation of liquids and solids. During an energetic core-disruptive accident, aerosols will form by all three of these mechanisms. For most scenarios, however, nucleation and condensation play the most significant role. The process of homogeneous nucleation will form nanometer-sized aerosol droplets. These aerosols will continue to grow by heterogeneous nucleation, yielding droplets in the submicrometer-to-micrometer diameter range. If there is a sufficient quantity of condensible vapor present such that large number concentrations of aerosols are formed ($\sim 10^8$ particles/cm^3) and if the aerosol transit time through the reactor system is sufficiently long (~ 1 s), further growth of the resident aerosols can occur by coagulation and agglomeration.

1.2 Aerosol Attenuation

After the aerosols have been formed, there are a number of aerosol attenuation mechanisms that can occur during aerosol transit within the region where the aerosols are formed and within the sampling system that can reduce the aerosol concentration and alter its size distribution. The primary attenuation of the aerosols occurs by their deposition onto structural surfaces. This deposition can occur by impaction, gravitational settling, diffusion, and thermophoresis. Brownian diffusion is dominant for small, submicrometer particles. As the aerosols grow to micrometer size by condensation and/or by agglomeration, gravitational settling and impaction become the dominant mechanisms.

2.0 Extractive Sampling Considerations

Most of the systems used to sample nuclear aerosols are extractive, i.e., they extract or withdraw a "representative" sample of the aerosol from the primary flow for subsequent analyses. These systems typically consist of a probe inserted into the primary flow that extracts a representative sample of particles and gas, a sampling line that connects the probe to the measurement instrumentation and/or collection devices, and a sampling flow control system. In the following, the considerations involved in the development of each of these components are presented. These include a presentation of the methods used to determine (1) the size range of particles travelling upward in the primary flow that will reach the sampling probe inlet, (2) the probe inlet diameter that achieves ideal sampling, and (3) the particle losses that occur in the sampling line during transit to the measurement instrumentation and/or collection devices.

2.1 Particle Transport in Vertical Flow

In most nuclear aerosol experiments extractive samples are obtained from primary flows that travel vertically upward. This is the result of the usual vertical orientation of the fuel pins and control rods. An initial consideration before determining the sampling probe dimensions and its sample flow rate is to estimate the size range of the particles that will reach the elevation at which the probe is located.

The largest particle based on its aerodynamic diameter that can be supported by a given vertical flow can be determined from the Stokes formula:

$$D_p = [18\eta CV/(\rho-\rho_g)g]^{1/2} \tag{1}$$

where D_p is the diameter of the particle, η the gas absolute viscosity, C the Cunningham slip correction factor, V the velocity of the gas, ρ the particle density, ρ_g the gas density and g the gravitational acceleration. For the range of pressures and temperatures that are typically of interest, the slip correction factor is near unity except for very small particles (<0.1 μm) and the gas density is small compared to the particle density. As seen in Table 1, only very large particles (supermicron in diameter) travelling with velocities that are typical of the primary flow in a reactor system are too large to be carried vertically upward.

2.2 Sampling Probe Inlet Design

Once the diameter range of particles in the primary flow that can travel upward to the sampling probe elevation is determined, calculations can be performed to arrive at the combination of gas flow rate and probe inlet diameter that provides efficient sampling for all particles of interest. This efficiency is defined as the ratio of the number of particles per unit volume of gas in a given diameter range that is sampled to the number of particles per unit volume of gas in the diameter range that is available to be sampled.

TABLE 1

Maximum Size of a Unit Density Spherical Aerosol Suspended
in a Vertical Flow as a Function of Flow Velocity at 2500K

Gas Composition	Flow Velocity (cm/s)	Aerodynamic Diameter (μm)
Steam	3.1	74
	15.7	166
	28.3	233
Hydrogen	2.5	41
	12.4	92
	22.4	124

The behavior of a particle travelling into the inlet of a sampling probe can be evaluated in terms of the particle relaxation time:

$$\tau = mC/(3\pi D_p \eta) \tag{2}$$

where m is the particle mass. Typical particle relaxation times encountered in a nuclear aerosol experiment are given in Table 2.

TABLE 2

Relaxation Times in Steam or Hydrogen
for a Unit Density Spherical Aerosol

	Particle Relaxation Times, τ(s)			
	Steam		Hydrogen	
D_p(μm)	1000 K	2500 K	1000 K	2500K
1	1.52×10^{-6}	5.89×10^{-7}	2.76×10^{-6}	1.49×10^{-6}
10	1.52×10^{-4}	5.89×10^{-5}	2.76×10^{-4}	1.49×10^{-4}
20	6.07×10^{-4}	2.36×10^{-4}	1.10×10^{-3}	5.96×10^{-4}
50	3.80×10^{-4}	1.47×10^{-3}	6.90×10^{-3}	3.72×10^{-3}
100	1.52×10^{-2}	5.89×10^{-3}	2.76×10^{-2}	1.49×10^{-2}

An analysis by Davies (1968) provides an upper and lower limit to the diameter of a small, thin-walled probe that will achieve ideal sampling in still air:

$$s[(2Q\tau)/\pi]^{1/3} < D_T < s^{-1}[(4Q)/(\pi g\tau)]^{1/2} \tag{3}$$

where D_T is the tube diameter, Q the volumetric flow rate, and s a nondimensional "range limit". Here, a "small, thin-walled probe" is one that has the same sampling efficiency at any orientation. The usual convention for ideal probe sampling is to make the range limit equal to 5. Smaller choices for the range limit result in sampling efficiencies less than 99%. Ideal probe sampling is obtained for a given probe flow rate when D_T is larger than the left side of the equation but smaller than the right side. Sample calculations using Equation (3) and some example sampling conditions are given in Table 3. The ideal sampling criterion is met only for the sizes in the table above the dashed lines. Thus, it is not possible to select one probe size to allow ideal sampling in a still gas for all of these conditions for 20 μm particles using the Davies criterion.

If necessary, a less restrictive criterion may be used to size the probe inlet. Work by Agarwal and Liu (1980) for the case of small, thin-walled probes has shown that a reasonable criterion for still air is:

$$(2\tau^2 g)/D_T < 0.1 \tag{4}$$

Meeting this criterion gives a still air sampling efficiency greater than 90%. Table 4 provides example calculations for the minimum probe diameter for representative sampling according to the Agarwal-Liu criterion for steam and for hydrogen.

For sampling situations where the primary flow is moving perpendicular (transverse) to the sampling probe inlet, the sampling efficiency can be estimated by [Davies (1968)]:

$$\text{Efficiency} = 1 - 0.8f + 0.08\,f^2 \quad \text{for } f < 1 \tag{5}$$

where

$$f = [(u^2 + V_f^2)/V_o^2]^{3/4},$$

$$V_o = [Q/(4\pi\tau^2)]^{1/3},$$

and u is the velocity transverse to the probe's inlet axial centerline, V_f the particle terminal settling velocity and V_o the "dynamical sampling" velocity as defined above. This equation gives a sampling efficiency of 90% for 10 μm particles in the case when the transverse velocity is 1 cm/s.

TABLE 3
Probe Diameter (cm) Limits from Davies (1968) Criterion

Steam:

Q(cm^3/s)	1.66	4.2	16.7	16.7
Temperature (K)	2500	2500	2500	1000

D_p(μm)	Probe Diameter (cm)			
1	.05-12.1	.05-19.3	.1-38.4	.13-23.9
10	.20-1.21	.27-1.93	.43-3.84	.58-2.39
20	.32-.61	.43-.96	.68-1.92	.93-1.20
50	----------	-----------	-----------	-----------

Hydrogen:

Q(cm^3/s)	1.66	4.2	16.7
Temperature (K)	2500	2500	2500

D_p(μm)	Probe Diameter (cm)		
1	0.57-7.52	.08-12.1	.13-24
10	.27-.76	.31-1.21	.58-2.4
20	----------	.58-.61	.933-1.21
50	----------	----------	------------

TABLE 4
Probe Diameter (cm) based on Agarwal-Liu (1980) Criterion
(any flow rate)

D_p(μm)	Steam (1000K)	Steam (2500K)	Hydrogen (2500K)
1	4.5×10^{-8}	3.6×10^{-9}	4.4×10^{-9}
10	4.5×10^{-4}	36.8×10^{-5}	4.4×10^{-4}
20	.01	1.1×10^{-3}	.01
50	.28	.04	.27
100	4.53	.68	4.35

In general, for primary flow velocities less than approximately a few centimeters per second, a sampling efficiency greater than approximately 90% can be achieved by a probe whose diameter is determined using analyses developed assuming that sampling occurs in "still air", i.e., from a stationary gas with no velocity.

2.3 Transport Losses in Tubes

Once the aerosols have been representatively sampled by the probe, they must be transported with minimal losses to the measurement devices. The ability of a sampling line to transport all particles sizes with acceptably low losses depends upon the settling and inertial effects of the particles travelling in the sampling line, which are dictated primarily by their size and the flow rate through the line. In general, the flow velocity within the sampling line must exceed the settling velocity for the largest particle of interest. However, the flow velocity must not be large enough to cause that particle not to follow its flow streamline and then possibly impact on the sampling line wall along any of its bends. Accepted theories have been developed that account for particle losses during transit as the result of gravitational settling, diffusion and losses in bends for tubes with either laminar or turbulent flow. These are presented below. Existing theory also describes the thermophoretic force on a particle that can occur in a sampling line under certain conditions. This theory, which is used to develop a method for calculating thermophoretic losses in a tube, is presented also.

2.3.1 Gravitational Settling

For laminar flow in a cylindrical, horizontal tube of length L and diameter D_T, the fraction of particles lost is given by Thomas (1958):

$$F_S = 2\pi^{-1}[2Z(1-Z^{2/3})^{1/2} + \sin^{-1}(Z^{1/3}) - Z^{1/3}(1-Z^{2/3})^{1/2}] \qquad (6)$$

where

$$Z = (3V_fL)/(4V_aD_T)$$

and V_a denotes the bulk velocity of the gas in the tube.

For turbulent flow, the fraction of particles lost in a cylindrical tube is given by Fuchs (1964):

$$F_S = 1-\exp[-A_gL_d] \qquad (7)$$

where

$$A_g = (4V_f)/(\pi V_a) \text{ is the dimensionless velocity}$$

and

$$L_d = L/D_T \text{ is the dimensionless length}$$

To account for the possibility of an inclined tube geometry, the tube length, L, should be replaced by Lcosθ, where θ is the angle in degrees that the inclined tube makes with respect to the horizontal plane

2.3.2 Diffusion

For laminar flow, the fraction of particles lost in a cylindrical tube is given by Sinclair, *et al.* (1976):

$$F_D = 1 - [0.819 \exp(-3.65\alpha) + 0.097 \exp(-22.3\alpha) + 0.032 \exp(-57.0\alpha)$$
$$+ 0.027 \exp(-123.0\alpha) + 0.025 \exp(-750.0\alpha)] \qquad (8)$$

where

$$\alpha = (\pi DL)/Q,$$

D is the diffusion coefficient and Q is the volumetric flow rate.

For turbulent flow, the fraction of particles lost in a cylindrical tube is given by Fuchs (1964):

$$F_D = 1 - \exp[-4V_dL/(D_TV_a)] \qquad (9)$$

where

$$V_d = D/\delta = D/(28.5D_T D^{1/4} Re^{-7/8}(\eta/\rho_g)^{-1/4})$$

and Re denotes the Reynolds number, which is equal to $\rho V D_T/\eta$.

In the region where both settling and diffusion are important, Heyder (1985) provides a method for predicting the total loss from the sum of both effects. The total loss in a tube is given by:

$$F_T = F_S + F_D - (F_D \times F_S)/(F_D + F_S). \qquad (10)$$

2.3.3 Bend Losses

For laminar flow, a number of different approaches for bend losses have been taken that all arrive at the same equation. Yeh (1976) transforms the equation to describe the fraction of particles lost in a tube:

$$F_{BL} = 1 - 2\pi^{-1}\cos^{-1}[(\phi STK)/2] + \pi\sin^{-1}(2\cos^{-1}[(\phi STK)/2]) \qquad (11)$$

where the Stokes number is defined in terms of the tube diameter and ϕ is the angle of the bend in radians.

For turbulent flow, the fraction of particles lost in a tube is given by the empirically determined deposition efficiency for a ninety degree bend developed by Pui *et al.* (1987):

$$F_{BL} = 1 - 10^{-0.963STK} \tag{12}$$

for Reynolds numbers between 1000 and 10,000. The theory of Cheng and Wang (1981) for Re=1000 agrees well with this expression. Calculations using such equations to determine the amount of material lost in a tube bend for angles other than ninety degrees and for Reynolds numbers less than 1000 should be considered to be only estimates. In general, no theory has been able to agree accurately with experimental measurements over the entire flow range that is usually of interest.

2.3.4 Thermophoretic Losses

Thermophoretic losses can be a potential problem for extractive sampling in nuclear systems. Generally, there is a large temperature gradient along the sampling line from the probe inlet to the collection device. Particle deposition onto the sampling line surface due to thermophoresis becomes significant when the ttemperature gradient between the particle and the surface is large and if the particle is small (submicrometer-size). This leads to particle losses along the sampling line.

The thermophoretic force is a force that a particle experiences when it is present in a large temperature gradient. This results from the difference in momentum that arises when "cold" gas molecules strike one of its sides and "hot" gas molecules strike the other side, thereby moving the particle in the direction of decreasing temperature [Waldmann and Schmitt (1966); Brock (1962)]. Particles of diameter $D_p < \lambda$, where λ is the mean free path of the gas molecules, experience a thermophoretic velocity:

$$V_{th} = (-0.55\eta\nabla T)/\rho_g T) \tag{13}$$

where ∇T is the temperature gradient in the gas surrounding the particle and T is the absolute temperature of the particle. Particles having diameter $D_p > \lambda$ have thermophoretic velocities:

$$V_{th} = (-3\eta CH\nabla T)/(2\rho_g T) \tag{14}$$

where

$$H = (1+6\lambda/D_p)^{-1}[(K_a/K_p + 4.4 \lambda/D_p)/(1+2K_a/K_p + 8.8 \lambda/D_p)]$$

and K_a and K_p are the thermal conductivities of the gas and of the particle, respectively. The mean free paths for the temperatures and pressures typical of reactor tests are on the order of 10^{-2} μm, so that Equation (13) can be applied to only the smallest submicrometer particles.

For laminar flow, the thermophoretic velocity can be used to determine the fraction of particles lost in a tube. The amount of time during which the thermophoretic force is acting is based on the length (or segment) of the tube, L, over which the wall temperature gradient can be assumed to be constant, divided by the velocity of the flow stream with which the particle is traveling. This force causes a radial displacement, ΔR, from the original trajectory given by:

$$\Delta R = (V_{th}L)/V' \tag{15}$$

where V' is the velocity of the gas along the particular streamline (trajectory). If the particle is initially within ΔR of the wall, then the particle will be lost to the wall before it can leave the tube or segment. The fraction of particles lost in the tube is then given by:

$$F_T = 1 - [(R_T - \Delta R)/R_T]^2 \tag{16}$$

or

$$F_T = 1 - [R_T - (V_{th}L)/V']^2/R_T^2 \tag{17}$$

where R_T is the tube radius.

Figure 1 indicates the dependence of the thermophoretic loss for a typical extractive sampling probe on the thermal conductivity of the aerosol, K_p. For very low particle Stokes numbers, less than approximately 2.5E-5, there is no dependence on K_p. As the Stokes number is increased, this dependence becomes significant. For Stokes numbers greater than approximately 0.04 the % transmission becomes independent of the Stokes number for a given particle conductivity. Figure 2 shows how the type of constituent gas that would be encountered in a typical nuclear aerosol sampling system line affects the quantity of aerosols lost due to thermophoresis.

All the preceeding particle loss equations are derived for monodisperse aerosols, i.e., particles of the same diameter. For aerosol distributions, the loss in each size fraction must be calculated individually. It should be pointed out that a broad aerosol distrubution can give rise to additional mechanisms such as kinetic and thermal coagulation. These mechanisms are important when large differential velocities are developed by the same force acting on masses that are orders of magnitude different. The impact of these mechanisms usually are assumed negligible for reactor aerosol sampling systems because (1) coagulation in the reactor plenum narrows the aerosol distribution , (2) the transport times are rapid, (3) the temperature gradients are reduced, and (4) dilution increases the time between particle collision.

The user of any type of transport system must also realize that while the equations for diffusion and settling are supported well experimentally, the experimental verification for bend losses and thermophoretic losses are practically non-existent. Therefore, it is recommended that a mock-up of any transport system be calibrated experimentally beforehand whenever possible. This will provide the most accurate data for calculating the sampled aerosol parameters from the measured parameters.

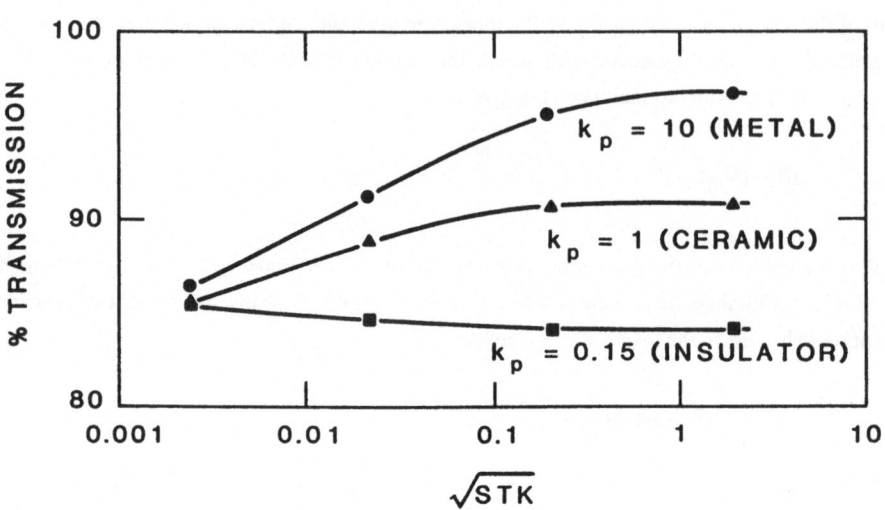

Figure 1. Thermophoretic Loss: Material Dependence

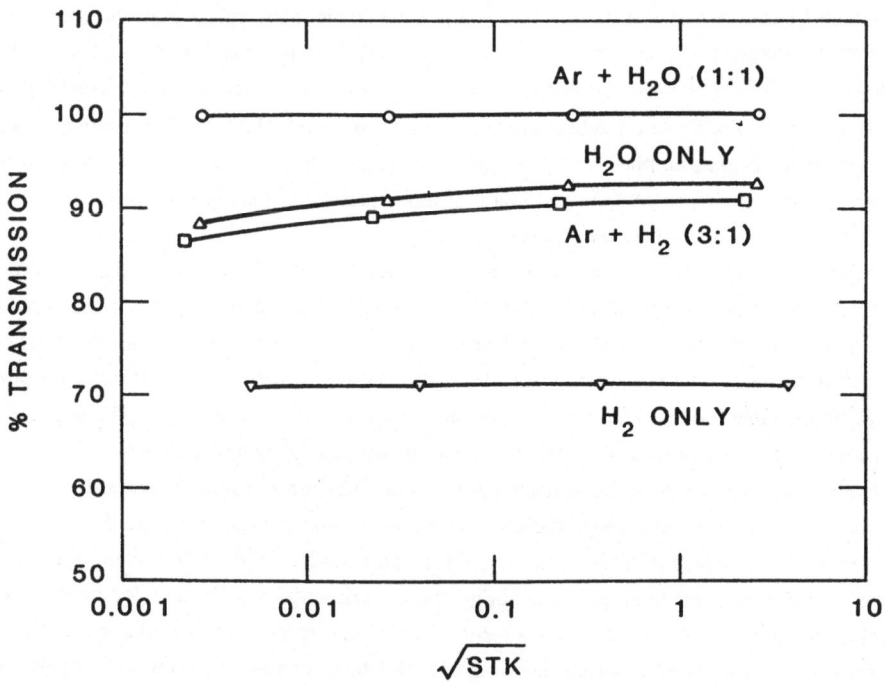

Figure 2. Thermophoretic Loss: Constituent Gas Effects

2.4 Probe Cooling and Dilution

For extractive sampling from nuclear or other high temperature systems, it is an accepted practice to cool and dilute the desired sample with additional gas supplied at the probe tip. This additional gas provides two major functions: (1) it cools the sampled gas rapidly to minimize thermophoretic transport losses and to meet maximum temperature limitations at the collector or reactor seal, and (2) it dilutes the sample sufficiently to 'fix' the aerosol size distribution and number concentration by reducing coagulation effects and eliminating condensation of additional material. Figure 3 gives an example of this type of probe, a variant of which was used in the Loss of Fluid Test (LOFT) experiments [Miller *et al.*(1984); McPherson and Hicks (1986)].

One of the major problems confronting the design of a probe tip is the temperature difference between the sampled gas and the probe sheath or dilution gas. This problem has two facets. One is a materials problem, in that the probe tip must withstand temperatures between 1000 and 2500 K and experience a gradient to approximately 700 K. The second is an aerosol problem, in that the sampled gas moving through the probe tip should not pass a surface cooler than the gas until it mixes with the dilution gas and is cooled by the dilution gas to the probe temperature. This is necessary to minimize the loss of particles by thermophoresis and to prevent possible condensation of vapors. To reduce or eliminate this problem, sheath gas can be introduced into the tip through a cylindrical annulus. The flow rate, tube diameter and cylinder length can be adjusted so that the sheath gas velocity equals or exceeds the thermophoretic velocity. Argon is considered a good choice for sheath gas because it is non-reactive, has high thermal conductivity and has a viscosity near that of steam, which is important for the control of the gas flow.

2.5 Sampling Flow Control System

The primary purposes of a gas flow control system are to control the amount of sample extracted and to provide a constant gas velocity through the aerosol measurement device. This seemingly simple task is complicated because sampling during typical simulated reactor accidents can be affected by changes in the sampling pressure, temperature and gas composition. This is in addition to the materials problems associated with the high-temperature steam that is present in these types of accidents.

These potential problems can be minimized by using a gas flow control system with a critical orifice located downstream of the sampling system that controls the sample flow rate. For this application, the orifice must be sized to allow the desired gas flow into the sampling system to adjust the sampling system pressure so that it is in equilibrium with the pressure of the plenum being sampled. If the size of the orifice is too small, less of a sample will be extracted than desired. If the rate of decrease in pressure in the plenum is sufficiently rapid, then the entire output of the orifice is needed solely to equalize the pressure in the sampling line with the plenum. More rapid rates of pressure decrease will allow only some of the gas contained in the sampling system to flow through the control orifice, while the remainder must flow through the tip into the plenum in order to equalize pressures. For experiments where the rate of change of pressure is known and constant, a properly sized simple orifice will function adequately. If the

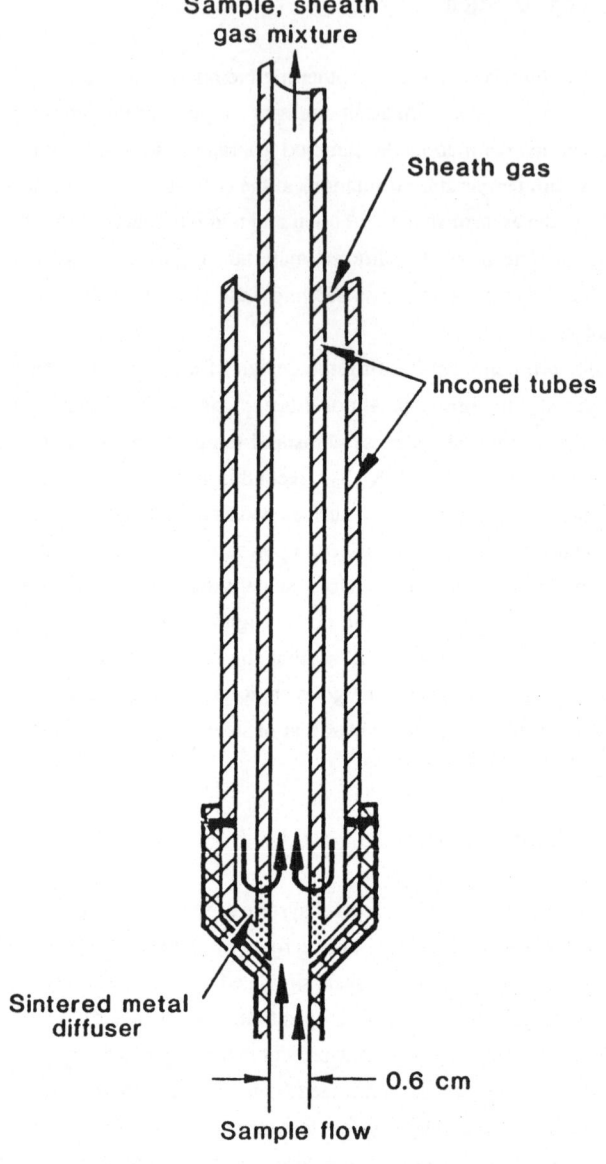

Figure 3. Sampling Probe

rate of change is unknown or not constant, an active pressure feedback mechanism is necessary to properly control the flow.

The use of an orifice also complicates the control of gas flow if the gas composition is changing, even if the pressure remains constant. This is because the gas flow through an orifice is sensitive to changes in the gas constant and specific heat ratios of the gas flowing through it. The specific heat ratio can change by as much as 30% during a typical reactor experiment and the gas constant by more than an order of magnitude. For the specific case of simulated reactor accident conditions, the sampled gas composition changes from steam to hydrogen during the course of the accident. One solution to this problem is to dilute the sampled gas greatly (steam-hydrogen) with another gas such as argon. Unfortunately, too large a dilution ratio results in no control, and thus a considerable uncertainty in the sample gas flow rate.

Another solution to the problem of changing composition is to use a catalyst to change one gas component into another, upstream of the orifice or other control device. For reactor systems CuO can be used as a catalyst to convert H_2 gas into steam. Because this recombination occurs on a mole for mole basis, the volumetric gas flow rate through the orifice remains unchanged. Converting the hydrogen back into steam reduces the impact of varying gas constants and specific heat ratios. Efficient recombination relies on the gas residence time in the recombination chamber and on the exposed surface area of catalyst. Condensation of the steam does not occur because the catalyst recombiner must be heated above 500° C to allow the H_2-to-steam reaction to proceed at a sufficient rate. From an engineering standpoint, the recombiner must be installed upstream of the controlling orifice, but downstream of the particle collection devices. The disadvantage in placing the recombiner downstream of the particle analysis devices is that changes in the viscosity of the gas mixture will introduce uncertainty in the data gathered from these devices. For typical steam-hydrogen-argon mixtures such as those used in the LOFT experiments, the resultant viscosity change was calculated to be less than 20%.

3.0 Aerosol Measurement Techniques

The aerosol measurements techniques that are commonly used in nuclear aerosol experiments are discussed in the following. This section is not intended to be a complete review of all techniques that have been used. Instead, it is intended to provide an overview of the types of systems that are available and how they can be applied for various types of measurements. These measurement techniques are divided into intrusive devices, which are located within the gas flow, and nonintrusive devices, which allow the gas to flow through without obstruction. Mass concentration and radioactivity measurements are emphasized because the goal of nuclear reactor experiments is to determine the total possible mass and activity that can be released for a given scenario.

3.1 Intrusive Techniques

In reactor safety experiments, a considerable number of different techniques have been used to determine the aerosol size distribution and number concentration. These have been primarily extractive, intrusive techniques requiring post-test analyses. Typically, particles are extracted isokinetically into the sampling system from the flow, i.e., with the same kinetic energy, hence velocity, they had in the flow. Then, they are passed into a collection device that separates the particles according to size by impaction, centrifugal force, gravitational settling and/or electrostatic precipitation. Samples acquired in this manner have often been analyzed to determine the number, mass, shape, chemical composition, and radioactivity of the particles.

Impactors of various designs have been used to measure size distributions over the particle diameter range from approximately 0.5 to 20 μm. These include, for example, size distribution measurements of UO_2, SrO_2 and NaI particles [Sauter and Schutz (1980)].

Spiral centrifuges have been used to determine particle size distributions over the diameter range from approximately 0.1 to 20 μm, including measurements of aerosols produced from simulant nuclear reactor core materials [Albrecht et al. (1980)] and of U_3O_8 aerosols [Parker (1980)]. A rotating wheel collection device has been used to determine the velocity and size distribution of aerosols in the 30 to 3000 m/s and 0.003 to 10 μm diameter ranges [Elrick (1980,1982)] produced by neutronically heating fresh UO_2 fuel. Cyclones have been used to size classify aerosols > 1μm diameter [Chyssler et al. (1983)].

Elutriators or gravitational settling chambers with collection plates have been developed to size classify simulant reactor core aerosols [Chyssler et al. (1983)] and volatile fission product aerosols [Dunn et al. (1983)] in approximately the 1 to 100 μm diameter range. Such a device that also incorporates fine wires to collect particles by impaction and diffusion is described in detail in Section 3.5 Another device based upon gravitational settling is a sequentially operated sedimentation sampler [Chyssler et al. (1983)].

An electrostatic precipitator has been used to collect yttrium oxide [Kanapilly et al. (1980)], sodium oxide and fuel oxide aerosols [Kanapilly et al. (1980); Schock (1980)] in approximately the 0.02 to 10 μm diameter range. An electrostatic aerosol analyzer (EAA) and diffusion battery also were used in experiments [Kanapilly et al. (1980)], yielding size information in the 0.003 to 1 μm diameter range. Number concentrations of submicrometer-size sodium oxide and fuel oxide aerosols using condensation nuclei counters (CNCs) have been measured, as reported in Nuclear Aerosols in Reactor Safety (1979). Note that the commercially available EAA and CNC devices cannot be adapted to large-scale reactor experiments due to pressure, temperature and gas composition contraints.

Intrusive optical techniques that can charaterize the size distribution of an extracted aerosol include a variety of light-scattering techniques and an aerodynamic particle sizer [Baron (1986)]. An on-line spectrometer also has been patented that can be used to monitor in real-time the size and chemical composition of aerosols [Sinha et al. (1982)]. To date, however, none of these optical techniques have been used to sample nuclear aerosols.

The collection of particles onto filters for post-test size distribution analyses has been used extensively. Examples include the collection of PuO_2 and UO_2 aerosols [Bunz and Schock (1979)], simulant reactor core aerosols [Albrecht et al. (1980); Chyssler et al. (1983)], and fission product and fuel aerosols [Buescher et al. (1982); Lorenz et al. (1971)]. UO_2 size distributions have also been determined through sieve analysis [Sauter and Schutz (1980)] and collection onto grids for electron microscope analysis [Wright et al. (1980)].

3.2 Nonintrusive Techniques

Nonintrusive techniques for particle analysis typically are optical and based upon either imaging or nonimaging techniques. Imaging techniques primarily involve photography of the particles or particle images. Analysis is accomplished following the test. Nonimaging techniques exploit either the extinction or scattering properties of the aerosol. In the former, the attenuation of a light source through a cloud of particles is measured. In the latter, the amount of light scattered from a single or countable number of particles is measured. Scattering techniques require a knowledge of the dependence between incident light wavelength, particle diameter, complex refractive index, scattering angle and polarization. Only in the limits of small or large particle diameter-to-incident wavelength ratios do simple relationships hold. Scattering methods yield a particle size distribution when sufficient parameters are known or measured. They can yield the particle volume-to-surface area mean (Sauter) diameter and number concentration. Also, most scattering techniques are applicable only to particle diameters >0.2 μm. In general, a single extinction measurement yields no information concerning the particle size distribution unless *a priori* assumptions are made about the shape of the distribution.

Forward scattering has been used in the laser aerosol spectrometer for sizing in the 0.4 to 2 μm diameter range [Schock (1980)], and in an optical particle counter in the 0.3 to 25 μm diameter range [Nuclear Aerosols in Reactor Safety (1979)]. The extinction of a single-wavelength HeNe laser light has been used to determine relative changes in the number concentration or mean concentration in sodium mist experiments [Himeno and Takahashi (1980)]. By using two disparate wavelengths of light, the Sauter mean diameter and number concentration can be measured in real-time [Ariessohn *et al.* (1980)]. The extinction method has been used in a series of simulant reactor core experiments [Kanapilly *et al.* (1980)]. This method has been extended by Novick (1988) to using two extinction cells in series to determine the particle number concentration as a function of time, which is discussed in detail in Section 4.4.3. Commercial devices such as a phase/doppler particle analyzer [Bachalo and Houser (1984)] and a particle-counter-sizer-velocimeter [Holve (1986)] may be applicable also. In general, optical techniques are subject to temperature limitations of the window material. Also, electronic components and other materials may be subject to degradation by radiation.

3.3 Mass Concentration

The measurement of mass concentration in reactor safety experiments is best accomplished by the extraction of representative aerosol samples and the subsequent collection in impactors, centrifuges, precipitators, settling chambers or on filters. Post-test analysis (typically weighing) is required for these techniques. The use of filters has been the most frequent approach [Albrecht *et al.* (1980); Bunz and Schock (1979); Buescher *et al.* (1982); Lorenz *et al.* (1971)]. Filter media able to withstand high temperatures include "saffil", an alumina oxide fiber matte, and sintered porous stainless steel. One technique that yields real-time mass concentrations is to direct the extracted aerosol samples onto a piezoelectric crystal microbalance [Woods (1979); Sem and Daly (1979)]. A virtual impactor designed specifically for mass concentration measurements in a large scale accident simulation experiment is described in Section 3.5

3.4 Radioactivity

Gamma spectrometry can be used in reactor safety experiments to determine the amount and disposition of the condensible and noncondensible radioactive fission products. Both real-time and post-test approaches have been used. For real-time measurements, the aerosol is scanned with stationary Ge(Li) or NaI detectors as it passes them (if there is sufficient activity during transit) or as material is collected onto filters. An on-line fission product detection system has been used to monitor the activity of fission product aerosols [Dunn et al. (1983)]. Post-test measurements can be performed after the radioactive aerosols are collected onto filters or other substrates, in condensate traps or in collection tanks. Examples of applying this technique include a system for on-line detection of gaseous and condensed (liquid) fission products during severe fuel damage experiments [Buescher et al. (1982)], for on-line detection of corium aerosols [Albrecht et al. (1980)], and for post-test determination of fission product and fuel materials [Lorenz et al. (1971)].

3.5 Examples of In-Pile Aerosol Measurement Systems

The following systems were designed for use in three different large-scale nuclear accident simulation experiments. They are described in detail in order to provide examples of systems that have been successfully applied for nuclear aerosol measurements. Included with the descriptions are the equations used for their analysis.

3.5.1 The Source Term Experiments Program (STEP)

The STEP experiments consisted of a series of four tests which were conducted in Argonne National Laboratory's Argonne-West TREAT reactor during 1983-84 [Herceg et al. (1984)]. The resultant aerosols were collected in aerosol canisters that were designed to characterize the potential releases in both low pressure (on the order of 0.2 MPa) and high pressure (on the order of 8 MPa) environments. Because prior information regarding the characteristics of these aerosols was limited, the system was designed to provide adequate sampling over a relatively wide particle size range. It was also designed to handle a wide range of particle loadings. Two canisters were attached to the STEP experimental test vehicles, as shown in Figure 4. Electrical heaters with temperature-regulated feedback control systems were used to maintain the temperatures of the canisters at the same temperature as that of the primary vessel's upper portion. Signals from the control thermocouples as well as additional thermocouples on the canisters were monitored during each test. The canister entrance ports were located at 0.5 m and 2.1 m above the top of the fuel. These locations allowed for sampling at the minimum and maximum distances above the fuel, given the space limitations of the test vehicles. The canister entrances

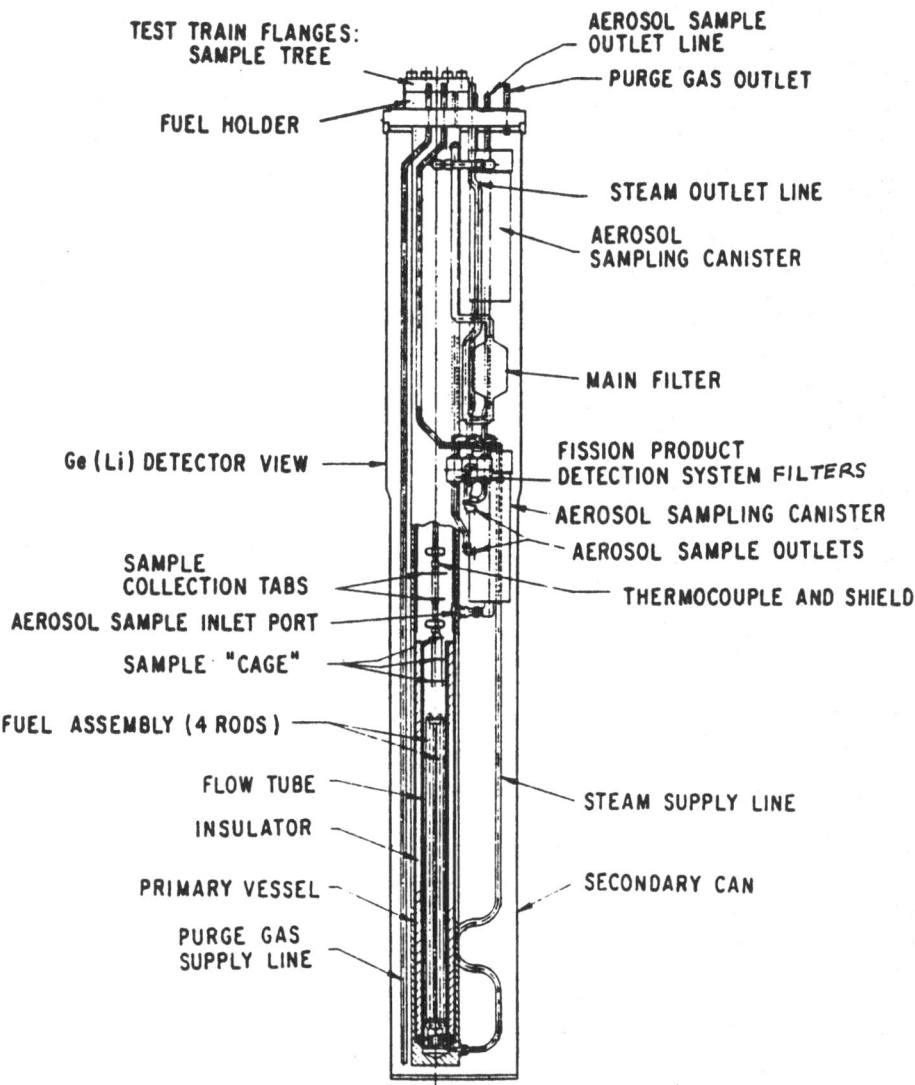

Figure 4. STEP In-Pile Test Vehicle

were flush with the internal surfaces of the primary vessels; space and mobility constraints prohibited extending the sampling lines into the flow streams.

Each canister contained three chambers. Each chamber consisted of a series of stages that were configured in such a way to create a long (3.2m), winding flow channel in a space that was only 0.16 m high. A schematic diagram of the stage arrangement within a canister is shown in Figure 5. Figure 6 shows the stages and collection devices. Significant fractions of the particles were not deposited by impaction around the bends because the velocities were low. The effect on quantitative results by such losses is nevertheless incorporated into the analyses, as described later. The channel cross-section was square. A portion of the steam/hydrogen mixture passed through one or more of the chambers in each canister, and particles were collected on collection devices located in the stages. The individual chambers were exposed to flow during different periods of the tests, allowing for temporal separation of the collected material.

Particles were collected on two types of sampling devices. Settling plates were positioned on the floors of the stages to collect larger particles by gravitational settling and smaller particles by diffusion. Here the chambers functioned essentially as horizontal elutriators. Particles were collected also on wires suspended perpendicular to the flow. The wires collected larger particles by impaction and interception, and smaller particles by diffusion. The system had adequate collection efficiencies for particles over a relatively large size range, from submicron to supermicron in diameter. At least one settling plate and one set of wires were positioned in each stage, with 14 stages per chamber. Therefore, particles were deposited on a large number of collection devices that were positioned along the channel length. Also, because of this 3.2 m long path length, there was sufficient time for many particles to settle gravitationally, leading to a considerable stratification of deposits along the channel length. This allowed for adequate sampling over a wide aerosol concentration.

The settling plates were 0.95 cm square and composed of stainless steel. They were held in position by stainless steel frames that meshed with the walls of the channels. A set of four wires were suspended together on one stainless steel frame, which was identical to the type of frame that supported a settling plate. The wires ranged in diameter from 2.5 to 250 μm (0.1 to 10 mil). These different wire diameters resulted in different collection efficiencies, providing some of the required versatility of the system. The wires were composed of a number of different materials. Most of the wires were chemically inert. Some, however, were not. Therefore, they could react with the fission products and provide information on their chemical composition. In addition, coupons of various materials were attached to selected settling plates to provide additional surfaces for chemical interaction. The material of the stages, entrance tubes, and canister housings was stainless steel.

The collection devices from the canisters were examined post-test using several microanalytical techniques. The purpose of the examinations was to obtain information on the size distributions, morphology and composition of the deposited materials. The majority of the information was obtained by scanning electron microscopy (SEM). The SEM generated images of the collection surfaces. The morphology of the deposits was determined from these micrographs. The SEM was equipped with energy dispersive X-ray analysis capability which was used to identify the elemental constituents of the deposits. Additional chemical and morphological information about the samples was obtained by electron

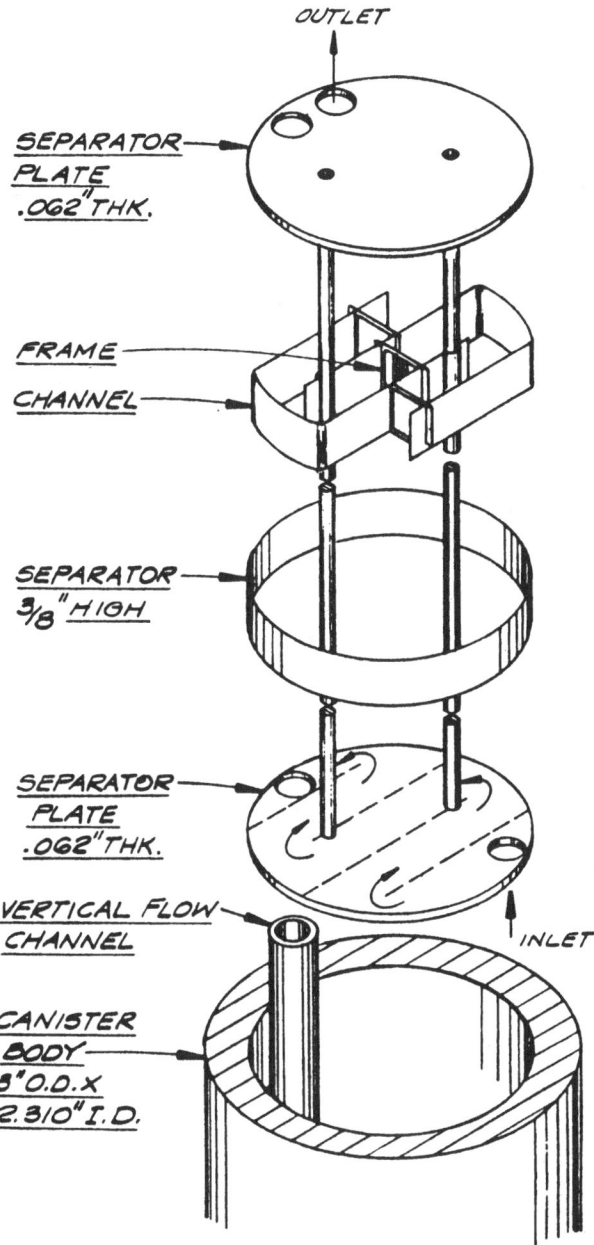

OUTLET

SEPARATOR
PLATE
.062"THK.

FRAME

CHANNEL

SEPARATOR
3/8" HIGH

SEPARATOR
PLATE
.062"THK.

VERTICAL FLOW
CHANNEL

INLET

CANISTER
BODY
3" O.D. X
2.310" I.D.

Figure 5. STEP Canister Stage Arrangement

592

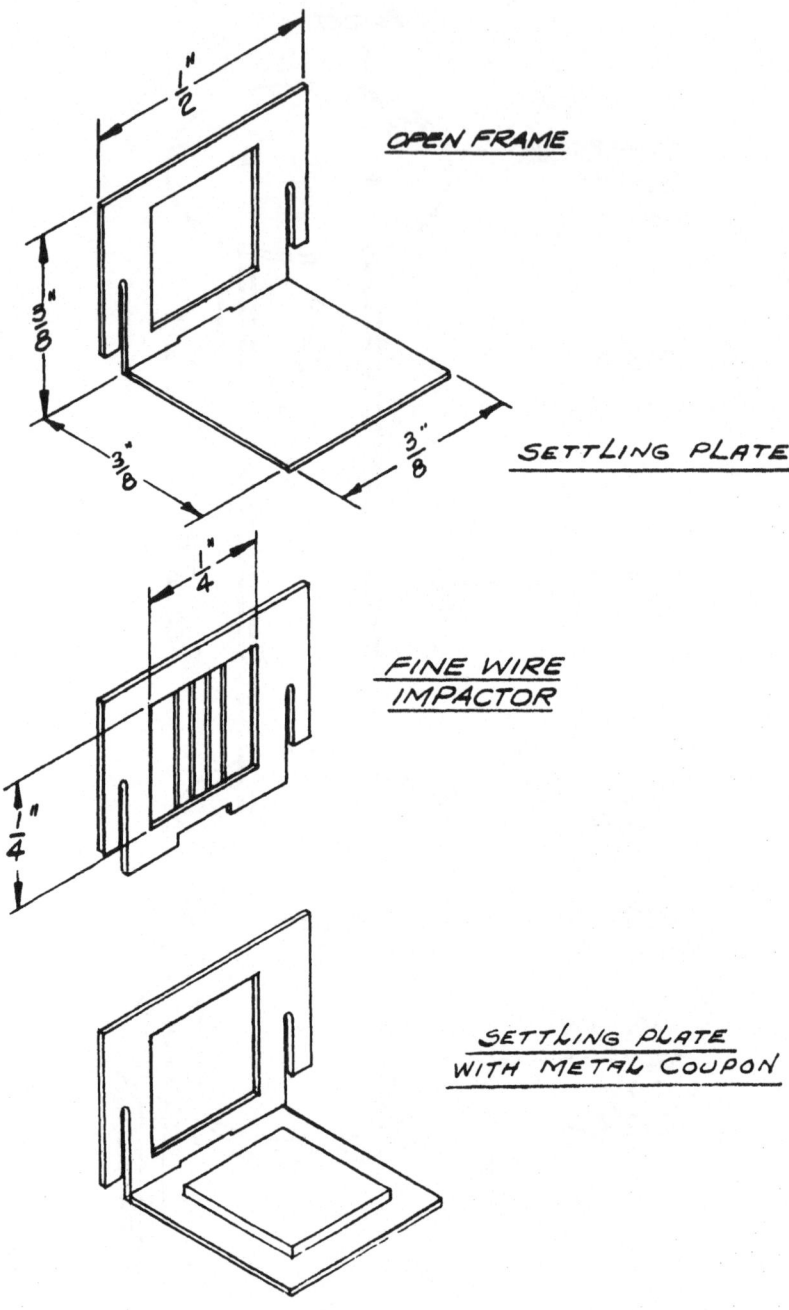

OPEN FRAME

SETTLING PLATE

FINE WIRE IMPACTOR

SETTLING PLATE WITH METAL COUPON

Figure 6. STEP Canister Collection Devices

microprobe analysis, secondary ion mass spectrometry, bulk analysis and X-ray diffraction. There were sufficient numbers of samples to allow for the destructive sample preparation required by some of these techniques.

The procedure for the aerosol size distribution and number concentration analyses consisted first of measuring the diameter of each particle in a selected area on a SEM micrograph and then of sorting the results by size to determine the number of particles in different size ranges on the surface of the settling plate or wire. Collection efficiencies of the wires and plates were calculated for the mean diameters of the particle size intervals using the equations listed below. The collection efficiencies then were applied to the particle count data to calculate the number of particles in the different size intervals that passed the particular collection device. Transport losses in the chamber upstream of the collection device were calculated according to the equations listed below. These losses then were applied to the number of particles that passed by the collection devices, yielding the number of particles in each size interval at the entrance to the chambers. These values were divided by the total gas flow to determine particle concentrations. The particle data was converted to a mass basis using the particle density. The thermal-hydraulic conditions used in these calculations were measured during the tests.

The analysis of the particle data from each micrograph corresponding to collection devices in a particular chamber theoretically should yield the same particle distribution function. In practice, however, there were variations in the results due to nonuniformity in the samples, errors introduced by the particle counting, and the approximate nature of the efficiency equations. By examining many micrographs from different collection devices in a chamber, sufficient data was generated to determine statistically significant values of the distribution functions.

All particles that were deposited were assumed to adhere to the collection surfaces. Forces of adhesion, although difficult to calculate, are known to be strong for particles of the size encountered in these tests. Experience with handling these samples supported the assumption that the material strongly adhered to the collection surfaces.

The collection efficiency equations used in the analysis were compiled from a review of the literature. The selected equations were determined to be the most applicable for the range of flow conditions and system geometry. Although the equations do not exactly model the conditions in the canisters, they are satisfactory, given the required precision of the results. Canister calibration experiments verified the application of the equations selected.

The equations used for calculating the collection efficiency of particles on the settling plates are as follows.

The efficiency of collection by gravitational settling in a rectangular duct is [Ingham (1977)]:

$$F_S = \frac{V_f}{Vh}. \tag{18}$$

where V_f is the particle vertical settling velocity, V the particle horizontal velocity, and h the channel height.

The efficiency of collection by diffusion is [Ingham (1975)]:

$$F_D = \left(\frac{4}{\sqrt{\pi}}\varepsilon^{1/2} - \varepsilon\right) \cdot 0.25 \text{ for } \varepsilon < 0.02, \tag{19}$$

$$F_D = [1 - [0.694\exp(-5.76\varepsilon) + 0.132\exp(-30.25\varepsilon)$$
$$+ 0.053\exp(-74.8\varepsilon)]] \cdot 0.25 \text{ for } \varepsilon > 0.02, \tag{20}$$

where D is the diffusion coefficient, L the length of the duct, r the radius of the cylindrical duct and

$$\varepsilon = \frac{DL}{Vr^2}.$$

These diffusion equations were developed for plug flow through a duct. Of the geometries for which solutions existed in the literature, this was thought to be the best approximation to the conditions in the chambers.

The total collection efficiency for a plate is approximated by the following equation [Heyder *et al.* (1985)]:

$$F_T = F_S + F_D - F_S F_D/(F_S + F_D) \tag{21}$$

In order to calculate the collection efficiencies of the wires, a model of the flow around the wires is required. This is incorporated into the efficiency equations through the hydrodynamic factor. The hydrodynamic factor, K_H, for a single row of parallel cylinders is [Kirsch and Stechkina (1978)]:

$$K_H = -\ln\left(\frac{\pi d_w}{4h'}\right) - 1.33 + \frac{1}{3}\left(\frac{\pi d_w}{4h'}\right) + \tau Kn, \tag{22}$$

where $\tau \cong 1$, h' is the distance between wires, d_w the wire diameter and Kn the Knudsen number, which equals $2\lambda/D_p$.

The collection efficiency due to diffusion on a wire is [Natanson (1957); Stechkina and Fuchs (1966)]:

$$F_D = 2.94^{-1/3}Pe^{-2/3} + 0.624Pe^{-1}, \tag{23}$$

where Pe is the Peclet number, which equals the product of the Prandtl, Pr, and Reynolds, Re, numbers.

The efficiency of collection due to interception is [Chen (1955); Kirsch and Stechkina (1978)]:

$$F_R = (2K_H)^{-1} \left[2(1+R) \ \text{L}n(1+R)^{-1} + \frac{2\tau Kn(2+R)R}{(1+R)} \right], \qquad (24)$$

where R is the interception parameter, which equals D_p/d_w.

A term to account for the combined effects of diffusion and interception is [Stechkina and Fuchs (1966)]:

$$F_{DR} = 1.23 K_H^{-1/2} Pe^{-1/2} R^{2/3}. \qquad (25)$$

The efficiency of collection due to impaction is [Brewer and Goren (1984)]:

$$F_{ST} = \frac{Stk^3_{10}}{Stk^3_{10} + 0.77 Stk^2_{10} + 0.22}, \qquad (26)$$

where Stk_{10} is defined in terms of the Reynolds number and a function A, where

$$A(K_H, Re) = 4(K_H)^{-1} + 4.95 \ (Re)^{1/2} \ , \text{ and} \qquad (27)$$

$$Stk_{10} = \frac{A(K_H, Re)Stk}{A(K_H, 10)}, \qquad (28)$$

where Stk is the Stokes number, defined for this case as $\rho D_p^2 CV/18\eta d_w$. The function $A(K_H, Re)$ was developed for a hydrodynamic factor for a different wire geometry; it is assumed to be valid within the desired accuracy for this application.

The total collection efficiency for wire is approximated by the following sum:

$$F_T = F_D + F_R + F_{DR} + F_{ST}. \qquad (29)$$

The fraction of particles of each size range attenuated by gravitational settling and diffusion along the flow path upstream of a particular collection device were calculated using the same equations as those used for a settling plate, except that the factor of 0.25 in the diffusion equation is neglected (to account for collection on all four surfaces).

The flow passes around a number of 180° bends, n, prior to reaching a collection device. The loss around one bend by inertial impaction is calculated by the following equation [Crane and Evans (1977)]:

$$F_{BL} = \frac{\phi \rho D_p^2 VC}{18\eta h}, \qquad (30)$$

where ϕ, the angle of the bend, is expressed in radians.

The total attenuation upstream of a collection device is approximated by the sum:

$$F_T = F_S + F_D + F_{BL} \cdot n \tag{31}$$

Thermophoretic losses were not considered because the aerosol streams were in thermal equilibrium with the canisters. Electrostatic effects also were not considered because any particle charge would have been neutralized by the ionized atmosphere in the test vehicle produced by radiation and because the canisters were electrically grounded to the reactor structure.

3.5.2 The Loss of Flow Test (LOFT) FP-2 Experiment Virtual Impactor

The sampling and measurement systems for the LOFT FP-2 experiment [McPherson and Hicks (1986)] are shown schematically in Figure 7. In this experiment the aerosol mass concentration was determined using a virtual impactor. This type of impactor, like the cascade impactor, uses inertial impaction to size classify particles. Cascade impactors are probably the most widely used devices that seprate particles according to their aerodynamic particle size. Cascade impactors can be designed for a wide range of flow rates, pressures, and temperatures, providing particle classification for up to 20 size ranges. Major drawbacks to cascade impactors are particle bounce and the limited amount of mass that can be collected on each substrate.

The virtual impactor is a device that allows particles to impact upon a void rather than a substrate while maintaining a minor flow through the void. Virtual impactors size particles by directing a particle-laden gas jet toward a collection probe that allows only a small fraction of the flow of the impinging jet to pass. This forces a majority of the flow to turn and exit the probe. The larger particles cannot negotiate the turn but follow the minor flow in the probe and are carried to a collection device. The smaller particles follow the streamlines of the major flow and are collected separately. A multistage impactor iterates this process with the major flow within practical limits.

A virtual impactor does not suffer from particle bounce, and the sample is collectecd by filtering the major and minor flow, thus collecting larger amounts of mass than a cascade impactor given the same external dimension. Internal losses are typically greater in virtual impactors, and the larger particle size fraction contains a percentage of the small particle size fraction roughly equivalent to the percentage of minor flow. The traditional virtual impactor separates particles into only two size ranges. The device developed for the LOFT experiment consists of a two-cut point, three-collectable fraction virtual impactor [Novick and Alvarez (1987)]. This device retains the advantages of large mass collection and reasonably sharp cut points while providing definition of particle size distribution via the three size ranges. The small physical size of the impactor permits its application to a variety of environments.

The two-stage impactor is shown in Figure 8. It was designed to be small in size and to handle a 2 L/min flow rate. The minor flow through most virtual impactors ranges between 5 and 15%. However,

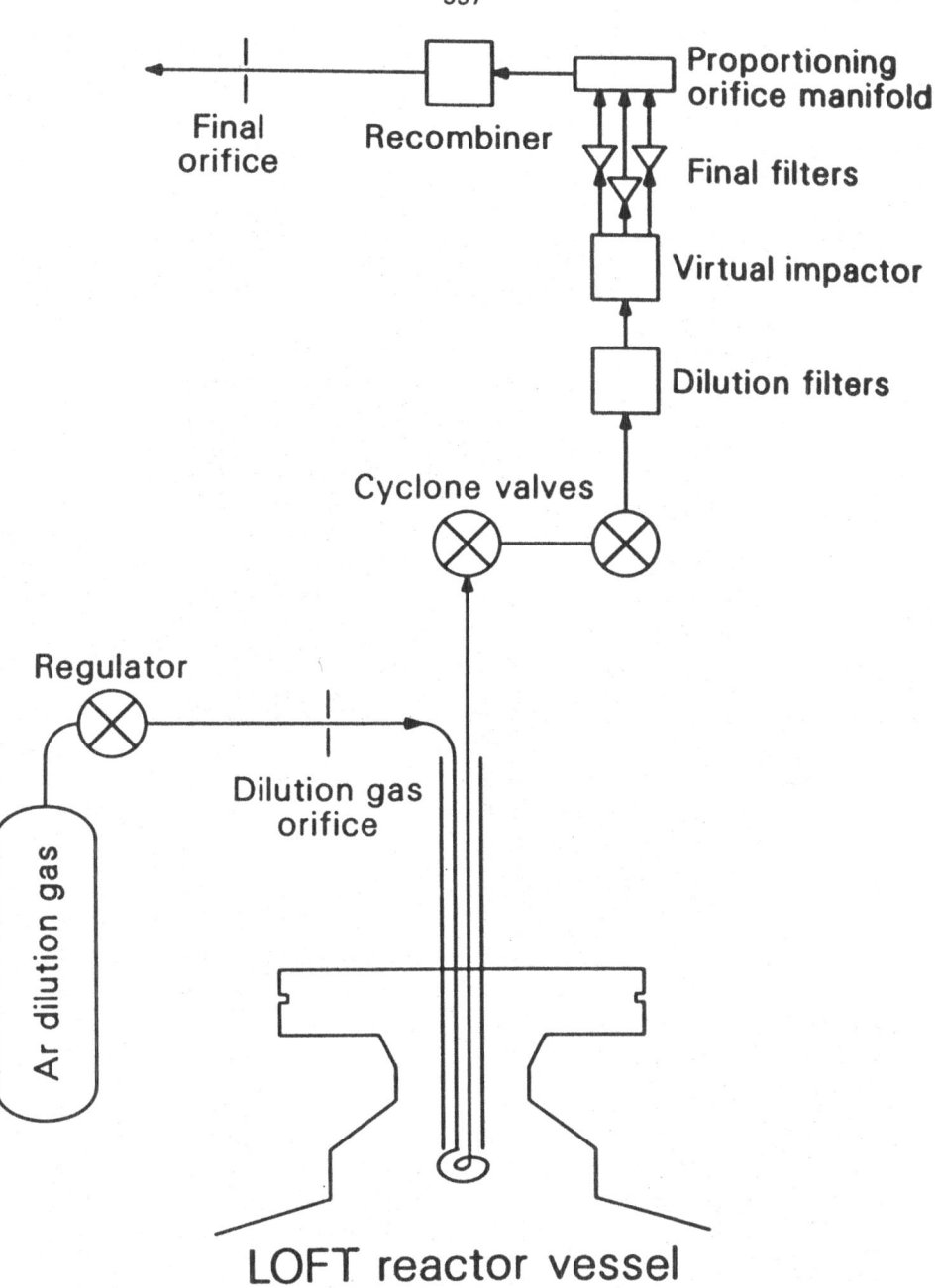

Figure 7. LOFT FP-2 Sampling System

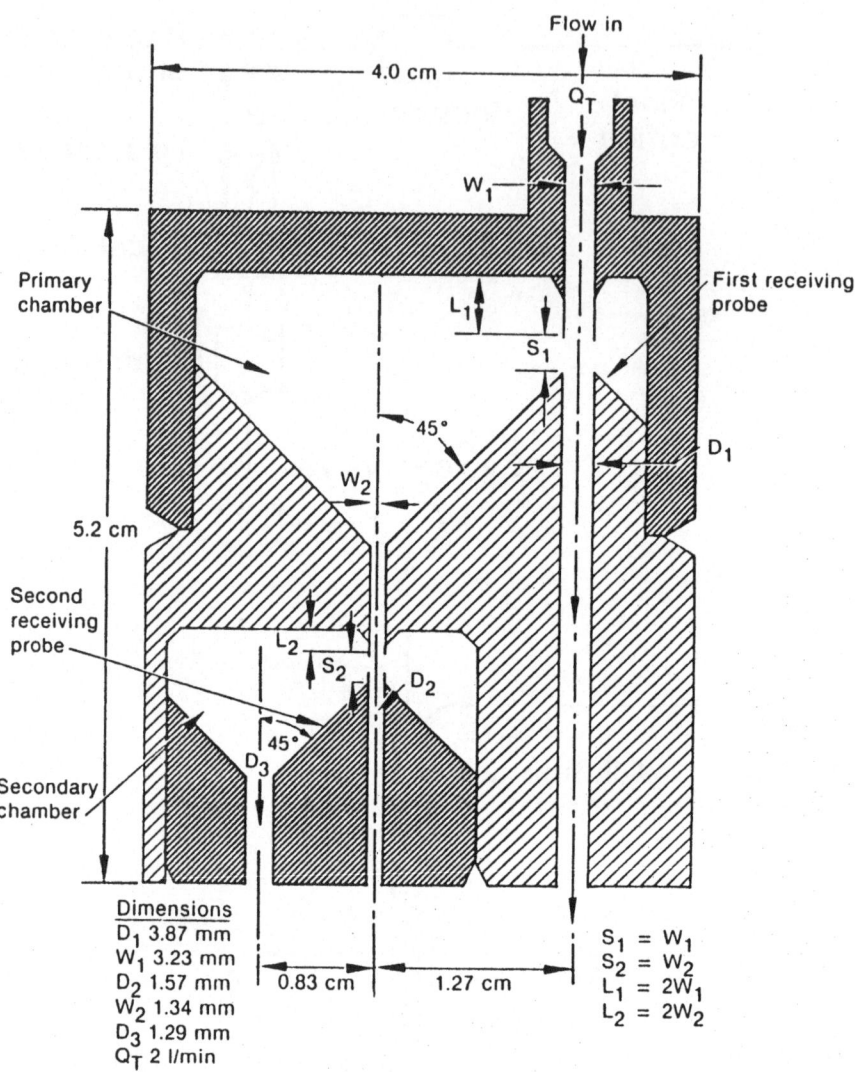

Figure 8. Two-Stage Virtual Impactor

this prototype was designed for minor flows between 15 and 20% in order to keep particles larger than 15 μm suspended in a vertical upward flow, required by the experiment for which the impactor was originally designed and tested. At this flow rate and a nominal temperature of 650 K, the first stage cut point occurs near 10 μm and the second stage cut point is near 3 μm. Flow through each stage is controlled by an orifice. The sum of the orifice areas can be used to control the total flow rate, Q, through the impactor, while the ratios of the orifice areas are used to control the relative amounts of major and minor flows. Filters are placed upstream of the orifices to collect the three particular size fractions. The operating characteristics of the impactor are governed by the Stokes number:

$$Stk = \frac{4pQCD_p^2}{9\pi\eta W^3} \tag{32}$$

where the characteristic dimension is the diameter of the nozzle, W, for a given stage.

A virtual impactor allows a fraction of the total flow to pass through the inlet probe of each stage. This fractional flow transports the larger particles to a collector but also entrains a fraction of low inertia particles (those smaller than the cut point diameter). This results in an efficiency curve that does not approach zero for sufficiently small values of particle Stokes number as do efficiency curves for cascade impactors. Virtural impactor efficiency curves instead approach a value based on the precentage of flow through the separating stage. This requires a correction to be applied to the values of the mass sampled by each stage to give the actual mass in the particular size class. The first-order corrections to the measured mass in each size class for a two-stage impactor are given below. If x, y, and z are the measured values of the collected mass, and X, Y, and Z are the corrected values for size classes X, Y, and Z, then:

$$Z = \frac{z}{1-q_1-q_2} \tag{33}$$

$$Y = \frac{y - q_2Z}{1-q_1} \tag{34}$$

$$X = x - q_1(Y+Z) \tag{35}$$

where q_1 is the fractional flow through the first stage, q_2 the fractional flow through the second stage, X and x the first stage values, largest class size, Y and y the second stage values, middle class size, and Z and z the third stage values, smallest size class.

Experimental laboratory testing has confirmed that the virtual impactor is able to accurately describe particle size distributions by determining the mass mean aerodynamic diameter and the geometric standard deviation. The impactor operates at close to theoretical efficiencies and sharpness of cut points. The prototype routinely collects 20 to 40 mg of aerosol mass with the total internal losses less than 9%. The impactor is extremely versatile. A wide range of cut points are available with a single design by simply changing the flow rate through the impactor.

3.5.3 The Power Burst Facility (PBF) SFD 1-4 Experiment

Most optical aerosol monitors for reactor experiments only measure the extinction of a beam of light. The extinction measurement can be interpeted as a history of the mass flux of aerosols during the experiment if both a particle size history and extinction coefficient history are assumed. A similar extinction device was added to the PBF SFD 1-4 experiment, as is shown in Figure 9. A schematic of the aerosol sampling and measurement systems for the PBF SFD 1-4 experiment [Adams et $al.$ (1986); Osetek (1987)] is shown in Figure 10. This device differed from other extinction measurement devices by using two extinction cells in series. For sufficiently high concentrations of aerosols, a difference in the extinction between the two cells can be detected and explained by coagulation, which tends to decrease the total cross-sectional area of the aerosol in the second cell, hence reducing the amount of extinction [Novick (1988)].

For a polydisperse aerosol, the extinction coefficient, Q, must be written as an integral over particle size, particle extinction coefficient, and size distribution. Let the value of this integral be designated by Q_{ext}. An equation describing the light extinction can be written for each cell. Hence,

$$I/I_o = \exp[-N\pi D_m^2 L Q_{ext}/4] \tag{36}$$

and

$$I'/I_0 = \exp[-N'D'_m{}^2 L'Q'_{ext}/4] \tag{37}$$

where N is the particle concentration, D_m the diameter of average mass, L the path length, I the measured light intensity and I_o the intensity of the light incident upon the particles. The unprimed and primed designations denote the first cell and the second cells, respectively. Because the cells are in series, the transit time is short, and therefore Q_{ext} equals Q'_{ext} and D_m equals D'_m. These assumptions are valid provided the coagulation process does not appreciably increase the particle size. D_m is used instead of the geometric size to describe the particle size because the total mass remains constant. Using the above assumptions and dividing Equation (37) by Equation (36) the following equation is obtained:

$$\frac{\ell n[I/I_o]}{\ell n[I'/I'_o]} = \frac{LN}{L'N'} \tag{38}$$

This equation is dependent only on the measured values of the intensity ratios for each cell and the number concentration in each cell.

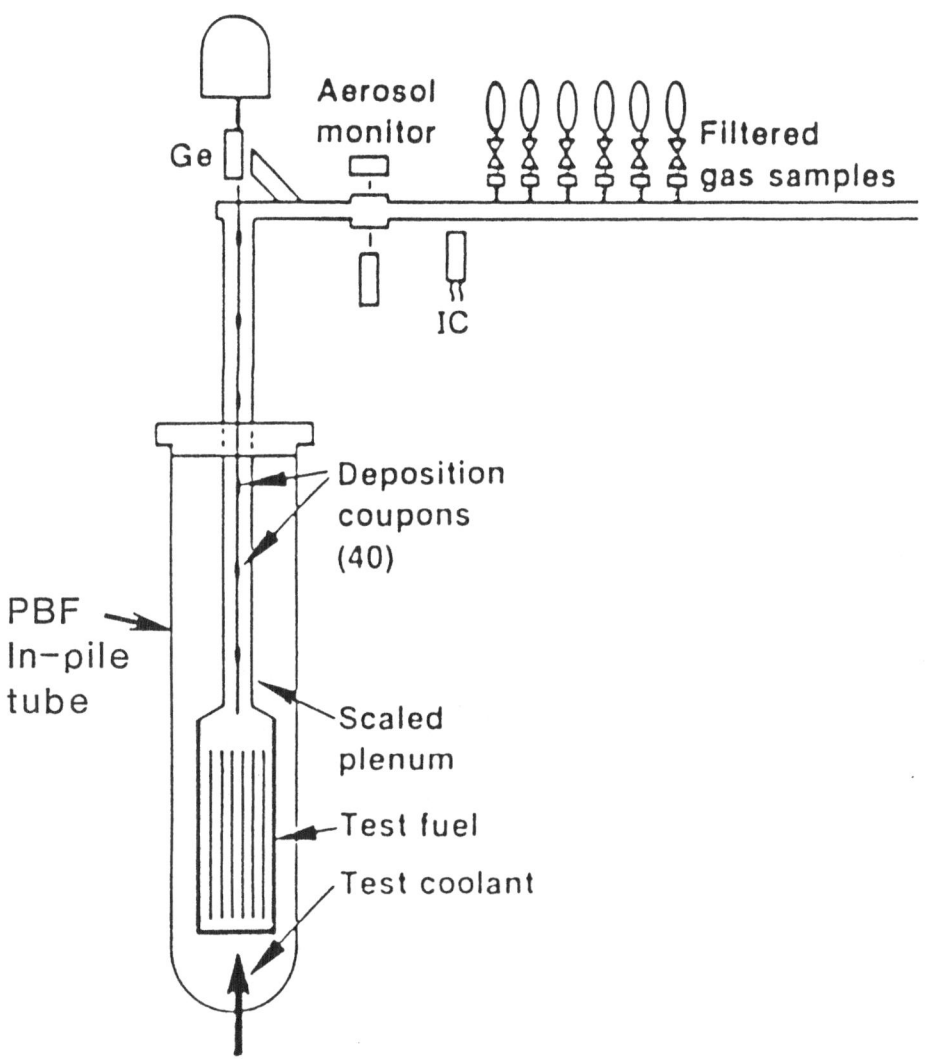

Ge

Aerosol monitor

Filtered gas samples

IC

Deposition coupons (40)

PBF In–pile tube

Scaled plenum

Test fuel

Test coolant

Figure 9.   PBF SFD-14 Experiment

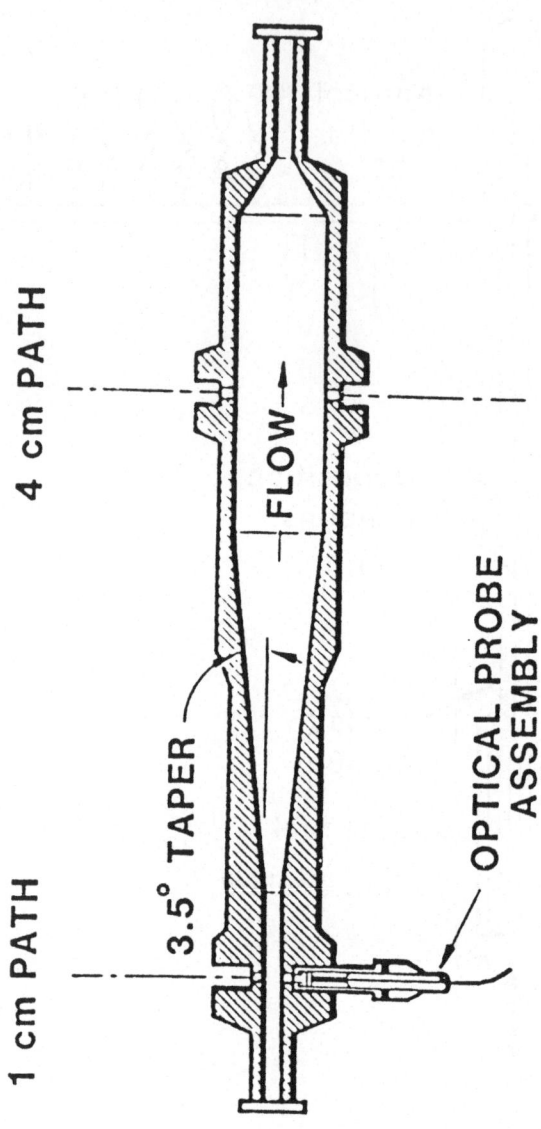

Figure 10. Dual Cell Extinction Monitor

The difference in number concentration is assumed to be due to only the coagulation process. As a result, the change in number concentration as a function of time for an aerosol undergoing coagulation is [Fuchs (1964)]:

$$dN/dt = 4\pi D_p D N^2 [1 + D_p(\pi Dt)^{-1/2}], \tag{39}$$

where D is the diffusion coefficient for particle of diameter, D_p, and t the coagulation time in seconds.

Depending on experimental conditions, the second term in the bracket is usually less than 1 and considered negligible. The solution of the simplified form of Equation (39) is given by:

$$1/N_0 = 1/N - kt, \tag{40}$$

where N_0 is the initial number concentration (particles/cm^{-3}), N the number concentration after time t (particles/cm^{-3}), and k the coagulation coefficient (cm^3/particles·s). The coagulation coefficient can be calculated for particles of a given size by:

$k = 4BTC/3\eta$,

$B =$ Boltzman's constant $= 1.38 \times 10^{-16}$ erg/K,

$T =$ temperature (K),

$C =$ Cunningham slip correction factor, and

$\eta =$ gas absolute viscosity (g/cm·s).

Based on the work of Gillespie (1963), Riest (1984) gives the following expression for the coagulation coefficient of log-normally distributed polydisperse aerosols:

$$k = 2BT[1 + exp(\ln^2 GSD) + 4.52[exp(0.5 \ln^2 GSD)$$
$$+ exp(2.5 \ln^2 GSD]\lambda/D_p]/3\eta, \tag{41}$$

where GSD is the geometric standard deviation of the particle size distribution and λ the mean free path of the gas molecules.

Changing from monodisperse to polydisperse aerosols does not alter Equation (40). The difference between the coagulation coefficients calculated by Equation (41) and that used in Equation (40) is less than 15% when $\lambda/d < 0.02$ and GSD < 1.5 In terms of the aerosol monitor, Equation (40) can be written with the initial concentration equal to the number concentration in the first cell, N, and the final concentration equal to the number concentration in the second cell, N':

$$N/N' = 1 + ktN. \tag{42}$$

Substituting into Equation (38):

$$\frac{L'\text{ℓn}[I/I_0]}{L\ \text{ℓn}[I'/I'_0]} = 1 + ktN \qquad (43)$$

Because of the high temperatures and pressures encountered in PBF-type experiments, the Cunningham correction factor reduces to near unity for particle sizes greater than 0.1 μm. This, in turn, yields a constant coagulation coefficient, k. Because the path lengths are known, the intensity ratios are measured and the transit time is determined from flow rate measurements, Equation (43) can be immediately solved for N. The number concentration as a function of time is obtained directly from extinction measurements with no *a priori* knowledge of particle size or extinction coefficient history.

References

Adams, J.P., Partin, J.K., Petti, D.A., and Reed, T.R., "Development and Calibration of an Aerosol Monitor Used in the PBF SFD 1-4 Experiment", Proceedings of the 4th Miami International Symposium on Multiphase Transport and Particulate Phenomena, Miami Beach, Florida, December 1986.

Agarwal, J. K., and Liu B. Y. H., "A Criterion for Accurate Aerosol Sampling in Still Air", American Industrial Hygiene Journal, Vol.41, p. 191-197, 1980.

Albrecht, H., Matschoss, V., and Wild, H., "Investigation of Activity Release during Light Water Reactor Core Meltdown", p. 278-283, NUREG/CR-1724, 1980.

Ariessohn, P.C., Self, S. A., and Eustis, R. H., "Two Wavelength Laser Transmissometer for Measurements of the Mean Size and Concentration of Coal Ash Droplets in Combustion Flows", Applied Optics, p. 3775-3781, 1980.

Bachalo, W.D., and Houser, M.J., "Development of the Phase/Doppler Spray Analyzer for Liquid Drop Size and Velocity Characterizations", Paper No. AIAA-84-1199, AIAA/SAE/ASME 20th Joint Propulsion Conference, Cincinnati, OH, June 1984.

Baron, P.A., "Calibration and Use of the Aerodynamic Particle Sizer (APS 3300)", Aerosol Science and Technology, Vol. 5, p. 55-67, 1986.

Brewer, J. M., and Goren, S. L., "Evaluation of Metal Oxide Whiskers Grown on Screens for Use as Aerosol Filtration Media", Aerosol Science and Technology, Vol. 3, p. 411-429, 1984.

Brock, J.R., "On the Theory of Thermal Forces Acting on Aerosol Particles", J. Colloid Science, Vol. 17, p. 768-780, 1962.

Buescher, B. J., Osetek, D. J. and Ploger, S.A., "Power Burst Facility Severe Fuel Damage Test Series", EG&G Idaho, Inc. Paper No. EGG-M-083382, 1982.

Bunz, H., and Schock, W., "Measurements of the Condensation of Steam on Different Aerosols under LWR Core Melt Down Conditions", p. 171-180, NUREG/CR-1724, 1980.

Chen, C. Y., "Filtration of Aerosols by Fibrous Media", Chem. Rev., Vol. 5, p. 595-623, 1955.

Cheng, Y.S., and Yang, C.S., "Motion of Particles in Bends of Circular Pipes", Atmos. Environ., Vol. 15, p. 301-306, 1981.

Chyssler, J., Hesbol, R., Jansson, E., Piispaner, W., Sandstrom, R., and Strom,"Measurement System, Tests 1 and 2", Marviken Project Report MX5-34M, May 1983.

Crane, R. I., and Evans, R. L., "Inertial Deposition of Particles in a Bent Pipe", Journal of Aerosol Science, Vol. 8, p. 161-170, 1977.

Davies, C. N., "The Entry of Aerosols into Sampling Tubes and Heads", British J. Applied Physics Ser. 2, Vol. 1, p. 921-932, 1968.

Dunn, P. F., Herceg, J. E., and Johnson, C.E., "A Sampling System for the Physicochemical Characterization of Fission-Product Aerosols Formed During Light-Water-Reactor Experiments", Aerosol Science and Technology, Vol. 2, p. 257, 1983.

Elrick, R. M., "A First Study of Aerosols Produced by Neutronic Heating of Fresh UO_2 Fuel Under Core-Disruptive Accident Conditions", NUREG/CR-2296, October 1982.

Elrick, R. M., "A Time-Resolving Sampler to Determine Initial Fuel Aerosols Under CDA Conditions", p. 73-83, NUREG/CR-1724, 1980.

Fuchs, N. A., The Mechanics of Aerosols, Pergammon Press, New York, p. 264, 1964.

Gillespie, T., "The Effect of Size Distribution on the Rate Constants for Collisions in Disperse Systems", J. Colloid Science, Vol. 18, p. 562-567, 1963.

Herceg, J. E., Blomquist, C. A., Chung, K. S., Dunn, P. F., Johnson, C. E.,Kraft, D. A., Schlenger, B. J., Shaftman, D. H., and Simms, R., "TREAT Light Water Reactor Source Term Experiments Program", American Nuclear Society Topical Meeting, Fission Product Behavior & Source Term Research, Snowbird, Utah, July 15-19, 1984.

Heyder, J., Gebhart, J., and Scheuch, G., "Interaction of Diffusional andGravitational Particle Transport in Aerosols", Aerosol Science and Technology, Vol. 4, p. 315-326, 1985.

Himeno, Y. and Takahashi, J., "Sodium Mist Behavior in Cover Gas Space of LMFBR, Out-Pile Experiment", J. of Nuclear Science and Technology, p. 404-412, 1980.

Holve, D.J., "In Situ Measurements of Flyash Formation from Pulverized Coal", Combust. Sci. and Tech., Vol. 44, p. 269-288,1986.

Ingham, D.B., "Gravity Settling of Aerosol Particles in Horizontal Rectangular   Tubes with Some Diffusion", Journal of Aerosol Science, Vol. 8, p. 139-147, 1977.

Ingham, D. B., "Diffusion of Aerosol From a Stream Flowing Through a Cylindrical Tube, Journal of Aerosol Science", Vol. 6, p. 125-132, 1975.

Kanapilly, G.M., Cheng, Y.S., Gray, R.H., and Yeh, H.C., "A Comparative Instrumental Study on the Size Characteristics of Yttrium Oxide Aggregate Aerosols", p. 302-317, NUREG/CR-1724, 1980.

Kirsch, A. A., and Stechkina, I. B., "The Theory of Aerosol Filtration with Fibrous Filters", Fundamentals of Aerosol Science, D. Shaw, Editor, John Wiley & Sons, New York, p. 165-256, 1978.

Lorenz, R. A., Hobson, D. O., and Parker, G. W., "Fuel Rod Failure Under Loss-of-Coolant Conditions in TREAT", Nuclear Technology, Vol. 11, p. 502-520, August 1971.

McPherson, G.D., and Hicks, D., "LOFT Goes out with a Severe Accident", Nuclear Engineering International, p. 24, April 1986.

Miller, R.W., Applehans, A.D., Bolstad, J.O., Courtright, E.L., Novick, V.J., Alvarez, J.L., and Deason, V.A., Instrument Development Summary Report for the Power Burst Facility Severe Fuel Damage Series 2, EG&G Report FIN No. A6305, March 1984.

Natanson, G. L., "Diffusion Precipitation of Aerosols on a Streamlined Cylinder with a Small Capture Coefficient", Dokl. Akad. Nauk USSR, Vol. 11, p. 100, 1957.

Novick, V. J., "The Use of Series Light Extinction Cells To Determine Aerosol Number Concentration", Aerosol Science and Technology - In Press, 1988.

Novick, V. J., and Alvarez, J. L. "Design of a Multistage Virtual Impactor", Aerosol Science and Technology, Vol. 6, p. 63-70, 1987.

Nuclear Aerosols in Reactor Safety, Nuclear Energy Agency, Organization for Economic Co-Operation and Development, June 1979.

Osetek, D.J., "Results of the Four Severe Fuel Damage Tests", NUREG/CP-0090, Fiftheenth Water Reactor Safety Information Meeting, Gaithersburg, Maryland, October 26-30, 1987.

Parker, G. W., "Experimental Techniques of the Characterization of Nuclear Aerosols", p. 278-301, NUREG/CR-1724, 1980.

Pui, D.Y.H., Romay-Novas, F., and Liu, B.Y.H., "Experimental Study of Particle Deposition in Bends of Circular Cross Section", Aerosol Science and Technology, Vol. 7, p. 301-315, 1987.

Riest, P.C., Introduction of Aerosol Science, MacMillan Publishing Company, New York, New York, p. 259, 1984.

Sauter, H. and Schutz, W., "Aerosol Release from a Hot Sodium Pool and Behavior in Sodium Vapor Atmosphere", p. 84-94, NUREG/CR-1724, 1980.

Schock, W., "Application of Optical Methods in Nuclear Aerosol Measurements", p. 221-231, NUREG/CR -1724, 1980.

Sem, G. J., and Daley, P. S., "Performance Evaluation of a New Piezoelectric Aerosol Sensor", Aerosol Measurement, D.A. Lundgren et al., Editors, University Presses of Florida, p. 672-686, 1979.

Sinclair, D., Countess, R. J., Liu, B.Y.H., and Pui, D.Y.H., "Experimental Verification of Diffusion Battery Theory", APCA Journal, Vol. 26, p. 661-663, 1976.

Sinha, M. P., Griffin, C. E., Norris, D. D., and Friedlander, S. K., "Continuous Monitoring of Aerosols", p. 54, NASA Tech Brief, Vol. 7, No. 1, Item #40, Fall 1982.

Stechkina, I. B., and Fuchs, N. A., "Studies on Fibrous Aerosol Filters-I.Calculation of Diffusional Deposition of Aerosols in Fibrous Filters", Ann. Occup. Hyg., Vol. 9, p. 59-64, 1966.

Thomas, J.W.,"Gravity Settling of Particles in a Horizontal Tube", APCA Journal,Vol. 8, p.32-34, 1958.

Waldmann, L., and Schmitt, K. H., in Aerosol Science, C. N. Davies, Editor, Academic Press, London, p. 137-161, 1966.

Woods, D. C., "Measurement of Particulate Aerosol Mass Concentration Using Piezoelectric Crystal Microbalance", Aerosol Measurement, D.A. Lundgren et al., Editors, University Presses of Florida, p. 119-130, 1979.

Wright, A.L., Kress, T.S., and Smith, A.M., "ORNL Experiments to Characterize Fuel Release from the Reactor Primary Containment in Severe LMFBR Accidents NUREG/CR-1724, 1980.

Yeh, H. C., "Use of Heat Transfer Analogy for a Mathematical Model of Respiratory Tract Deposition", Bulletin of Mathematical Biology, 36, No. 2, p. 105-116, 1976.

Lecture Notes in Engineering

Edited by C.A. Brebbia and S.A. Orszag

Lecture Notes in Engineering

Lecture Notes in Engineering

Edited by C.A. Brebbia and S.A. Orszag